2nd European Conference
on the
Mathematics of Oil Recovery

AMONG OUR BOOKS

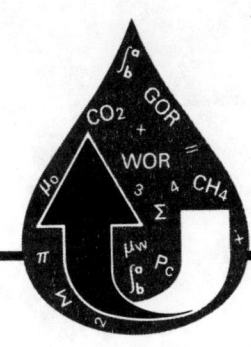

2nd European Conference on the Mathematics of Oil Recovery

Proceedings of Presentations
held in Arles, France,
September 11-14, 1990

Edited by

Dominique Guérillot
Institut Français du Pétrole

Olivier Guillon
Elf Aquitaine

Organized by
Institut Français du Pétrole

1990

ÉDITIONS TECHNIP 27 RUE GINOUX 75737 PARIS CEDEX 15 **techniP**

SCIENTIFIC COMMITTEE:

F. Da Silva, Petrofina
C. Descalzi, Agip
C. L. Farmer, AEA Winfrith
F. J. Fayers, BP
A. Grant, Shell
D. Guérillot, IFP
O. Guillon, Elf Aquitaine (President of the Scientific
    Committee)
Z. Heinemann, Mining University Leoben
O. Jensen, Maersk
B. Legait, IFP
B. Maitin, RWE-DEA
M. Rioche, Total CFP
P. R. Thomassen, Statoil

SCIENTIFIC SECRETARIAT:

D. Guérillot, IFP

CONFERENCE SECRETARIAT:

Y. Rondot, IFP
M.-F. Baltus, IFP

The papers included in these proceedings were selected following review of information contained in an abstract submitted by the author(s). Contents of the papers have not been reviewed.

ISBN 2-7108 – 0589-8

Printed in France
by Imprimerie Nouvelle, 45800 Saint-Jean-de-Braye

Foreword

Back in May 1988, John Fayers (from BP) initiated the process which led to the European Conference on the Mathematics of Oil Recovery (ECMOR), held in July 1989 in Cambridge (UK). This first conference was a clear success, by the quality of the presentations and by the number of participants (about 150).

This success encouraged us to pursue this initiative which meets the need for high quality conferences held in Europe, with a somewhat more mathematical approach to problems than existing conferences and focusing on specific problems that we Europeans are facing (North Sea as an example). The recent developments in East-European countries and the level of participation from East Europeans to this conference is a sign that the future conferences will get richer with new problems, new visions and new people and this is certainly to be welcome.

What will remain from ECMOR 90 is this book:

More than a hundred abstracts have been reviewed by the committee. Only a little over thirty of these could be included for full paper presentation in the 3 days conference. These papers, organized by themes, each introduced by a keynote paper given by individuals of the highest caliber, make up the largest part of this book.

All aspects of Reservoir Description and Simulation are covered : the theme of Reservoir Characterization is dominated by geostatistical methods which allow to generate possible reservoir descriptions compatible with whatever knowledge is available. The theme on Effective Properties groups problems in simulating the behaviour of heterogeneous reservoirs as if they were homogeneous. The next themes, Numerical Schemes and Reservoir Simulations cover some aspects of simulating flow in (deterministically or stochastically) heterogeneous reservoirs. Analytical methods dealing mainly with gas or solvent flooding and dispersion come next. The last theme on gridding and grid refinement addresses the problem of minimizing the number of grid-blocks while describing accurately moving fronts, large scale heterogeneities or well regions.

Many good abstracts that had been excluded from the formal presentations were rescued for the poster session and are available in an abridged version in this volume.

On behalf of the committee, I wish to thank the Institute of Mathematics and its Applications (IMA) and especially the Society of Petroleum Engineers (SPE) for their help in highly publicizing this conference; the financial support of Association de Recherche sur les Techniques d'Exploitation du Pétrole (ARTEP) and ELF AQUITAINE has been greatly appreciated. The efficient organization of the conference and the timely edition of this book should both be attributed to the enthusiasm and competence of Dominique Guérillot and Yolande Rondot from the Institut Français du Pétrole.

We all hope this conference will be a great success and will be the second of a long list of European Conferences!

For the Scientific Committee

O. GUILLON
President of ECMOR 1990

Contents

Foreword VII

List of Authors XV

Reservoir Characterization

Stochastic Simulation of Lithofacies: an Improved Sequential Indicator Approach 3
V. Suro-Pérez and A. G. Journel

Combining Geology, Geostatistics and Multiphase Fluid Flow for 3D Reservoir Studies 11
A. Galli, D. Guérillot, C. Ravenne and Heresim Group

Development of Geostatistical Methods Dealing with the Boundary Conditions Problem Encountered in Fluid Mechanics of Porous Media 21
A. Dong, S. Ahmed and G. de Marsily

Large-Scale Barriers in Extensively Drilled Reservoirs 31
J. Høiberg, H. Omre and H. Tjelmeland

Effective Properties

Accurate Calibration of Empirical Viscous Fingering Models 45
F. J. Fayers, M. J. Blunt and M. A. Christie

Effective Permeability of Heterogeneous Reservoir Regions 57
L. J. Durlofsky and E. Y. Chung

Application of Analytical Methods in Predicting Waterflood Performance of Reservoirs with Stochastic Sand Bodies 65
O. B. Abu-elbashar, T. S. Daltaban, C. G. Wall and J. S. Archer

Numerical Simulation and Homogenization of Diphasic Flow in Heterogeneous Reservoir 75
B. Amaziane, A. Bourgeat and J. V. Koebbe

Stochastic Characterization of Grid-Block Permeabilities: from Point Values to Block Tensors 83

J. J. Gomez-Hernandez and A. G. Journel

Large-Scale Properties for Flow through a Stratified Medium: a Discussion of Various Approaches 91

A. Ahmadi, A. Labastie and M. Quintard

Numerical Schemes

The Use of Second-Order Godunov-Type Methods for Simulating EOR Processes in Realistic Reservoir Models 101

K. Holing, J. Alvestad and J. A. Trangenstein

Modelling Flow through Heterogeneous Porous Media with Boundary Integrals Using Higher-Order Surface Singularities 113

D. W. Wong, J. S. Archer and J. M. R. Graham

A Finite Element Method for Calculating Transmissibilities in N-point Difference Equations Using a Non-Diagonal Permeability Tensor 121

P. Samier

Implicit Flux Limiting Schemes for Petroleum Reservoir Simulation 131

M. Blunt and B. Rubin

The Use of Boundary Element Method in Front Tracking for Composite Reservoirs 139

J. Kikani and R. N. Horne

Reservoir Simulations

Control Volume Method to Model Fluid Flow on 2D Irregular Meshing 149

I. Faille

Heterogeneous Porous Media and Domain Decomposition Methods 157

M. S. Espedal, R. Hansen, P. Langlo, O. Sævareid and R. E. Ewing

Parallel Simulation of Petroleum Reservoirs 165

J. Larsen and N. Bech

Comprehensive Mathematical Modeling of Horizontal Wells 169

M. R. Islam

Analytical Methods

Curvilinear Grid Generation Techniques 179
C. L. Farmer and D. E. Heath

An Analytical Investigation by the Method of Characteristics of Gravity
Stabilised Gas Injection 187
R. W. S. Foulser

Dispersive Mixing in Unstable Displacement 197
L. J. T. M. Kempers

Fluid Flow in Porous Media and Related Rock Mechanics Problems 205
M. Boutéca and J. P. Sarda

Composition Paths in Binary CO_2-C_{10} Displacements: Effects of Reservoir
Heterogeneity and Crossflow on Displacements with Limited Solubility 211
K. K. Pande and F. M. Orr, Jr

Mathematical and Numerical Analysis of a Hyperbolic System Modeling
Solvent Flooding 219
T. Johansen and R. Winther

Gridding

Mixed Methods, Operator Splitting, and Local Refinement Techniques for
Simulation on Irregular Grids 237
M. S. Espedal, R. E. Ewing and T. F. Russell

Domain Decomposition Methods in Reservoir Simulation Coupling Well and
Full Field Models 247
O. Gosselin and J. M. Thomas

A Characteristic Finite Element Method for Solving Non-Linear Convection-
Diffusion Equations on Locally Refined Grids 255
R. H. J. Gmelig Meyling

A Coordinate System for Local Grid Refinement Close to Wells 263
S. Ekrann

Data Structure and Algorithms for Adaptive Mesh Refinement 271
T. Hermitte and D. Guérillot

Poster-Conferences

Two Dimensional Stochastic Modelling of Flow in Non-Uniform Confined Aquifers. Correction of the Systematic Bias Introduced by Numerical Models when They Are Used Stochastically 283
P. Lachassagne, E. Ledoux and G. de Marsily

An Estimator for the Effective Permeability 287
L. Holden, J. Høiberg and O. Lia

Numerical Simulation of Hydraulic Fracturing in a Discrete Element System 291
S. Thallak, L. Rothenburg, M. Dusseault and R. Bathurst

A 3-D Network Simulating Two Phase Immiscible Displacements in Porous Media 297
D. Zhou and E. H. Stenby

An Analytical and Numerical Study of the Three-Phase Surfactant Displacement Problem 301
J. W. Barker

A Triangular Model for Three-Phase Flow 305
L. Holden and A. Tveito

A Practical Front Tracking Technique for Control of Numerical Diffusion 309
M. Halilu and R. I. Issa

FAC Method for Reservoir Simulation 315
R. Boyer, B. Martinet and K. Saïkouk

Mixed Finite Elements for Multiphase Flow in Porous Media Consisting of Different Rock Types 319
Ø. Bøe and G. E. Fladmark

Radial Transport in Porous Media with Dispersion and Adsorption 323
L. Ci-Qun and Y. Jie

Constant-Time Step Deconvolution Model for Variable-Rate Well Test Pressure Data 327
S. Buitrago, G. Gedler and R. Manzanilla

Inverse Modeling for Compressible Flow. Application to Gas Reservoirs 331
B. Bréfort and V. Pelcé

Computer Geological Simulation in Oil Recovery 335
V. A. Badyanov

A New Formulation for Generalised Compositional Simulation 339
R. E. Mott and C. L. Farmer

CONTENTS

P3D Modeling of Vertical Hydraulic Fracture Growth 343
P. Valko and B. Pertik

Irregular Averaging of Filtration Transfer Processes in Heterogeneous Media 347
M. B. Panfilov

History Matching Problems of Filtration Theory. Complex Adaptative Geological Models of Fields 351
B. Palatnik, I. Zakirov and G. Agaev

Multilevel Methods in Porous Media Flow 355
R. Teigland and G. E. Fladmark

List of Authors

Abu-Elbashar O.B., 65

Agaev G., 351

Ahmadi A., 91

Ahmed S., 21

Alvestad J., 101

Amaziane B., 75

Archer J.S., 65, 101

Badyanov V.A., 335

Barker J.W., 301

Bathurst, R., 291

Bech N., 165

Blunt M.J., 45, 131

Bøe ø., 319

Bourgeat A., 75

Boutéca M., 205

Boyer R., 315

Bréfort B., 331

Buitrago S., 327

Christie M.A., 45

Chung E.Y., 57

Ci-Qun L., 323

Daltaban T.S., 65

Dong A., 21

Durlofsky L.J., 57

Dusseault M., 291

Ekrann S., 263

Espedal M.S., 157, 237

Ewing R.E., 237

Faille I., 149

Farmer C.L., 179, 339

Fayers F.J., 45

Fladmark G.E., 319, 355

Foulser R.W.S., 187

Galli A., 11

Gedler G., 327

Gmelig Meyling R.H.J., 255

Gomez-Hernandez J.J., 83

Gosselin O., 247

Guérillot D., 11, 271

Halilu M , 309

Hansen R., 157

Heath D.E., 179

Hermitte T., 271

Høiberg J., 31, 287

Holden L., 287, 305

Holing K., 101

Horne R.N., 139

Islam M.R., 169

Issa R.I., 309

Jie Y., 323

Johansen T., 219

Journel A.G., 3, 83

Kempers L.J.T.M., 197

Kikani J., 139

Koebbe J.V., 75

Labastie A., 91

Lachassagne P., 283

Langlo P., 157

Larsen J., 165

Ledoux E., 283

Lia O., 287

Manzanilla R., 327

Marsily (de) G., 21, 283

Martinet B., 315

Mott R.E., 339

Orr (Jr.) F.M., 211

Palatnik B., 351

Pande K.K., 211

Panfilov M.B., 347

Pelcé V., 331

Pertik B., 343

Quintard M., 91

Ravenne C., 11

Rothenburg L., 291

Rubin B., 131

Russel T.F., 237

Sævareid O., 157

Saïkouk K., 315

Samier P., 121

Sarda J.P., 205

Stenby E.H., 297

Suro-Pérez V., 3

Teigland R., 355

Thallak S., 291

Tjelmeland H., 31

Trangenstein J.A., 101

Tveito A., 305

Valko P., 343

Wall C.G., 65

Winther R., 219

Wong D.W., 101

Zakirov I., 351

Zhou D., 297

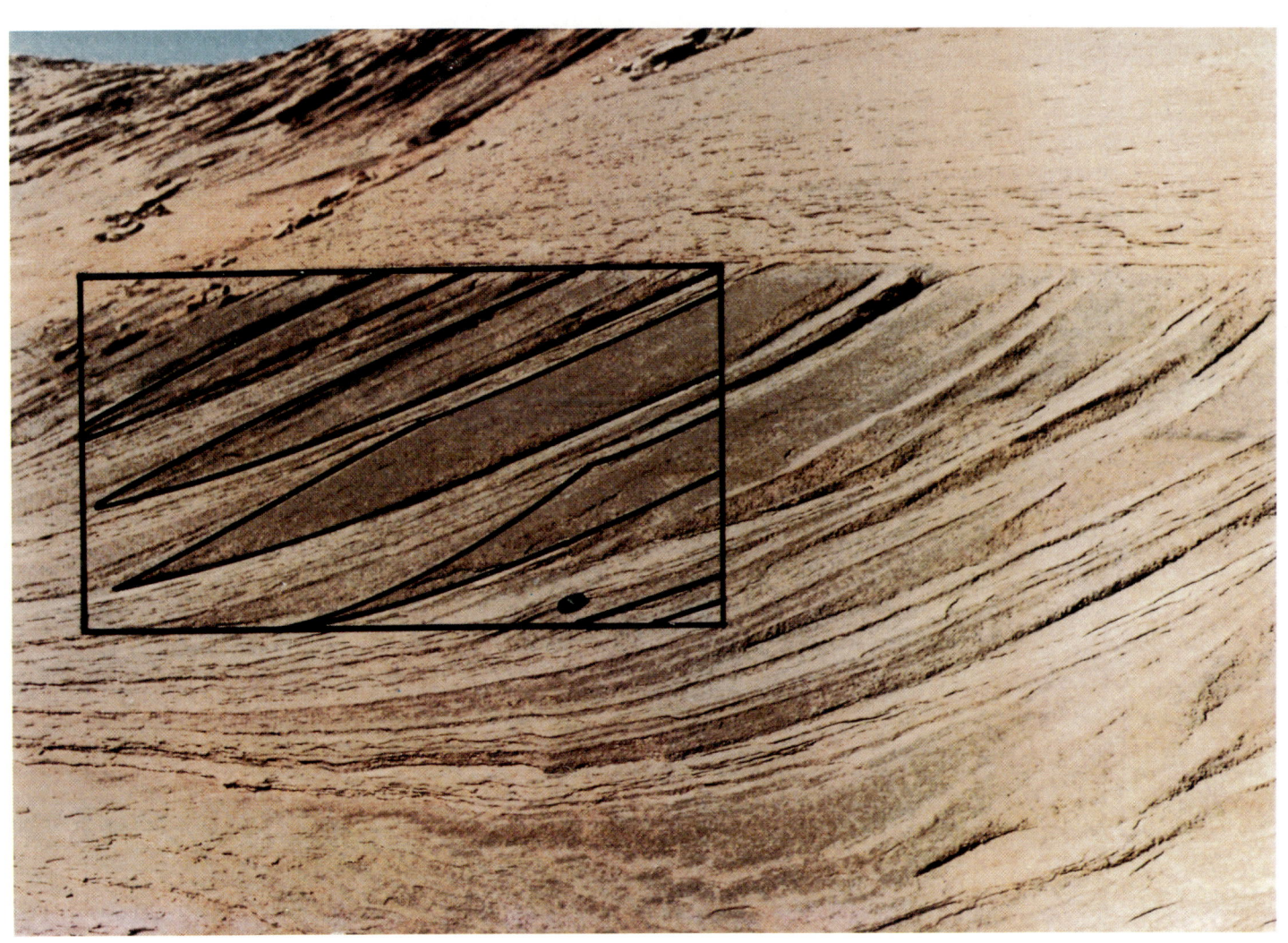

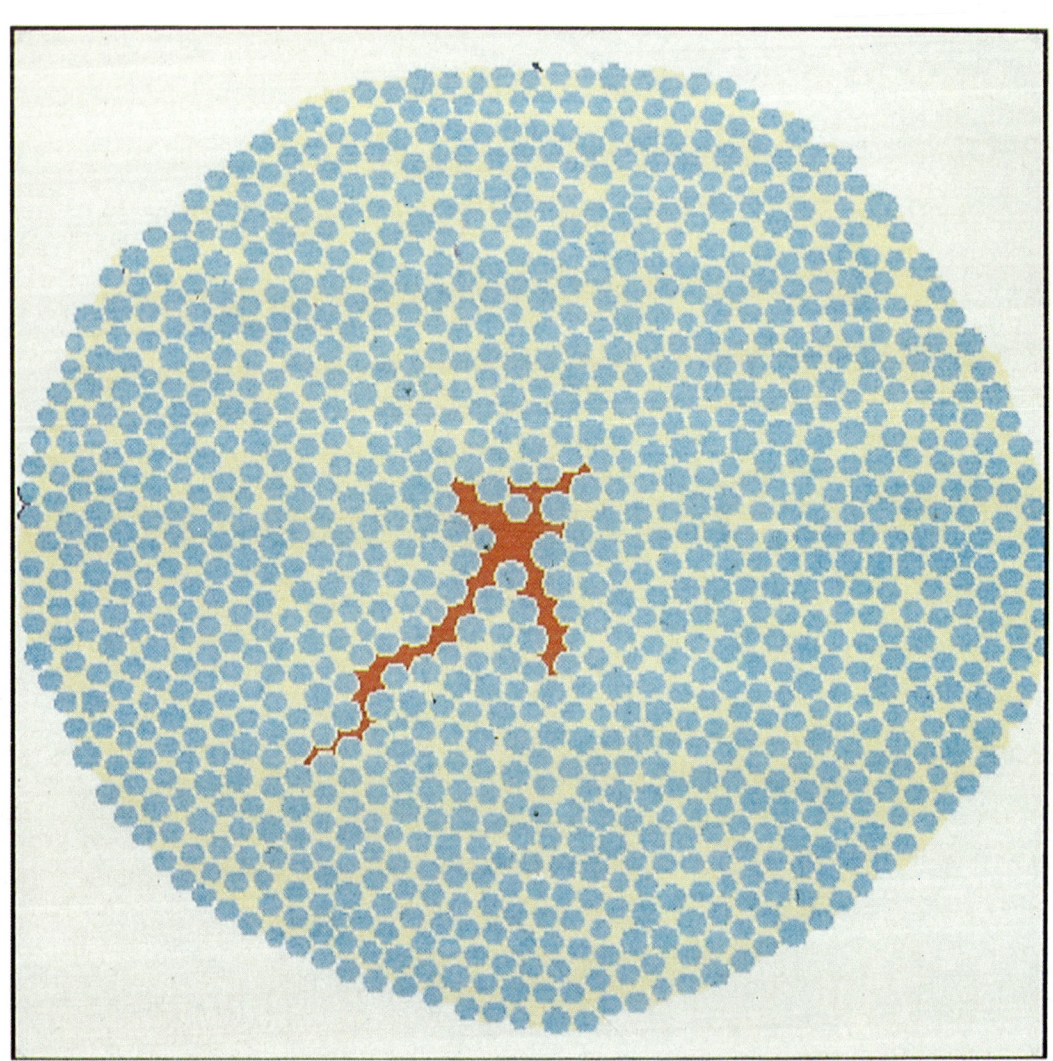

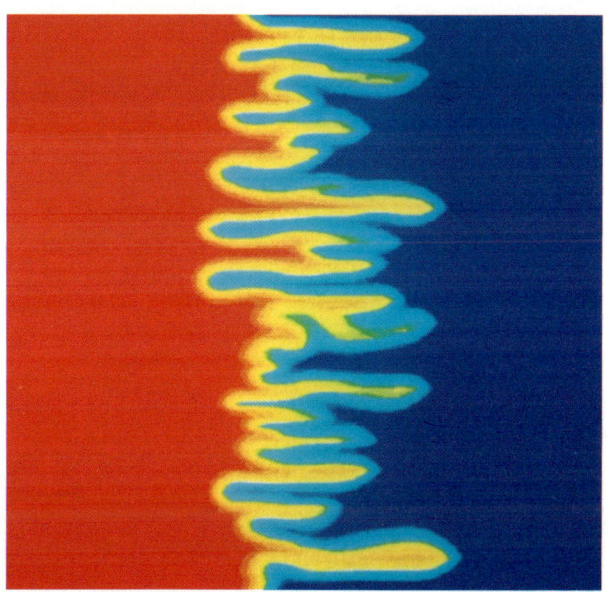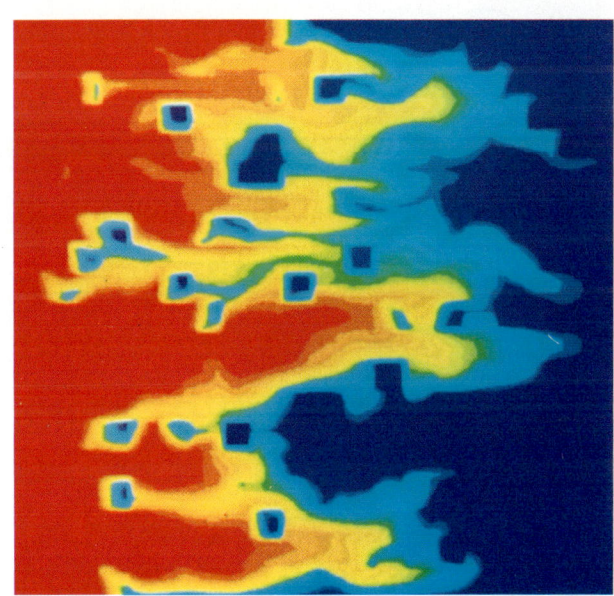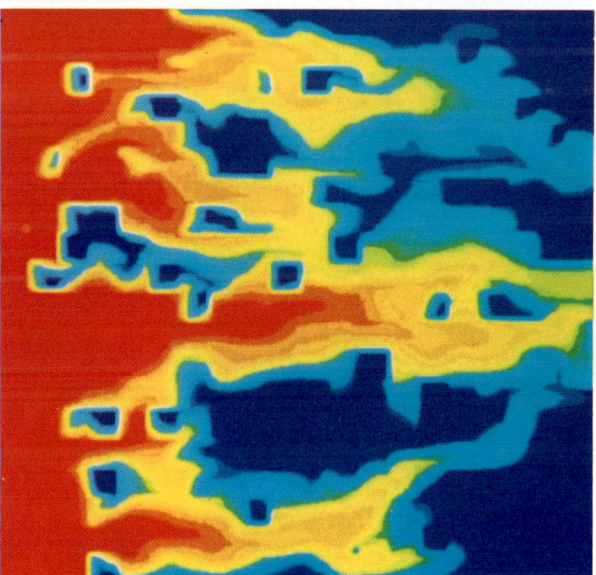

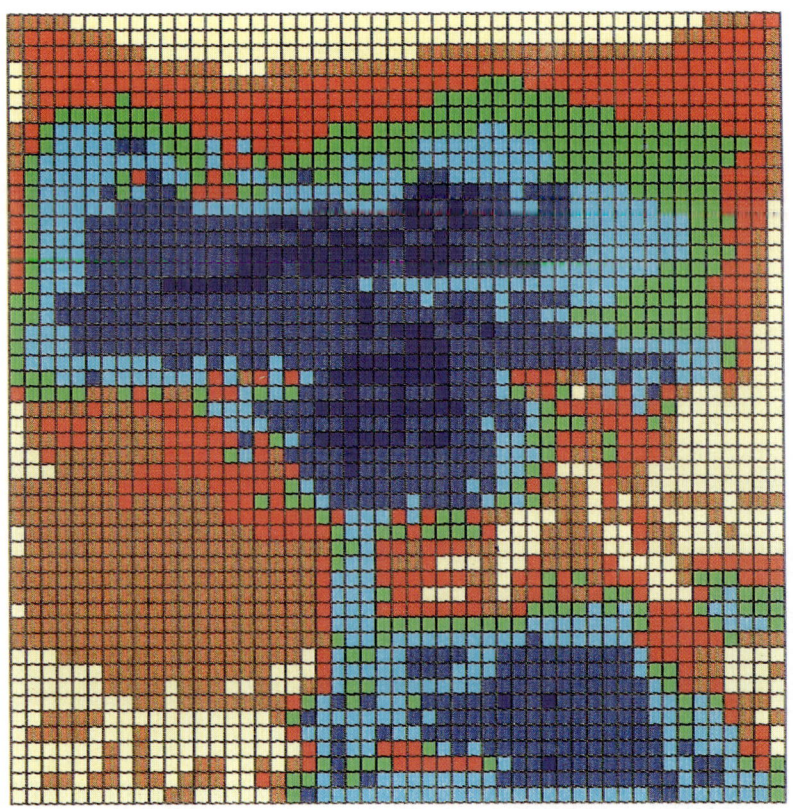
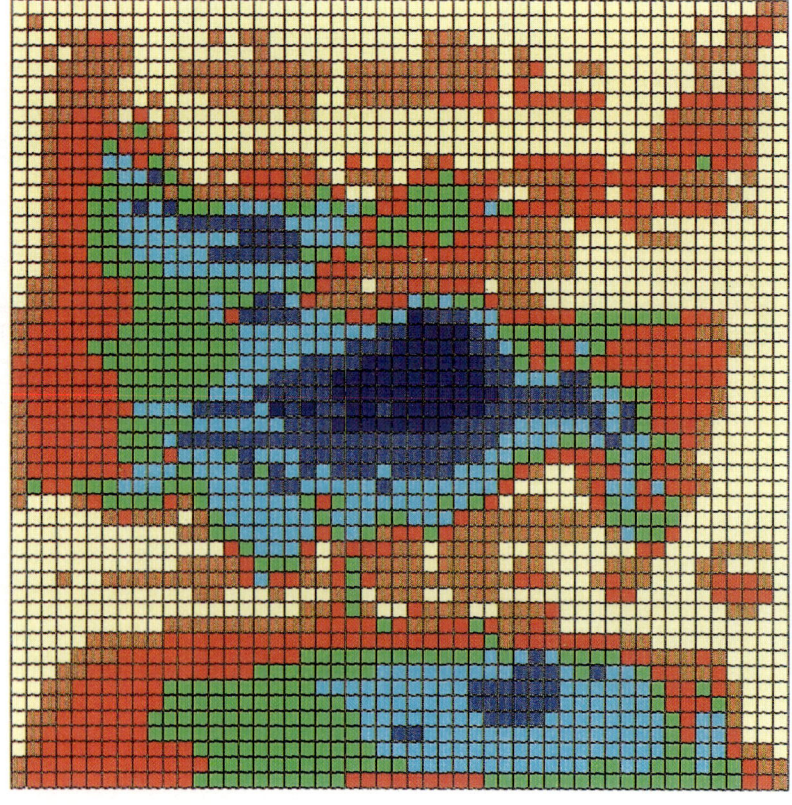

Reservoir
Characterization

2nd European Conference on the Mathematics of Oil Recovery
© D. Guérillot, O. Guillon (Editors) and Éditions Technip, Paris 1990, pp. 3-10
27 rue Ginoux, 75015 Paris

Stochastic Simulation of Lithofacies:
an Improved Sequential Indicator Approach

V. Suro-Pérez and A. G. Journel[1]

Abstract

Reservoir characterization demands the estimation or simulation of important reservoir parameters such as permeability and porosity. Often, reservoir geology is accounted for via continuous values of permeability or porosity, without explicit reference to the reservoir facies. The approach proposed starts by modeling the spatial distribution of the reservoir facies and, only then, conditions the generation of permeability/porosity values to the simulated facies geometry. It allows accounting for spatial relationships between different lithologies (covariances and crosscovariances), and uses such relations to estimate (or simulate) the most probable lithology at any specific location. The ARCO data set is used to build stochastic simulations of six different lithofacies over a particular vertical section, considering only three conditioning wells out of ten actually available. Each stochastic simulation is a lithological reservoir image which reproduces the patterns of continuity of the lithofacies considered and honor the data values at the conditioning well locations. Repeated generation of such lithofacies images, allows assesment of the geological heterogeneities impact on the oil recovery.

Introduction

Description of the spatial distribution of reservoir properties is critical to the correct prediction of recovery performance. A sophisticated flow simulator is not enough to obtain accurate predictions; a good characterization of the reservoir heterogeneities is also necessary to obtain a reliable forecast.

The approach proposed divides the integration of information in two major steps. The first step emphasizes quantitative use of basic geological information (lithofacies) to provide the geometric architecture of the reservoir. The second step consists of filling each lithofacies with values of permeability, porosity and saturations. These latter values can be derived from well logs, seismic data or core data and their analysis is made conditional to lithofacies.

There had been numerous attempts in the literature to simulate lithofacies or categorical variables. Journel and Isaaks (1984), Haldorsen et al (1988), Matheron et al (1988) and Journel and Gomez-Hernandez (1989) have proposed different schemes to handle this problem. However, none of these techniques considers explicitly the reproduction of cross-statistics relating two different facies.

(1) Department of Applied Earth Sciences, Stanford University, 310 Mitchell Building, Stanford, California 94345-2225, USA.

This paper proposes a method capable of reproducing both the facies proportions (volumes) and their relative spatial variabilities (covariances and crosscovariances). The goal here is the simulation of geological (categorical) variables such as lithofacies codes. Once the lithofacies spatial distribution is characterized and visualized through stochastic images, a flow unit discrimination can be done allowing a better permeability-porosity-saturation simulation. Flow simulation can then focus on the analysis of the extreme permeability connected bodies, whether flow barriers or flow paths.

Facies Structural Information:

A typical situation in an oil reservoir consists of a certain number of wells or locations $\mathbf{x}$ where the lithology $L(\mathbf{x})$ is known. Let $I(\mathbf{x}; l_k)$ be an indicator random variable which is assumed stationary and defined by:

$$I(\mathbf{x}; l_k) = \begin{cases} 1 & \text{if } L(\mathbf{x}) \in l_k \\ 0 & \text{otherwise} \end{cases} \quad (1)$$

with l_k being one of the K lithologies present in the reservoir. The corresponding expected value is the lithofacies volume proportion p_k.

Assuming a second order process, the facies auto(cross)covariance is:

$$C_I(\mathbf{h}; l_k, l_{k'}) = E[I(\mathbf{x}; l_k)I(\mathbf{x} + \mathbf{h}; l_{k'})] - p_k p_{k'} \quad (2)$$

The facies covariances and crosscovariances describe their relative spatial variability, that is their relative geometry within the reservoir. Correlation range, nugget and shape of the covariances are all linked to that geometry.

Conditional Probabilities:

The conditional probability:

$$Prob(L(\mathbf{x}) \in l_{k_0} | L(\mathbf{x}_\alpha) = l(\mathbf{x}_\alpha) \ , \alpha = 1, \dots, n) \quad (3)$$

requires knowledge of the $(n+1)$-variate joint probability density function (pdf) which is never accessible in a practical situation. If a training image is available, limited statistical information of order two such as the facies covariances and crosscovariances can be extracted from it. Therefore, a model for (3) using K^2 facies covariances and crosscovariances would be:

$$I^*(\mathbf{x}; l_{k_0}) = p_{k_0} + \sum_{k=1}^{K} \sum_{\alpha=1}^{n} \lambda_{k\alpha}[I(\mathbf{x}_\alpha; l_k) - p_k] \quad (4)$$

However, in a practical situation such number K^2 of covariances and crosscovariances would represent a too large modeling effort.

Stochastic Simulation of Lithofacies:

Lithofacies geometry is captured in part by the facies covariances and crosscovariances, therefore the simulation goal will be the reproduction of those measures of two-points spatial continuity. The proposed approach is an extension of the sequential indicator simulation method proposed by Journel (1989) and generalized in Suro-Perez and Journel (1990b) where details can be found.

This sequential algorithm here proposed ensures reproduction of the proportions p_k and all facies auto and crosscovariances $C_I(\mathbf{h}, l_k, l_{k'})$. The key step is the obtention of a model of the conditional distribution (3) by *simple* indicator cokriging, then using MonteCarlo for drawing from (4) a simulated lithofacies for each unsampled node. The procedure finishes when all nodes have been visited and informed with a simulated lithofacies.

Indicator Principal Component Kriging (IPCK)

The approach consists of transforming the original indicator variables (1) into their linear principal components (Suro-Perez and Journel, 1990a). This transformation is done by orthogonalizing the covariance matrix $\Sigma_I(\mathbf{h'})$

$$Cov(\mathbf{I}(\mathbf{x}; l), \mathbf{I}(\mathbf{x} + \mathbf{h'}, l)) = \Sigma_{\mathbf{I}}(\mathbf{h'}) \quad (5)$$

of the indicator vector

$$\mathbf{I}(\mathbf{x}; l) = [I(\mathbf{x}; l_1) \dots I(\mathbf{x}; l_K)] \ . \quad (6)$$

4

That orthogonalization is accomplished by deriving an orthogonal matrix $\mathbf{A}$ from the spectral decomposition of the covariance matrix (5). The indicator vector (6) can then be transformed into the vector of principal components:

$$\mathbf{Y}(\mathbf{x}) = \mathbf{A}^{\mathbf{T}} \mathbf{I}(\mathbf{x}; \mathbf{l}) \qquad (7)$$

with elements $Y_k(\mathbf{x})$, called hereafter facies principal components. After transformation (7), $\mathbf{Y}(\mathbf{x})$ and $\mathbf{Y}(\mathbf{x} + \mathbf{h}')$ are exactly orthogonal. The constitutive hypothesis is that the covariance matrix $\Sigma_Y(\mathbf{h})$ will also be aproximately diagonal for all $\mathbf{h}$.

Conditional Probabilities under IPCK:

If the facies principal component covariance matrix is approximately diagonal for any $\mathbf{h}$, then the estimation of the facies component $Y_k(\mathbf{x})$ can be done independently of $Y_{k'}(\mathbf{x}_\alpha)$, with $k \neq k'$. Practically, $Y_k^*(\mathbf{x})$ is obtained by *simple* kriging instead of the much heavier *simple* cokriging algorithm.

An inverse transformation of expression (7) provides an estimate for the indicator vector (6):

$$\mathbf{I}^*(\mathbf{x}; \mathbf{l}) = \mathbf{A}\mathbf{Y}^*(\mathbf{x}) \qquad (8)$$

Expression (8) can be read as a model for the conditional distribution (3).

Remarks:

- The sum of all K elements of vector (6) is always 1, therefore one of the eigenvalues of the indicator covariance matrix (5) must be 0. This implies that kriging is performed only for the first $K - 1$ components of $\mathbf{Y}^*(\mathbf{x})$ with the last component estimated by its mean:

$$Y_K^*(\mathbf{x}) = p_{Y_K}$$

- The number of covariance models required for IPCK is only $K - 1$ compared to K^2 for the cokriging estimator (4).

- The motivation in using Principal Component Analysis is not dimensionality reduction, but to account indirectly for indicator crosscovariances.

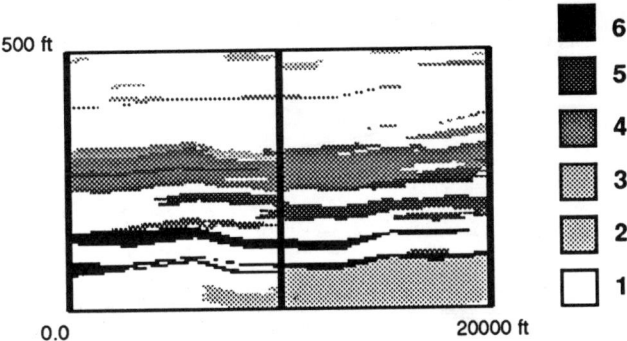

Figure 1: Geological section showing the spatial lithofacies distribution interpreted from 10 actual wells. The solid lines refer to wells to be used as conditioning data for the stochastic simulation.

The Arco Data Set

This data set consists of several interpreted geological sections of an oil reservoir owned by ARCO. The various grey scales of figure 1 correspond to different lithofacies, most often shale lenses intersecting the reservoir sands. These shales act as vertical permeability barriers, isolating production intervals. Permeability profiles as read from the logs are extremely irregular due to rapid changes in lithofacies. Therefore, a prior proper lithofacies characterization is essential to correctly map the permeability field. The procedure consisting of simulating permeability across all facies would not be appropriate in this case. Conditional statistics did show the dependence of permeability statistics on the lithofacies.

Figure 1 shows one of the interpreted geological sections as provided by Arco geologists. Six different lithofacies are observed along almost horizontal layers. This particular section is intersected by 10 different wells. Therefore, an abundance of information derived from both core data and well logs is available.

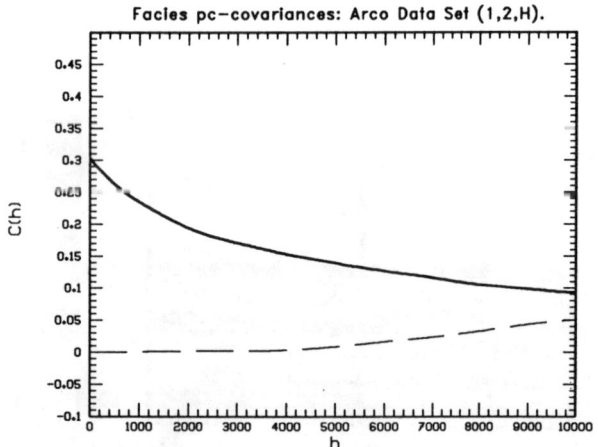

Figure 2: PC-covariances in the horizontal direction. The solid line refers to the first component covariance and the dashed line to the crosscovariance between the first and second component.

Facies PC-Covariances

Figures 2 to 5 show some of the pc covariances and crosscovariances in the horizontal and vertical direction. The pc covariance matrix in the horizontal direction is seen to be well approximated by a diagonal matrix. None of these pc covariances seems to show a significant nugget effect. The vertical correlation range for the last three components are quite smaller. Within the autocorrelation ranges the level of crosscovariance compared to that of autocovariance is seen to be insignificant.

Two nested exponential structures were considered to model each facies covariance. A large horizontal to vertical anisotropy is adopted for all cases.

Conditional Simulation of Lithofacies

Three of the many equiprobable stochastic simulations of the facies generated are presented. The conditional information consists of three wells located at the edges and at the middle of figure 1. Figures 6 to 8 show the corresponding realizations. The patterns of connectivity are seen to be well-reproduced and images are "close" to the reference image of figure 1; in such comparison one should account for the fact that the exhaustive ge-

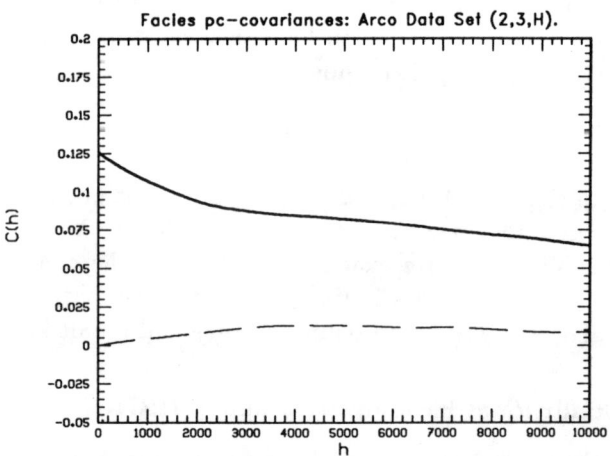

Figure 3: PC-Covariance in the horizontal direction for the second component (solid line) and crosscovariance between the second and third component (dashed line).

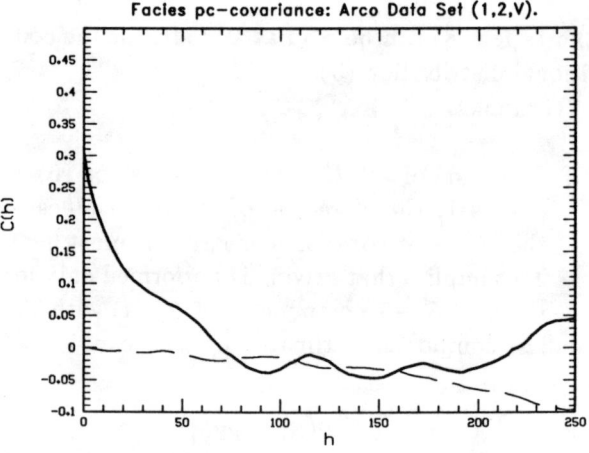

Figure 4: PC-Covariance in the vertical direction for the first component (solid line) and crosscovariance between first and second component (dashed line).

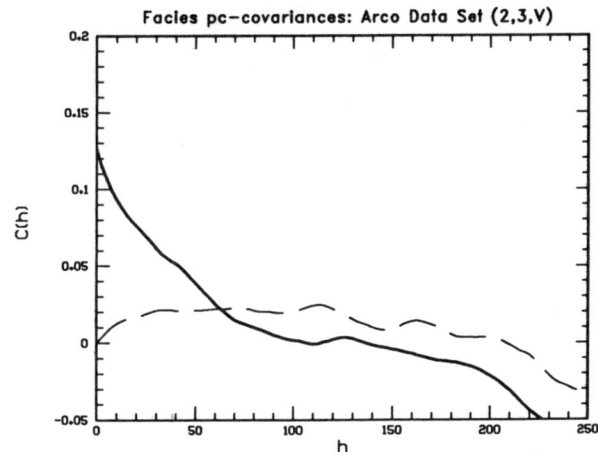

Figure 5: PC-Covariance in the vertical direction for the second component (solid line) and crosscovariance between second and third component (dashed line). Observe that in all the cases the level of crosscovariance compared to the covariance level is insignificant.

ological section (figure 1) was built considering 10 wells whereas the conditional simulations considered only *three* wells.

The proportions are also well reproduced, even for those lithologies in small proportions. Figures 9 to 15 present a comparison of the exhaustive facies indicator covariances and crosscovariances as calculated from figure 1 and the corresponding covariances of the simulation of figure 8, for the horizontal and vertical directions. In all the cases, covariances and crosscovariances were well reproduced.

Analysis of the different stochastic images provides information about heterogeneities, flow paths and flow barriers.

Conclusions

Reservoir description can be improved if lithological information is used explicitly prior to simulation of continuous attributes such as as porosity, permeability and saturations. Also prior consideration and simulation of lithofacies geometries allows relaxing the stationarity hypothesis: different models for the spatial distributions of the continu-

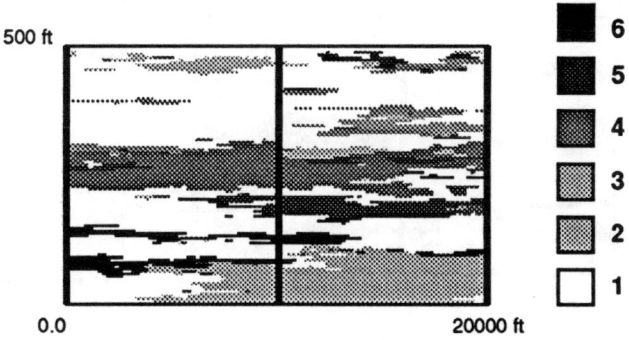

Figure 6: First lithofacies simulation considering three conditioning wells. The volume of each lithofacies is reproduced together with the corresponding facies covariances and crosscovariances.

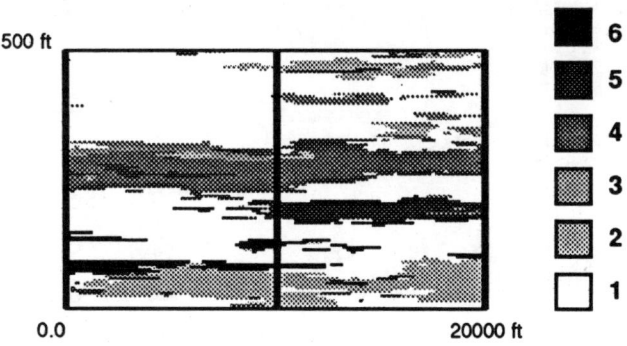

Figure 7: Second lithofacies simulation. Note how the change in behavior of the lithofacies 2 at the bottom of the section.

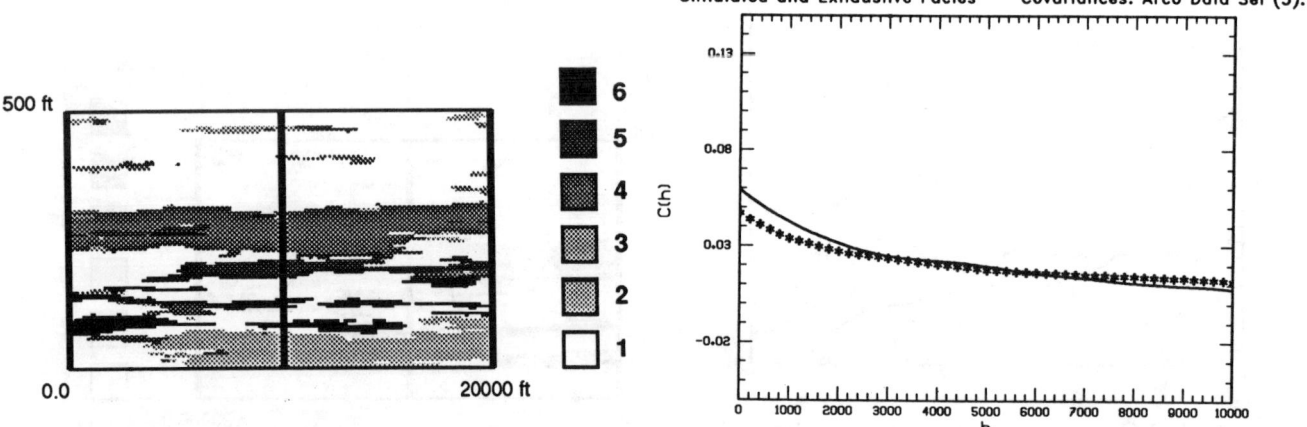

Figure 8: Third lithofacies simulation. In all simulations, lithofacies 4 showed the least variation since the three conditioning wells did intersect it.

Figure 10: Comparison between the exhaustive horizontal covariance (solid line) for lithofacies 5 and the simulated covariance (asterisks) derived from figure 8. Note the reproduction of the large correlation range.

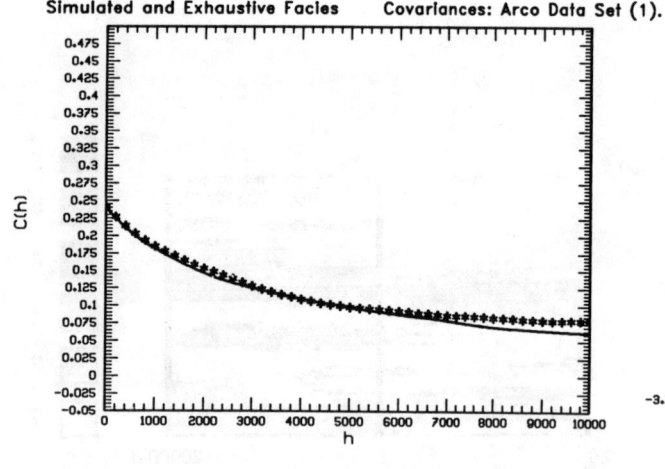

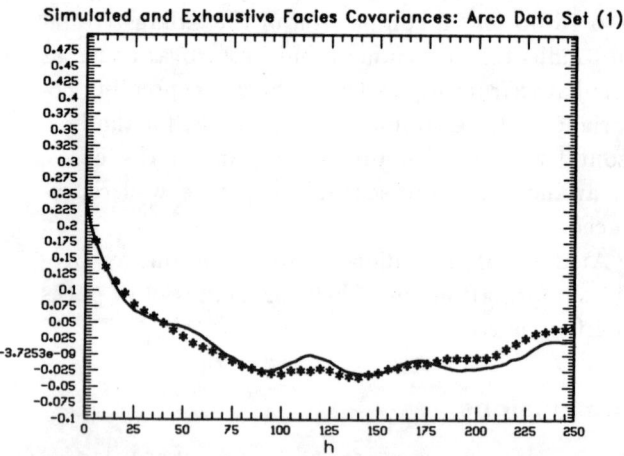

Figure 9: Comparison between the exhaustive horizontal covariance (solid line) for lithofacies 1 and the simulated covariance (asterisks) derived from figure 8.

Figure 11: Comparison between the exhaustive vertical covariance (solid line) for lithofacies 1 and the simulated covariance (asterisks) derived from figure 8.

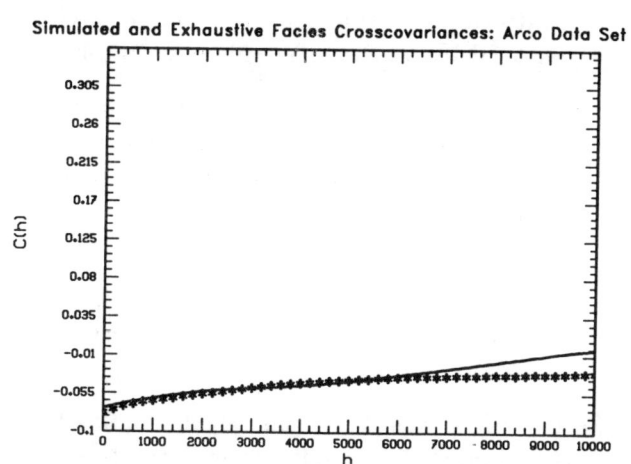

Figure 12: Facies crosscovariances for lithofacies 1 and 2 in the horizontal direction. The exhaustive crosscovariance (solid line) shows a shorter correlation range than that derived from the simulation of figure 8.

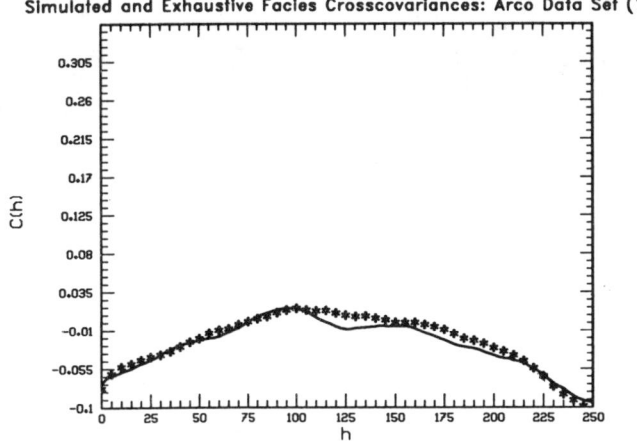

Figure 13: Facies crosscovariances for lithofacies 1 and 2 in the vertical direction. The exhaustive crosscovariance (solid line) is well reproduced by that crosscovariance extracted from figure 8.

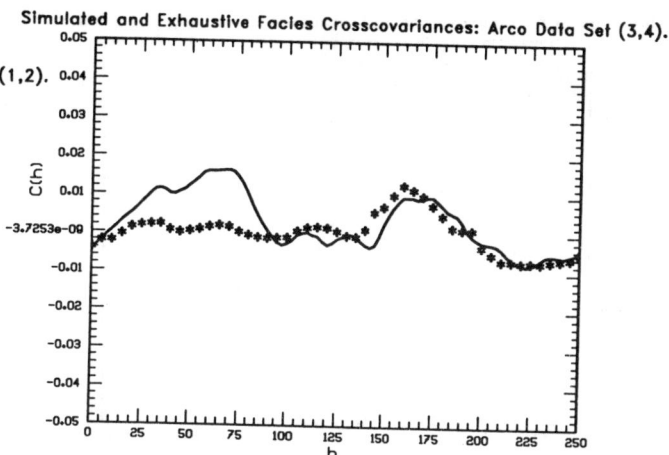

Figure 14: Facies crosscovariances for lithofacies 3 and 4 in the vertical direction. The exhaustive crosscovariance (solid line) is not well reproduced for the shorter distances by that crosscovariance extracted from figure 8 (asterisks).

ous variables may be considered for different facies.

Upscaling of continuous properties done within lithofacies-homogeneous blocks is also made easier (King, 1988). The subsequent flow simulation will reflect the geometric architecture (or flow piping) of the reservoir which is, possibly, the most critical aspect of that reservoir heterogeneity. Repetitive generation of equiprobable lithofacies realizations provides quantitative information about the impact of such heterogeneities on oil recovery. It can point out areas of large uncertainty where additional information might be required.

IPCK, as applied to lithofacies simulation, appears as an effective tool for describing heterogeneities in oil reservoirs. There seems to be no significant loss of structural information by considering that the pc-covariance matrix is diagonal. This diagonal character in practice needs to be verified. Current, although limited experience, has not shown any case where this diagonal condition was not honored to a sufficient degree of approximation.

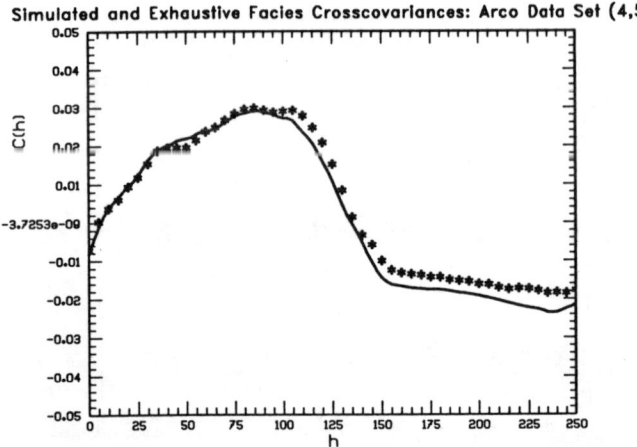

Figure 15: Facies crosscovariances for lithofacies 4 and 5 in the vertical direction. The exhaustive crosscovariance (solid line) is well reproduced by that crosscovariance extracted from figure 8 (asterisks).

Acknowledgements

ARCO Oil and Gas Co. is gratefully acknowledged for providing the data used in this project. Special thanks to Steve Crane of ARCO Plano who provided much needed advise.

References

King, P. R., 1988, Effective values in averaging, in **Mathematics in Oil Production**, eds. S. Edwards and P. R. King, Oxford Science Publications.

Haldorsen H. H., P. J. Brand and C. J. Macdonald, 1988, Review of the stochastic nature of reservoirs, in **Mathematics in Oil Production**, eds. S. Edwards and P. R. King, Oxford Science Publications.

Journel A. G. and E. Isaaks, Conditional Indicator Simulation: Application to a Saskatchewan uranium deposit, Mathematical Geology, v. 16, pp. 685-718.

Journel A. G. and J. Gomez-Hernandez, 1989, Stochastic imaging of the Wilmington clastic sequence, SPE paper # 19857.

Journel, A. G., 1989, **Fundamentals of Geostatistics in Five Lessons**, American Geophysical Union Press, Washington DC, 40 p.

Matheron G., H. Beucher, Ch. de Fouquet, A. Galli and Ch. Ravenne, 1988, Simulation conditionnelle a trois facies dans une falaise de la formation du Brent, Sci. de la Terre, Ser. Inf., v. 28, pp. 213-249.

Suro-Perez V. and A. G. Journel, 1990a, Indicator Principal Component Kriging: Theory, submitted to Mathematical Geology.

Suro-Perez V. and A. G. Journel, 1990b, Stochastic Simulation of Lithofacies for Reservoir Characterization, Report 3, SCRF, Stanford University.

2nd European Conference on the Mathematics of Oil Recovery
© D. Guérillot, O. Guillon (Editors) and Éditions Technip, Paris 1990, pp. 11-19
27 rue Ginoux, 75015 Paris

Combining Geology, Geostatistics and Multiphase Fluid Flow for 3D Reservoir Studies

A. Galli[1], D. Guérillot[2], C. Ravenne[2] and HERESIM Group[1,2]

ABSTRACT

A geostatistical model is used to generate a 3D lithofacies model between wells. Then fluid flow simulations are performed in this 3D geometry. The geological investigations and the main characteristics of the geostatistical model are recalled and comments on the simulation and the fluid flow behavior are made.

INTRODUCTION

In most of the reservoir engineering studies, heterogeneities must be taken into account and their influence on fluid flow behavior has been known to be important for about 30 years (Warren and Price, 1961).

New geological studies based mainly on outcrop observations have opened new perspectives on reservoir characterization. Most of the outcrop studies use new sequential stratigraphy concepts. Moreover the use of geostatistical tools (Matheron, 1987) has made it possible to quantify natural observation. The geological work done by the HERESIM Group is partially reviewed here.

We then recall the main features of the *geostatistical model* which was elaborated. A new challenge is to integrate such 3D imaging of reservoir heterogeneities in the process of production forecast using a classical "reservoir simulator".

As an illustration we propose a 2 phase fluid flow displacement in a five spot study. Even in such a simple case, the analysis of the results open new perspectives for research.

1. GEOLOGICAL METHODOLOGY

To develop a modelling method that is applicable to reservoirs of the fluvio–deltaic type, two different problems must be dealt with : (1) finding an algorithm capable of properly modelling the geology in terms of lithofacies, and then testing it on a series of well–known cases to prove or disprove it, (2) determining the essential sedimentological data to be incorporated in such a model to obtain as realistic images of the reservoir as possible.

At the start of the HERESIM project, it was obvious that, in order to solve the two problems mentioned above, for the methodological part it was not sufficient to work with subsurface reservoirs. A larger geological databank had to be available to obtain perfectly coherent images for use in working out and testing the method and also to gain access to a maximum degree of geological knowledge so as to assess its validity.

To achieve this, it was decided to analyse outcrops and associated core samples to gain 3D information.

The great extent of vertical and lateral shale and sand interbedding in the fluvio–deltaic series in the North Sea reservoirs led us to examine Middle Jurassic outcrops from cliffs in Yorkshire, Great Britain. The cliff chosen gives a remarkable display of a great variety of fluvial to fluvio–deltaic deposits over a length of about 10 km. The first studies concerned two formations in the Ravenscar group, i.e. the Cloughton formation and the Scalby formation (Ravenne et al., 1987, 1988).

This first study as well as ongoing ones demonstrate as a general rule the need for breaking the entire reservoir into

---

(1) Centre de Géostatistique, École des Mines de Paris, 35, rue Saint-Honoré, 77305 Fontainebleau, France.
(2) Institut Français du Pétrole, 1 et 4, avenue de Bois-Préau, 92506 Rueil-Malmaison, France.

sedimentary units that are genetically linked. This means :
(1) determining the unit boundaries, (2) determining
reference level in each unit, (3) obtaining the sequential
order of each unit. This kind of study must be carried out
with sequential stratigraphic technics.

2. THE GEOSTATISTICAL MODEL

Although a reservoir is a purely deterministic object,
a partial understanding of it from both borehole and seismic
data leads to a description of it by using probabilistic
methods. These data coupled with an understanding of the
sedimentological environment cannot produce a single
image of the reservoir. Hence it is preferable to aim for a
probabilistic tool since simulations can lead to obtaining
representations of the reservoir that are not the only
possible ones but that all verify two prime features :

A. They respect the heterogeneity of the reservoir.
Such models must respect the spatial distribution of the
heterogeneities from a statistical viewpoint. One tool for
this is the variogram, which enables this spatial distribution
to be quantified.

B. Such representations of the reservoir may vary
from well to well, but all of them must represent the reality
known in the wells. They are called conditional simulations.

The model developed by the Heresim Group for the
conditional simulation of the lithofacies of the reservoir is
based on a Gaussian random function model. This will be
decribed briefly below (Matheron et al., 1987, de Fouquet
et al., 1988).

The basic idea behind this model is to describe the n
lithofacies using one indicator function per facies. The
indicator for facies i is defined by using a Gaussian random
function $Y(x)$ by : "the point x belongs to facies i if $Y(x)$ is in
the interval (a_{i-1}, a_i)". The a_i values are directly related to
the proportion of each lithofacies by means of the
proportion curves. For the case described here, a_i is a
function of depth z only and is directly related to what is
called the vertical proportion curve $p_i(z)$, which is the
proportion of lithofacies i at level z. These proportion curves
are the first parameters of the model. The other parameters
are the variograms and cross-variograms of the lithofacies.
It can be shown that they are directly related to the
variogram of the Gaussian random function $Y(x)$ (Matheron
et al., 1988). Hence to fit a model for the variograms and
cross-variograms of the lithofacies, we choose a variogram
for $Y(x)$, and from this variogram we compute the
corresponding variograms and cross-variograms for the
lithofacies. We then check whether it fits with the
experimental variograms and cross-variograms for the
lithofacies. If they do not, we select another variogram for
Y and continue until a correct fit is reached.

3 THE CASE STUDY

Through this case study, we follow the general
methodology defined above. The data used reflect actual
situations (§ 3.1). The variogram model (§ 3.2) is then
described. We then have all the paramaters for simulating
the 3D geometry (§ 3.3). and fluid flow behavior can be
calculated (§ 3.4).

3.1 GEOLOGICAL BACKGROUND

For the five spot study, we used wells from the
Ravenscar area in Yorkshire. These wells have crossed the
entire Middle Jurassic, but for the purpose of the study we
have concentrated on the 30 m corresponding roughly to the
Saltwick formation, which is dated Aalenian to Bajocian.

These data have been slightly modified :

a) the location of the wells is slightly different,

b) the lithologies of the basal part, which belongs to
another formation displaying different geological
properties were modified, and

c) the sandstone proportion was increased by
approximately 10%.

Four of the five wells are located at the corners of a
400 m square, and the fifth one at the center. The thickness
of the unit under study is 30 m. The wells were described
with four facies : channel-fill sandstone, slightly
argillaceous channel-fill sandstone, argillaceous channel-
fill or crevasse splay sandstone, delta-plain shale (Fig. 1).

The main environment here is an aggrading
delta-plain. In such an environment, the distribution of the
channels (sandstone reservoir) is quite "stochastic".

For the purpose of the simulation, we have used a
marker (coal-bed) as a reference level for the flattening of
the sedimentary unit. Such a reference level is important
when computing variograms and proportion curves. This
marker is located in the overlying sequence (sequence 3, see
Eschard, 1989, Ravenne et al., 1988).

These two sequences are assumed to be deposited
conformably to this reference level.

In the sequence under study, we can see on the
proportion curves computed from the five wells (Fig. 2) a
major sandstony peak located in the middle (according to
the reference level). It corresponds to a sandy layer
attributed to a major channeling level (probably due to a
decrease or a stop in the aggrading rate). Some peaks
appear, mainly in the upper part; they correspond to
sandstone channels disseminated in the delta-plain. On the
main sandstony peak in the middle, we observe an upward
increase of shale proportion, at a decametric scale. This is
quite typical of such channels in a delta- plain. The
proportion curves were computed with only five wells but by
our own experience, they do not entirely reflect the
variability of this type of environment. This is why we have
modified these proportions. The new proportion curves,

presented on Fig. 3, follow the same broad lines as those defined by the wells, and according to the geologist's know-how, they reflect the general environment.

3.2 THE MODEL PARAMETERS

The model used for the Gaussian random function is a product of 3 exponential models, one per direction. The ranges are 150 m in both the N–S and the E–W directions, and 5 m along the vertical direction.

These were selected experimentally from the Ravenscar outcrop. The outcrop being roughly oriented North–South, and furthermore, given the scale of the study, there was no evidence of anisotropy, and we therefore decided to use the same range in the E–W direction. The proportion curves used are those modified as shown on Fig. 3.

3.3 THE LITHOFACIES SIMULATION

We performed a geostatistical simulation in lithofacies on a 600 m × 600 m × 30 m block, centered on the central well and discretized into 20 m × 20 m × 1 m cells. The depth is –100 m at the bottom and –70 m at the top. This simulation was conditioned by the five wells.

For the presentation of the simulations, we have corrected the general structural dip (NE–SW), but we have kept the minor deformation. This simulation is therefore not entirely parallel to the flattened reference level, and it takes the minor structuration into account.

In the middle of the two vertical cross–sections (Figs. 4, 5), we observe a major layer with high sand content. This is in agreement with the idea of the geologist and it reflects the large extension (at the scale of the study) of the meandered belt. Above and below, we can see small patches of sandstone, corresponding to the channels disseminated in the deltaic plain. On the NE–SW section, we observe the effect of the minor structuration of the main sand layer dipping south–eastward. This effect can also be seen on the horizontal section (Figs. 6, 7, 8, 9).

On the simulation, we computed the connex components of sandstone and shaly sandstone. (Two grid nodes are said to be connected if they belong to sandstone or to shaly sandstone and if they have a common edge. A connex component is the collection of the connected grid nodes). Figs. 10, 11, 12, 13, 14, 15 show the largest connex component found. Its volume is 28% of the volume of the block. Looking at this component, it is interesting to see that some of the disseminated channels are connected to the main channel layer. This is probably due to the short vertical distance between these and the major sand unit. Fig. 11 clearly shows how important the third direction is for connexity.

3.4 FLUID FLOW SIMULATION

After the 3D imaging of the reservoir, the three following steps must be considered :

Step 1 : Assigning petrophysical values to the lithofacies.

Step 2 : Discretizing the reservoir for the fluid flow simulator and averaging fine scale petrophysical values for the grid cells of the reservoir simulators.

Step 3 : Simulating the fluid flow behavior with suitable source terms, initial and boundary conditions (flow rates and pressure at the wells, fluids in place, type of aquifer, no–flow boundaries, etc.)

Here we have considered a simple case of secondary recovery : water injection in an oil reservoir. Let us describe this two–phase non–missible displacement following the above steps.

Step 1 : The same petrophysical properties are associated to the two lithofacies with the highest content in shale. Table 1 sums up some petrophysical characteristics, and Fig. 16 illustrates the relative permeability curves (capillary effects are neglected). Table 2 contains the viscosities and the densities of water and oil. More generally, the number of lithofacies to be considered for the fluid flow does not need to be the same as for the lithofacies simulations. Few petrophysical data are generally available, and it is not necessary to quantify the lithofacies in great detail. However, to check the geological sequences of the deposit and to assess the validity of the simulation, it is important to consider all the facies.

It is in this step that diagenesis effects can be taken into account, giving for instance the trends in absolute permeabilities along the vertical direction (Rudkiewicz et al., 1989). The dissociation between (1) the lithofacies simulation and (2) the quantification of the geometry gives a great flexibility when dealing with actual case studies.

Step 2 : The same mesh has been used for the reservoir simulator. (The multipurpose reservoir model $\Sigma CORE$ is from *ELF*, *IFP*, *GDF* and *CFP*). We have restricted the grid used for fluid flow to the part containing the wells. The total number of grid cells is 13,320 which is usual for fluid flow simulations. It is not therefore necessary to average the petrophysical values to assign them to the reservoir simulator mesh.

For heterogeneities with average lengths smaller than the reservoir grid cells, the petrophysical values need to be averaged, which is a complex process for non–additive parameters such as absolute permeability (Matheron, 1967, 1968, 1979, Le Loc'h, 1989, Guérillot, 1989). To average the relative permeability and capillary pressure curves, two different aspects must be considered : (1) the modification of the parametrers of the generalized Darcy's equations and even their transformation (Quintard, 1988, Amaziane and Bourgeat, 1988), and (2) the modification of relative permeability curves for numerical dispersion which may occur even with an homogeneous porous medium.

Step 3 : The constant injection rate is 400 m³ per day in the central well, and the production rates are 100 m³ per

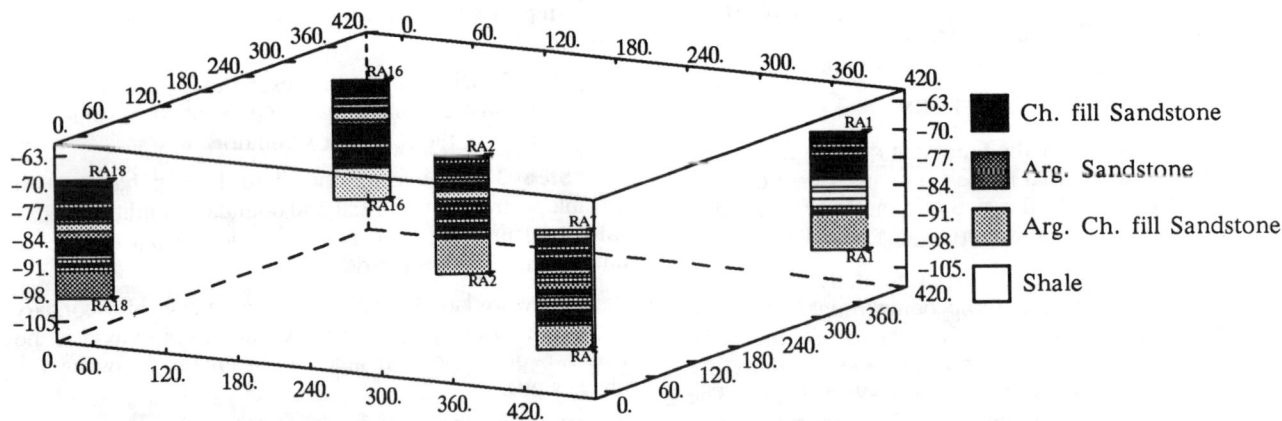

Figure 1 : Five spot

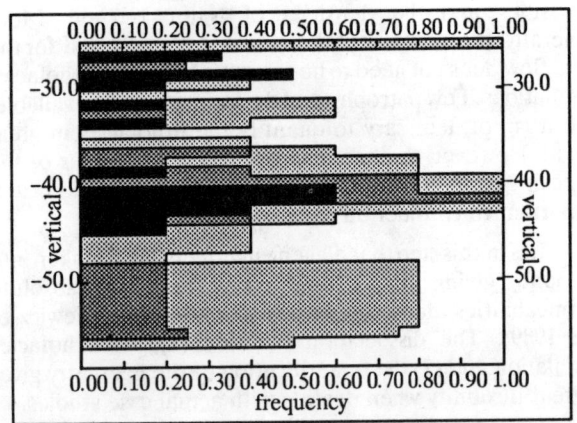

Figure 2 : Raw vertical proportions
Depth from the reference level

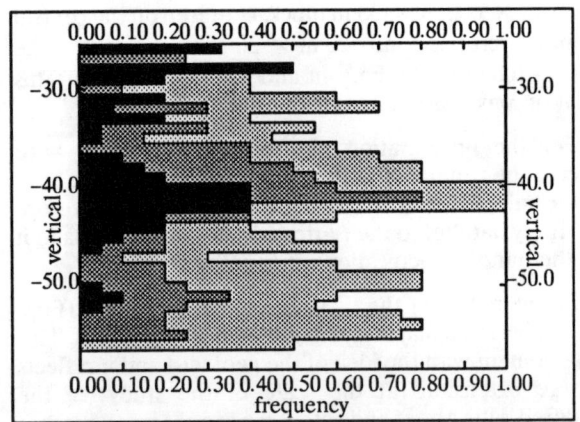

Figure 3 : Modified vertical proportions
Depth from the reference level

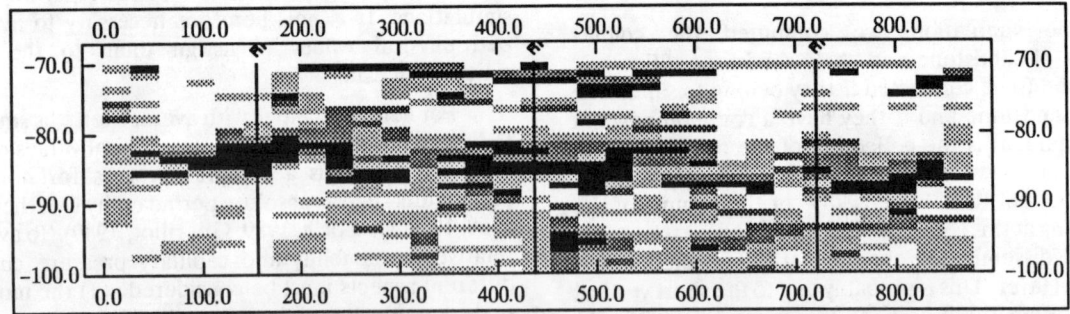

Figure 4 : Lithofacies simulation – Section RA16 – RA7

14

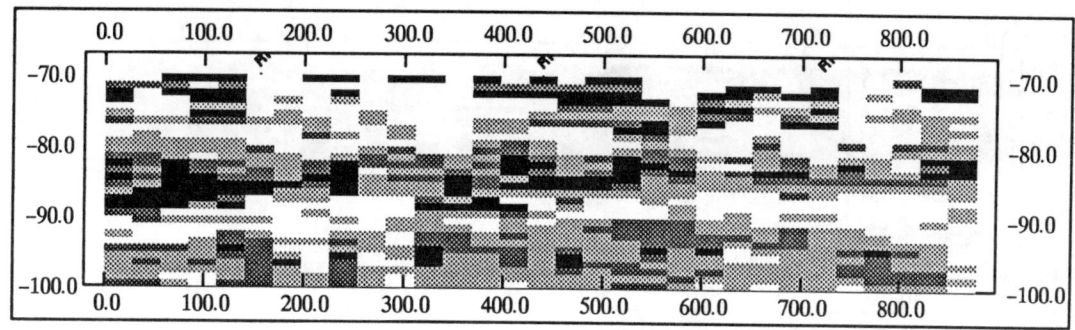

Figure 5 : Lithofacies simulation – Section RA18 – R17

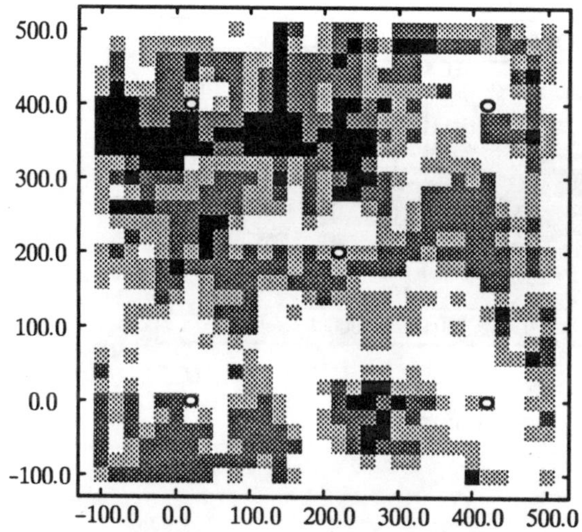

Figure 6
plane
z = 81

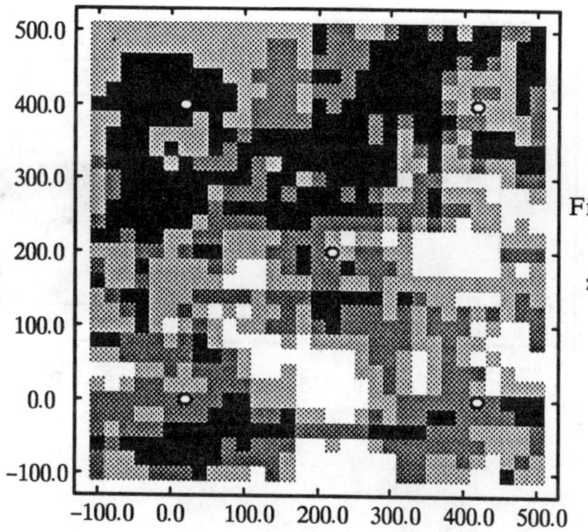

Figure 7
plane
z = 84

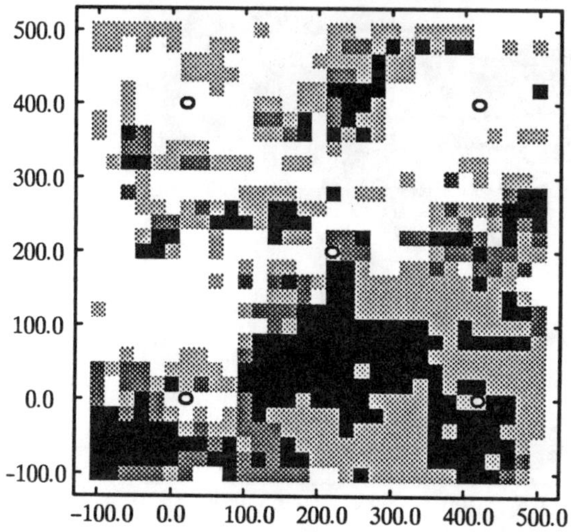

Figure 8
plane
z = –88

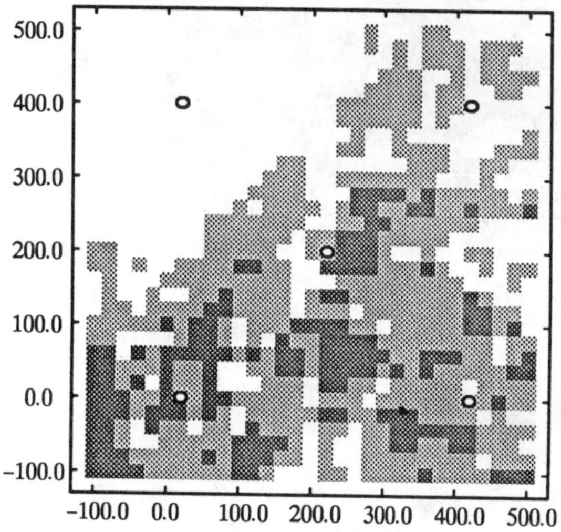

Figure 9
plane
z = 91

Comments for Figs 6,7,8,9 : Horizontal sections from the 3–D simulation
The o represents the wells

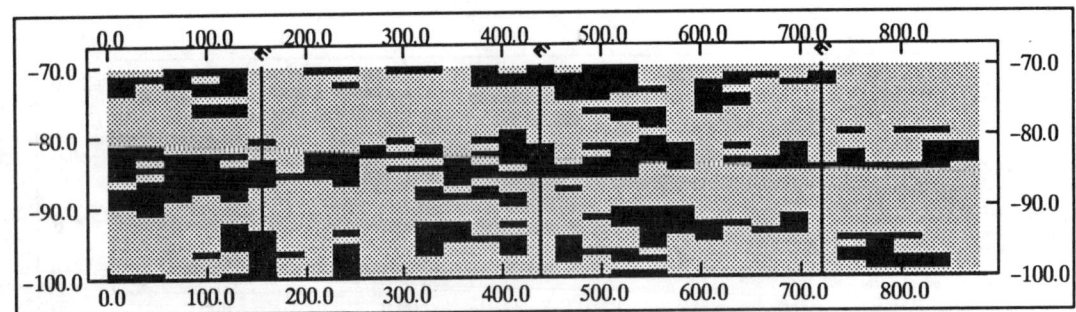

Figure 10 : Section RA18 – RA1 . The largest connex component
The black points belong to the the <u>same</u> connex component

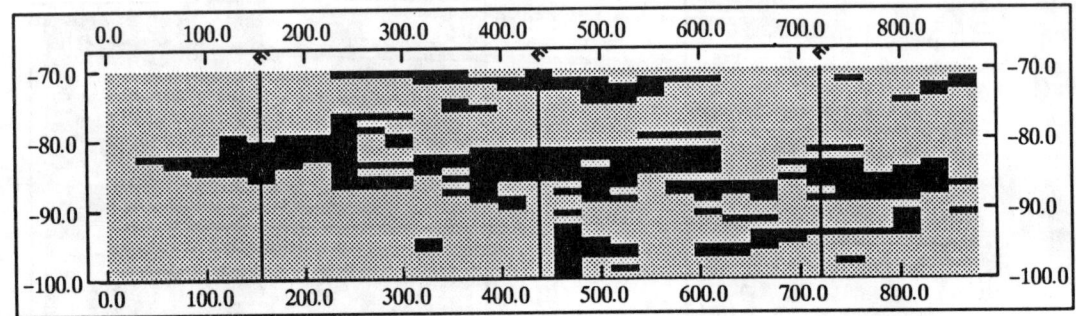

Figure 11 : Section RA16 – RA7 . The largest connex component
The black points belong to the the <u>same</u> connex component

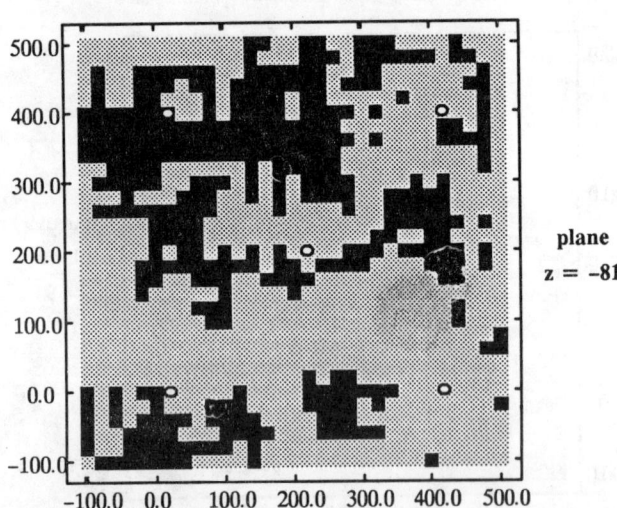

Figure 12 : The largest connex component
for simulated plane at level z = −81

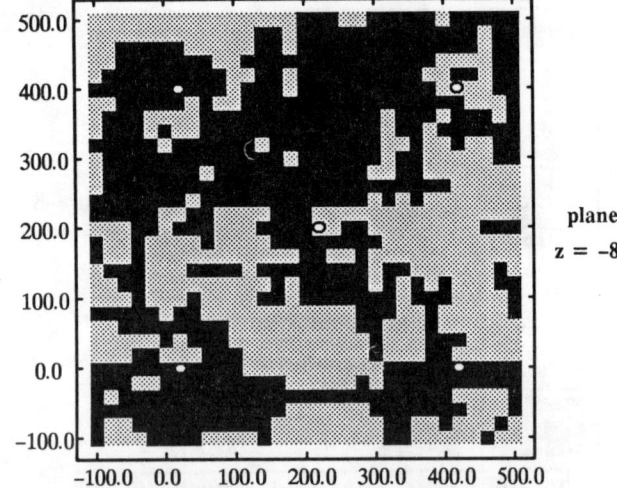

Figure 13 : The largest connex component
for simulated plane at level z = −84

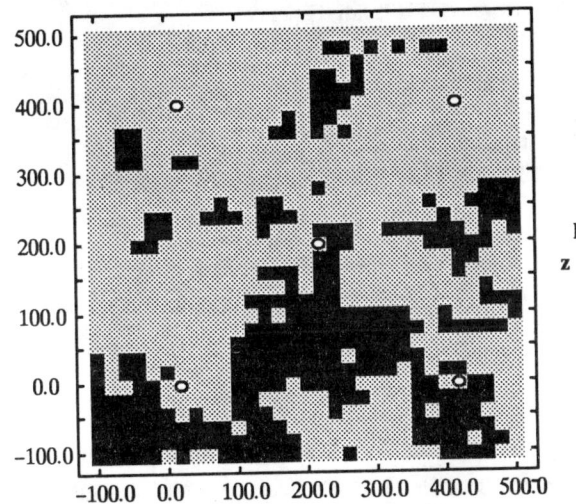

Figure 14 : The largest connex component
for simulated plane at level z = −88

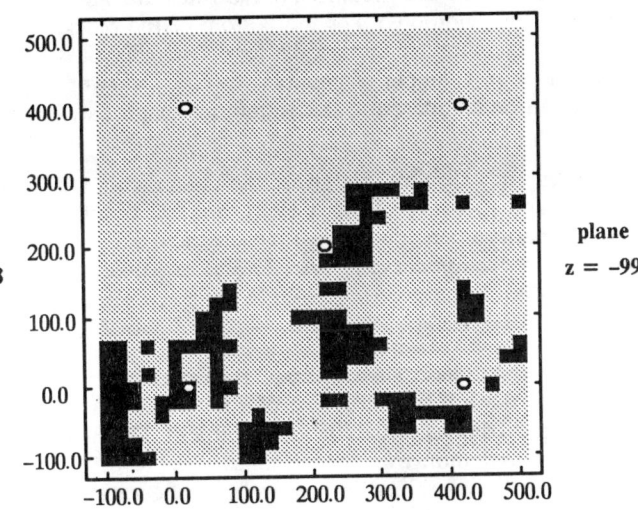

Figure 15 : The largest connex component
for simulated plane at level z = −99

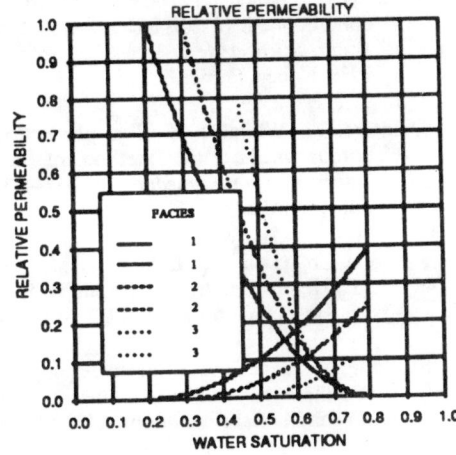

Figure 16 : Relative permeability

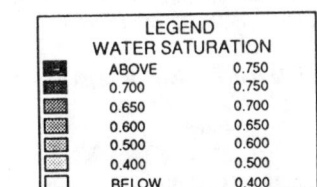

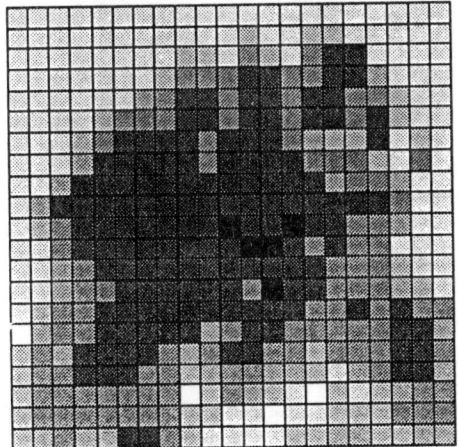

Figure 17 : Water saturation at the level
z = −99 after 400 days

Figure 18 : Water saturation at the level
z = −99 after 800 days

day for each well in the corners. (These rates are given under reservoir conditions). For the bottom of the block, water saturation maps are displayed after 400 and 800 days of injection as an example of possible results. The swept efficiency can be compared to the lithofacies map of this layer (Fig. 16,17).

Figures 18, 19, 20, 21 gives the water–cut, oil production and recovery of the 4 wells. We can observe differences between each breakthrough time and between cumulative production curves.

CONCLUSION

This article has shown how the geostatistical description of the internal structure of a reservoir can be used for a typical diphasic fluid flow displacement. The fluid flow calculations demonstrate the influence of the reservoir heterogeneities, constrained by the wells configurations, correlation lengths and the proportions of lithofacies, and the contrast of petrophysical values, on the well productions.

There is a new challenge to integrate the knowledge of quantified geology to improve the production forecasts. The methodology described here is not yet proven, but the results have been encouraging so far.

REFERENCES

Amaziane, B., Bourgeat, A., Numerical simulation in oil recovery, The IMA vol in Mathematics and Applications, Vol. 11, 1988, pp. 1–22, ed. by Wheeler, Springer Verlag.

Eschard, R., *Géométrie et dynamique de séquences de dépots dans un système deltaïque*, Thèse de 3ème Cycle, Juin 1989, Université Louis Pasteur, Strasbourg.

de Fouquet, Ch., Beucher, H., Galli, A., Ravenne, Ch., Conditional simulation of random sets - Application to an argillasceous sandstone reservoir, in *Geostatistics*, Proc. 3rd International Geostatistics Congress, ed. by M. Armstrong, Vol. 2, pp. 517–530, Kluwer Academic Publ., Dordrecht, 1989.

Guérillot, D., Rudkiewicz, J.L., Ravenne, C, Renard, G., An integrated model for computer aided reservoir description: from outcrop study to fluid flow simulations, Fifth European

Symp. on Improved Oil Recovery, Budapest, April 25–27, 1989, Revue de l'IFP, Jan./Fév. 1990.

Le Loc'h, G., An efficient stategy for combining the permeabilities: practical applications on a simulated reservoir, in *Geostatistics*, Proc. 3rd International Geostatistics Congress, ed. by M. Armstrong, Vol. 2, pp. 557–568, Kluwer Academic Publ., Dordrecht, 1989.

Matheron, G., *Eléments pour une théorie des mileiux poreux*, Masson, Paris, 1967.

Matheron, G., Composition des perméabilités en milieux pôreux hétérogènes: critique de la règle de pondération géométrique. *Revue de l'IFP*, 1968, 2, pp. 201–218.

Matheron, G., L'émergence de la loi de Darcy, 1979, CGMM, ENSMP, Fontainebleau.

Matheron, G., Beucher, H., de Fouquet, Ch., Galli, A., Guérillot, D., Ravenne, Ch., Conditional simulations of the geometry of fluvio–deltaic reservoirs, 1987, SPE 16753.

Matheron, G., Beucher, H., de Fouquet, Ch., Galli, A., Simulations conditionnelles à trois facies d'une falaise de la formation du Brent, *Sciences de la Terre*, Série Informatique Géologiuqe, 1988, Nancy, France.

Quintard, M, Whitaker, S., Two–phase flow in heterogeneous porous media: the method of large–scale averaging, *Transport in porous media 3*, pp. 357–413, 1988, Kluwer Acad. Publ., Dordrecht.

Ravenne, Ch., Galli, A., Mathieu, Y., Montadert, L., Rudkiewicz, J.L., Heterogeneities and geometry of sedimentary bodies in a fluvio–deltaic reservoir, 1987, SPE 16752.

Ravenne, Ch., Beucher, H., Recent developments in the decription of sedimentary bodies in a fluvio–deltaic reservoir and their 3D conditional simulations, 1988, SPE 18310.

Rudkiewicz, J.L., Guérillot, D., Galli, A. and Group Heresim, High definition reservoir lithology and property prediction with an integrated software - Application to the Yorkshire Middle Jurassic Formation, 2nd International Conf. on North Sea Oil and Gas Reservoirs, Trondheim, 1989.

Warren, J.E., Price, H.S., Flow in heterogeneous porous media, Soc. Pet. Eng. J., 1961, Vol. 1, pp. 153–169.

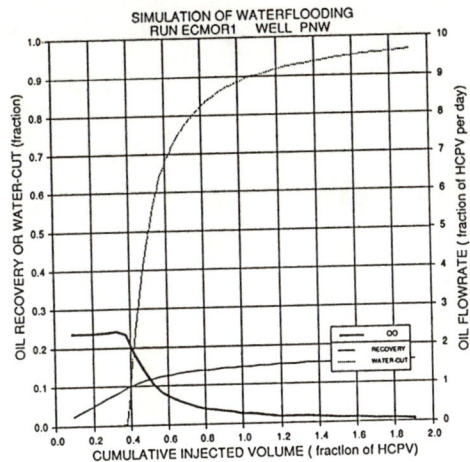

Figure 19 : Water–cut, oil production and recovery for well RA16

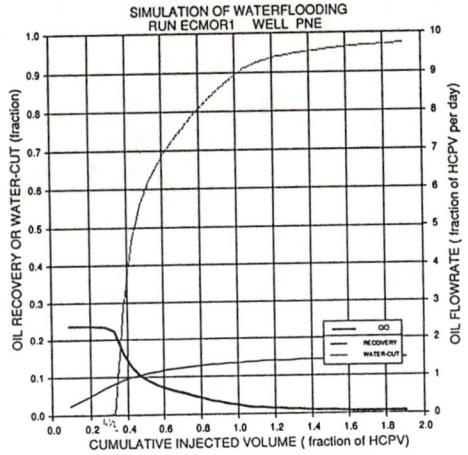

Figure 20 : Water–cut, oil production and recovery for well RA1

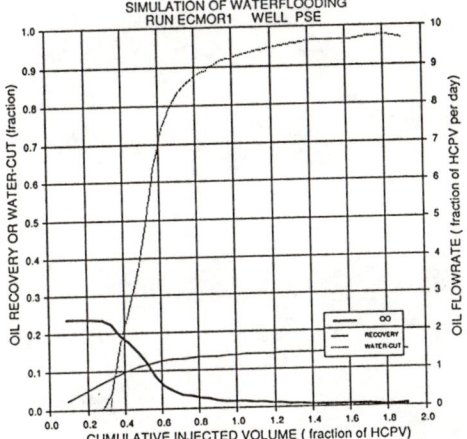

Figure 21 : Water–cut, oil production and recovery for well RA18

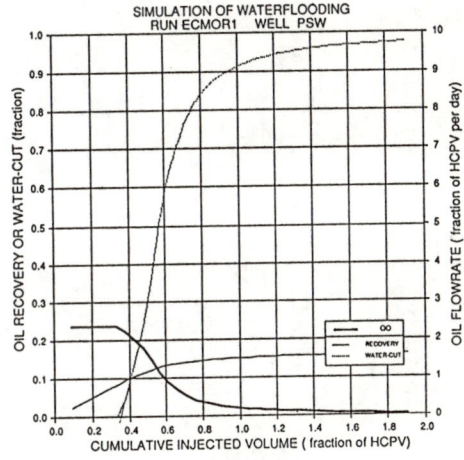

Figure 22 : Water–cut, oil production and recovery for well RA7

Facies	1	2	3 et 4
Porosity	0.20	0.10	0.05
Horizontal permeability (md)	1000	10.00	1.00
Vertical permeability (md)	100	1.00	0.10
Irreducible water saturation Swi	0.2	0.30	0.45
Residual oil saturation Sor	0.2	0.20	0.25
Kro (Swi)	1.	1.00	0.80
Krw (Sor)	0.4	0.25	0.05

Table 1 : Petrophysical characteristics

	Water	Oil
Viscosity (cp)	1.00	2.00
Density	1.00	0.85

Table 2 : Fluid characteristics

19

2nd European Conference on the Mathematics of Oil Recovery
© D. Guérillot, O. Guillon (Editors) and Éditions Technip, Paris 1990, pp. 21-30
27 rue Ginoux, 75015 Paris

Development of Geostatistical Methods Dealing with the Boundary Conditions Problem Encountered in Fluid Mechanics of Porous Media

A. Dong[1], S. Ahmed[2] and G. de Marsily[3]

ABSTRACT

Until recently, partial differential equations (pde) and geostatistics were seldom associated. Yet, in earth sciences, examples of phenomena controlled by these equations are numerous: fluid flow or heat transfer...The presence of a pde implies three main points:

- **the covariances used to describe the variables involved in such an equation must be consistent with it.** If we consider two variables Z and Y linked by the Poisson equation (1) where Δ is the Laplacian operator and if we want to cokrige Z using data on Z and Y then we need to know their simple and cross–covariances. **If Y has a given covariance, that of Z is linked to this covariance by a relation deriving from (1).**

$$\Delta Z = Y \qquad (1)$$

- as it seems natural to constrain an estimator to honour the data, **one might insist that the estimator satisfy the same pde.** In the case of equation (1), this means that the estimator Z^* is expected to satisfy the following equation at any point where Y is known:

$$\Delta Z^* = Y$$

- the third point appears if we consider the direct problem (for instance estimation of Z using data on Y). **How can we involve the boundary conditions in the geostatistical estimation?**

In this paper, we study equation (1) and the modified Poisson equation (x is one direction of the space)

$$\Delta Z = \frac{\partial Y}{\partial x} \qquad (2)$$

In most of the papers published in literature, a common hypothesis was considered: Z and Y were often supposed to have the same "degree of stationarity" i.e. they were assumed to be both stationarity random functions (SRF) with real stationary covariances or both intrinsic random functions of order 0 (IRF–0) with variograms. Thus a whole set of solutions of the problem have been ignored because they do not obey the previous hypothesis.

Here, we make a brief review on how different authors have involved pdes in the structural analysis and in the estimation. Then we point out how some non–stationary solutions can be obtained using the theory of Intrinsic Random Functions of order k (IRF–k). Structural analysis in this theory is exposed in detail for (1) and we place special emphasis on the boundary conditions and on the fact that the estimator satisfies the initial pde. Equation (2) is studied from the inverse problem point of view. Z and Y are variables related to transmissivity and hydraulic head. The structural analysis is deduced from the analysis made for equation (1). Lastly, we deal with cokriging in the case of the inverse problem; our aim is to estimate transmissivity using data of transmissivity and hydraulic head. A case study is presented: the aquifer comes from a chalk area in the north of France.

(1) Centre de Géostatistique, École des Mines de Paris, 35, rue Saint-Honoré, 77305 Fontainebleau, France.
Now with DETN, Gaz de France, 361, avenue du Président Wilson, 93211 La Plaine-Saint-Denis Cedex, France.
(2) National Geophysical Research Institute, 416, Rachana Nagar, Bhopal, MP 462 023 India.
(3) Université Pierre et Marie Curie Paris VI, 4, Place Jussieu, 75006 Paris, France.

1 REVIEW OR PREVIOUS WORKS

Papers published in literature dealing with the estimation of variables linked by pde generally cover two main fields: geothermy and fluid mechanics.

In geothermy, Vasseur and Singh (1986) used a probabilistic approach to study the relationship between heat production and heat flow at the earth's surface: in fact, an equation linking these variables was empirically established by Roy, Blackwell and Birch (1968), equation from which we deduce an estimation of the thickness of the upper part of the earth's crust. Since, other observations do not confirm this relation and Vasseur and Singh's probabilistic modelization of the lateral variations of the heat production tends to propose a new interpretation of the phenomenon. For a homogeneous medium (conductivity K is constant), if A and T denote the heat production and the temperature, the heat equation for a steady state is

$$K \Delta T = - A$$

The heat flow q which is supposed to be radial (along the vertical axis z) is given by the following equation

$$\text{div } q = A$$

where div is the divergence operator. Having modeled A as a stationary stochastic process, Vasseur and Singh's main aim is to deduce the covariance of the heat flow q from the one of A. The method proposed is the **harmonic analysis of stationary random functions.** Such an approach has been used by Mizell (1980) in fluid mechanics to study the relation linking variables related to transmissivity and hydraulic head. The starting equation is a modified Poisson equation; it is obtained after linearization of the flow equation in porous media using the perturbation method when the global flow is supposed to be along one direction. By denoting T the transmissivity and ψ the hydraulic head, the original flow equation without recharge is

$$\text{div } (T \nabla \psi) = 0$$

As T often has a logarithmic distribution we rather consider its logarithm denoted Y. According to the perturbation method, Y (resp. ψ) is splitted into a deterministic "mean value" F (resp. Φ) and a random function f (resp. ϕ). If x the direction of flow, the final equation is given by equation (3). In this case, $\partial \Phi / \partial x$ is a constant and we obtain something comparable to equation (2) with ϕ as Z and f as Y.

$$\Delta \phi = - \frac{\partial \Phi}{\partial x} \frac{\partial f}{\partial x} \qquad (3)$$

From this equation, we can derive the link between the covariances of Y and Z. The harmonic analysis of SRF approach used by Vasseur and Singh (1986) and Mizell (1980) consists of establishing the relationship between the spectral measures associated with the simple and cross covariances of Y and Z (i.e. their Fourier transform). For a given covariance of Y, we deduce the spectral measure of Z and by calculating its inverse Fourier transform, we derive the covariance of Z.

We must underline the fact that such **an approach is based on the hypothesis that Y and Z are simultaneously SRF, which is very restrictive.** There are stationary covariances of Y for which equations (1) and (2) admit non–stationary solutions Z that this approach cannot yield. Such a situation has been encountered by Mizell (1980) when he considered the following stationary covariance for Y

$$C_Y(h) = h_W K_1(h_W)$$

where K_1 is the modified Bessel function and h_W is a variable proportional to the norm of the vector h. Here the harmonic analysis of SRF approach yields a solution Z whose "variance" is infinite, which is in contradiction with the initial hypothesis "Z is a SRF".

Another approach was considered by Kitanidis and Vomvoris (1983) and Hoeksema and Kitanidis (1984). The authors' aim is to find a rapid and convenient way to fit simultaneously the covariances of Z and Y. Y is assumed to be an IRF–0 with a linear variogram γ (equation (4)) and the basic asumption is that the "experimental covariance matrix" between Z and Y can be modeled by (5) where A,B and C are matrixes involving cross–structures between Z and Y.

$$\gamma(h) = \theta_1 + \theta_2 |h| \qquad (4)$$

$$A + \theta_1 B + \theta_2 C \qquad (5)$$

The starting flow equation has been linearized in order to determine explicitly the elements of A,B and C. The coefficients θ_1 and θ_2 have been fitted by maximum likelihood criterion.

Special attention must be focused on the works of Dagan (1985), Rubin and Dagan (1987a, 1987b) whose basic hypotheses are that Y and Z are SRF in the wide sense and that their probability density is gaussian. The first step is the determination of the joint probability density of Y and Z. Then using the Bayes relation, the conditional probability density (conditional to the measures) can be derived easily. As the estimator chosen is the conditional expectation, we need to know the simple and cross covariances of Y and Z. The majority of the papers cited above are dedicated to this determination which has been carried out by direct resolution of the flow equation using the Green kernel corresponding to an infinite domain.

Two remarks are necessary: the first point deals with the gaussian density asumption for Y and Z. The authors specified that this approach was possible for non gaussian distributions but in that case the conditional expectation is rather complex and it is no longer a linear combination of measurements. As for the cokriging approach, no asumption on the distributions of the variables is necessary. If the distribution is gaussian, the cokriging estimator is identical to the conditional expectation; in other cases, cokriging is certainly not the best estimator but its determination is relatively simple. The second remark deals the asumption of stationarity (in the wide sense) on Y and Z. In the present paper, we will see examples of functions of different degrees of stationarity that satisfy equations (1) or (2); the linear model for Y is such an example. The approach proposed by Dagan would have yielded a non–stationary covariance for Z whose main drawback is that it cannot be fitted. The case when Y has a variogram with a finite sill is very particular (spherical,...). These models can be considered as real stationary covariances and we will see later that Z is then an IRF–0; so we have the illusion of not having moved the stationarity degree from Y to Z. Note that in these cases, we obtain the same result (Table 2) as Dagan (1985, equations 4d,4e).

In the next paragraph, a coherent structural analysis of the variables Y and Z linked by equation (1) is carried out within the IRF–k theory.

2 STRUCTURAL ANALYSIS

The IRF–k theory is a generalization of the notion of IRF–0 (variogram). A thorough presentation of this theory is in Matheron (1971, 1973) and Delfiner(1979). A brief review of the principal lines is presented in Dong (1989, 1990, chapter III and appendix A). One of the reasons why we have chosen to work in the scope of this theory is the existence of a theorem established in (Matheron, 1971, chapter 5) that states: "If Y is a continuous SRF then there exists a unique IRF-1 Z (up to a linear drift) twice differentiable that satisfies equation (1)". In fact, this theorem is more general (if Y is an IRF-k then Z will be an IRF-(k+2)); it is somewhat a probabilistic consequence of the Liouville theorem which states that a bounded harmonic function is necessarily constant.

By denoting σ the stationary covariance of Y, C the non–stationary covariance of Z and K a generalized covariance of Z (K is known up to an even polynomial of degree ≤ 2 as Z is an IRF–1). Estimation involves K, σ and the cross structure between Z and Y. Which is the link between them? If x and y denote two elements of $\mathbf{R}^n$ whose coordinates are x_i, y_j i,j = 1...n. The definition of σ and (1) imply

$$\sigma(x-y) = E[\Delta Z(x)\,\Delta Z(y)]$$

where E is the mathematical expectation. As Z is twice differentiable, its non–stationary covariance C is fourth differentiable and by denoting Δ_x and Δ_y the Laplacian operator with respect to x and y, we have

$$\sigma(x-y) = \Delta_x\,\Delta_y\,C(x,y)$$

C and K are linked by the following equation where a_l are continuous functions, f^l monomials of degree up to 1 and p the number of functions f^l:

$$C(x,y) = K(x-y) + \sum_{l=1}^{p} a_l(y)f^l(x) + \sum_{l=1}^{p} a_l(x)f^l(y)$$

The last two terms are filtered out by Δ. We finally get (6) where Δ^2 is the Laplacian operator iterated twice:

$$\boxed{\Delta^2 K(h) = \sigma(h) \text{ where } h = x-y} \qquad (6)$$

For a given covariance σ, the resolution of (6) yields a set of solutions K which differ from each other by an even polynomial of degree up to 2. The cross–structures involved in the cokriging matrix can be derived from the expression $\Delta K(h)$; thus the resolution of (6) is vital for the derivation of the estimation systems. Here we relate in detail two methods of resolution of (6) in the case when Y is a SRF but it can be performed in other cases for instance if Y is an IRF–0.

If Y has an isotropic structure then (6) can be transformed into a differential one before being solved. r denotes the norm of the vector h and H its square. We consider the functions K_1 and σ_1 such that $K_1(H) = K(h)$ and $\sigma_1(H) = \sigma(h)$. In $\mathbf{R}^2$, (6) becomes (7) where the derivative of order m of K_1 with respect to H is denoted $K_1^{(m)}$:

$$H^2 K_1^{(4)}(H) + 4H K_1^{(3)}(H) + 2K_1^{(2)} = \sigma_1(H)/6 \qquad (7)$$

In $\mathbf{R}^3$, we consider the function σ_2 such that $\sigma_2(r) = \sigma(h)$, equation (6) becomes equation (8)

$$\frac{1}{r}\frac{\partial^4 (rK)}{\partial r^4}(r) = \sigma_2(r) \qquad (8)$$

If Y has an anisotropic structure, we rather use the **harmonic analysis of SRF and IRF–k**. The Fourier transform of (1) (denoted by $\mathcal{F}$) gives (9) where χ_0 is the spectral measure associated with σ; thus, it is positive, summable and without an atom at the origin.

$$(4\pi^2|u|^2)^2 \, [\mathcal{F}(K)](du) = \chi_0(du) \qquad (9)$$

Unfortunately, the inverse Fourier transform of $\mathcal{F}(K)$ does not always exist because the following integral where (u,h) is the scalar product of the vectors u and h of $\mathbf{R}^n$ does not converge near 0.

$$\int \frac{\cos[2\pi \,(u,h)]}{(4\pi^2|u|^2)^2}\chi_0(du)$$

We can get round this problem (Matheron,1971) by subtracting from the numerator the first two terms of the expansion of $\cos[2\pi(u,h)]$ (which are filtered out by the Laplacian iterated twice), that is :

$$P_1[2\pi(u,h)] = 1 - [2\pi(u,h)]^2/2$$

The solutions of (6) are given in (10) where the spectral measures χ_0 required can be found in the literature.

$$\boxed{K(h) = \int_{\mathbf{R}^n} \frac{\cos[2\pi(u,h)] - P_1[2\pi(u,h)]}{(4\pi^2|u|^2)^2}\chi_0(du)} \qquad (10)$$

3 ESTIMATION AND BOUNDARY CONDITONS PROBLEM

Now we know how to deduce the structure of Z from that of Y using (7), (8) or (10); the problem is to carry out the estimation of Z (the "direct problem") and to study the properties of the estimator. We develop in detail the case when Y is a SRF, Z is an IRF–0.

Suppose that data on Z and Y are available. The natural reflex is to use cokriging to estimate Z with those data. The derivation of the cokriging system is straightforward; we denote the estimator of Z

$$Z^*(x_0) = \sum_{\alpha=1}^{n} \lambda_\alpha Z(x_\alpha) + \sum_{\beta=1}^{m} \lambda_\beta Y(x_\beta) \qquad (11)$$

As Y and Z are linked by (1) the condition guaranteeing that the estimation error would have a finite variance does not involve the coefficients λ_β (because Δ filters out polynomials of degree up to 1 and Z is an IRF–1). Finally, the estimation system is given in (12) where $\alpha' = 1...n$ and $\beta' = 1...m$. The term $F_{\alpha\beta}$ denotes the function F calculated at the value $(x_\alpha - x_\beta)$.

$$\begin{cases} \sum_{\alpha=1}^{n} \lambda_\alpha K_{\alpha\alpha'} + \sum_{\beta=1}^{q} \lambda_\beta \Delta K_{\alpha'\beta} + \sum_{l=1}^{p} \mu_l f^l(x_0) = K_{0\alpha'} \\[2mm] \sum_{\alpha=1}^{n} \lambda_\alpha \Delta K_{\alpha\beta'} + \sum_{\beta=1}^{q} \lambda_\beta \sigma_{\beta\beta'} \quad\quad = \Delta K_{0\beta'} \\[2mm] \sum_{\alpha=1}^{n} \lambda_\alpha f^l(x_\alpha) - f^l(x_0) \end{cases} \qquad (12)$$

This is a classical cokriging system but the simple and cross covariances involved are linked by (10); in order to establish this system, one might refer to the results of Table 1 of the appendix. As usual, the estimation variance is the scalar product of the R.H.S. of system (12) and the kriging coefficients vector (equation (13)).

$$V = K(0) - \sum_{\alpha=1}^{n} \lambda_\alpha \Delta K_{\alpha 0} - \sum_{\beta=1}^{m} \lambda_\beta \Delta K_{0\beta} - \sum_{l=1}^{p} \mu_l f^l(x_0) \quad (13)$$

As for the boundary conditions linked to the direct problem, we consider the Dirichlet conditions which state that at some points x_α, Z must be equal to fixed values Z_α (this happens at the frontier between an aquifer and a river for instance). If we consider these boundary conditions Z_α as ordinary data, (12) yields an estimator which is an **exact interpolator**:

$$Z^*(x_\alpha) = Z_\alpha$$

So the Dirichlet conditions are automatically fulfilled. Moreover as Δ and cokriging are linear operators

$$\Delta Z^*(x_0) = (\Delta Z)^*(x_0)$$

Z and Y are linked by (1), the R.H.S. of the previous equation is the estimator of Y using the same data as in (11) e.g. the cokriging estimator of Y. Finally

$$\boxed{\Delta Z^*(x_0) = Y^*_{CK}(x_0)} \quad \text{at any point } x_0 \qquad (14)$$

$$\boxed{\Delta Z^*(x_\beta) = Y(x_\beta)} \quad \begin{array}{l}\text{at any point } x_\beta \\ \text{where Y is known}\end{array} \qquad (15)$$

Pratically equation (15) means that if we estimate Z at points that enable the calculations of the discrete version of $\Delta Z^*(x_0)$, this quantity converges to $Y(x_\beta)$ when the distances between those points decrease.

The approach is similar for the Neumann conditions which constrain the values of the normal derivative of Z denoted $\partial Z/\partial \vec{n}$ at some points x_β of the frontier (this occurs when there is no flow at the frontier for instance). In this case we can build an estimator for Z by considering the Neumann boundary conditions as ordinary data (R.H.S. of (11)) e.g.

$$Z^*(x_0) = \sum_{\alpha=1}^{n} \lambda_\alpha Z(x_\alpha) + \sum_{\beta=1}^{m} \lambda_\beta Y(x_\beta) + \sum_{\eta} \frac{\partial Z}{\partial \vec{n}}(x_\eta) \quad (16)$$

Equation (15) is still satisfied and as $\partial/\partial\vec{n}$ is linear

$$\boxed{\frac{\partial Z^{*}}{\partial\vec{n}}(x_\eta) = \frac{\partial Z}{\partial\vec{n}}(x_\eta)} \quad \text{at points constrained by the Neumann conditions} \quad (17)$$

The cokriging system is simply larger than (12) and we must add terms related to $\partial Z/\partial\vec{n}$ e.g. $\partial K/\partial\vec{n}$, $\partial\Delta K/\partial\vec{n}$ and $\partial^2\Delta K/\partial\vec{n}^2$ for the cross structures.

It is obvious that **any other type of linear boundary conditions such as periodical or free surface ones can be taken into account if they are considered as ordinary data and incorporated in the building of the estimator of Z as in (11) and (16).**

There are some differences between this method and other usual techniques of resolution of pdes such as finite differences methods... A structural analysis is necessary before starting the cokriging and the matrixes involved generally do not contain many zero values, which might slow down the inversion. On the other hand, cokriging yields an estimation variance and can be applied directly to any pattern of data : no intermediate stage of estimation of Y on the nodes of the final grid is necessary unlike the other methods cited above. Lastly, **estimation of Z is still possible by cokriging even if all the boundary conditions are not known along the entire frontier.** That might seem suspicious as we know that this condition is vital to establish the existence and uniqueness of a solution of (1). This points out that the cokriging formalism within the IRF-k theory considers boundary conditions **as a posteriori conditioning data** and **not as a priori conditions** applied to Z. This is a fundamental point and further investigation should be focused on the consistency between the different possible forms of boundary conditions and the degree of the IRF Z.

4 STUDY OF THE MODIFIED POISSON EQUATION

4.1 Covariance Determination

In the case of equation (2), there is no preexisting theorem (such as the one of Matheron quoted in §2 for (1)) that helps us in the structural analysis. The first question that appears is the existence of an IRF-k Z solution of (2) when Y is a SRF. If such a solution exists in the IRF-k theory, what is the degree k and how can we deduce its generalized covariance from the stationary covariance of Y ?

Equation (2) involves the partial derivative with respect to x of Y which, in fluid mechanics, is the perturbation of the logarithm of transmissivity. As this variable is rather erratic, it is not smooth enough to be derivable. Instead of interpreting (2) in terms of functions, we consider it as a distribution (Schwartz, 1950). Then Y can be modeled by non differentiable covariances (spherical, ...) and still the expression $\partial Y/\partial x$ (derivative of Y with respect to x) has a meaning in the distribution theory.

Let us consider two distributions X and Y linked by (1). According to §2, if Y is a SRF then there exists a unique IRF-1 twice differentiable X solution of (1). Now consider the distribution

$$Z = \frac{\partial X}{\partial x}$$

As we know that X is an IRF-1 then Z is an IRF-0 characterized by a variogram. Using the ordinary properties of operators applied to distributions and test functions and by denoting $<D,g>$ the distribution D applied to the test function g,

$$<\Delta Z, g> = -<\Delta X, \frac{\partial g}{\partial x}> = -<Y, \frac{\partial g}{\partial x}>$$

We get the following relation which is the weak formulation of (2). Finally, **if Y is a SRF then there exists a unique IRF-0 Z satisfying equation (2).**

$$<\Delta Z, g> = <\frac{\partial Y}{\partial x}, g> \quad \text{for any test function}$$

The structural analysis of Z knowing that of Y can be deduced from the approach described in §2. By denoting σ the stationary covariance of Y, K a generalized covariance of X and γ the variogram of Z, according to (2), we have equation (3) where h_1 is the 1st coordinate of the vector h. As X and Y are linked by (1) we can use the results of §2 and of the appendix to determine K for (18). The results of the determination of the variograms γ corresponding to usual stationary covariances σ of Y are presented in the appendix.

$$\boxed{\gamma(h) = \frac{\partial^2 K}{\partial h_1^2}(h)} \quad (18)$$

4.2 Estimation

Our aim is to consider (2) from the inverse problem point of view e.g. to estimate Y using data on Y and Z. As Z is an IRF-0 when Y is a SRF, the variance of the estimation error will be finite only if the latter is an authorized linear combination of order 0 e.g. this combination must filter out constants (polynomials of degree 0). We consider an estimator built with data on Z and Y.

$$Y^*(x_0) = \sum_{\alpha=1}^{n} \lambda_\alpha Z(x_\alpha) + \sum_{\beta=1}^{m} \lambda_\beta Y(x_\beta)$$

The previous condition is fulfilled if

$$\sum_{\alpha=1}^{n} \lambda_\alpha = 0 \qquad (19)$$

The coefficients λ_β are not involved because Y is equal to ΔZ and Δ filters out constants. After minimization of the estimation variance with constrain (19), we obtain the system (20) which yields a non–biased and optimal estimaton whose variance is given by (21) (the notations are the same used for (12)).

$$\begin{cases} -\sum_{\alpha=1}^{n} \lambda_\alpha \, \gamma_{\alpha\alpha'} + \sum_{\beta=1}^{m} \lambda_\beta \frac{\partial \Delta K}{\partial x} \alpha'\beta + \mu_0 = \frac{\partial \Delta K}{\partial x} 0\alpha' \\[2ex] \sum_{\alpha=1}^{n} \lambda_\alpha \frac{\partial \Delta K}{\partial x} \alpha\beta' + \sum_{\beta=1}^{m} \lambda_\beta \sigma_{\beta\beta'} \qquad = \sigma_{0\beta'} \qquad (20) \\[2ex] \sum_{\alpha=1}^{n} \lambda_\alpha = 0 \end{cases}$$

$$\text{var} = \sigma(0) - \sum_{\alpha=1}^{n} \lambda_\alpha \frac{\partial \Delta K}{\partial x} \alpha 0 - \sum_{\beta=1}^{m} \lambda_\beta \sigma_{0\beta} \qquad (21)$$

The cokriging approach is classical but the models involved in (20) are determined after a **coherent structural analysis between Y, Z and (2)**. This guarantees that the final estimator Y^* satisfies (2). Indeed

$$\frac{\partial Y^*}{\partial x}(x_0) = (\Delta Z)^*(x_0) = \Delta \, Z^*(x_0)$$

As Z is an exact interpolator so

$$\boxed{\frac{\partial Y^*}{\partial x}(x_0) = \Delta Z(x_0)} \text{ at points where Z is known}$$

4.3 Example

4.3.1 Description of the data

The aquifer under study comes from a chalk area of 3200 km^2 in the North of France and has been studied by many authors from the geological and hydrogeological point of view (Ahmed (1987),...). Fig. 1, 2 and 3 are quoted from Ahmed (1987). Only 72 values of transmissivity are available (fig. 1); the values of hydraulic head ψ and the "mean values" Φ of ψ have been simulated by Ahmed (1987) with the boundary

conditions represented in fig. 2. The black cells are areas where the potential hydraulic head has an imposed value, the hatched cells are zones where there is exchange between a river and the aquifer and the cells with a point at the centre represent contact between the aquifer and a river without exchange.

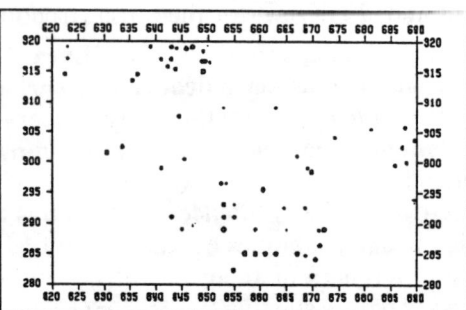

FIG. 1. Location of the 72 values of transmissivity (Y for equation (2)).

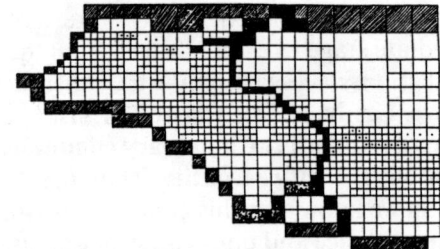

FIG. 2. Boundary conditions used to simulate the hydraulic heads ψ.

The procedure adopted is the following:

- first, a geostatistical conditional simulation of Y (logT) is performed using the 72 known values. The mesh of the simulation grid is a square cell of 1km on each side.

- then the results of this operation are used to initialize a flow simulation using the boundary conditions of fig. 2. The values obtained are considered as realizations of the hydraulic head ψ (see §1).

- a similar flow simulation is performed with the same boundary conditions but using the mean values of the transmissivity: we obtain realizations of the mean of the hydraulic head Φ (§1).

- in order to get the values of the perturbation of the hydraulic head ϕ (3) we subtract Φ from ψ. Contour lines of Φ are presented in fig. 3.

In order to test the methodoly described in § 4.1 and §4.2, the data retained for the structural analysis and the cokriging belong to two groups: first the 72 values of Y and among the values of ϕ deduced from the simulated ones of Φ and ψ, only 143 points have been retained (fig. 4).

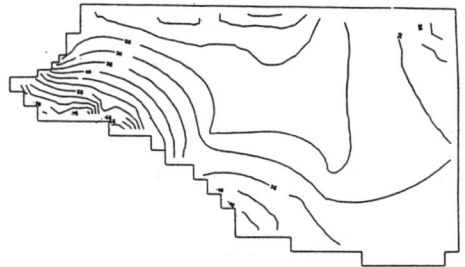

FIG. 3. Contour lines of the variable Φ "mean value" of the hydraulic head obtained after flow simulation using the boundary conditions described in fig. 2.

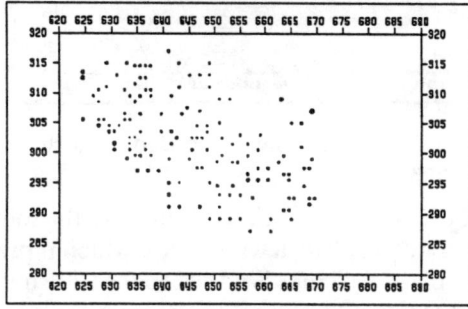

FIG. 4. Location of the 143 values of the perturbations ϕ (Z for equation (2)).

4.3.2. Structural analysis of the data

In §2, the theoretical structural analysis has been performed for (2). We can transpose that to (3) which controls the flow of our case study (though the presence of exchange between the aquifer and a river makes things become complex, we suppose that the flow is along one direction Ox (fig. 3)). For that, we multiply the generalized covariances K_Z of Z (Table 2) by J_1^2 and the cross structure between Z and Y by J_1.

Now the problem is to find the covariances that fit the best the hydraulic head and transmissivity data. According to §2, the choice of a model for one of these two variables determines the structure we must use to model the other; the only parameter that can be moved "independantly" in order to have the best fit of the two variables is the constant J_1 which is unknown (for a given covariance for Y, different values of J_1 yield struc-

tures of different shapes for Z). Generally this parameter is one of the first characteristics determined by the hydrogeologist after the study of hydraulic head maps. Usually, we cannot make it vary "independantly". Still, here the real flow is complex (fig. 3) and it is difficult to evaluate a mean value for J_1 in the whole domain; we have deliberately chosen to use this coefficient as an extra degree of liberty.

As only few data are available for the transmissivity, we assume that this variable is isotropic and the mean variogram for all directions is represented in fig. 6a. The variogram looks noisy and we are tempted to include a nugget effect in the fitting model. But as we only work with continuous models, the nugget effect must be banned. We rather use continuous models. In fig. 6a, the fitting is performed with an exponential covariance whose range and sill are equal to 1.6 and 0.4578. The first two points of the experimental variogram are well fitted. What are the consequences of this on the fitting of the experimental variograms of ϕ.

According to (18), we know that ϕ is anisotropic. The four directions of calculations for the variograms of ϕ (referred to the flow direction) are plotted in fig. 5

FIG. 5. The 4 directions of calculations for the variogram of ϕ

Its variograms in these four directions are represented in fig. 6b (direction 0°), in fig. 6c (directions 45° and 135°), in fig. 6d (direction 90°). Note that for directions 2 and 4 which are symmetrical compared to the perpendicular of the flow direction, (18) gives the same expression for γ; so we join the variograms in these two directions in the same figure. Another remarkable point is the shape of the variogram in the direction perpendicular to the flow (fig. 6d); after a distance of 20 km (which corresponds to the height of the informed zone (fig. 7), the variogram drops. As the hydraulic head decreases in the direction of flow, the drop might be explained by the fact that the flow is not really along one direction but there must be parasite flows (fig. 8). This is not surprising if we recall fig. 4. We note that the models γ read in the appendix (Table 2, 5th line) that correspond to the exponential covariance σ fitted for Y enable a good fit of the experimental variograms of the hydraulic heads. Other structures have been tested but this model yields the best simultaneous fit of all the 5 variograms. In fig. 9

are presented the results of the cokriging approach described in § 4.2 using data on transmissivity and hydraulic head and the structural analysis of fig. 6. The variable is the perturbation of the logarithm of transmissivity: thick, dashed and plain lines are zero, negative and positive lines and the step is equal to 0.25. The map of transmissivity is plotted in fig. 10. Dashed lines delimit areas of specially great values. Their location is the exchange area between the river and the aquifer.

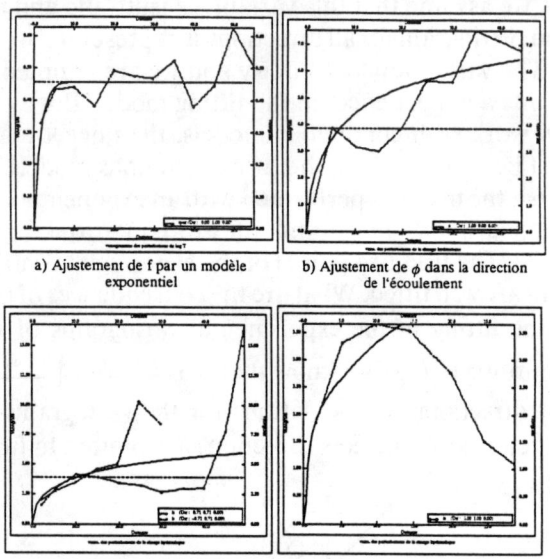

a) Ajustement de f par un modèle   b) Ajustement de ϕ dans la direction
         exponentiel   de l'écoulement

FIG. 6. Simultaneous fit of the experimental variograms of the transmissivity Y and hydraulic head Z:
for Y : exponential model (range = 1.6, sill = 0.4578)
   (a :upper left)
for Z : corresponding variogram (Table 2)
   (b: upper right) direction 1
   (c: lower left) directions 2 and 4
   (d: lower right) direction 3.
The constant J_1 is chosen equal to 1.5.

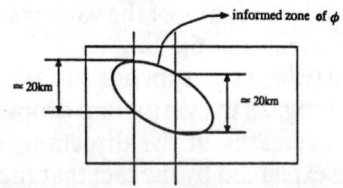

FIG. 7. Shape of the informed zone

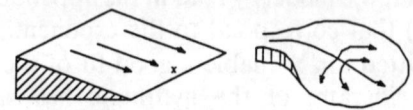

FIG. 8. Two hypotheses of flow : along 1 direction (left) and with parasite flows (right)

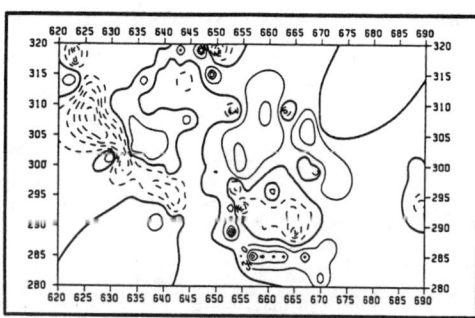

FIG. 9. Cokriging of the perturbation of the logarithm of transmissivity

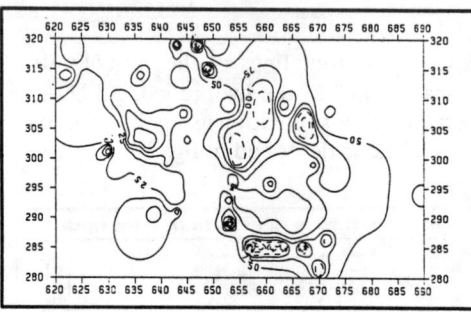

FIG. 10. Transmissivity deduced from the estimation by cokriging presented in fig. 9

In order to examine the instability of the solutions of the inverse problem, a white noise which represents 3 and 6% of the variance is added to the transmissivity data. We perform structural analysis and cokriging as in the previous case. In fig. 11 we plot the correlation clouds between these results and those of fig. 9.

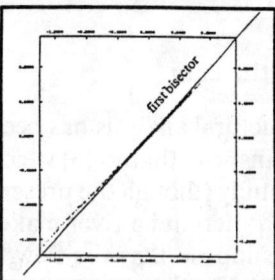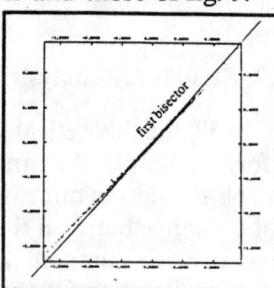

FIG. 11. Correlation clouds between the "original estimator" and the cokriging estimators obtained after addition of a white noise representing 3% of the variance (left) and 6% of the variance (right)

We note that the clouds are very close to the first bisector and it seems that no symptom of instability has been observed. Does it mean that this approach has managed to get round the istabilities inherent to the inverse problem? These results are encouraging but further investigation is necessary before we can draw any conclusions on that subject.

CONCLUSION

Two elliptic pdes have been studied in this paper: the Poisson equation (1) and one of its modified version (2). The IRF-k approach makes it possible to get out of the set of SRF and **can reach non-stationary solutions**: when Y is a SRF, Z is an IRF-1 (for (1)) or an IRF-0 (for (2)). In order to establish the estimation systems, a coherent structural analysis has been carried out and the results in the case when Y is a SRF are presented in the appendix. Equation (1) has been considered from the "direct problem" point of view: cokriging yields an optimal, non-biased estimator which, moreover, **satisfies the initial pde. Any type of <u>linear</u> boundary conditions can be taken into account** by considering them as a posteriori conditioning data and not as a priori conditions on the variable. This method can be directly applied to any pattern of data Y without intermediate estimation stage of Y on the nodes of the final grid, unlike other techniques such as the finite elements method for instance. A case study is presented to illustrate the **"inverse problem"** for equation (2): the aim is to estimate transmissivity using data of transmissivity and hydraulic head. A practical and coherent fit of the variograms of both variables is presented. One notes that the results of cokriging are relatively stable even if a nugget effect is added to the initial data: **no instability symptom has been observed**. This is an encouraging clue which would deserve further investigation.

REFERENCES

Ahmed, S., 1987, Estimation des transmissivités des aquifères par des méthodes géostatistiques multivariables et résolution indirecte du problème inverse, Ph. D. Thesis defended at the Ecole des Mines de Paris, 133 p.

Ahmed, S. and de Marsily, G., 1987, Comparison of Geostatistical Methods for Estimating Transmissivity using Data on Transmissivity and Specific Capacity, Water Res. Research, vol. 23, n° 9, pp. 1717-1737.

Dagan, G., 1985, Stochastic Modeling of Groundwater Flow by Unconditional and Conditional Probabilities: the Inverse Problem, Water Res. Research, vol. 21, n° 1, pp. 65-72.

Delfiner, P., 1979, The Intrinsic Model of Order k, course note C-77, Ecole des Mines de Paris, Centre de Géostatistique, Fontainebleau.

Dong, A., 1988, Cokrigeage et Krigeage de variables liées par l'équation $\Delta Z = Y$, internal report N-11/88/G, Ecole des Mines de Paris, Centre de Géostatistique, Fontainebleau.

Dong, A., 1989, Kriging Variables that satisfy the Partial Differential Equation $\Delta Z = Y$, Proceedings of the IIIrd International Geostatistics Congress, Avignon, sept. 1988, part 1, *Geostatistics*, Armstrong M. ed., Kluwer Pub. Co., Dordrecht, Netherlands, pp. 237-248.

Dong, A., 1990, Etude de l'équation $\Delta Z = Y$. Une solution pour les conditions aux limites du type Dirichlet, séminaire de Géostatistique "Journées de Juin", Fontainebleau, June 12-13 1989, to be published in Sciences de la Terre, série Informatique Géologique.

Dong, A., 1990, Estimation Géostatistique des Phénomènes régis par des Equations aux Dérivées Partielles, Ph. D. Thesis defended at the Ecole des Mines de Paris, 262 p.

Hoeksema, R. J. and Kitanidis, P. K., 1984, an Application of the Geostatistical Approach to the Inverse Problem in Two-dimensional Groundwater Modeling, Water Res. Research, vol. 20, n° 7, pp. 1003-1020.

Kitanidis, P. K. and Vomvoris, E.G., 1983, A Geostatistical Approach to the Inverse Problem in Groundwater Modeling (steady state) and One-dimensional Simulations, Water Res. Research, vol. 19, n° 3, pp.677-690.

Matheron, G., 1965, Les Variables Régionalisées et leur Estimation, Masson, Paris, 305 p.

Matheron, G., 1971, La Théorie des Fonctions Aléatoires Intrinsèques Généralisées, internal report N-252, Ecole des Mines de Paris, Centre de Géostatistique, Fontainebleau, 64 p.

Matheron, G., 1972, Leçons sur les Fonctions Aléatoires d'ordre 2, course note C-53, Ecole des Mines de Paris, Centre de Géostatistique, Fontainebleau, 80 p.

Matheron, G., 1973, The Intrinsic Random Functions and their Applications, Adv. Appl. Prob., n° 5, pp. 439-468.

Mizell, S. A., 1980, Stochastic Analysis of Spatial Variability in two–dimensional Groundwater Flow with Implications for Observations–well–network design, Ph. D. Thesis defended at the New Mexico Institute of Mining and Technology, Socorro, New Mexico, U.S.A.

Mizell, S. A., Gutjahr, A. L. and Gelhar, L.W., 1982, Stochastic Analysis of Spatial Variability in Two–dimensional Steady Groundwater Flow assuming Stationary and Non–stationary Heads, Water Res. Research, vol. 18, n° 4, pp. 1153–1167.

Roy, R. F., Blackwell D. D. and Birch F., 1968, Heat Generation of Plutonic Rocks and Continental Heat Flow Provinces, Earth and Planetary Science Letters 5, pp. 1–12.

Rubin, Y. and Dagan, G. 1987(a), Stochastic Identification of Transmissivity and Effective Recharge in Steady Groundwater Flow. 1. Theory, Water Res. Research, vol. 23, n° 7, pp. 1185–1192.

Rubin, Y. and Dagan, G., 1987(b), Stochastic Identification of Transmissivity and Effective Recharge in Steady Groundwater Flow. 2. Case Study, Water Res. Research, vol. 23, n° 7, pp. 1193–1200.

Schwartz, L., 1950, Théorie des Distributions, Hermann.

Vasseur, G. and Singh, R.N., 1986, Effects of Random Horizontal Variations in Radiogenic Heat Source Distribution on its Relationship with Heat Flow, Journal of Geophysical Research, vol. 91, n° B10, pp. 397–404.

APPENDIX
DETERMINATION OF THE STRUCTURE OF Z WHEN Y IS A SRF AND WHEN Z AND Y ARE LINKED BY EQUATION (1) OR (2)

The results of the determination of the structures of Z knowing that of Y are summed up in the Tables 1 and 2 where E is the space of work. We denote

C : sill   a : range

h_1 : 1st coordinate of the vector h

$h_a = h/a$ $dx_a = h_1/a$

TABLE 1

Study of equation (1): determination of the generalized covariance K_Z of the IRF-1 Z when Y is a SRF whose stationary covariance σ is spherical or exponential.

σ	E	GENERALIZED COVARIANCES K_Z OF Z	
S P H E R I C A L	R	$\frac{Ca^4}{4}\left[\frac{1}{6}h_a^4 - \frac{1}{20}h_a^5 + \frac{1}{420}h_a^7\right]$	if $h_a \leq 1$
		$\frac{Ca^4}{4}\left[\frac{1}{4}h_a^3 - \frac{1}{5}h_a^2 + \frac{1}{12}h_a - \frac{1}{70}\right]$	if $h_a \geq 1$
	R^2	$\frac{Ca^4}{16}\left[\frac{1}{4}h_a^4 - \frac{8}{75}h_a^5 + \frac{8}{1225}h_a^7\right]$	if $h_a \leq 1$
		$\frac{Ca^4}{16}\left[\frac{2}{5}h_a^2\log h_a + \frac{3}{35}\log h_a + \frac{1}{75}h_a^2 + \frac{669}{4900}\right]$	if $h_a \geq 1$
	R^3	$\frac{Ca^4}{240}\left[2h_a^4 - h_a^5 + \frac{1}{14}h_a^7\right]$	if $h_a \leq 1$
		$\frac{Ca^4}{240}\left[\frac{18}{7} - 5h_a + 4h_a^2 - \frac{1}{2h_a}\right]$	if $h_a \geq 1$
E X P O N E N T I A L	R	$Ca^4\left[e^{-h_a} + h_a + \frac{h_a^3}{6}\right]$	
	R^2	$\frac{Ca^4}{2}[-3Ei(-h_a) - \frac{1}{2}h_a^2 Ei(-h_a) + \frac{5}{2}e^{-h_a}$ $- \frac{1}{2}h_a e^{-h_a} - \frac{1}{4}h_a^2 + (\frac{1}{2}h_a^2 + 3)\log(ah_a) + \frac{3}{2}]$	
	R^3	$Ca^4\left[-\frac{4}{h_a} + 3 - h_a + \frac{1}{6}h_a^2 + \frac{4}{h_a}e^{-h_a} + e^{-h_a}\right]$	

TABLE 2

Study of equation (2): determination of the variogram γ of the IRF-0 Z when Y is a SRF whose stationary covariance σ is spherical or exponential.

σ	E	VARIOGRAMS γ OF Z	
S P H E R I C A L	R	$\frac{Ca^2}{4}\left\{\left[\frac{2}{3}h_a^2 - \frac{1}{4}h_a^3 + \frac{1}{60}h_a^5\right] + dx_a^2\left[\frac{4}{3} - \frac{3}{4}h_a + \frac{1}{12}h_a^3\right]\right\}$	if $h_a \leq 1$
		$\frac{Ca^2}{4}\left\{\left[\frac{3}{4}h_a - \frac{2}{5} + \frac{1}{12h_a}\right] + dx_a^2\left[\frac{3}{4h_a} - \frac{1}{12h_a^3}\right]\right\}$	if $h_a \geq 1$
	R^2	$\frac{Ca^2}{16}\left\{\left[\frac{8}{175} + h_a^2 - \frac{8}{15}h_a^3\right] + dx_a^2\left[2 - \frac{8}{5}h_a + \frac{8}{35}h_a^3\right]\right\}$	if $h_a \leq 1$
		$\frac{Ca^2}{16}\left\{\left[\frac{32}{75} + \frac{3}{35h_a^2} + \frac{4}{5}\log h_a\right] + dx_a^2\left[\frac{4}{5h_a^2} - \frac{6}{35h_a^4}\right]\right\}$	if $h_a \geq 1$
	R^3	$\frac{Ca^2}{240}\left\{\left[8h_a^2 - 5h_a^3 + \frac{1}{2}h_a^5\right] + dx_a^2\left[16 - 15h_a + \frac{5}{2}h_a^3\right]\right\}$	if $h_a \leq 1$
		$\frac{Ca^2}{240}\left\{\left[8 - \frac{5}{h_a} + \frac{1}{2h_a^3}\right] + dx_a^2\left[\frac{5}{h_a} - \frac{3}{2h_a^3}\right]\right\}$	if $h_a \geq 1$
E X P O N E N T I A L	R	$Ca^2\left\{\left[\frac{1-e^{-h_a}}{h_a} + \frac{1}{2}h_a\right] + dx_a^2\left[\frac{e^{-h_a}-1}{h_a^3} + \frac{e^{-h_a}}{h_a^2} + \frac{1}{2h_a}\right]\right\}$	
	R^2	$\frac{Ca^2}{2}\left\{\left[\frac{3}{h_a^2} - 3\frac{e^{-h_a}}{h_a^2} - 3\frac{e^{-h_a}}{h_a} + \log(ah_a) - Ei(-ha)\right] + dx_a^2\left[6\frac{e^{-h_a}}{h_a^4} + + 6\frac{e^{-h_a}}{h_a^3} - \frac{6}{h_a^4} + 2\frac{e^{-h_a}}{h_a^2} + \frac{1}{h_a^2}\right]\right\}$	
	R^3	$Ca^2\left\{\left[\frac{1}{3} - \frac{1}{h_a} - \frac{e^{-h_a}}{h_a} - 4\frac{e^{-h_a}}{h_a^2} - 4\frac{e^{-h_a}}{h_a^3} + \frac{4}{h_a^3}\right] + dx_a^2\left[\frac{e^{-h_a}}{h_a^2} + \frac{1}{h_a^3} + 5\frac{e^{-h_a}}{h_a^3} + 12\frac{e^{-h_a}}{h_a^4} + 12\frac{e^{-h_a}}{h_a^5} - \frac{12}{h_a^5}\right]\right\}$	

2nd European Conference on the Mathematics of Oil Recovery
© D. Guérillot, O. Guillon (Editors) and Éditions Technip, Paris 1990, pp. 31-41
27 rue Ginoux, 75015 Paris

Large-Scale Barriers
in Extensively Drilled Reservoirs

J. Høiberg, H. Omre and H. Tjelmeland [1]

ABSTRACT

Shale and carbonate semented barriers in petroleum reservoirs are of major concern due to its influence on flow characteristics. A stochastic model for shale distribution in a sandstone reservoir is established. A Markov field model combined with marked point processes are applied. Based on the model and constrained by the observations in wells, several realizations of the reservoir description are generated. The description is homogenized and transfered to a format suitable as input to ECLIPSE. Reservoir production simulations are performed on the realizations, hence the uncertainty in the production profiles can be evaluated.

1 INTRODUCTION

The stochastic model for reservoir description to be presented was developed during an evaluation of an oil field in the North Sea. The objective of the study was to establish a reservoir description suitable for evaluation of the production potensial including uncertainty assessments. Hence the description must be transferable to a reservoir production simulator, in this study ECLIPSE, and only the most important features for fluid flow can be included. An overview over stocahstic models in reservoir descrip-

tion can be found in Haldorsen and Damsleth (1990).

The reservoir comprises a complex sequence primarily of sandstone, but with several subordinate interbedded shales. The reservoir sand is of excellent quality, while the shales can be considered as non-permeable. The shales could be divided into two classes from the observations in core samples: Firstly, interchannel floodplain of large areal extent containing traces of soil production. These could in many instances be interpreted to correlate between wells which were several kilometers apart. Secondly, abandoned channels, bar-form drape and mudstone breccia facies of small areal extent, much less than the well spacing. A conseptual drawing by the geologist is presented in Fig. 1.

The geologist and engineers decided that modeling the heterogeneity in the shales would be sufficient for a reliable evaluation. The sand quality is considered to be relative homogeneous and good. The sand/shale models in Haldorsen and Lake (1984) and Begg et. al. (1989) are based on the assumption of relatively small shales. This can not be assumed in this study, hence an alternative model must be established.

The next section contains a presentation of the stochastic model for reservoir description and references to the corresponding simulation procedure. Section three is devoted to the procedure for ho-

---

(1) Norwegian Computing Center, PO Box 114, Blindern N, 0314 Oslo 3, Norway.
(*) Presently at Østfold Regional College, Halden, Norway.

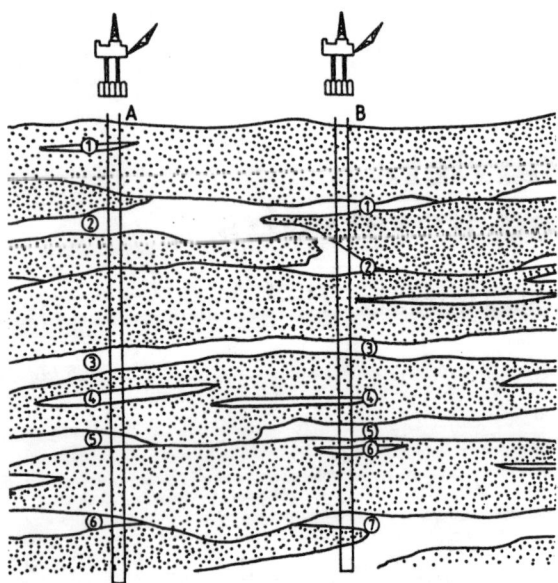

FIG. 1. Sketch of geologic setting

(by A.MacDonald, 1988)

mogenization. In section four, some results representative for the reservoir study are presented. Emphasize is put on the description, but some preliminary results for the production in a skematic setting are given. Lastly, some concluding remarks are forwarded.

2 STOCHASTIC MODEL AND SIMULATION PROCEDURE

The modeling has to be based on the general geologic knowledge of the geologist and observations from cores in a limited number of wells.

The stochastic model is based on the assumption that the reservoir architecture consists of two facies classes – sand and shale. The former with homogeneous porosity, saturation and permeability characteristics. The shales appear as two types, corresponding to the ones observed by the geologists.

The stochastic model defines the sand as a background facies into which the shales have been sedimented. In the following, the model is specified, but a more thorough discussion can be found in Høiberg et al. (1990). The shale model is as follows:

- Type I - consists of large shales which usually are intersected by more than one well.

The shales are assigned a thickness and are defined to be located in some user defined curved surfaces in the reservoir. These surfaces are thought of as time horizons, hence they are non-intersecting. They are denoted correlation surfaces. The parametrization of the shales in each correlation surface is:

- areal coverage ratio, specified as the proportion of the correlation surface covered by sand

- thickness distribution, specified as the pdf of shale thicknesses

- well-to-well continuity, specified as the probability for shales to be continuous on a straight line between a given pair of wells

- general appearance of shales is modeled by a 2D Markov random field model in the correlation surface. It is defined on a rectangular net of pixels over the correlation surface, $(i, j) \in$ CS. In each pixel a shale thickness discretized into $(K + 1)$ classes of length a, $t_{ij} \in \{0, a, ..., Ka\}$, is defined. If $t_{ij} = 0$, the pixel contains sand.

The Markov model (Ripley 1987) is specified by the conditional probability for shale thickness in an arbitrary location (i, j) given the thicknesses in a defined neighbourhood, ∂_{ij}, and the actual expression in this study is:

$$\text{Prob}\{t_{ij}|t_{kl}; (k, l) \in \partial_{ij}\}$$

$$= C\exp\Big\{ -\sum_{(k,l)\in\partial_{ij}} \Big[\frac{\beta_{k-i,l-j}}{a} \cdot |t_{ij} - t_{kl}| +$$

$$\Delta_{k-i,l-j} \cdot \delta(t_{ij} = 0 \neq t_{kl} \vee t_{ij} \neq 0 = t_{kl})\Big] \Big\}$$

where

- C — a normalizing constant

- $\beta_{\Delta i, \Delta j}$ — smoothness parameter for thickness in shale unit

- $\Delta_{\Delta i, \Delta j}$ — shape parameter for size of shale units and transaction with sand

- $\delta(x) = \begin{cases} 1 & \text{if } x \text{ is true} \\ 0 & \text{else} \end{cases}$

This model is fully specified, but cannot be solved analytically. Realizations are usually generated by the Metropolis algorithm. In the present setting, the simulations should be constrained to reproduce the shale thicknesses observed in the wells, the areal coverage ratio, the thickness distribution and the probabilities for well-to-well continuity. An extension of the Metropolis algorithm is applied for generating realizations of the shales in the correlation surfaces.

- Type II - consists of small shale units, each of them never penetrated by more than one well. The shale units may be located anywhere in the reservoir. The parameterization of these shales is:

  - net-gross ratio, specified as the proportion of sand in the reservoir.

  - shale frequency, specified as the average number of shales hit per meter in the wells

  - general location and size of type II shales is modeled by a 3D marked point process. Each element in the process is representing a shale unit, $B : \{\underline{x}, \underline{s}\}$. Here $\underline{x}$ refers to location and $\underline{s} = (w, l, t)$ to size specified by width, length and thickness, respectively. The size distribution is specified by a trivariate cdf, $F_{\underline{s}}(w, l, t)$.

The marked point process (Stoyan et al. 1987) is specified by the joint probability for N elements within the reservoir, and the actual expression in this study is:

$$\text{Prob}\{B_1, B_2, \ldots, B_N\}$$

$$= C \exp\left\{-\sum_{\substack{i,j=1 \\ i \neq j}}^{N} c(\underline{x}_i - \underline{x}_j)\right\} \cdot \prod_{\substack{i,j=1 \\ i \neq j}}^{N} \delta(B_i \cap B_j)$$

$$= C \prod_{\substack{i,j=1 \\ i \neq j}}^{N} \left[g(\underline{x}_i - \underline{x}_j) \cdot \delta(B_i \cap B_j)\right]$$

where

- C — a normalizing constant

- $g(\Delta \underline{x}) = \exp\{-c(\Delta \underline{x})\}$ — location interaction function for elements in the process

- $\delta(A) = \begin{cases} 1 & \text{if } A = \emptyset \\ 0 & \text{else} \end{cases}$

The model allows for pairwise repulsion between shale units and it guarantees no overlap. This model is complicated to treat analytically, but realizations from it can be generated by the Ripley/Kelly algorithm.

The reservoir description is based on the stochastic model defined above. Based on this model, specified parameter values and constrained by observations in the wells, a set of reservoir realizations can be generated. Each realization will reproduce the inherent heterogeneity in the model and the set of realizations will reflect the uncertainty.

The quality of the results will be dependent on to what extent the model is representative for the true reservoir. The verification of the model must primarily be based on visual inspection of the realizations by geologists. Comparison of the shale pattern in the wells with the pattern in fake wells arbitrarily located in the realizations has proved particularly helpful.

3 HOMOGENIZATION PROCEDURE

The reservoir realizations appear as detailed descriptions of the reservoir well suited for verification by the geologist. In order to use it as input parameters to the reservoir production simulator, an averaging of properties into a considerably coarser system of blocks must be performed.

The block system is regular but with varying density horizontally. Vertically the block borders coinsides with the curved correlation surfaces. The reservoir characteristics in each block is computed from the realizations as follows:

- net-gross ratio, computed by dividing sand volume by total block volume. The thickness of the type I shales is also included

- porosity, assigned the value of sand porosity

- absolute permeability, computed from the type II shales in the block by the turtuosity algorithm (Haldorsen and Lake, 1984)

Flow characteristics are also assigned to the block borders as transmissibility multipliers:

- vertically, computed as the areal fraction of the block side not covered by shale. Recall that the block side coinside with the correlation surface

- horizontally, computed for each of the four sides by taking the thickness of the shales in the associated part of the correlation surfaces into consideration.

4 RESERVOIR STUDY

The actual reservoir is located in the North Sea. The study area and the location of the 19 wells are presented in Fig. 2. All data is transformed to make them non-recognizable, but not beyond representability of the problem. The geologists have classified the core observations in the wells as either sand or one of the two shale types. Well-to-well interpretations have defined 12 correlation surfaces containing shales of large extent, type I. Based in this, a 13 layer description, outlined vertically by top and base of reservoir and the 12 correlations surfaces, is defined, see Fig. 2.

The type I shales are constrained to be located in the correlation surfaces, although with a thickness. The areal coverage ratio for each surface is estimated from the well observations by the relative number penetrating shales. The thickness distribution in each surface is also estimated from well observations by a smoothed version of the empirical thickness histogram. The well-to-well continuity values, in total, are based on geologic interpretation. In the model, a pixel net of 48×118 and a thickness discretization of 0.33m in 45 classes are used. A pixel neighbourhood of (5x5) is defined. The smoothness and shape parameters can hardly be estimated from the observations, and they were set by trial and error with respect to the realizations of shales in the correlation surfaces. The actual values of the parameters for type I shales of general interest are listed in Fig. 2.

The type II shales may appear everywhere in the reservoir. The net-gross ratio is estimated from the observations in the wells by the sand to total ratio, corrected for the volume of type I shales. The upper most layer is defined to be shale-free, the remaining

have constant net-gross ratio. Similar assumptions were made for shale frequency, which also is estimated from well observations. The probability distribution for (width, length, thickness) of shale units is specified, based on thickness observations in core and studies of available litterature. The location interaction function can hardly be estimated from observations, it is defined from inspection of consequences in the realizations. No overlap of type II shales are allowed. The actual values of the parameter for type II shales of general interest are listed in Fig. 2.

Based in the 3D stochastic reservoir description model, the assigned parameter values and constraints by the observations in the wells; a set of reservoir realizations can be generated. In Figs. 3 and 4 various displays of one realization are presented. Note the following:

- the inherent heterogeneity in the reservoir characteristics is reproduced, see Fig. 3.

- the observations in the wells are reproduced, see Figs. 3 and 4.

- the extensive type I shales may be penetrated by several wells and their thickness may have considerable impact on production, see Figs. 3 and 4.

- the small type II shales are numerous, hence they are expected to have impact on production, see Fig. 3.

- the set of reservoir realizations will reflect the uncertainty due to lack of information.

The reservoir realizations are subject to verification by visual inspection by geologists. Support statistics like empirical, horizontal, length distribution of type I shales in various direction and displays of fake wells can easily be produced, see Fig. 5. The former can be compared to outcrop equivalents, the latter to the observations in wells.

In order to evaluate the reservoir production potential, the fluid flow is simulated by ECLIPSE. The reservoir description realizations have to be homogenized to suit as input to ECLIPSE. This is done automatically and a file in ECLIPSE-format is produced for each realization. The system of blocks is 20x21 horizontally and 13 vertically. Vertically

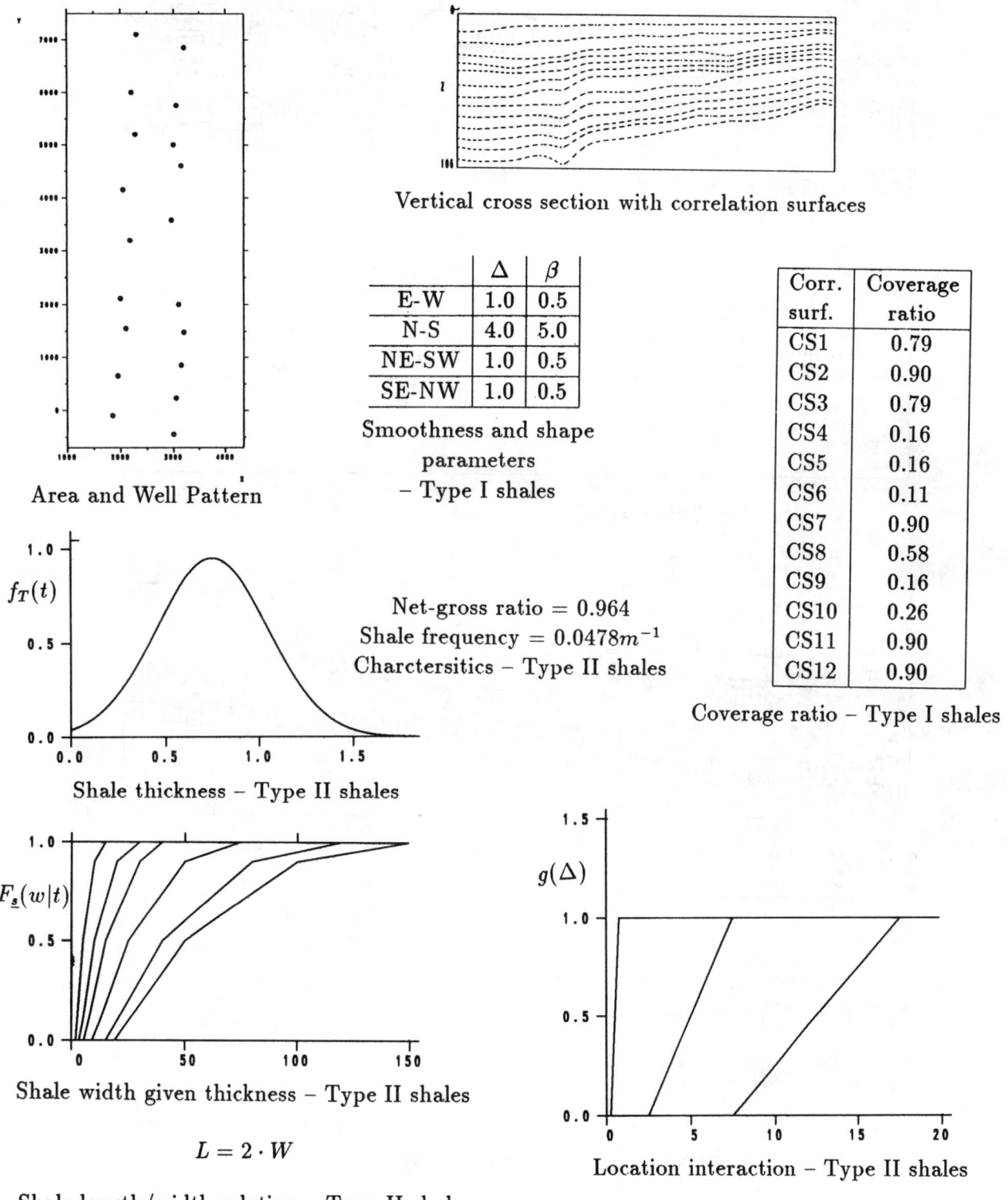

Area and Well Pattern

Vertical cross section with correlation surfaces

	Δ	β
E-W	1.0	0.5
N-S	4.0	5.0
NE-SW	1.0	0.5
SE-NW	1.0	0.5

Smoothness and shape
parameters
– Type I shales

Corr. surf.	Coverage ratio
CS1	0.79
CS2	0.90
CS3	0.79
CS4	0.16
CS5	0.16
CS6	0.11
CS7	0.90
CS8	0.58
CS9	0.16
CS10	0.26
CS11	0.90
CS12	0.90

Coverage ratio – Type I shales

Net-gross ratio $= 0.964$
Shale frequency $= 0.0478 m^{-1}$
Charctersitics – Type II shales

Shale thickness – Type II shales

Shale width given thickness – Type II shales

$$L = 2 \cdot W$$

Shale length/width relation – Type II shales

Location interaction – Type II shales

FIG. 2. Outline and parameter values for reservoir study.

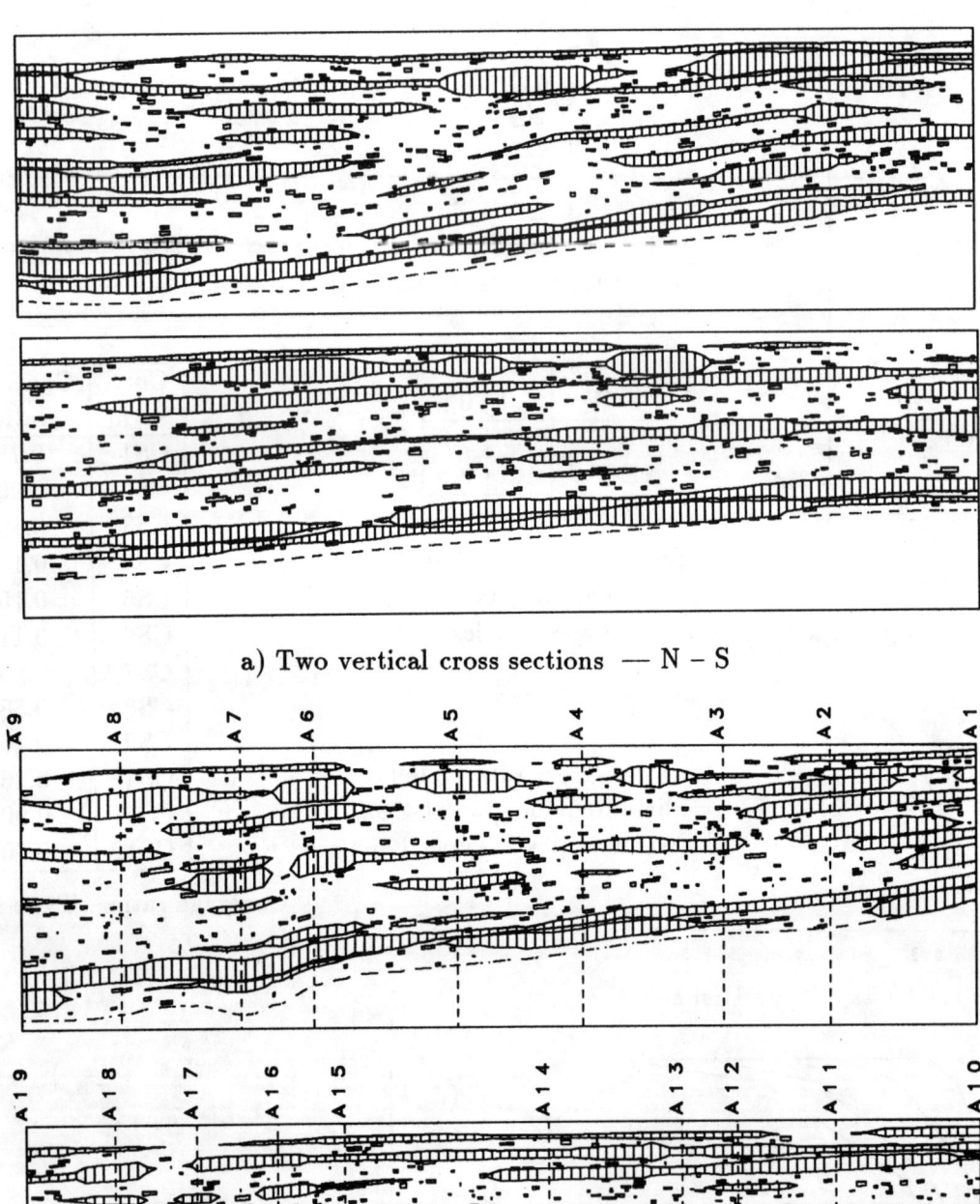

a) Two vertical cross sections — N – S

b) Two vertical fence diagrams.

FIG. 3. Realization of reservoir Description.

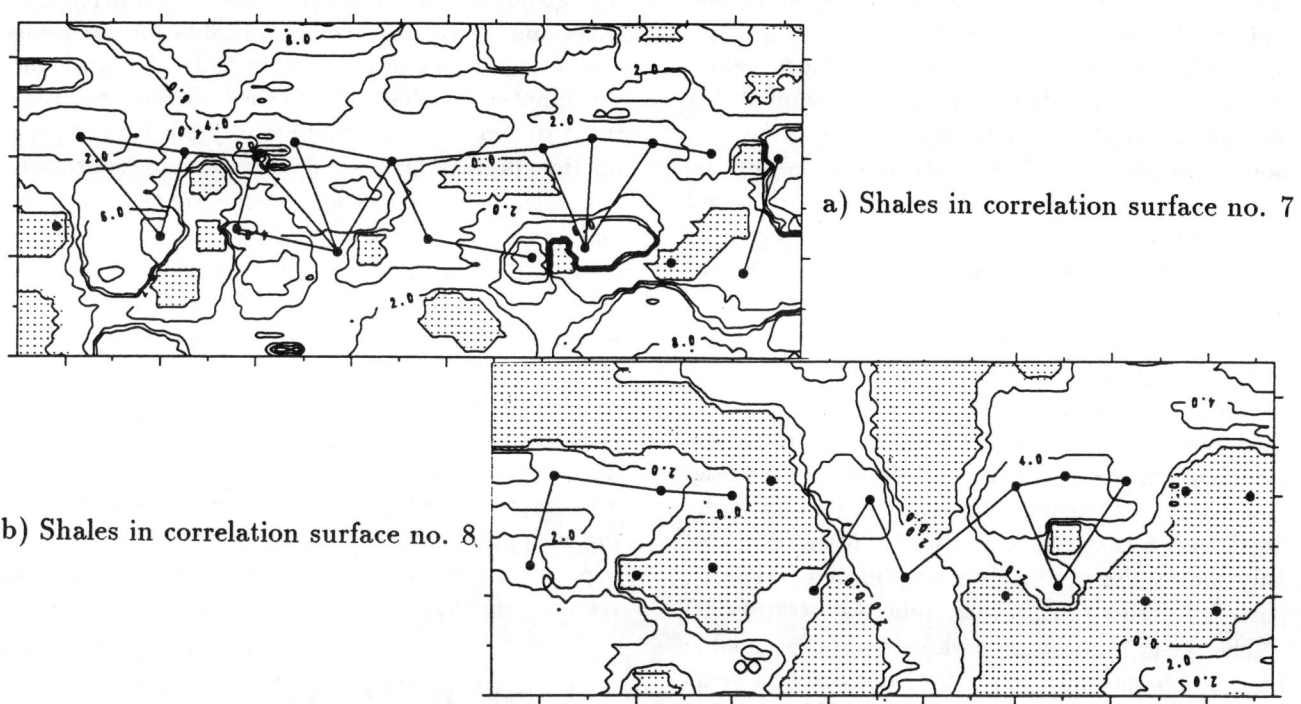

a) Shales in correlation surface no. 7

b) Shales in correlation surface no. 8

FIG. 4. Realization of reservoir description.

X-DIRECTION:.. ,Y-DIRECTION:.
SHALE
P
1.0
0.8
0.6
0.4
0.2
0.0
0
500
1000
1500
2000
2500
3000
3500
4000
4500
LENGTH

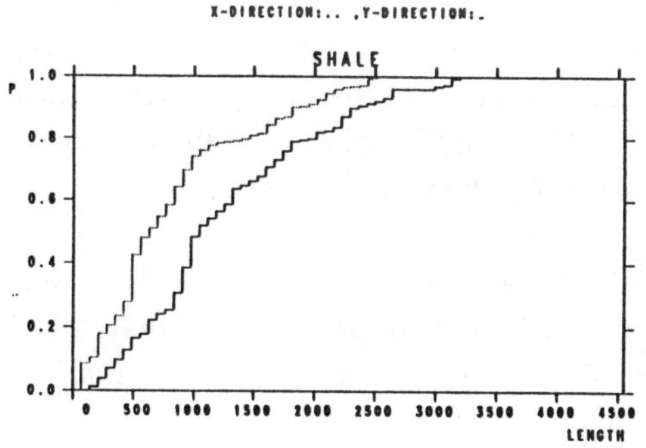

FIG. 5. Statistics from reservoir realization.
a) Length distribution of shales.
b) Well comparison.

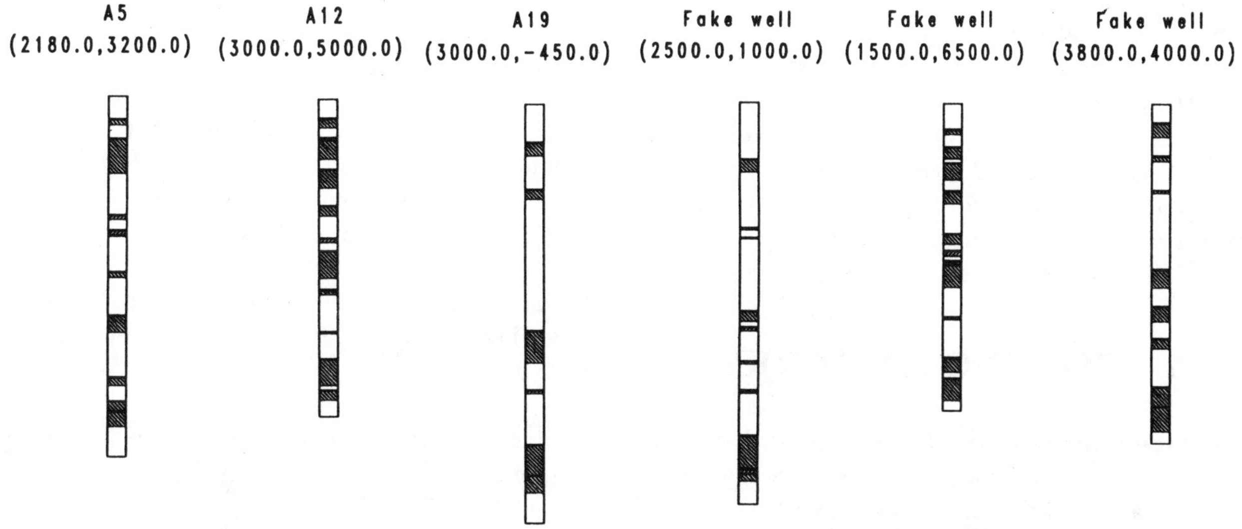

A5
(2180.0,3200.0)
A12
(3000.0,5000.0)
A19
(3000.0,-450.0)
Fake well
(2500.0,1000.0)
Fake well
(1500.0,6500.0)
Fake well
(3800.0,4000.0)

the block borders coinside with the top and base, and the 12 correlation surfaces. The porosity assigned the sand is 0.21. The net-gross ratio, absolute permeability and vertical/horizontal transmissibility multipliers are computed from the reservoir description realizations. In Fig.6 various displays of one realization are presented. The displays correspond to the display in Figs. 3 and 4. Note the general correspondance between the description and the homogenized values, and the correlations between the various homogenized values.

The objective is to evaluate production potential for which production profiles with time is a widely used measure. To obtain this, various production technical parameters have to be defined. In the original study, a realistic production strategy was defined, and a thorough evaluation is planned. Here, only the results from a skematic, limited case is reported, in order to show the evaluation strategy.

The reservoir technical parameters are specified in Fig. 7. The production strategy the two first years is: No injection, and produce 10000 stb/d from each well as long as pressure is above 1000 psi. Thereafter production at constant pressure, 1000 psi. The strategy the thirteen following years: injection of 10000 stb/d water per injection well and production at constant pressure, 1000 psi.

In this skematic case, seven realizations of the reservoir description are generated, transfered to ECLIPSE and the productions are simulated. This provides seven realizations of possible production profiles. The results are presented in Fig. 8. Note than the seven realizations have a similar appearance, but that their dispersion indicates the uncertainty. The break through of water can for example be expected to occur after 10 to 12 years.

In the reservoir study, a considerably more thorough evaluation will be made. In the system, geological parameters and well data can be given as input and the corresponding production profiles for several realizations can be generated completely automatically under a given production strategy.

5   CLOSING REMARKS

The reservoir description has a stochastic base and is obtained by combining two model classes, namely Markov random fields and marked point processes.

The geologists have confirmed the reservoir realizations and accepted them as suitable for reservoir evaluation. A line of production, fully automized on a computer, has been established all the way from input of geological parameters to prediction of production profiles for oil. Each reservoir realization will reflect the inherent heterogenity of the reservoir and hence give rise to improved prediction of production profiles. A set of reservoir realizations will provide a set of production profiles which dispersion will reflect the uncertainty due to lack of information about the reservoir under the specified model.

Development of petroleum reservoirs must be based on decisions made under uncertainty. The uncertainties concerning the resource base and possible depletion are considerable. Stochastic modeling with possibilities for quantifying uncertainties is expected to be an important component in future reservoir management procedures.

6   ACKNOWLEDGEMENT

Eva Halland, Boye Flood and Anne Markhus at Norwegian Petroleum Directorate and Alister MacDonald Institute for Energy Technology Research have all contributed to the geological interpretations. The IMB/Europe, Middle East and Africa Petroleum Application Centre in Stavanger, Norway has gently provided computer facilities for the study. Special thanks to Yuanchang Qi IBM/EPAC for reservoir technical support. The research is founded by Norwegian Petroleum Directorate and Norwegian Computing Center.

REFERENCES

Begg, S.H; Carter, R.R. and Dranfield, P.; 1989: Assigning Effective Values to Simulator Gridblock Parameters for Heterogeneous Reservoirs; SPE Reservoir Engineering; Nov. 1989, pp 455-463.

Haldorsen, H.H. and Damsleth, E.; 1990: Stochastic Modeling; Journal of Petroleum Technology; April 1990; pp 404-412.

Haldorsen, H.H. and Lake, L.W.; 1984: A new Approach to Shale Management in Field Scale Models, Soc. of Petr. Eng. Journ., August 1984.

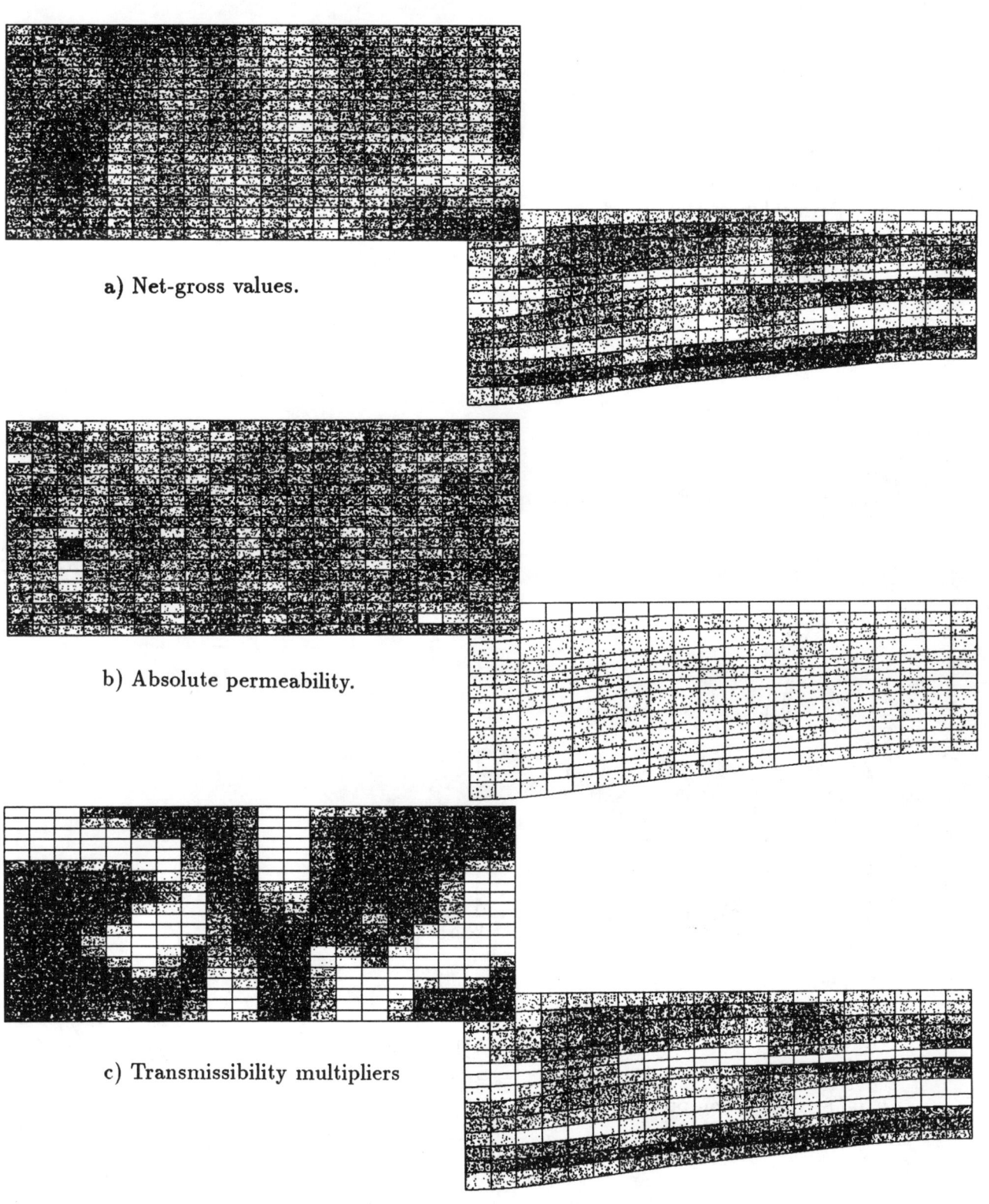

a) Net-gross values.

b) Absolute permeability.

c) Transmissibility multipliers

FIG. 6. Input to reservoir production simulator.

Sand Characteristics:
Porosity: 0.21
Permeability: 800 mD

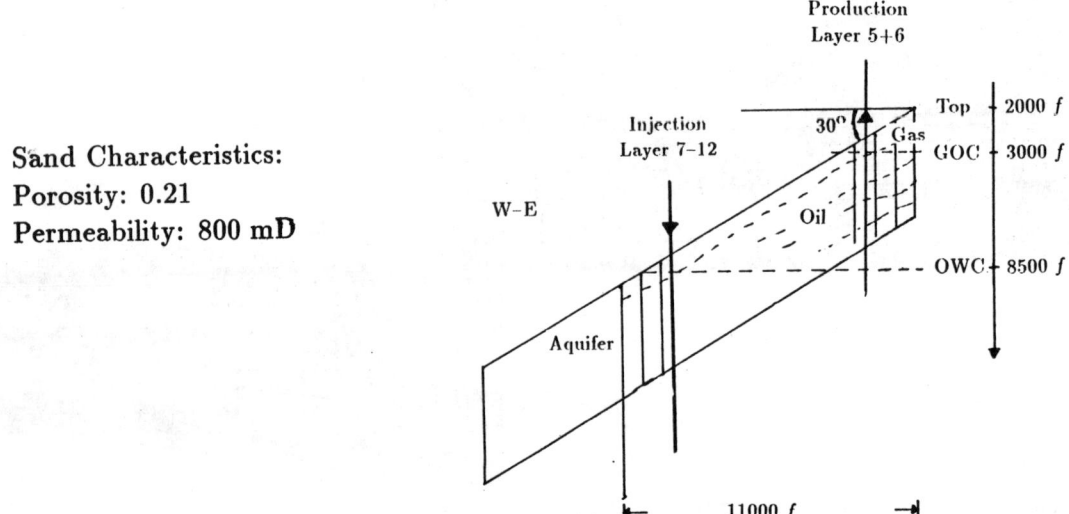

FIG.7. Reservoir technical parameters.

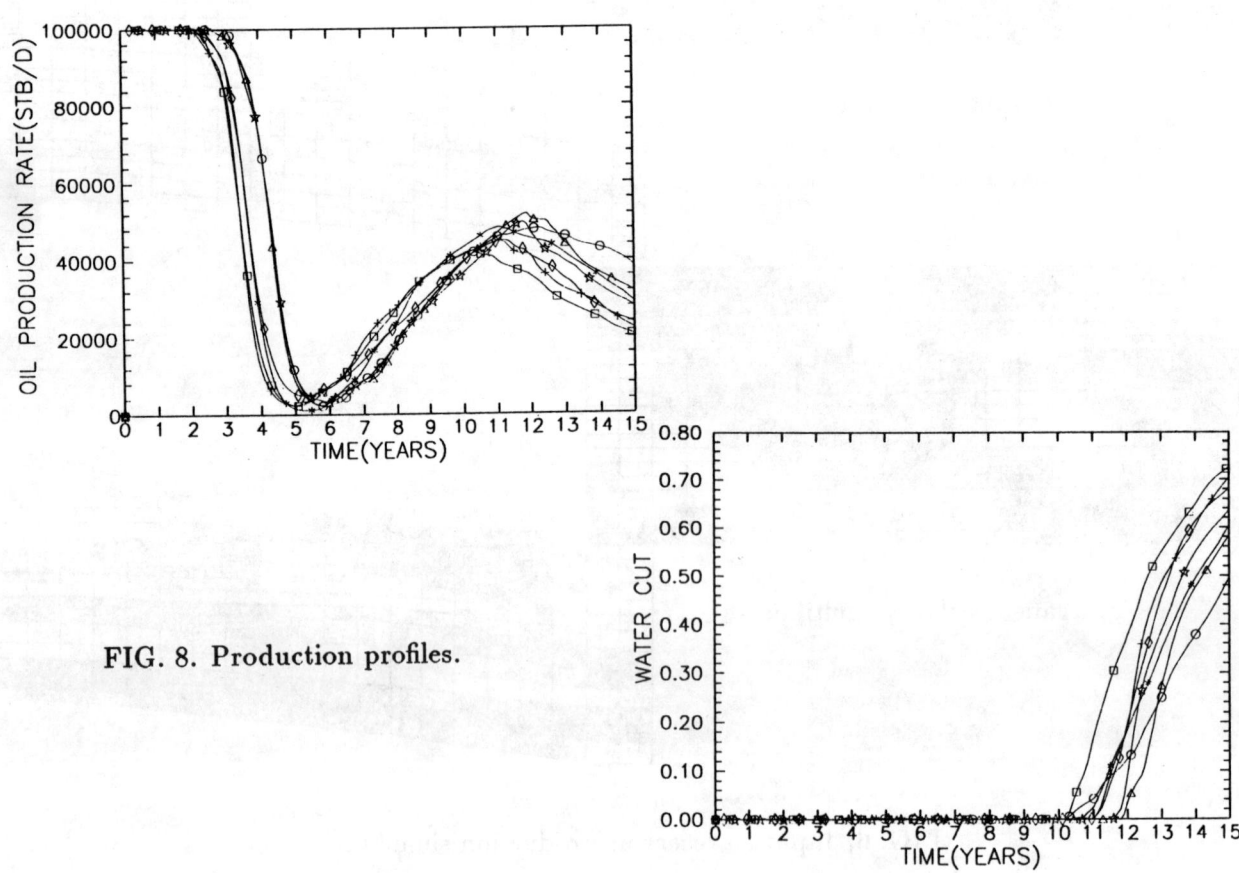

FIG. 8. Production profiles.

Høiberg, J.; Omre, H. and Tjelmeland, H.; 1990: A Stochastic Model for Shale Distribution in Petroleum Reservoirs; Proceedings from The Second Codata Conference on Geomathematics and Geostatistics, (in Sciences de la Terre), Leeds U.K.; Sept. 10.-14. 1990, to appear.

Ripley, D.D.; 1987: Stochastic Simulation, John Wiley & Sons Inc., 1987.

Stoyan, D.; Kendall, W.S. and Mecke, J.; 1987: Stochastic Geometry and its Applications, John Wiley & Sons Inc., 1987.

Effective Properties

2nd European Conference on the Mathematics of Oil Recovery
© D. Guérillot, O. Guillon (Editors) and Éditions Technip, Paris 1990, pp. 45-55
27 rue Ginoux, 75015 Paris

Accurate Calibration of Empirical Viscous Fingering Models

F. J. Fayers, M. J. Blunt and M. A. Christie[1]

ABSTRACT

We review the use and calibration of empirical models for viscous fingering. The choice of parameters for the three principal approaches (Koval, Todd and Longstaff, and Fayers methods) is outlined. The methods all give similar levels of accuracy when compared with linear experiments, but differ in performance in two-dimensional applications. This arises from differences in the formulation of the total mobility terms. The superiority of the Todd and Longstaff and Fayers methods is demonstrated for two-dimensional and gravity influenced flows by comparison with experiments and high resolution simulation.

The use of high resolution simulation to calibrate empirical models in a systematic manner is described. Results from detailed simulation demonstrate the sensitivity of empirical model parameters to viscous to gravity ratio, recovery process (secondary, tertiary or WAG), and geological heterogeneity. It is shown that for large amplitude heterogeneities with short correlation lengths, the accuracy of the empirical models is not satisfactory, but is improved by the addition of a diffusive term.

1 INTRODUCTION

Miscible gas flooding of petroleum reservoirs has a number of attractions as a technique for enhancing oil recovery. Miscible gas injection can achieve very low residual oil saturations in laboratory scale experiments. However, on a field scale, the process is subject to instabilities which decrease the sweep efficiency and reduce the oil recovery.

Numerical solution of the full set of two-component miscible displacement equations offers a promising way of characterising field scale process efficiency. Such calculations for a secondary displacement have received much attention in recent years. See, for example, Christie and Bond (1987), Christie (1989), Araktingi and Orr (1988), Moissis *et al* (1989). The two dimensional solutions obtained exhibit reproducible behaviour for near homogeneous porous media. Similar results for three dimensional miscible flows are also now becoming available (Christie *et al*, 1991).

The fingering behaviour can also be strongly influenced by gravity segregation as shown by Fayers and Newley (1988) and Fayers and Muggeridge (1989). The implementation of miscible gas flooding is usually undertaken in association with water injection (i.e. a water/alternating gas or WAG scheme) which partially stabilises the fingering. Gravity segregation again affects the behaviour since water separates in the reverse direction to gas. All of these effects can now be studied using 3D-high resolution simulation (Christie *et al*, 1991).

The calculation of fingering in the presence of heterogeneities is a difficult challenge, and work on quantifying the effects of non-uniform geological properties is still in its infancy. For example, fine scale random heterogeneities will perturb the front, while wide permeability variations with a large correlation length may dominate the overall displacement pattern.

To plan field scale applications of miscible flooding, it is necessary to use empirical fractional flow models, which in a macroscopic sense describe the consequences of viscous fingering within the grid blocks of coarsely gridded models. There are three different models available, namely those of Koval (1963), Todd and Longstaff (1972) and Fayers (1988). All of these models alter the fractional flow of solvent to describe the effects of heterogeneities and fingering. A comparison of the three

(1) BP Research Centre, Chertsey Road, Sunburry-on-Thames, Middlesex, TW16 7LN, UK.

models has been reported by Newley (1987).

The Todd and Longstaff model is probably the most widely used in conventional field simulators; one of its principal attractions lies in the fact that the model requires the selection of only one parameter ω whose value must encompass the combined effects of all the factors influencing fingering described above.

Fitting the Todd & Longstaff model to experiments in uniformly packed sandpacks, such as those of Blackwell *et al* (1959) showed that $\omega = 2/3$ was a reasonably good choice for all mobility ratios. However, the constancy of the value of ω required for homogeneous secondary displacements is somewhat illusory, since rather different values are required for tertiary, WAG, heterogeneous, and gravity influenced flows.

2 TWO-FLUID MODELS

The fractional flow models introduced in the previous section are all examples of 'two-fluid' models for which flow is represented by two pseudo-phases, one more oil dominated and the other more gas dominated. The equations of motion are:

Darcy's law for the two fluids ($i =$ oil, gas):

$$\mathbf{v}_{ie} = -K \frac{k_{rie}}{\mu_{ie}} (\nabla P + \rho_{ie} \mathbf{g}) \qquad (1)$$

The conservation of gas component is:

$$\phi \frac{\partial S_g}{\partial t} + \nabla . (S_{go} \mathbf{v}_{oe} + S_{gg} \mathbf{v}_{ge}) = 0 \qquad (2)$$

Also $S_{oo} + S_{go} = 1$, $S_g = S_{go} S_{oe} + S_{gg} S_{ge} = 1 - S_o$ where S_{oo} is the oil concentration in the pseudo oil phase, S_{og} is the oil concentration in the pseudo gas phase etc. The other symbols are defined in the nomenclature. It is usual to solve this system with a fixed injection flow rate such that

$$\nabla . (\mathbf{v}_{oe} + \mathbf{v}_{ge}) = 0 \qquad (3)$$

for incompressible flow, which results in $v_{oe} + v_{ge} = q$ in one dimension.

The gas conservation eqn (2) can be expressed as

$$\phi \frac{\partial S_g}{\partial t} + \nabla . f_g (\mathbf{v}_{oe} + \mathbf{v}_{ge}) = 0 \qquad (4)$$

where f_g is the fractional flow of gas.

2.1 Koval Model

To obtain the Koval model, we assume that $k_{rge} = S$, $k_{roe} = 1 - S$, $\mu_{oe} = \mu_o$ and

$$\mu_{ge} = (0.22/\mu_g^{1/4} + 0.78/\mu_o^{1/4})^{-4} \qquad (5)$$

i.e. the pseudo-gas behaves as though it had a fixed composition $S_{gg} = 0.22$.

Neglecting gravity, the Koval fractional flow is given by:

$$f_g = \frac{S}{S + (1 - S)/M_e} \qquad (6)$$

where the effective viscosity ratio $M_e = \mu_o/\mu_{ge}$. Koval did not consider total mobility or gravity influenced flows, but consistent with his model we may write $\rho_{oe} = \rho_o$, $\rho_{ge} = 0.22\rho_g + 0.78\rho_o$, and $\lambda_t = S/\mu_{ge} + (1 - S)/\mu_o$.

2.2 Todd and Longstaff Model

For the Todd & Longstaff model, the same assumptions are made on relative permeabilities, but the effective viscosities of both phases are obtained by considering them to be modifications of the pure mixture viscosity,

$$\mu_m = (S/\mu_g^{1/4} + (1 - S)/\mu_o^{1/4})^{-4} \qquad (7)$$

in the form $\mu_{ge} = \mu_g^{1-\omega} \mu_m^{\omega}$, $\mu_{oe} = \mu_o^{1-\omega} \mu_m^{\omega}$. The mixing parameter ω can be chosen between the limits $\omega = 1$ for complete mixing, and $\omega = 0$ for complete separation. Neglecting gravity the Todd & Longstaff fractional flow is again given by eqn (6) except that now $M_e = (\mu_o/\mu_g)^{1-\omega}$. Like the Koval model, the effective mobility ratio in the Todd & Longstaff model is independent of S.

The inclusion of gravity requires definition of an effective oil concentration in the pseudo oil phase in the form:

$$S_{oo} = \frac{(\frac{\mu_o}{\mu_g})^{1/4} - (\frac{\mu_o}{\mu_{oe}})^{1/4}}{(\frac{\mu_o}{\mu_g})^{1/4} - 1} \qquad (8)$$

with a similar expression for S_{og}, and then the effective densities are given by $\rho_{oe} = S_{oo}\rho_o + (1 - S_{oo})\rho_g$, $\rho_{ge} = S_{og}\rho_o + (1 - S_{og})\rho_g$. The consistent total mobility in the Todd & Longstaff method is

$$\lambda_t = S/\mu_{ge} + (1 - S)/\mu_{oe} \qquad (9)$$

2.3 Fayers Model

In the Fayers model a more physically based attempt is made to describe the relative permeabilities of the pseudo phase in terms of a finger width function, $\Lambda(S_{gg})$, such that $k_{rge} = \Lambda$ and $k_{roe} = 1 - \Lambda$. In common with Koval, it is then assumed that $\mu_{oe} = \mu_o$ but

$$\mu_{ge} = (S_{gg}/\mu_g^{1/4} + (1 - S_{gg})/\mu_o^{1/4})^{-4} \qquad (10)$$

An empirical form for Λ was suggested, $\Lambda = a + (1 - a)S_{gg}^{\alpha}$, in which a is the initial finger width parameter, and α is the growth exponent. S_{gg} is related to S through the relation $\Lambda = S_{gg}/S$, which highlights the difference n relative permeability assumption in this model. The ractional flow is given by:

$$f_g = \frac{\Lambda S_{gg}}{\Lambda + (1 - \Lambda)\mu_{ge}/\mu_o} \qquad (11)$$

Gravity is introduced using the consistent relations $\rho_{oe} = \rho_o$, $\rho_{ge} = S_{gg}\rho_g + (1 - S_{gg})\rho_o$, and the total mobility is given by $\lambda_t = \Lambda/\mu_{ge} + (1 - \Lambda)/\mu_o$.

All three empirical models can fit the recovery curves of the benchmark experiments of Blackwell *et al* (1959) successfully. These authors demonstrated that the magnitude of fingering in laboratory experiments is dependent on length to width ratio (L/W) and that representative behaviour is only obtained when $L/W < 3.0$. It is the asymptotic results for which the parametric values in the three empirical models are confirmed. The success of the Koval model is perhaps the most surprising, since the assumption of a constant effective gas viscosity associated with the choice $S = 0.22$ is unphysical. Nevertheless having chosen this value, the resulting fractional flows accurately fit experiments for a range of mobility ratios. Comparison of the effective mobilities indicate that the Todd & Longstaff model will reproduce the Koval results if

$$\omega = 1 - 4log\left(0.78 + 0.22\left(\frac{\mu_o}{\mu_g}\right)^{1/4}\right)/log\left(\frac{\mu_o}{\mu_g}\right) \qquad (12)$$

In fact use of this equation to set ω gives a somewhat better overall fit to the Blackwell experiments than the choice $\omega = 2/3$. Fayers found for his fingering function that the choice of parameters $a = 0.1$ and $\alpha = 0.42(\mu_o/\mu_g)^{0.4}$ gave a good fit to these experiments. His results are illustrated in Fig. 1. It is not possible to choose a physically acceptable Λ function which will cause eqn (11) to reproduce exactly the Koval fractional flow in eqn (6).

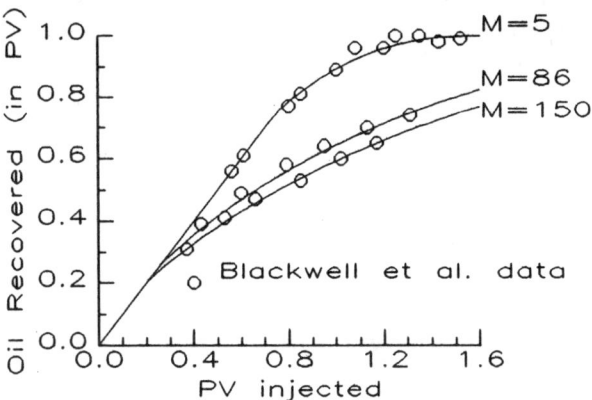

FIG. 1 Comparison of Fayers Model with Blackwell's experiments

3. EXTRA PHYSICAL EFFECTS

3.1 Total Mobility

The calibration of empirical models against standard laboratory experiments usually fails to examine the total mobility or pressure drops, which are rarely measured in miscible displacement experiments. However, high resolution simulation (Christie and Bond, 1987) provides accurately computed pressure fields, as well as the details of the composition fields, so that appropriate averages may be taken across the y-lines of two-dimensional solutions (or across planes in 3D) to examine the effective 1D properties. This approach was adopted by Fayers and Newley (1988). Fig. 2 shows a comparison between the three types of empirical model and the average total mobilities determined from high resolution simulation. The Todd & Longstaff and Fayers models give almost identical and correct results, but the Koval model significantly underestimates the effective mobility. The high resolution simulation pressure solution also shows that the average pressures in the bypassed oil and solvent invaded finger regions are very close, which confirms an important basic assumption in the two-fluid models.

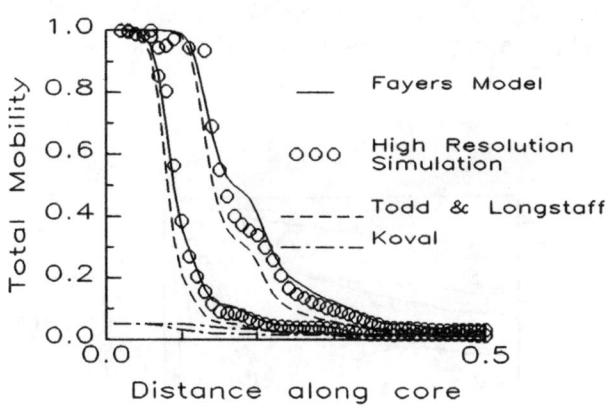

FIG. 2 Comparison of Total Mobilities with M=86

The effects of errors in total mobility come into play in multi-dimensional applications. For the two-dimensional problem of a repeated 5-spot pattern, comparisons with two sets of measurements of oil recovery against the three empirical models are shown in Fig. 3, for a mobility ratio, $M = 10$. The Lacey *et al* (1961) data refer to a beadpack and Habermann's (1960) data to a sandpack. The Habermann results are plotted with small corrections applied to compensate the systematic errors in his measurements of sweep efficiency (see Newley 1987). The Todd & Longstaff and Fayers results agree with the 2D porous media experiments, without the need for further adjustment of the model parameters from their one-dimensional values. The Koval result shows a systematic error in the 2D-case, since the constant effective gas viscosity is too large. Adjustments to model parameters will change both the front shape and the areal sweep pattern, so it is possible to obtain apparently good oil recovery response by cancelling errors. Differences in sweep patterns for the three models are shown in Fig. 4 for $M = 40$.

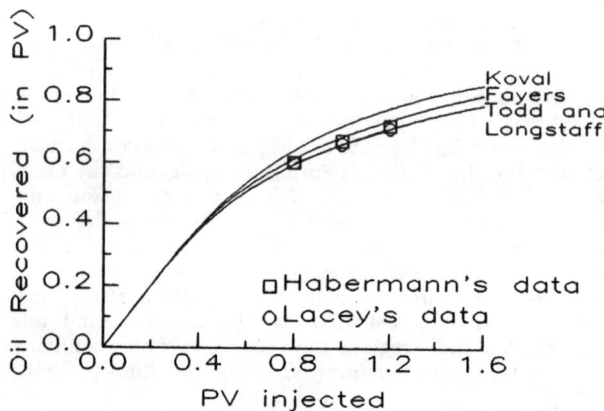

FIG. 3 Recovery Predictions for 5-spot experiments, $M = 10$

dimensional flow for Todd & Longstaff and Koval models:

$$G = f_g \left(1 - \frac{K(\rho_{oe} - \rho_{ge})g}{q\mu_{oe}}(1 - S) \right) \qquad (13)$$

The scaling group $K(\rho_{oe} - \rho_{ge})g/q\mu_{oe}$ measures the ratio of gravity to viscous terms. For the Fayers model $(1 - S)$ is replaced in eqn (13) by $(1 - \Lambda)$. For each of the models it is possible to define a critical flow rate q_c, for which the flow has overall stability, i.e. gravity stabilises fingers at all wavelengths. The performance of the three empirical models in terms of gravity effects in one-dimensional flow is discussed by Newley (1987).

A difficult challenge for the empirical models relevant to reservoir applications is two-dimensional flow in a vertical cross section, where gravity acts normal to the principal flow direction. The displacement behaviour will now vary from a viscous fingering pattern at high flow rates to a gravity override tongue at the lower rates. It is usual to categorise these flows by the viscous to gravity parameter, N_{gr}. Our preferred definition of N_{gr} has been developed for slightly tilted vertical sections (Fayers and Muggeridge, 1989) in the form:

$$N_{gr} = 2 \left(\frac{(M - 1)q}{\lambda_g(\rho_o - \rho_g)g} - \theta \right) \frac{W}{L} \qquad (14)$$

where θ is the dip angle of the section. Fayers and Muggeridge presented results which suggested that for this type of problem fingering will disappear when $N_{gr} < 1$.

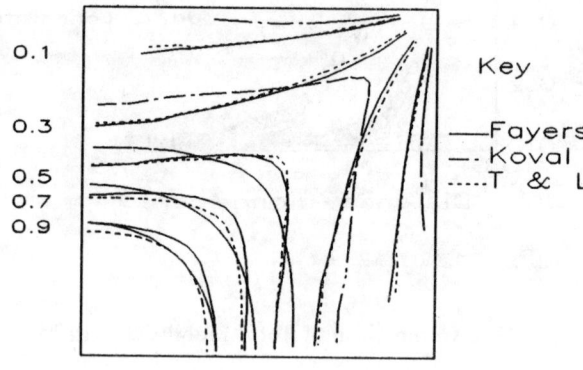

FIG. 4 Solvent Concentration at 0.4 pv injected, $M = 40$

3.2 Gravitational Segregation

Gravity terms in eqs (1) and (2) lead to modified fractional flows, with the general form for downward one-

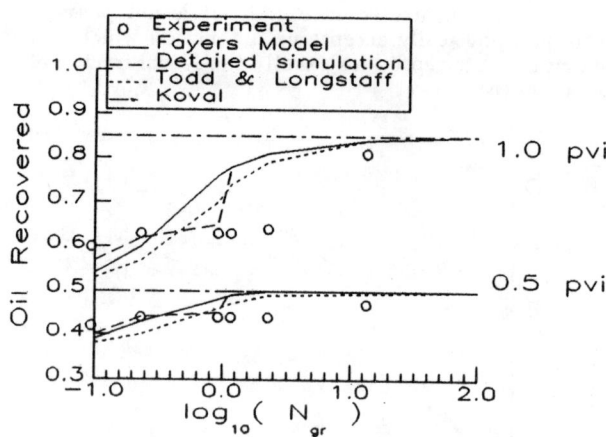

FIG. 5 Effect of Viscous to Gravity Ratio on Recovery for van der Poel Experiments (M=7.9)

In Fig. 5 we show results for the three empirical models and high resolution simulation compared with experimental oil recoveries (at 0.5 and 1.0 pore volumes of solvent injected) for the vertical section experiments ($\theta = 0$) of van der Poel (1962). High resolution simulation, Todd & Longstaff and Fayers models achieve good agreement with the measurements at the high and low

end of the range in N_{gr}, but the changeover of behaviour in the intermediate N_{gr} region is incorrect. This has been attributed to the effects of a finite third dimension in the experiments, not represented in the two-dimensional simulations. More recently, carefully executed experiments by Christie *et al* (1990) with a very small thickness between the plates of the section show a different response in the mid-range of N_{gr}. The resulting excellent agreement between high resolution simulation and measurements is shown in Fig. 6. Three dimensional high resolution simulation results are being used to calibrate empirical models accurately in reservoir applications where gravity effects are important (Christie *et al*, 1991). Both the Todd & Longstaff and Fayers models cope reasonably well with gravity effects, but the extension of the Koval model for these problems is unsuccessful.

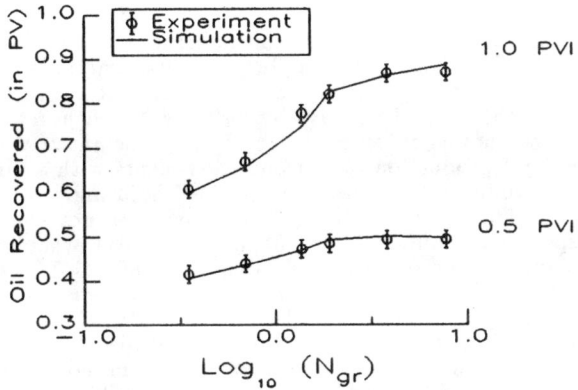

FIG. 6 Effect of Viscous to Gravity Ratio on Recovery for Christie *et al* Experiments (M=5.0 and 6.8)

3.3 Water Alternating Gas (WAG)

None of the discussion in the preceeding Sections has been concerned with the practical reservoir problems associated with tertiary displacement or WAG performance. We now have two phases, water and hydrocarbon, with gas and oil components in the hydrocarbon phase. The presence of water implies the need to define the water fractional flow function:

$$f_w(S_w, S_g) = \frac{\lambda_w}{\lambda_w + \lambda_h} \qquad (15)$$

where λ_w and λ_h are the mobilities of the water and hydrocarbon and phases respectively. The gas fractional flow is:

$$F_g(S_w, S_g) = (1 - f_w)f_g \qquad (16)$$

where f_g was defined in Section 2 for each empirical model.

The conservation equations in one-dimension with a constant flow rate q for water and gas become:

$$\phi \frac{\partial S_w}{\partial t} + q \frac{\partial f_w}{\partial S_w} \frac{\partial S_w}{\partial x} + q \frac{\partial f_w}{\partial S_g} \frac{\partial S_g}{\partial x} = 0 \qquad (17)$$

$$\phi(1 - S_w)\frac{\partial S_g}{\partial t} + q(S_g - f_g)\frac{\partial f_w}{\partial S_w}\frac{\partial S_w}{\partial x} +$$
$$q\left((1 - f_w)\frac{\partial f_g}{\partial S_g} + (S_g - f_g)\frac{\partial f_w}{\partial S_g}\right)\frac{\partial S_g}{\partial x} = 0 \quad (18)$$

This system of equations is hyperbolic provided $f_g > S_g$. For the Koval and Todd & Longstaff methods the fractional flow function is concave and satisfies this condition, but this is not true everywhere for Fayers model, i.e. this solution involves a leading shock front.

The presence of water in the tertiary displacement partially stabilises the flow behaviour, because the total mobility of the generated oil bank can be larger than for oil alone, and similarly the effective mobility of the gas plus water is less than for gas alone. Stalkup (1983) has shown for WAG that optimum stabilisation occurs when the water and gas advance at the same rate. This occurs when, $V_g/S_g^- = V_w/(1 - S_g^-)$, i.e. the gas saturation S_g^- in the drive region is given by solving the implicit equation, $S_g^- = f_g^-$. In the oil bank the oil saturation S_o^+ is determined by the speed of the leading characteristic being coincident with the velocity of the leading shock, $df_o/dS_o^. = f_o^+/(S_o^+ - S_{orw})$. Thus the effective mobility ratio of the system in the equal velocity case becomes:

$$M_e^* = \frac{\lambda_g(S_g^-) + \lambda_w(S_g^-)}{\lambda_o(S_o^+) + \lambda_w(S_o^+)} \qquad (19)$$

In the Todd and Longstaff method we use $M_e = (\mu_o/\mu_g)^{1-\omega}$, so that to obtain correct results we should use for the optimal WAG case:

$$M^{1-\omega^*} = M_e^{*1-\omega} \qquad (20)$$

where ω is the normal value ($\omega = 2/3$), thus ω can be recalibrated using

$$\omega^* = 1 - (1 - \omega)log(M_e^*)/log M \qquad (21)$$

A similar analysis applies for secondary WAG displacement. High resolution simulations of WAG displacement have been described by Christie (1988), and these confirm the stabilisation ideas expressed above. Fig. 7 shows a high resolution simulation of fingering in miscible tertiary flooding in a line drive compared with a WAG flood at optimum WAG ratio. The viscosity ratios used were $\mu_o/\mu_g = 40$ and $\mu_o/\mu_w = 10$, with simple quadratic relative permeability functions.

Figure 8 shows identical WAG oil recovery curves for high resolution simulation and Todd & Longstaff computed using $\omega = 0.98$, as obtained from eqn (21). The shift in the required value of ω is large, and this value is confirmed by the high resolution simulation results. Fig. 8 also gives a similar comparison for tertiary miscible flooding, where $\omega = 0.81$ gives the correct Todd & Longstaff recovery profile. Further extensions of this work are needed to analyse total mobilities and gravity override in WAG processes.

miscible flood

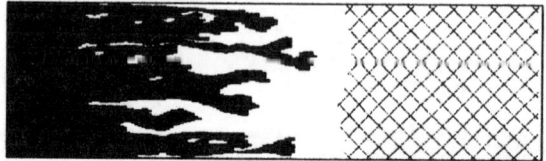

WAG

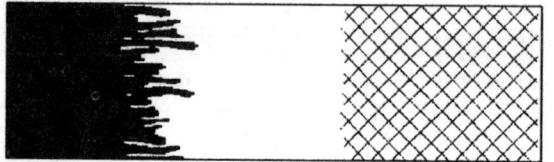

FIG. 7 High Resolution Simulations of Tertiary Miscible and WAG recovery (solvent in black, water hatched)

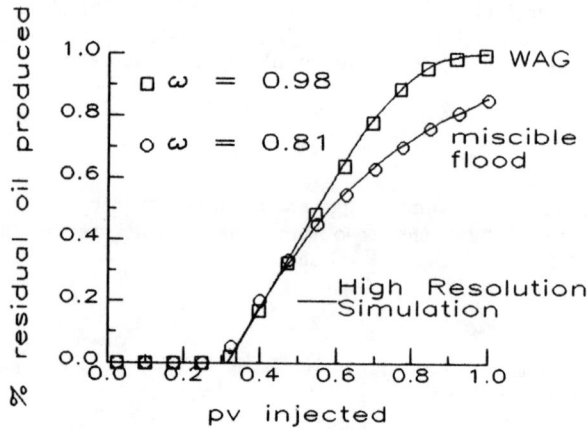

FIG. 8 Production Curves for Tertiary Recovery, compared with Todd & Longstaff

4 HETEROGENEITIES

4.1 Extension of Empirical Models

Koval extended his fractional flow model to include heterogeneous effects by introducing the heterogeneity parameter H as a multiplier on effective viscosity ratio, thus the overall Koval mobility parameter M_e is given by:

$$M_e = H \frac{\mu_o}{\mu_{ge}} \qquad (22)$$

with the fractional flow given by eqn (6). Koval analysed experimental results by fitting H so that the oil recovery from a test at unit viscosity ratio would be matched by his model at 1 pore volume injection.

This formulation can give a reasonable approximation to the tracer production behaviour observed from tracer experiments in slightly heterogeneous cores. Koval also analysed the behaviour of H as a function of the Dyksta-Parsons coefficient V_{DP}, and showed that a fractional flow behaviour of the form (6) was also obtained from layered systems with randomly chosen permeabilities. A correlation for H is given by:

$$log H = \frac{V_{DP}}{(1 - V_{DP})^{0.2}} \qquad (23)$$

The resulting values of H can be large, e.g. for $V_{DP} = 0.75$, $H = 9.8$. This implies that the fractional flow behaviour for layered systems is considerably modified from the normal effective viscosity result with $H = 1$. Koval showed evidence that using the constant H value, solvent production rates from experiments with a range of mobility ratios were well predicted by using eqn (22).

The Todd and Longstaff model does not contain specific recommendations for heterogeneity, other than using an appropriately adjusted value of ω. This implies in the limit $M = 1$, that $f_g = S$ (see eqn (6)), i.e. no elongation of the displacement profile. This result is clearly not acceptable. For non-unit viscosity ratios, the Todd & Longstaff parameter ω can be adjusted to match the Koval model, using eqn (12), which requires a $log H$ term in the brackets for the heterogeneous case. In terms of the normal ω (e.g. $\omega = 2/3$) this leads to the expression:

$$\omega^* = \omega - \frac{log H}{log(\mu_o / \mu_g)} \qquad (24)$$

When H is large, there is the unacceptable possibility that $\omega^* < 0$. Thus we do not expect the conventional Todd & Longstaff to be satisfactory in the more extreme cases of heterogeneity. An alternative modification of the Todd & Longstaff method would be to follow Koval's procedure and to define the effective Todd & Longstaff mobility ratio by the equation:

$$M_e = H \left(\frac{\mu_o}{\mu_g} \right)^{1-\omega} \qquad (25)$$

where ω is chosen as the homogeneous value. Recent papers have shown that the H factor in Koval's method is not satisfactory (Giodarno et al, 1985), but nevertheless this remains the only formal procedure for compensating for heterogeneities in reservoir simulation of miscible displacement.

The Fayers model has not been investigated in terms of its extension to heterogeneous problems. For mild heterogeneities the parameters a and α in the fingering function can be adjusted. Fayers hypothesised that for more extreme heterogeneities, semi-permanent bypassing of oil can occur, so that the fingering function never occupies the full width. This led to the definition $\Lambda = a + bS_{gg}^{\alpha}$ with $a + b < 1$.

4.2 Numerical Studies

We use high resolution simulation to determine the average fingering behaviour in a heterogeneous two dimensional system. Firstly we generate correlated permeability maps where the logarithm of K has a normal distribution. To achieve this, uncorrelated random numbers with a normal distribution are assigned to each grid block. The logarithm of K at a point is then the average of these numbers in a circle of radius λ_D centred on that point. This produces a log-normal distribution of K values with a dimensionless correlation length λ_D, where λ_D is the circle radius as a fraction of the length of the system. The nature of the heterogeneity is specified by both λ_D and σ_{logK} which is the variance in the logarithm of the generated permeability.

We then simulate the constant-rate injection of miscible solvent into oil. We have averaged the saturations S along the direction perpendicular to mean flow. The calculations have been executed on grids of 300×100 and 150×50 with σ_{logK} in the range 0 to 5 and various values of λ_D from $1/300$ to 1. We have chosen $L/W = 3.0$, which conforms with the ratio suggested by Blackwell *et al* (1959). Explicit microscopic diffusion has not been included, so that the consequent large Peclet numbers are governed by low levels of numerical dispersion associated with the FCT high order method in the high resolution simulation code (Christie and Bond, 1987).

4.3 Unit Viscosity Ratio

Single phase flow of a tracer is believed to be governed by a convection/diffusion equation of the general form:

$$\frac{\partial S}{\partial t} + \frac{\partial f(S)}{\partial x} = \frac{\partial}{\partial x} D(S) \frac{\partial S}{\partial x} \qquad (26)$$

Distance x and time t are dimensionless units: x is a distance measured in a system of unit length, and t is measured in pore volumes injected. The dispersion $D(S)$ arises because of the perturbations introduced by the variations in permeability on the pressure field, which satisfies the equation $\nabla . K \nabla P = 0$. For the simple case $f = S$ and $D = const$, the solution is given by:

$$S = \frac{1}{2} \left(1 - erf \left(\frac{x-t}{\sqrt{2Dt}} \right) \right) \qquad (27)$$

The mean position of the front moves with constant speed, but the width of the profile increases only as $t^{1/2}$, due to dispersion.

It is possible to use perturbation theory to determine the dispersion coefficient for small variations in permeability with a correlation length λ_D (see for instance the review paper by Lake (1987)) which gives the result:

$$D = C_v^2 \lambda_D (1 - e^{-t/\lambda_D}) \qquad (28)$$

where C_v is the coefficient of variation of K. For a log-normal distribution $log(C_v^2 + 1) = \sigma_{logK}^2$. For large correlation lengths, or short times $D \to C_v^2 t$. In this case a convective-like solution is obtained:

$$S = \frac{1}{2} \left(1 - erf \left(\frac{x-t}{\sqrt{2C_v t}} \right) \right) \qquad (29)$$

where the front broadens linearly with time, which is applicable to, for instance, layered systems of the type considered by Koval, which we will not investigate here. Notice that the solution for S may be written as a function of $v = x/t$. This is also the case for all purely convective solutions of eqn (26), when $D = 0$, which would be obtained from the fractional flow models discussed earlier.

The limit $D \to C_v^2 \lambda_D$ is of considerable interest, since it describes the long time behaviour in a reservoir with small scale heterogeneities. The solution for S is given by eqn (27) and is a function of $(x - t)/t^{1/2}$.

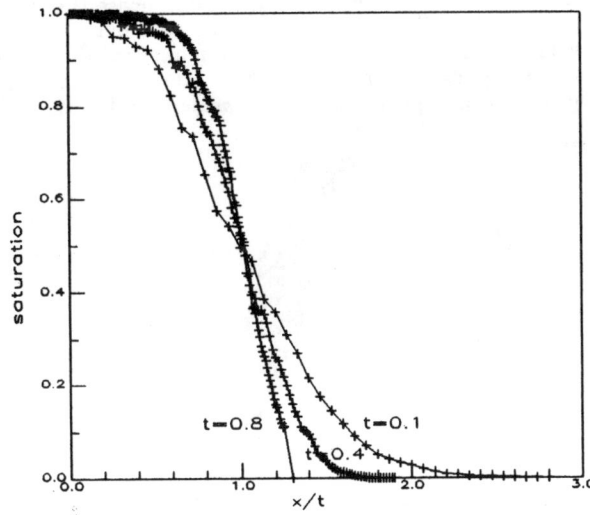

FIG. 9 Saturation Profiles for Unit Viscosity Ratio in a Heterogeneous System as a Function of x/t

Figure 9 shows the saturation profile for an $M = 1$ flood with $\lambda_D = 1/150$ and $\sigma_{logK} = 2$ at different times. S is plotted as a function of $v = x/t$. If a fractional flow formulation were correct the curves should all lie on top of one another. This is clearly not the case, and Koval's method is therefore not valid at this short correlation length.

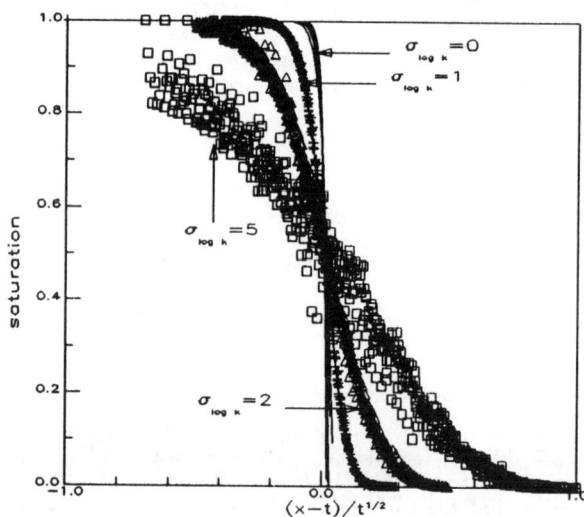

FIG. 10 Saturation Profiles for Unit Viscosity Displacement in a Heterogeneous System as a Function of $(x-t)/\sqrt{t}$

Figure 10 shows the same saturation profiles measured at 20 times from $t = 0.1$ to $t = 2.0$, plotted as a function of $(x-t)/t^{\prime}$. If a dispersive formulation were correct the curves would be coincident. This is so for the various values of σ_{logK} illustrated ($\lambda_D = 1/150$ in all cases). The curves for $\sigma_{logK} = 0$ indicate the degree of numerical dispersion. The high resolution simulation solution method gives very little numerical dispersion, and the physically valid dispersion caused by the heterogeneities in the other cases is considerably larger.

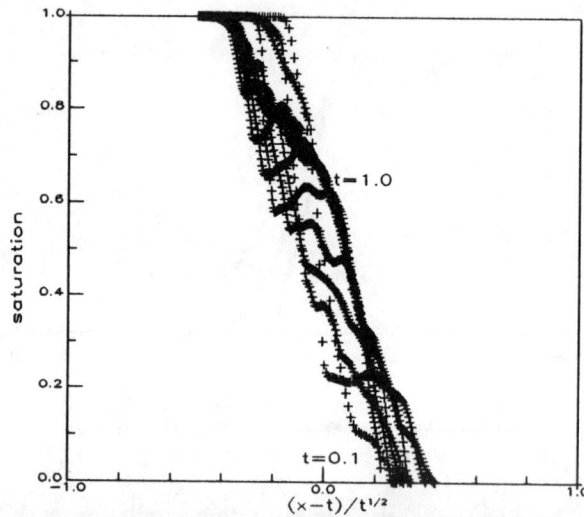

FIG. 11 Saturation Profiles for Unit Viscosity Ratio in a Correlated Heterogeneous System

Figure 11 shows saturation profiles for a larger correlation length, $\lambda_D = 1/10$ with $\sigma_{logK} = 1$. The at-

tempt to collapse all the data onto a single curve has been less successful. This is for two reasons. Firstly, the large scale heterogeneities perturb the profile, with channelling through regions with large K, leading to a fingered front. This means that the average saturation profile varies widely depending on where the average is taken: S is sensitive to the details of the pattern of permeability. Secondly the profiles at later times are systematically broader (as a function of $(x-t)/t^{\prime}$) than those at earlier times, which indicates that a fractional flow formulation is becoming appropriate.

Figure 12 illustrates the displacements for $\sigma_{logK} = 1$, $\lambda_D = 1/300$ and $\sigma_{logK} = 1$, $\lambda_D = 1/10$. In the former case the front experiences random perturbations about a smooth profile; for a larger correlation length the pattern is noticeably channelled through streaks of high permeability.

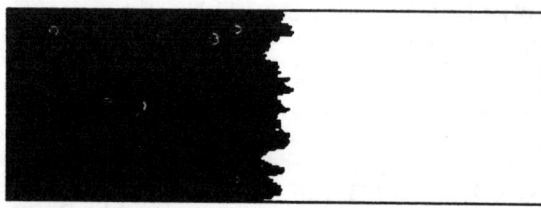

(a) $\lambda_D = 1/300$

(b) $\lambda_D = 1/10$

FIG. 12 Unit Mobility Ratio Displacements at $t = 0.5$ with $\sigma_{logK} = 1$. Areas with $s > 0.5$ are shaded

We have seen that heterogeneities have a dispersive effect on the front, if the correlation length λ_D is small. The table below gives the effective dispersion constant D for the simulations we have performed. D is measured from the best fit of the solution for S in eqn (27) to the profiles shown in Figs. 10 and 11. D is also compared with the perturbation theory prediction. This is only valid for $\sigma_{logK} < 1$ and as can be seen it greatly overestimates D for large σ_{logK}.

We have indicated that when $\lambda_D \sigma_{logK} \simeq 0.1$ the front begins to look fingered and the overall shape of the pattern is dominated by the permeability distribution. Also the average behaviour differs widely between different realizations. Only for a well characterized geometry, such as a layered system, would an averaged description be possible and then a fractional flow model is appropriate. Araktingi and Orr (1988) found similar results, with a cross-over to heterogeneity dominated flow when $\lambda_D \sigma_{logK} > 0.25$.

Table 1: Effective Diffusion Coefficients for $M = 1$

σ_{logk}	λ_D	D (measured)	D (from eqn (28))
0	1/150	$2 \times 10^-$	0
1	1/150	$6 \times 10^-$	$1.1 \times 10^-$
2	1/150	$3 \times 10^-$	0.36
5	1/150	0.2	147
1	1/10	0.2	0.17

The measured values are only accurate to 1 significant figure.

4.4 Adverse Mobility Ratios

As we discussed earlier, if a displacement is performed in a nearly homogeneous system at adverse viscosity ratio, the saturation profiles are functions of $v = x/t$ only. For instance, for the example $M = 10, \lambda_D = 1/150$ and σ_{logK} is only 0.2, we may average $S(v)$ at different times to obtain the smoothed profile shown in Fig. 13. The dispersion is negligible.

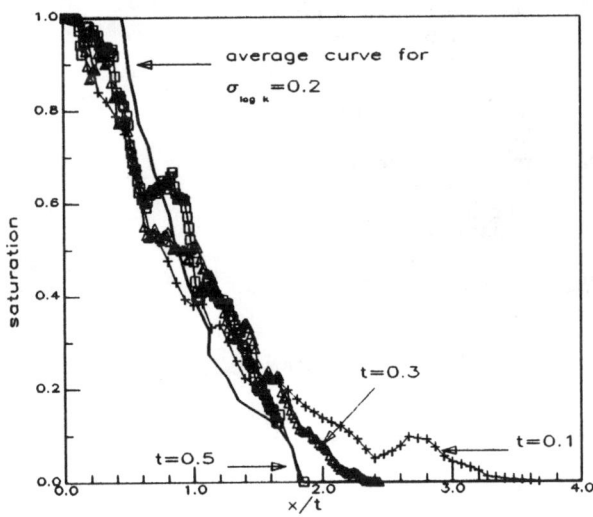

FIG. 13 Saturation Profiles for $M = 10$

Figure 13 also shows results for $M = 10, \sigma_{logK} = 2$ and $\lambda_D = 1/150$, where the saturation profiles at different times are not functions of v only. The solid line is an average profile for a nearly homogeneous case. The quality of the results is not very good, since the displacement profile is fingered, which gives large fluctuations in the average saturation. However, it would appear that for large times the plots are converging towards the saturation, $S(v)$, measured from the nearly uniform flood. Thus the general solution may be interpreted as a displacement modelled by $f(S)$ from a homogeneous system with a non-zero dispersion constant D, i.e. the general

form of eqn (26).

In general one may expect D to be a function of both M and S, and to be related to $f(S)$. However, our preliminary results are insufficiently detailed to determine D with any precision. Since the diffusive term in eqn (26) is a correction to the convective behaviour for small scale heterogeneities, a good prediction of the observed displacement for adverse mobility ratios might be possible if the homogenous $f(S)$ is used together with a reasonable estimate of D.

The easiest approach would be to use a constant D obtained from a unit mobility ratio flood and a Koval or Todd & Longstaff fractional flow from a homogeneous, adverse mobility displacement. We may test the accuracy of this method. The nearly-homogeneous averaged saturation profiles for $M = 10$ and $M = 30$ were measured (as in Fig. 13) and from them the fractional flow function consistent with the observed $S(v)$ is found. For $M = 10$ this was fitted to an $f(S)$ described by eqn(6), for which we find $M_e = 1.93$ (indicating a Todd & Longstaff ω parameter of 0.71). The Koval model would predict $M_e = 1.88$ from eqn (5), which agrees well with the high resolution simulation result. For $M = 30$ we found $M_e = 2.3$, corresponding to a Todd & Longstaff ω of 0.76. The Koval prediction is $M_e = 2.81$, which is an over-estimate. $S(x,t)$ is then predicted from a numerical finite difference solution of eqn (26). For $M = 100$ we found that M_e was a function of saturation for the nearly homogeneous flood, which contradicts the Koval and Todd & Longstaff formulations. The validation of the empirical models for very large M requires further study.

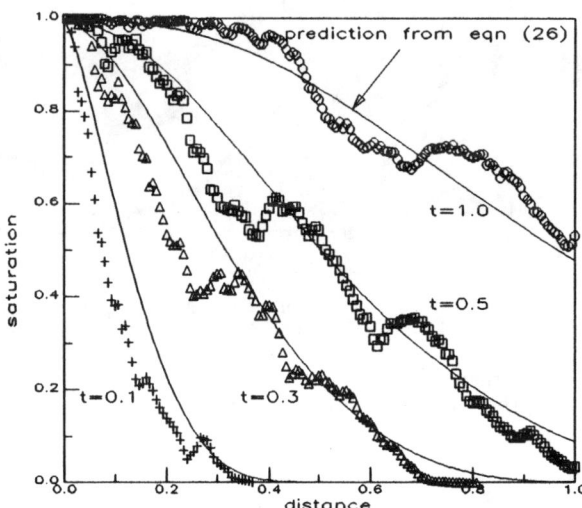

FIG. 14 Saturation Profiles for $M = 10$, compared with Todd & Longstaff plus Diffusion Term

Figure 14 shows predicted and measured solutions for $\sigma_{logK} = 2, \lambda_D = 1/150$ and $M = 10$, and Fig. 15 illustrates the case $M = 30$. In each case, the high resolution simulation results are compared with predictions from a Todd and Longstaff model with a diffusion term. Within the considerable fluctuations due to fingering, the predicted profiles are reasonable estimates. This is not surprising, since the dominant influence comes from $f(S)$

and we have ensured that this is approximately correct.

In reservoir simulators, the effective dispersion constant D may be included as a correction to the fractional flow, by writing:

$$F(S, \frac{\partial S}{\partial x}) = f(S) - D\frac{\partial S}{\partial x} \qquad (30)$$

then if F is considered to be the flux, $D\partial S/\partial x$ may be an estimate of the correction to a Todd & Longstaff form. For instance, if $F(S)$ in eqn (30) is matched to a Koval fractional flow at a given time t we may estimate an effective H parameter, which will decrease to 1 with time. For the results presented in Fig. 15, the correction to $f(S)$ in eqn (30) at $S = 0.5$ leads (after some algebra) to an effective $H = 1.4$ at $t = 0.1$ and $H = 1.04$ at $t = 1$, using the homogeneous $M_e = 2.3$. Notice that for large times this correction is small: for short correlation lengths eqn (23) would greatly over-estimate H. The Dyksta-Parsons coefficient V_{DP} is $1 - e^{-\sigma}$. For the results in Fig. 15, this gives $V_{DP} = 0.865$ and from eqn (23) $H = 19.5$ which is far too high.

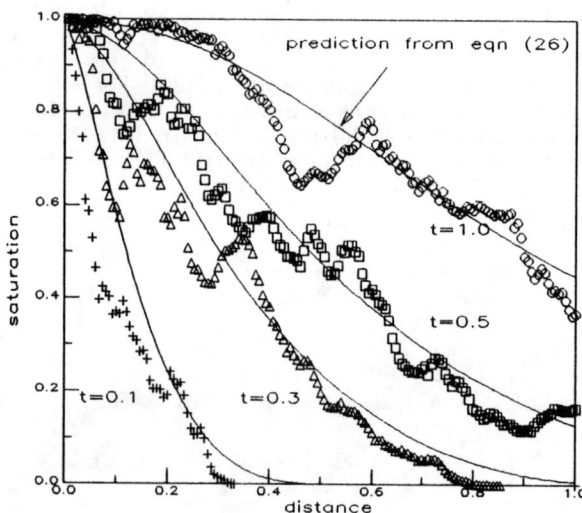

FIG. 15 Saturation Profiles for $M = 30$, compared with Todd & Longstaff plus Diffusion Term

Further improvements would stem from finding an appropriate analytical form for $D(S, M)$.

5 CONCLUSIONS

(i) All three empirical models give satisfactory fractional flows for pseudo one dimensional homogeneous problems at the laboratory scale. However, the Koval model inadequately describes total mobilities and gravity influenced flows. The Fayers and Todd and Longstaff models appear to be similar in performance for multi-dimensional homogeneous problems. In spite of its more clearly defined physical basis, there is no strong case for preferring use of the Fayers model as a replacement for the Todd & Longstaff approach in reservoir simulators. We have also illustrated the strong effect that a WAG process can have on the effective ω value in the Todd & Longstaff method.

(ii) The effect of small scale heterogeneities is to introduce a diffusion-like or dispersive broadening of the displacement, which is inconsistent with a fractional flow formulation. For moderate mobility ratios, a good approximation to the observed behaviour was obtained by using the homogeneous fractional flow and a diffusive term with a dispersion constant measured from a unit viscosity ratio flood.

(iii) Large scale heterogeneities produce channelling and the behaviour is dependent on the exact pattern of permeability.

(iv) Further work on three dimensional systems, and to determine a reliable diffusivity function $D(S, M)$ should be performed.

Nomenclature

a	parameter in Fayers model
b	parameter in Fayers model
C_v	coefficient of variation
D	dispersion coefficient
f	fractional flow
g	gravity acceleration
G	fractional flow under gravity
H	heterogeneity factor
k_r	relative permeability
K	absolute permeability
L	length
M	mobility ratio
M_e	effective mobility ratio
N_{gr}	viscous to gravity ratio
P	pressure
q	total flow rate
S	saturation or concentration
q	total flow rate
v	Darcy velocity
V_{DP}	Dykstra-Parsons coefficient
W	width
α	growth rate parameter
λ_D	dimensionless correlation length
λ_t	total mobility
Λ	finger width function
μ	viscosity
ϕ	porosity
σ_{logK}	standard deviation in $logK$
ρ	density
θ	inclination
ω	Todd & Longstaff mixing parameter

Subscripts

g	gas
o	oil
w	water
e	effective

Acknowledgements

We are grateful to Dr P.R. King for useful discussions and to Dr G. Verly for generating the correlated permeability distributions used in the paper. We thank the British Petroleum Co. plc for permission to publish this paper.

References

Araktingi, U. G. and Orr, R. F. M., 1988 *Viscous Fingering in Heterogeneous Porous Media* SPE 18095.

Blackwell, R. J., Rayne, J. R. and Terry, W. M., 1959 *Factors Influencing the Efficiency of Miscible Displacement* Trans AIME **216** 1.

Christie, M. A. and Bond, D. J., 1987 *Detailed Simulation of Unstable Processes in Miscible Flooding* SPERE **2** 514-522.

Christie, M. A., 1988 *Application of High Resolution Simulation to Modelling Fluid Instabilities* Proc IMA conf on Mathematics of Flow in Porous Media, eds. S. F. Edwards and P. R. King, Clarendon Press, Oxford.

Christie, M. A., 1989 *High Resolution Simulation of Unstable Flows in Porous Media* SPERE **4** 297-304.

Christie, M.A ., Jones, A. D. W. and Muggeridge, A. H., 1990 *Comparison between Laboratory Experiments and Detailed Simulations of Unstable Miscible Displacement Influenced by Gravity* In *North Sea Oil and Gas Reservoirs II* to be published by Graham and Trotman.

Christie, M. A., Muggeridge, A. H. and Barley, J. J., 1991 *3D Simulation of Viscous Fingering and WAG Schemes* SPE 21238 to be presented at 11th SPE Symposium on Reservoir Simulation, Anaheim, CA, Feb 1991.

Fayers, F. J., 1988 *An Approximate Model with Physically Interpretable Parameters for Representing Miscible Viscous Fingering* SPERE **3** 551-558.

Fayers, F. J. and Newley, T. M. J., 1988 *Detailed Validation of an Empirical Model for Viscous Fingering with Gravity Effects* SPERE **3** 542-550.

Fayers, F. J. and Muggeridge, A. H., 1989 *Extension to Dietz Theory and Behaviour of Gravity Tongues in Slightly Tilted Reservoirs* SPE 18438. Presented at 5th SPE Symposium on Reservoir Simulation, Houston, 1989.

Giodarno, R. M., Salter, S. J. and Mohanty, K. K., 1985 *The Effects of Permeability Variations on Flow in Porous Media* SPE 14365, Presented at 60th Annual Meeting of SPE, Las Vegas.

Habermann, B., 1960 *The Efficiency of Miscible Displacements as a Function of Mobility Ratio* Trans AIME **219** 264.

Koval, E. J., 1963 *A Method for Predicting the Performance of Unstable Miscible Displacement in Heterogeneous Media* Trans AIME **228** 145.

Lacey, J. W., Faris, J. E. and Brinkman, F. H., 1961 *Effect of Bank Size on Oil Recovery in the High-Pressure Gas Driven LPG-Bank Process* J Pet Tech 806.

Lake, L. W., 1987 *A Marriage of Geology and Reservoir Engineering* in *Numerical Simulation in Oil Recovery* IMA Vol.3 in Mathematics and its Applications, ed. M.F. Wheeler. Springer-Verlag, Berlin 177-198.

Moissis, D. E., Miller, C. A. and Wheeler, M. F., 1989 *Simulation of Miscible Viscous Fingering Using a Modified Method of Characteristics: Effects of Gravity and Heterogeneity* SPE 18440, presented at 5th SPE Symposium on Reservoir Simulation, Houston, TX, 1989.

Newley, T. M. J., 1987 *Comparison of Empirical Models for Unstable Miscible Displacement* Proc European Symposium on EOR, Hamburg, Oct 27-29.

Stalkup, F. I., 1983 *Miscible Displacement* SPE monograph series, Vol.8, New York.

Todd, M. R. and Longstaff, W. J., 1972 *The Development, Testing and Application of a Numerical Simulator for Predicting Miscible Flood Performance* Trans AIME **253** 984.

van der Poel, C., 1962 *Effect of Lateral Diffusivity on Miscible Displacement in Horizontal Reservoirs* Trans AIME **225** 317.

2nd European Conference on the Mathematics of Oil Recovery
© D. Guérillot, O. Guillon (Editors) and Éditions Technip, Paris 1990, pp. 57-64
27 rue Ginoux, 75015 Paris

Effective Permeability of Heterogeneous Reservoir Regions

L. J. Durlofsky and E. Y. Chung[1]

ABSTRACT

Most reservoir heterogeneities, such as detailed cross-bedding, are on a scale that is too fine to be directly included in reservoir simulation or reservoir engineering calculations. Therefore, 'averaging' procedures are required to scale up permeabilities from the fine scale, over which heterogeneities occur, to the coarser scales appropriate for reservoir engineering computation. The purpose of this paper is to describe the numerical implementation of such an averaging technique and to apply this technique to the scale up of a representative region of a crossbedded eolian depositional system. Toward this end, a triangle based nonconforming finite element approximation of the fine scale pressure equation, subject to appropriately formulated periodic boundary conditions, is solved to give the effective coarse scale permeability. It will be seen that the magnitudes of the cross terms of the effective permeability tensor are significant for the eolian system considered and that the accurate determination of these terms is essential for modeling flow through crossbedded systems.

1. INTRODUCTION

The improved characterization of heterogeneous reservoirs has resulted from advances in a variety of related areas. These advances include improved borehole measurement capabilities (*e.g.*, formation microresistiv-ity scanning), the development of conditioned geostatistical reservoir descriptions and detailed outcrop studies. Through use of these and other techniques, it is possible, given sufficient data, to generate realistic, detailed reservoir descriptions. However, if a highly detailed reservoir description is used directly as input to a flow simulation, computing requirements are prohibitive. Therefore, methods are needed to scale up (or average) the fine scale reservoir description to coarser scales more appropriate for reservoir simulation. Properties on this coarser scale are referred to as effective properties.

Our intent here is to present a general formulation and numerical method applicable to the determination of the effective absolute permeability of heterogeneous reservoir regions. We shall apply this method to the determination of the effective permeability of a representative portion of an eolian depositional system. Eolian depositional systems, formed by wind blown sand dune deposits, are of both practical and theoretical interest. Their practical interest derives from the fact that many reservoirs contain eolian strata (*e.g.*, reservoirs in the Western Overthrust Belt, the Gulf of Mexico and the North Sea); they are of theoretical interest because of their complex cross stratification, which has a significant effect on their effective permeability, as we shall see.

Many previous investigators have considered the problem of computing the effective permeability of heterogeneous reservoir regions. Most of these studies, some of which are discussed by Desbarats (1987), address rather specific situations (*e.g.*, small permeability variations, particular types of sequences). A more general approach was

---

(1) Chevron Oil Field Research Company, PO Box 446, La Habra, California 90633-0446, USA.

suggested by White and Horne (1987). These investigators recognized that, because effective permeability must be treated as a tensor quantity, the boundary conditions for the fine scale problem (which must be solved to obtain the coarse scale effective permeability) must be specified with care. Otherwise, these boundary conditions can artificially influence the effective permeability result. The approach of White and Horne entails the use of several different sets of boundary conditions, but does not assure the symmetry of the resulting effective permeability tensor, as required by Onsager reciprocity. Thus its use for complicated systems may be suspect.

Kasap and Lake (1989) developed an analytical method for effective permeability calculations that is able to treat locally tensorial permeabilities. This method was applied to the modeling of flow through eolian systems, such as those considered here. The method is particularly suitable for layered systems; it is only approximate for more complex scenarios.

Several investigators have applied homogenization theory to the determination of effective permeability (*e.g.*, Bourgeat, 1984; Saez *et al.*, 1989; Mei and Auriault, 1989). These approaches establish the validity of the effective permeability (under certain assumptions) and yield equations for its determination. To date, solutions to these equations have only been obtained for simple heterogeneity fields, though there is no limitation in principle to their application to more complex systems. Homogenization approaches yield, in all cases, symmetric effective permeability tensors.

The general numerical method presented in §2 derives from the homogenization approaches discussed above, though the equations we actually solve differ from those that result from the homogenization procedures. The method entails a nonconforming triangular finite element discretization of the fine scale pressure equation. The use of triangular elements allows for the modeling of complex geometries; this capability is necessary for the cross strata we wish to consider. Periodic boundary conditions are imposed on the fine scale problem. This specification provides unambiguous boundary conditions and always results in symmetric effective permeability tensors. A brief description of eolian depositional systems and the permeabilities associated with the various strata is provided in §3. In §4 we present results for the effective permeability of a portion of an eolian outcrop. There it will be seen that the accurate determination of the cross terms of the effective permeability tensor is essential for a valid scale up of the eolian region under study. Additional conclusions are drawn in §5.

2. DETERMINATION OF EFFECTIVE PERMEABILITY

In this section we outline the general formulation and numerical procedure used for the determination of the effective permeability of heterogeneous reservoir regions. A more complete description of the method, including implementation details, will be published separately.

2.1 General Formulation

Single phase, incompressible flow through porous media is described by combining Darcy's law and continuity to yield the pressure equation (given in dimensionless form):

$$\nabla \cdot (\mathbf{k} \cdot \nabla p) = 0, \tag{1}$$

where p is pressure and $\mathbf{k}$ is the permeability tensor. Here, $\mathbf{k}$ is considered to vary on at least two scales, with the finest variation on a scale finer than that which can be resolved by a large scale reservoir model. Our intent is to replace (1) with an equation of identical form but in which $\mathbf{k}$ is replaced with an effective permeability tensor, designated $\mathbf{k}^*$; *i.e.*,

$$\nabla \cdot (\mathbf{k}^* \cdot \nabla p) = 0, \tag{2}$$

where $\mathbf{k}^*$ now varies on a coarse (simulation grid block) scale. Solution of (2) is clearly less costly computationally, once $\mathbf{k}^*$ is determined, than is solution of (1), because many fewer grid blocks are required to resolve variations in $\mathbf{k}^*$ than are required for $\mathbf{k}$.

The homogenization approaches discussed above have demonstrated that the simplification from (1) to (2) is theoretically possible when two disparate length scales exist. An example of such a case is a spatially periodic porous medium, represented schematically in Fig. 1. Here, two distinct length scales are evident: a fine scale $\mathbf{y}$ and a coarse scale $\mathbf{x}$. For such cases, we have devised a general formulation which yields the effective coarse scale permeability. Though this procedure is based on the homogenization result which assumes two disparate length scales, it can be expected to be applicable to much more general cases with more complicated variation in the permeability field. This point will be demonstrated explicitly in another publication.

Our method entails solution of the fine scale pressure equation (1) over the region for which the effective permeability is to be determined. This region is considered to be imbedded in a larger domain which is comprised of periodic replications of the region under study. Thus the imposition of periodic boundary conditions is appropriate. We further approximate the global pressure field locally as

$$p = p_0 + \mathbf{G} \cdot \mathbf{x}, \tag{3}$$

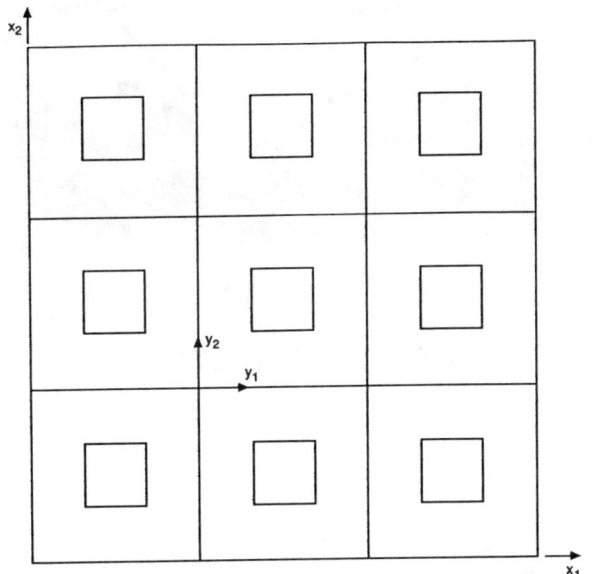

FIG. 1. Spatially periodic porous medium. **k** varies rapidly in **y** and does not vary in **x**.

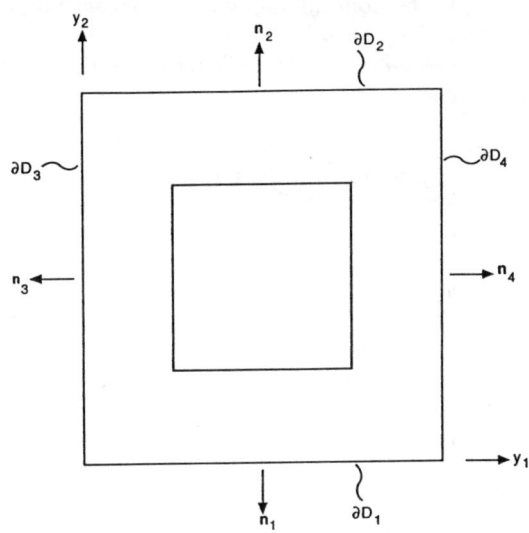

FIG. 2. Unit cell for effective permeability calculation.

where **G** represents the local pressure gradient and **x** the global coordinate. These two specifications, *i.e.*, periodicity and a locally linear pressure field, allow us to unambiguously impose boundary conditions on the fine scale problem. We first resolve **G** into its two components in the **y** coordinate system; *i.e.*, write $\mathbf{G} = G_1\mathbf{i}_1 + G_2\mathbf{i}_2$. To compute **k***, two problems must be solved, one for which $G_1 = 0$ and one for which $G_2 = 0$. For the latter case, the boundary conditions are as follows (with reference to Fig. 2):

$$p(y_1, y_2 = 0) = p(y_1, y_2 = 1) \quad \text{on } \partial D_1 \text{ and } \partial D_2, \quad (4a)$$

$$\mathbf{u}(y_1, y_2 = 0) \cdot \mathbf{n}_1 = -\mathbf{u}(y_1, y_2 = 1) \cdot \mathbf{n}_2 \\ \text{on } \partial D_1 \text{ and } \partial D_2, \quad (4b)$$

$$p(y_1 = 0, y_2) = p(y_1 = 1, y_2) - G_1 \quad \text{on } \partial D_3 \text{ and } \partial D_4, \quad (4c)$$

$$\mathbf{u}(y_1 = 0, y_2) \cdot \mathbf{n}_3 = -\mathbf{u}(y_1 = 1, y_2) \cdot \mathbf{n}_4 \\ \text{on } \partial D_3 \text{ and } \partial D_4. \quad (4d)$$

Upon specification of pressure at any single point in the domain, the boundary value problem (1) subject to (4) can be solved.

The effective permeability is determined by equating the integrated fluxes from the fine scale solution with those which would obtain from the global pressure gradient **G** and the effective permeability tensor **k***. Specifically, we write

$$\langle u_1 \rangle = -(k_{11}^* G_1 + k_{12}^* G_2), \quad (5a)$$

$$\langle u_2 \rangle = -(k_{21}^* G_1 + k_{22}^* G_2), \quad (5b)$$

where $\langle u_1 \rangle$ and $\langle u_2 \rangle$ are the total fluxes across the faces of the unit cell with normal in the $\mathbf{i}_1$ and $\mathbf{i}_2$ directions respectively. Now, for the case $G_2 = 0$, k_{11}^* and k_{21}^* are readily determined from G_1, $\langle u_1 \rangle$ and $\langle u_2 \rangle$. Solving for the case $G_1 = 0$ (with appropriate modifications to (4)) allows for the determination of the remaining two components of **k***. In all cases the resulting values for the cross terms of **k*** are identical, as required by the symmetry of **k*** (the requirement that **k*** be symmetric is proved by Mei and Auriault, 1989).

2.2 Numerical Solution Procedure

Many different numerical procedures could be applied to the solution of (1) subject to (4). Because we require accurate fluxes for the calculation of **k*** and because we need to resolve complex geometries, mixed or nonconforming finite element methods are most appropriate (see Thomasset, 1981, for descriptions of both these methods). For the triangular elements we shall use, Marini (1985) has proved that the lowest order nonconforming finite element method (P1 triangles) yields fluxes that are identical to those of the lowest order mixed method, which is computationally more intensive. In both cases, fluxes are computed to the same degree of accuracy as the pressure field (*i.e.*, to $O(h^2)$, where h is the length of a typical element edge), in contrast to standard finite element methods, for which fluxes are one order less accurate than the pressure field. Thus, the solution of (1) subject to (4) will be accomplished via a triangle based nonconforming finite element method.

To apply the nonconforming finite element method, we first discretize the solution domain D into a set of M triangular elements. Pressure is linearly approximated over each element, $i.e.$,

$$p(y_1, y_2) = \sum_{i=1}^{N} \gamma_i \Phi^i(y_1, y_2), \qquad (6)$$

where N is the total number of element edges. The Φ^i basis functions are defined over each element such that $\Phi^i = 1$ at the midpoint of edge i and zero at the midpoints of the other two edges of the element.

The residual equations for the nonconforming finite element method are formed by weighting the pressure equation with the Φ^k basis functions and integrating over the domain D:

$$\int_D [\nabla \cdot (\mathbf{k} \cdot \nabla p)] \Phi^k dS = 0. \qquad (7)$$

Integrating by parts, applying the divergence theorem and introducing the expansion for p (6) gives

$$\gamma_i \int_D (\mathbf{k} \cdot \nabla \Phi^i) \cdot \nabla \Phi^k dA + \int_{\partial D} (\mathbf{u} \cdot \mathbf{n}) \Phi^k dl = 0, \qquad (8)$$

where summation is implied by a repeated index. The boundary conditions are imposed by specifying the correspondences between pressure and flux unknowns on opposite boundaries stated in (4). For the flux specifications, this entails writing out ∇p terms via differentiation of (6). The resulting equation set is very sparse though the periodicity requirement relates unknowns on one boundary to those on the opposite boundary, complicating the matrix structure. Solution can be achieved through either direct or iterative techniques.

This completes our description of the general formulation and numerical procedure for the determination of $\mathbf{k}^*$. We now turn to a brief description of eolian depositional systems and then apply the general method to the calculation of the effective permeabilities for such systems.

3. EOLIAN DEPOSITIONAL SYSTEMS

Eolian depositional processes (wind blown sand dune deposits) display a variety of common features, though there is a great degree of variability among eolian systems. The significant stratification features of eolian depositional systems, grainflow, wind ripple and grainfall deposits, are discussed by Goggin et $al.$ (1988) and Lindquist (1983). A representative eolian depositional system is depicted schematically in Fig. 3. $Grainflow$ strata

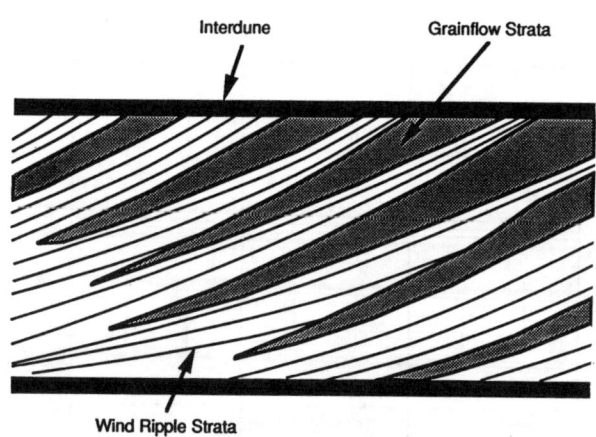

FIG. 3. Diagram of stratification features of eolian depositional system.

are steeply oriented strata formed by avalanching events which occur during dune migration. These regions display orientations, relative to the base of the dune, of angles as high as 30°. Grainflow strata are typically of high permeability relative to the other eolian strata (permeability ranges will be discussed below). Grainflow permeability is usually nearly isotropic.

$Wind$ $ripple$ deposits are especially prevalent in eolian systems. These strata are formed by the wind driven reworking of sand grains. Wind ripple deposits are comprised of thin, inversely graded laminae (coarsening upwards) with typical orientations relative to the base of the dune of 0° − 25°. Wind ripple permeabilities are almost always lower than those of the associated grainflows. Further, due to the inverse grading, they can display considerable permeability anisotropy, with principal directions parallel (the maximum permeability direction) and perpendicular (the minimum permeability direction) to the wind ripple strata orientation. These two principal values of the permeability tensor will be referred to as $k_\parallel$ and $k_\perp$, with the understanding that they apply to wind ripple regions. $Grainfall$ strata (not shown in Fig. 3) are seldom preserved in the rock record. These deposits result from wind blown, very fine scale sand grains accumulating on the upper portion of the dune. Individual dunes may be separated from one another by regions referred to as $interdune$ $areas$ or simply $interdunes$. Interdunes typically display the lowest permeabilities of the eolian strata.

Quantification of the permeabilites associated with the various eolian strata discussed above is a difficult task. To a large extent, progress in this area has come from detailed outcrop studies. In these studies, a minipermeameter is used to obtain fine scale permeability measurements (see $e.g.$, Goggin et $al.$, 1988, 1989). Comparison of outcrop and core data shows agreement between the two in the trends and relative magnitudes of the permeabilities of the various strata, though the absolute magnitudes of the permeabilities may differ considerably. Because the minipermeameter measurements of outcrops do not allow for the measurement of directional permeabilites, however, the permeabilities of the various strata must be treated as scalars. This scalar treatment is most problematic for the wind ripple regions, where permeability anisotropy is expected to be significant. Because the minipermeameter is positioned such that the bulk of the flow is along the high permeability direction (*i.e.*, along the strata), however, the reported wind ripple permeabilities are probably nearly the maximum principal value of the permeability tensor. This speculation is to some extent verified by the data of Lindquist (1988), discussed below.

Goggin *et al.* (1989) report mean permeability ratios for surface and subsurface measurements of the Page sandstone outcrop in Northern Arizona. For the surface data,

$$k_{grainflow}/k_{wind\ ripple} = 5.5,$$

$$k_{grainfall}/k_{wind\ ripple} = 1.6,$$

$$k_{grainflow}/k_{interdune} = 20.$$

For the subsurface data,

$$k_{grainflow}/k_{wind\ ripple} = 3.0,$$

$$k_{grainfall}/k_{wind\ ripple} = 1.5.$$

Under the speculation offered above, these numbers suggest that the ratio of grainflow permeability to the component of wind ripple permeability along the direction of the wind ripple strata ($k_\parallel$) is about 3-5. This is in approximate agreement with the values reported by Lindquist (1988). There, an example is given with grainflow permeabilities of 137-154 md in the horizontal direction and 200 md in the vertical direction (these measurements indicate the relative isotropy of grainflow permeability) and wind ripple permeabilities of 1.2-42 md in the horizontal direction (this corresponds to $k_\parallel$ as the angle of inclination of the wind ripple strata is $0°$) and 0.02 md in the vertical direction ($k_\perp$). These data indicate permeability ratios of roughly the following values:

$$k_{grainflow}/k_\parallel \sim 8,$$

$$k_\parallel/k_\perp \sim 1000,$$

where, again, $k_\parallel$ and $k_\perp$ refer to wind ripple regions. The $k_\parallel/k_\perp$ ratio is very large in this case; other data reported by Lindquist (1988) suggest ratios of order 1-100. In any event, the $k_{grainflow}/k_\parallel$ ratio is comparable to that reported by Goggin *et al.*

4. k* FOR EOLIAN DEPOSITIONAL SYSTEMS

We now turn to the calculation of the effective permeability of an eolian depositional system. We shall compute the effective permeability of a region of the Page sandstone outcrop, discussed above. From a photograph of the outcrop, the wind ripple/grainflow boundaries were digitized. A finite element mesh, shown in Fig. 4, was then constructed; the nodes and elements of the mesh were specified to conform to the stratigraphy evident in the outcrop. The grid contains 1760 elements; the shaded elements correspond to the grainflow regions and the unshaded elements to the wind ripple regions. No interdune exists in this portion of the outcrop.

A total of eleven different regions are present in the figure. The five grainflow regions are considered to have isotropic permeability of value unity; *i.e.*;

$$\mathbf{k}_{grainflow} = \begin{pmatrix} 1 & 0 \\ 0 & 1 \end{pmatrix}.$$

The wind ripple regions are assigned permeabilities as follows. The orientation of the wind ripple strata, as a function of position, is estimated from the outcrop photograph. These angles (designated θ) vary from nearly $0°$, in the lower left hand portion of the domain, to $24°$ in the upper right hand portion of the domain. Values are then assigned to $k_\parallel$ and $k_\perp$. These values are assumed to be constant throughout the entire domain. Now, given θ (as a function of position), $k_\parallel$ and $k_\perp$, the full permeability tensor of each element can be calculated through use of a simple tensor rotation. At this point, the problem is fully specified and the effective permeability of the eolian system under study can be computed.

Effective permeabilities for the region in Fig. 4 were computed for several different values of $k_\parallel$ and $k_\perp$. In these calculations, $k_{grainflow}$ is always taken to be 1, $k_\parallel$ was varied from 0.1 to 1 and $k_\perp$ was varied from 0.01 to

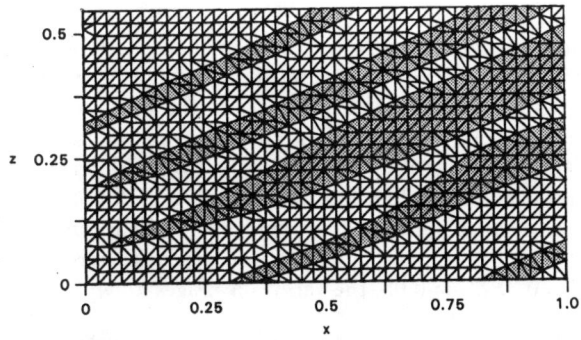

FIG. 4. Finite element grid of the region of the Page sandstone outcrop under study.

1. These ranges are consistent with the measurements reported in the references discussed above, though extreme values for the ratio $k_\parallel/k_\perp$ were not considered. The effective permeability results are summarized in Table 1 below. The table also lists the 'orientation' of $\mathbf{k}^*$, designated θ^*. This orientation is the angle through which the coordinate axes must be rotated to align with the principal directions of $\mathbf{k}^*$. Results for $\mathbf{k}^*$ using a coarser grid (520 elements) are within 10% of the results presented in Table 1 in all cases.

TABLE 1

$\mathbf{k}^*$ for eolian cross section

$k_\parallel$	$k_\perp$	k_{xx}^*	k_{xz}^*	k_{zz}^*	θ^*
1	1	1	0	1	-
1	0.1	0.8876	0.2660	0.2624	20.20
0.3	0.3	0.5106	0.04123	0.4315	23.10
0.3	0.03	0.4359	0.1471	0.1103	21.05
0.1	0.1	0.3138	0.06510	0.1856	22.72
0.1	0.01	0.2376	0.08727	0.05323	21.72

The results presented above display many interesting features. Most notable is the magnitude of the cross term of the permeability, k_{xz}^*. With the exception of the first and third cases, (the first case is completely isotropic and the third is nearly so), the magnitude of the cross term is comparable to or greater than the magnitude of the k_{zz}^* term. Further, in these cases both the diagonal terms of $\mathbf{k}^*$ are significantly less than 1, the value of the grainflow permeability. Also evident is the only slight variation in θ^*. Apparently, θ^* is most affected by the orientation of the strata and only minimally affected by the specific values of $k_\parallel$ and $k_\perp$. However, the actual components of $\mathbf{k}^*$ vary considerably for the different cases considered. This variation indicates the importance of careful quantification of the directional permeabilities of the various strata.

We now demonstrate the application of the effective permeability of a crossbedded region in a flow calculation and compare the results with those of commonly used, simpler approaches. In the absence of an effective permeability calculation, the permeability of the region would probably be approximated as either the arithmetic, geometric or harmonic mean of all the permeabilities in the region. Given that 37% of the region (by area) is grainflow strata ($k = 1$) and 63% of the region is wind ripple strata, these average permeabilities can be computed, component by component, once $k_\parallel$ and $k_\perp$ are specified. For the case $k_\parallel = 0.3$, $k_\perp = 0.03$, the various averages give:

$$\mathbf{k}_a = \begin{pmatrix} 0.5590 & 0 \\ 0 & 0.3889 \end{pmatrix},$$

$$\mathbf{k}_g = \begin{pmatrix} 0.4684 & 0 \\ 0 & 0.1098 \end{pmatrix},$$

$$\mathbf{k}_h = \begin{pmatrix} 0.4049 & 0 \\ 0 & 0.04679 \end{pmatrix},$$

where the a subscript indicates the arithmetic mean, g the geometric mean and h the harmonic mean. In computing these averages, the wind ripple permeabilites are treated as diagonal tensors, with $k_{xx} = k_\parallel$ and $k_{zz} = k_\perp$; the cross terms are taken to be zero, as is usual practice in reservoir simulation.

To determine the effects of using these simplified $\mathbf{k}'s$ rather than $\mathbf{k}^*$ computed via the method described in this paper, we compute the flow field for single phase flow through a square cross section comprised of many regions identical to the one discretized in Fig. 4. For such a region, the effective permeabilities given in Table 1 are strictly applicable, as the system is periodic. Two scenarios are considered: (1) injection in the lower left hand corner and production in the upper right and (2) injection in the lower right and production in the upper left. The flow field within the domain is computed using $\mathbf{k}^*$, $\mathbf{k}_a$, $\mathbf{k}_g$ and $\mathbf{k}_h$ for each of the two injection-production scenarios. The total fluxes through the system, given a prescribed pressure drop between the injector and producer of 1 in each case, as well as % error relative to the $\mathbf{k}^*$ result, are presented below in Table 2.

TABLE 2

Fluxes through crossbedded cross section

Scenario	$\mathbf{k}$	Flux	% error
1	$\mathbf{k}^*$	0.05498	-
1	$\mathbf{k}_a$	0.1293	+135
1	$\mathbf{k}_g$	0.06171	+12
1	$\mathbf{k}_h$	0.03381	-39
2	$\mathbf{k}^*$	0.02516	-
2	$\mathbf{k}_a$	0.1293	+414
2	$\mathbf{k}_g$	0.06171	+145
2	$\mathbf{k}_h$	0.03381	+34

As is evident from the table, the standard averaging techniques perform quite poorly for this problem. Arithmetically averaged permeability overpredicts the flux substantially for both scenarios; geometric averages overpredict as well, though only slightly for scenario 1. Harmonically averaged permeability appears to perform the best of the three averaged permeabilities, in the sense that the average error is the least. However, in one case harmonically averaged permeability overpredicts the flux while in the other case it underpredicts.

Because the wind ripple permeabilities are treated as diagonal tensors in computing $\mathbf{k}_a$, $\mathbf{k}_g$ and $\mathbf{k}_h$, none of these

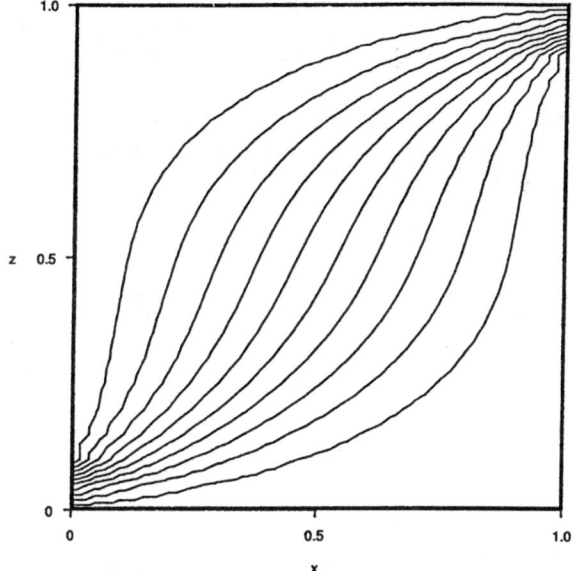

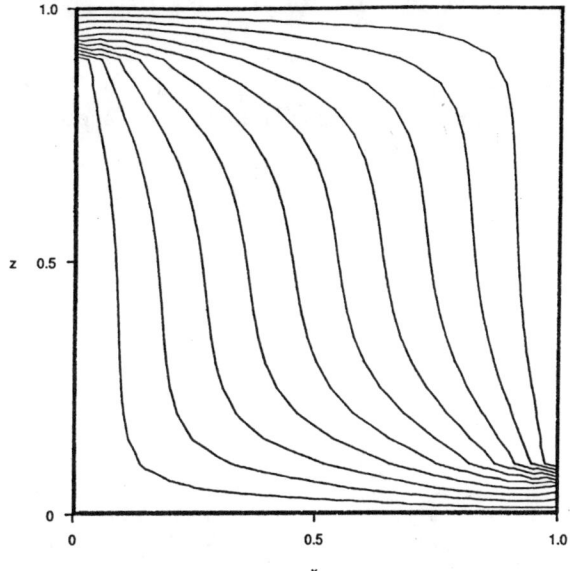

FIG. 5. Streamlines for flow through a cross-bedded region with the line connecting the injector and producer aligned with crossbedding.

FIG. 6. Streamlines for flow through a cross-bedded region with the line connecting the injector and producer nearly normal to cross-bedding.

averages is able to distinguish between the two scenarios, yielding identical fluxes for both cases and mirror image pressure and velocity fields for the two cases. Physically, given the same boundary (or bottom hole) pressures in the two cases, one would expect the solution to vary depending on the orientation of the injector and producer relative to the crossbedding direction, with the larger flux resulting when the line connecting the injector and producer is more nearly aligned with the crossbedding (*i.e.*, along the high permeability direction of the wind ripple strata). The effective permeabilities computed via the method described in this paper do in fact display the expected result; the flux is more than a factor of two greater when the flow is predominantly along the direction of the cross strata than when it is predominantly normal to the strata. Streamlines for the two scenarios, with permeability represented via $\mathbf{k}^*$, are shown in Figs. 5 and 6. The sweep is clearly less (though the flux is higher) in the case when the flow is along the bedding (Fig. 5) than when it is across the bedding (Fig. 6).

This completes our discussion of the effective permeability of a representative region of an eolian depositional system. The importance of a full tensor representation for effective permeability should now be clear. Other depositional systems, particularly those for which crossbedding is prevalent, would similarly be expected to require a full tensor effective permeability.

5. CONCLUSIONS

In this paper, we presented a general method for the calculation of the effective permeability of heterogeneous reservoir regions. Through the use of periodic boundary conditions, the method avoids artificial boundary condition dependence and always yields symmetric effective permeability tensors. The triangle based nonconforming finite element discretization of the fine scale pressure equation allows for the modeling of complex cross stratified systems, as illustrated by the eolian system considered in §4. The use of the effective permeability tensor in flow calculations clearly demonstrated the importance of the accurate determination of the cross terms of $\mathbf{k}^*$ for this case.

A scale up procedure such as that presented here may be of practical use in several ways. One application is simply the scale up, given fine scale measurements, of a portion of heterogeneous rock. Another use for such a procedure might be to scale up typical permeability data for particular types of beddings or cross strata, as presented in §4. This data, in conjunction with geological interpolation or geostatistical data, could then be used as input to a reservoir simulator. Such an approach might well qualitatively capture the effects of complex crossbedding and other heterogeneities.

ACKNOWLEDGMENTS

We are grateful to D.J. Goggin and T.B. Eschner for enlightening discussions regarding eolian depositional systems.

REFERENCES

Bourgeat, A., 1984, Homogenized Behavior of Two-Phase Flows in Naturally Fractured Reservoirs with Uniform Fractures Distribution, *Comp. Meth. Appl. Mech. Eng.*, **47**, p. 205-216.

Desbarats, A.J., 1987, Numerical Estimation of Effective Permeability of Sand-Shale Formations, *Water Resources Res.*, **23**, p. 273-286.

Goggin, D.J., Chandler, M.A., Kocurek, G. and Lake, L.W., 1988, Patterns of Permeability in Eolian Deposits: Page Sandstone (Jurassic), Northeastern Arizona, *SPEFE*, p. 297-306.

Goggin, D.J., Chandler, M.A., Kocurek, G. and Lake, L.W., 1989, Permeability Transects in Eolian Sands and Their Use in Generating Random Permeability Fields, SPE paper 19586 presented at the SPE Annual Technical Conference and Exhibition, San Antonio.

Kasap, E. and Lake, L.W., 1989, An Analytical Method to Calculate the Effective Permeability Tensor of a Grid Block and its Application in an Outcrop Study, SPE paper 18434 presented at the SPE Symposium on Reservoir Simulation, Houston.

Lindquist, S.J., 1983, Nugget Formation Reservoir Characteristics Affecting Production in the Overthrust Belt of Southwestern Wyoming, *JPT*, p. 1355-1365.

Lindquist, S.J., 1988, Practical Characterization of Eolian Reservoirs for Development: Nugget Sandstone, Utah-Wyoming Thrust Belt, *Sedimentary Geology*, **56**, p. 315-339.

Marini, L.D., 1985, An Inexpensive Method for the Evaluation of the Solution of the Lowest Order Raviart-Thomas Mixed Method, *SIAM J. Numer. Anal.*, **22**, p. 493-496.

Mei, C.C. and Auriault, J.-L., 1989, Mechanics of Heterogeneous Porous Media with Several Spatial Scales, *Proc. Roy. Soc. Lond. A*, **426**, p. 391-423.

Saez, A.E., Otero, C.J. and Rusinek, I., 1989, The Effective Homogeneous Behavior of Heterogeneous Porous Media, *Transport in Porous Media*, **4**, p. 213-238.

Thomasset, F., 1981, *Implementation of Finite Element Methods for Navier-Stokes Equations*, Springer-Verlag, New York.

White, C.D. and Horne, R.N., 1987, Computing Absolute Transmissibility in the Presence of Field-Scale Heterogeneity, SPE paper 16011 presented at the SPE Symposium on Reservoir Simulation, San Antonio.

2nd European Conference on the Mathematics of Oil Recovery
© D. Guérillot, O. Guillon (Editors) and Éditions Technip, Paris 1990, pp. 65-73
27 rue Ginoux, 75015 Paris

Application of Analytical Methods in Predicting Waterflood Performance of Reservoirs with Stochastic Sand Bodies

O. B. Abu-elbashar, T. S. Daltaban, C. G. Wall and J. S. Archer [1]

1 ABSTRACT

Due to the complex architecture of fluvial reservoirs composed of uncorrelatable (stochastic) sand bodies, straightforward application of simple one dimensional analytical methods to predict their waterflood performance is not possible. This is due to the inadequacy of the existing techniques to represent the complex connectivities between flow elements in one dimensional models.

In this paper, a new stochastic sand modelling method is presented. It simplifies sand description in grid blocks for simulation, and for averaging of the reservoir parameters such as permeability and porosity. Although the method is conditioned by the available seismic and well data which are extremely sparse, it employs only modest use of statistical methods. Also, contrary to many current techniques which are restricted to two dimensional realisation, the current method extends the realisation into three dimensions. This is carried out by using the special stochastic sand conditioning technique in which the generated flow elements are matched to a prescribed girded domain.

In the scaling up of the multi-dimensional reservoir parameters to represent a consistent one dimensional equivalent, a new hybrid averaging technique linked with the sand generation scheme is applied. The validity of this technique has been verified by comparing the one dimensional and three dimensional simulation results of a water/oil displacement process. A radial model of the generated three dimensional stochastic domain has also been created to simulate pressure response of a well to drawdown and build-up tests . In this case the averaged permeability predicted from the tests has shown a reasonable agreement with the pseudoised permeability predicted using the method of this paper.

This one dimensional model has been used for Buckley-Leverett model in a water/oil displacement process. The agreement achieved between detailed three dimensional simulation and the analytical results suggests that these techniques might be used successfully for waterflood performance prediction of complex reservoir systems with stochastic sands.

2 INTRODUCTION

Due to the enormous amount of data required to perform an explicit fluvial reservoir

---

(1) Department of Mineral Resources Engineering, Imperial College, London SW 7 2 BP, UK.

description study, the stochastic modelling techniques were adopted by many investigators (Haldorsen et al, 1987, Martin et al, 1988, Johnson and Krol, 1984, Knuston, 1976). These aim to generate possible realisations of fluvial reservoir structures based on well observations and analogue statistical data. Discretisation of these conceptual models in grid blocks for simulation becomes a real problem regarding the random distribution of the sand bodies and their variable sizes. The available techniques are inadequate to tackle the complex connectivities between the sand bodies. This is due to one or more of the following reasons:

(i) A large number of grid blocks is needed to accurately discretize the discontinuous sands.

(ii) Very large computer memory and time are needed for performance prediction calculations.

(iii) Inadequacy in describing the complex connectivities of the sand bodies.

Therefore, a new modelling scheme which can reduce the need for enormous computer memory and time to accurately describe these complex domains is applied in this project. This technique – which is thoroughly discussed by (Abu-elbashar, 1989, 1990) – can generate 3-D realisations of the reservoir. A sequential psedoisation process is then used to reduce the original three dimensional domain into two dimension. Particular care have been given to the spatial connectivities of sand bodies which cannot be reflected otherwise by the conventional two dimensional realisations. The new hybrid averaging technique is then applied to these 2-D pseudo models to generate representative 1-D homogeneous models which can be used by analytical prediction methods. Simulation studies are performed to compare the waterflood performance of the 3-D model, the 2-D pseudo model, and the 1-D model model generated by this averaging technique. The good similarity obtained between the performances of the three models justified the use of the 1-D model as a representative of the discontinuous reservoir domain.

Buckley-Leverett prediction method is then applied on this 1-D model to give a performance almost identical to the 3-D model.

These encouraging result suggest the use of the techniques presented in this paper to obtain simple models which can be used by simple prediction methods to obtain reasonably accurate prediction results.

A brief review of this new modelling technique is given below as it constructs the model which we reduce to 1-D by our hybrid averaging technique.

3 REVIEW OF THE RESERVOIR MODELLING TECHNIQUE

Based on well information and statistical outcrop data, a data base is set for each zone or layer of the reservoir under consideration. A cross-section normal to the paleocurrent direction of flow is divided into grid blocks having a thickness equal to the observed most-likely thickness of the channels (in wells), and a length calculated from the most-likely ratio of length to thickness and the most-likely thickness. These dimensions were found to be the optimum for minimising the number of grid blocks without reducing the accuracy of description. The sand bodies are generated as shown below.

3.1 Generation and Conditioning of the Sand Bodies

The dimensions of each sand body are sampled randomly from the corresponding data base, and the sand body is randomly positioned on the cross-section (of the zone or layer under consideration). The base of the sand body is then adjusted to the nearest grid layer boundary. After this spatial conditioning process, the horizontal co-ordinates of the sand body are adjusted to be multiples of the grid block length. This dimensional conditioning process simplifies the stochastic sand discretisation in grid blocks while keeping the sand dimensions within the boundaries specified by the data base. The conditioned sand bodies are then extended to the third dimension using random angles between 45 and 135 degrees.

The process is repeated until the fractional area occupied by the sand bodies in the cross-section equals the observed net-to-gross (NTG) in wells. At the end of this stage the generation of the 3-D model is completed. Fig. 1 demonstrates the conditioning process while Fig. 2 shows a cross-section of a fluvial reservoir generated by this technique.

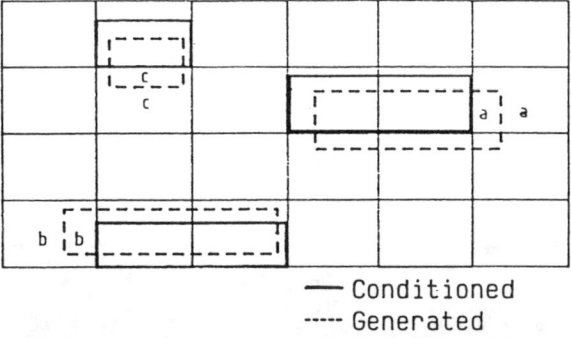

— Conditioned
---- Generated

FIG. 1. Generation and Conditioning of the Sand Bodies.

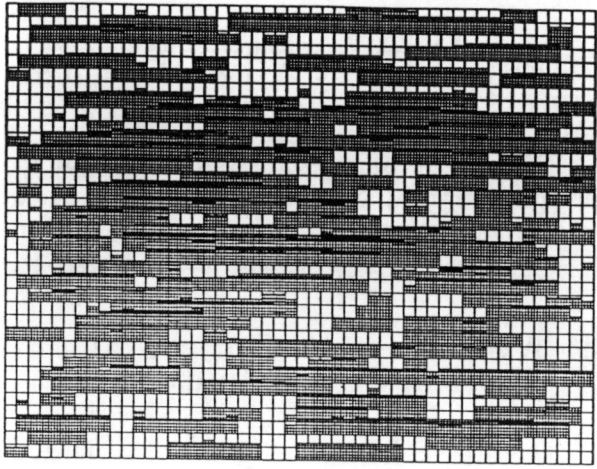

FIG. 2. Cross-section of a Fluvial Reservoir.

3.2 Creation of the 2-D Pseudo Models

Grid blocks apparently isolated in 2-D cross-sectional models may be connected through parallel grid sections. This can happen in both horizontal and vertical directions (see Fig. 3). To account for this phenomenon in 2-D models, an automatic parameter pseudoisation technique is

applied. This implies assignment of pseudo permeabilities to the inactive grid blocks separating the apparently isolated (active) grids. These are calculated by Darcy's law to give a pressure drop in the direct path equal to double the pressure drop in the indirect (rear) path (as the liquids of both grid sections share the available paths).

Grid blocks with pseudo horizontal permeability are assigned zero vertical permeability unless if they were assigned pseudo vertical permeabilities. Similarly, grid blocks with pseudo vertical permeabilities are assigned zero horizontal permeabilities unless if they were assigned pseudo horizontal permeabilities.

To declare them active – without changing the initial oil in place – grid blocks with pseudo permeabilities are assigned very small non-zero porosities as most of the simulation packages consider grid blocks with zero porosities as inactive.

A pseudo NTG of 1 is assigned to each grid block with pseudo properties. This value indicates the cross-sectional area used for each grid block in pseudo permeability calculations.

As fluids in a parallel grid section may share the paths of the front grid section resulting in reduction of their conductivity to fluids of the front section, an automatic back pseudoisation process is applied which results in reduction of the permeabilities of the grids constructing these paths by 50%.

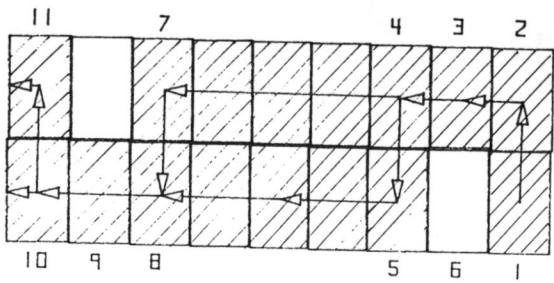

FIG. 3. Communication through parallel Grid sections.

3.3 Verification of the Pseudoisation Technique

A comparative simulation study was used to verify the pseudoisation technique. The performance of three models was performed: (i) a 3-D model; (ii) a conventional 2-D cross-sectional model; (iii) and a 2-D cross-sectional model with pseudo parameters.

Figures 4(a) and 4(b) show the waterflood performance of the the three models. In Fig. 4(a), the dimensionless oil production of the 3-D and the 2-D pseudo model remain fairly close with increasing time whereas the production of the conventional 2-D model decreases immediately after water breakthrough (BT) to give a conservative recovery factor.

The early BT of the 2-D conventional model is revealed by Fig. 4(b) which shows close BT times and water-cut behaviour for the 3-D model and the 2-D model with pseudos.

The good agreement between the performance of the 2-D model with pseudos and the 3-D model validates our pseudoisation technique, whereas the performance of the conventional 2-D model diverges from the 3-D behaviour immediately after breakthrough (BT).

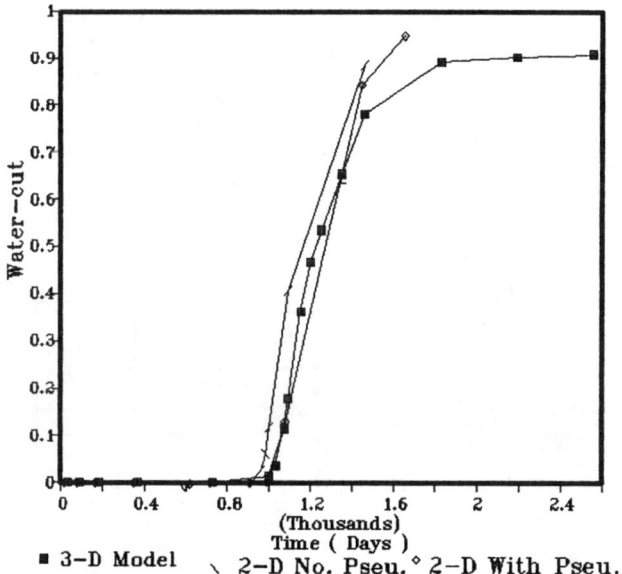

FIG. 4(b). Water-cut versus Time for the three Models.

4 AVERAGING THE RESERVOIR PARAMETERS

Different techniques are used to average the various reservoir parameters due to the particular characteristics of each parameter as demonstrated below.

4.1 Averaging the Horizontal and Vertical Permeabilities

Due to the directional nature of permeability, conventional averaging techniques such as arithmetic and geometric means may not give accurate results. Moreover, these techniques cannot handle systems with zero permeabilities. For systems with non-zero permeabilities, Cardwell and Parsons, 1945, compared the results of such techniques with the results of an electric analogy experiment. For various arrangements of permeability, the equivalent permeability of a four block model always fell between the harmonic and arithmetic means. The harmonic mean always gave the lower value.

King, 1987, introduced the renormalisation technique which is analogous to the electric net-

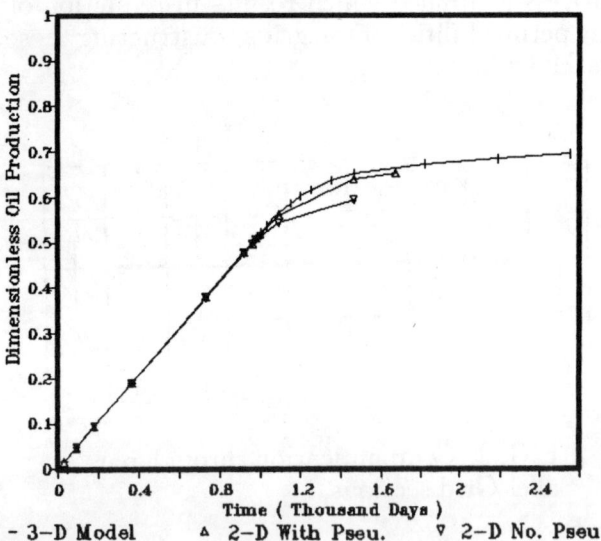

FIG. 4(a). Dimensionless Oil Production versus Time for the three models.

works. This new technique can handle systems with zero permeabilities. A modified form of the 2-D renormalisation method is introduced in this paper to account for permeability anisotropy.

Comparative simulation studies indicated that, the appropriate representative permeability of a domain with stochastic sand bodies, is the arithmetic mean of the value calculated by this modified form and the value obtained from a method using the arithmetic and harmonic means sequentially (the combination method).

Averaging of the permeability is performed in three stages:

Stage 1:

The 3-D connectivities of the reservoir are represented in 2-D models by introducing pseudo permeabilities as shown above.

Stage 2:

The horizontal permeabilities are weighted according to the respective volume of sand in each grid block. This is automatically done (at the end of the sand generation stage) by multiplying the permeability values by the NTG.

The active grid blocks with NTG less than 1 are assigned zero vertical permeabilities due to the thin shale layer at the top of the grid (see Fig. 2).

Stage 3:

The permeability values of each four grids starting from the end of the top two grid layers are read sequentially, the average value is calculated, and then assigned to a coarser grid. The process is repeated on the coarser grid permeabilities until the permeabilities of the domain are reduced to one value.

The averaging process is performed by four methods:

(a) The arithmetic mean, K_{ar}.
(b) The harmonic mean, K_{hr}, considering the shales to have small non-zero permeability (0.0001 millidarcy).
(c) The combination method, K_{co}: Divide the cross-section into slides (normal to the direction of permeability), find the arithmetic mean of the values in each slide, and then find the harmonic mean of the resulting values.

(d) The modified renormalisation method, K_{mr}: In this new form, the vertical resistors of the equivalent network are represented by the vertical permeability while the horizontal resistors are represented by the horizontal permeability as shown in Fig. 5.

The resulting resistor network is reduced in the way presented by King, 1987 to obtain the following expression for the effective permeability, K_{eff}.

$$K_{eff} = [2(NC \bullet K_1 K_2 + NB \bullet K_3 K_4) F1 \bullet F2] \bullet \\ [(NC \bullet NB) + (K_{1v} \bullet K_{3v}) F2 (NC \bullet K_1 K_2 + \\ NB \bullet K_3 K_4) + (K_{2v} K_{4v}) F1 \\ (NC \bullet K_1 K_2 + NB \bullet K_3 K_4)] \qquad (1)$$

Where,
$$NB = K_1 K_2 K_3 (K_{3v} + K_{1v}) F2 + F1 F2 K_2 \\ + F1 F2 K_1 + F1 K_1 K_2 K_4 \\ (K_{4v} + K_{2v}) \qquad (2)$$
$$NC = K_1 K_3 K_4 (K_{3v} + K_{1v}) F2 + F1 F2 K_4 \\ + F1 F2 K_3 + K_2 K_3 K_4 F1 \\ (K_{4v} + K_{2v}) \qquad (3)$$
$$F1 = K_1 K_3 K_{3v} + K_1 K_3 K_{1v} \\ + K_1 K_{1v} K_{3v} + K_3 K_{3v} K_{1v} \qquad (4)$$
$$F2 = K_2 K_4 K_{4v} + K_2 K_4 K_{2v} \\ K_2 K_{2v} K_{4v} + K_4 K_{4v} K_{2v} \qquad (5)$$

Subscript v denotes vertical permeability.

Table 1 shows the averaged horizontal and vertical permeabilities obtain for our model.

TABLE 1
Permeabilities averaged by
different techniques

Method	Hori. K	Ver. K
K_{hr}	375.3	21.69
K_{mr}	372.4	25
K_{co}	403.2	25
K_{ar}	485.5	108.8

The representative horizontal and vertical permeabilities obtained by this procedure are 387.8 md. and 25 md. respectively.

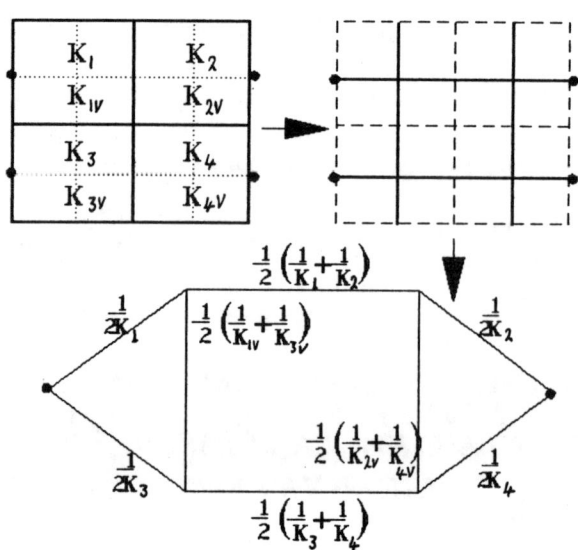

FIG. 5. Construction of the Equivalent Resistor Network.

4.2 Averaging the Porosity

Unlike permeability, averaging the porosity is simple and straightforward. The porosity of each grid block is multiplied by it's NTG (weighting according to the respective volume represented by the grid block). The arithmetic mean of the weighted values is considered as the effective porosity of the reservoir. This porosity value (0.133 fraction) conserves the initial oil in place.

4.3 Generation of the Averaged Relative Permeabilities

For heterogeneous reservoirs a great number of relative permeability tables may result, which is impractical for simulation purposes. Therefore, reservoir zonation (according to permeability) and averaging the permeability within each zone, became a common practice. For fluvial reservoirs, due to the presence of stochastic shales, the horizontal/vertical permeability ratio is substantial and the general direction of flow is effectively horizontal. Therefore, the average horizontal permeability which is calculated in (paragr. 4.1) is used in the correlation

chart (Daltaban, 1988) shown in Fig. 6 to generate the averaged relative permeabilities.

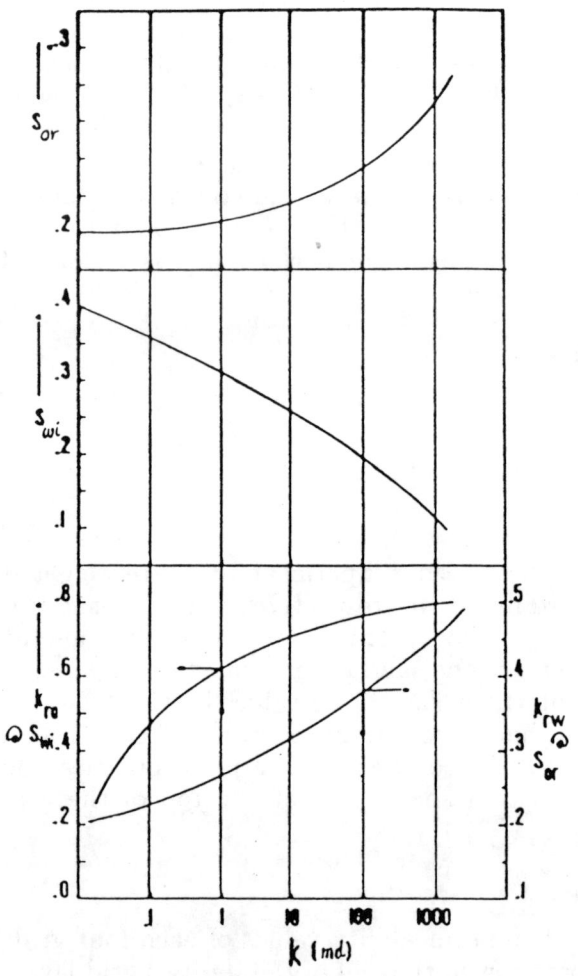

FIG. 6. General Relative Permeability Correlation Chart.

4.4 Validation of the averaged permeability

To validate the permeability averaging procedure presented in this paper, the 2-D model with pseudos is transformed into a radial cylindrical model (see Fig. 7), one well is completed at one end (the centre) and no flow boundaries are imposed on the top, bottom, and the outer boundary of the domain. The objective being to model and simulate typical drawdown and build-up tests (Abu-elbashar, 1990) and compare the permeability values calculated by the standard test analysis methods with the one obtained by the method of this paper.

The well is flown for 25 hours with an average flow rate of 6000 STB/Day followed by a 25 hour build-up period. The resulting drawdown and build-up permeabilities are 430.8 md. and 327.8 md. respectively. The average (arithmetic) of the permeabilities obtained from this study is 379.3 which is very close to the value calculated in (paragr. 4.1) which supports the validity of the averaging techniques presented in this paper. A reservoir limit test is used to justify using this test simulation approach. The reservoir volume obtained (1,669,280 barrel) is almost identical to the actual pore volume of the model, (1,725,931 barrel) .

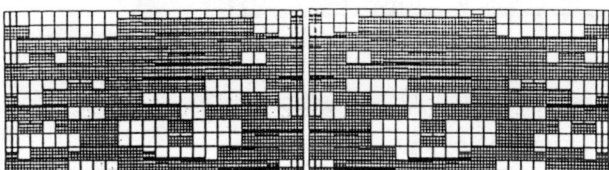

FIG. 7. The Radial Cylindrical Model.

5 COMPARATIVE ANALYTICAL AND SIMULATION STUDIES

A comparative simulation study has been carried out using the 3-D stochastic model and it's 2-D representation obtained by the averaging techniques of this paper to examine the water/oil displacement performance of each model. The pseudo 2-D cross-sectional model has uniform but anisotropic permeabilities.

For the 3-D model reservoir zonation is applied to generate representative permeabilities for each zone which can be used to generate a relative permeability table from (Fig. 6), whereas the horizontal permeability of the 2-D model is used for relative permeability calculation. Some of the relevant data base for the two models are shown in table 2.

TABLE 2
Data for the 3-D model and it's 2-D representation

Property	3-D	2-D
Res. length ft.	9600	9600
Res. width ft.	800	400
Res. thick. ft.	240	240
No. blocks in X	24	24
No. blocks in Y	2	1
no. blocks in Z	12	12
Oil visc. cp.	0.3	0.3
Initial Wat. Sat.	0.12 - 0.15	15
Wat. visc. cp.	0.3	0.3
Inj. rate RB/D	19800	9900
Prod. rate RB/D	19800	9900

The 2-D model has been assigned an injector at one end and a producer at the other while the 3-D model is assigned two injectors and two producers. Figure 8 shows plots of the pore volumes of oil produced versus pore volumes of water injected for the 3-D model, the 2-D pseudo model, the 2-D homogeneous model. It is clear from this plot that the 3-D and the homogeneous models show similar performances which verifies the averaging techniques of this paper. The homogeneous model gives recoveries slightly higher than the 2-D pseudo model and almost identical to the 3-D model due to it's higher accessibility and absence of the dead ends.

Due to the high horizontal permeability of the 2-D anisotropic model as compared with the vertical permeability (ratio of 15:1) the flow can be considered unidirectional (horizontal) and the model can simply be assumed 1-D by ignoring the vertical permeability.

At this stage, a simple analytical technique such as Buckley-Leverett can now be applied to predict the waterflood performance of this

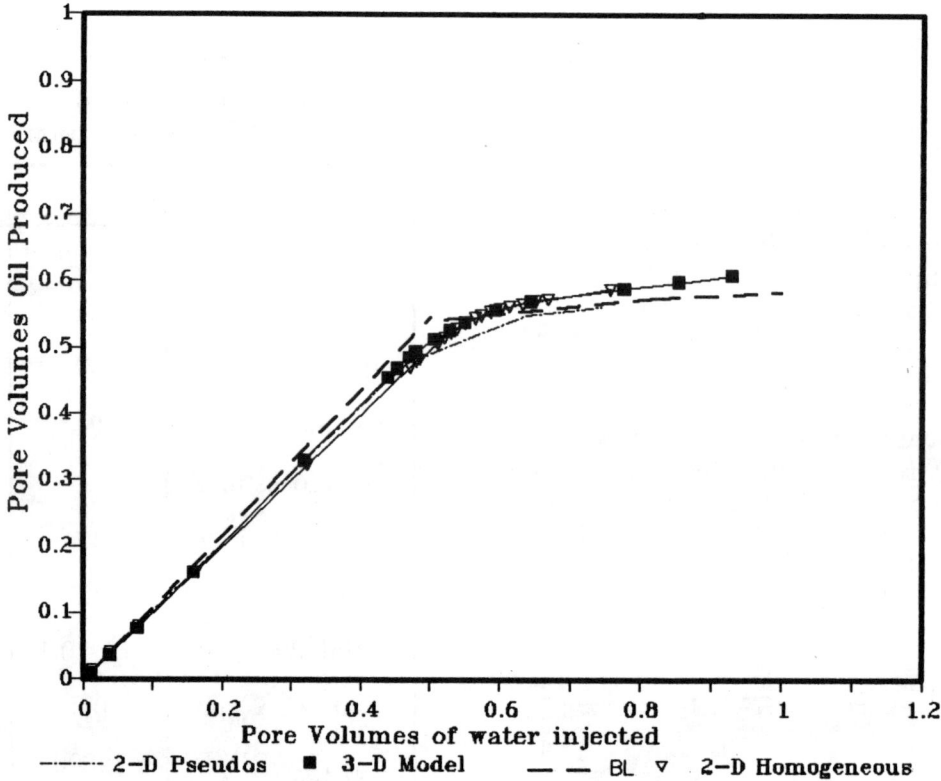

FIG. 8. 2-D and 3-D Simulation Performance versus Analytical Results.

1-D model. This is performed by utilising the horizontal permeability to generating the relative permeability curves using Fig. 6, and then applying the standard Buckley-Leverett procedure.

Figure 8 shows the good agreement between the 1-D analytical solution of Buckley-Leverett and the solution of a full scale 3-D model simulation. This remarkable agreement suggests that this technique might be used successfully for waterflood performance prediction of complex reservoirs with stochastic sands.

6 CONCLUSIONS

1 - A new stochastic sand modelling technique is presented. It simplifies sand Discretisation in grid blocks and reduces the computational time in simulation studies.

2 - A new reservoir averaging procedure is introduced. This procedure can generate pseudo 2-D models reflecting the 3-D reservoir characteristics.

3 - This averaging procedure can also generate representative 1-D models that can be used by analytical methods to give the 3-D reservoir performance.

4 - Buckley-Leverett technique appears to be well suited for reservoirs with stochastic sand bodies after averaging by the above techniques.

7 ACKNOWLEDGEMENTS

I would like to thank Sudan government for sponsoring this project. My thanks to DR. King, P.R. of BP for assistance and advice.

8 REFERENCES

Abu-elbashar, O.B., 1990, Characterisation and Modelling of Layered Discontinuous Reservoirs for Waterflood Predictions, Ph.D. Dissertation, Imperial College, London University, London, p. 140-240.

Abu-elbashar, O.B., 1989, Modelling and Simulation of Stochastic Sands in Stratified Reservoirs: 2-D Description by Pseudo Parameters, Presented at the Student Paper Competition, Offshore Europe, Aberdeen.

Cardwell, W.T. and Parsons, R.L., 1945, Average Permeabilities of Heterogeneous Oil Sands, T. P. 1853, Petroleum Technology, March, 1945.

Daltaban, T.S., Private communication.

Haldorsen, H.H. et al., 1987, Review of the Stochastic Nature of Reservoirs, Presented at the joint SPE/IMA seminar, Robinson College, Cambridge, U.K.

Johnson, H.D. and Krol, D.E., 1984, Geological Modelling of a Heterogeneous Sandstone Reservoir: Lower Jurassic Stratfjord Formation, Brent Field, SPE Paper No. 13050.

King, P.R., 1987, Effective Values in Averaging, Presented at the joint SPE/IMA seminar, Robinson College, Cambridge, UK.

Knuston, C.F., 1976, Modelling of Noncontinuous Fort Union and Mesaverde Sandstone Reservoirs, Piceance Basin, North-western Colorado, SPEJ, Aug. 1976, p. 175-188.

Martin, J.H. et al, 1988, Reservoir modelling of Low- Sinuosity Channel Sands: A Network Approach, SPE Paper No. 18364.

2nd European Conference on the Mathematics of Oil Recovery
© D. Guérillot, O. Guillon (Editors) and Éditions Technip, Paris 1990, pp. 75-81
27 rue Ginoux, 75015 Paris

Numerical Simulation and Homogenization of Diphasic Flow in Heterogeneous Reservoir

B. Amaziane[1], A. Bourgeat[2] and J. V. Koebbe[3]

1. INTRODUCTION

By mean of the so called homogenization theory, see for instance [16], we derive mathematically rigorous "effective" reservoir equations from exact local equations of incompressible two-phase flow (miscible or immiscible) in a heterogenous reservoir. The main result is that "effective" equations are exactly of the same type as the original ones. In general cases the effective permeability tensor is given only as a mathematical limit. In some special cases where there is some additional knowledge on the heterogeneities repartition as for instance a spatial periodic repartition, we may really compute the effective parameters and then numerically compare both behaviour in a heterogeneous or in a homogenized reservoir. In [1] and [12], we have presented some simulation of stratified medium; in [2] and [12] we have presented several simulations on spatially periodic heterogeneities. Herein we are presenting only one of such a simulation to illustrate our results.

2. THE TWO-PHASE FLOW MODEL

Standing governing equations modeling two immiscible fluid phases with no mass transfer between the fluids in a porous reservoir $\Omega \subset \mathbb{R}^n$, $1 \leq n \leq 3$, over a time period $J =]0,T[$, are made up of equations describing conservation of mass in each phase coupled via a capillary pressure law $P_{cap}(x,S)$, and relative permeabilities curves $k_{ri}(S)$ in each phase $i = w,0$. S_i and P_i are the saturation and pressure in each phase, where the subscripts w and 0 refer to water and oil respectively. The equations of conservation of mass and continuity for each phase $i = 0,w$ are given by:

$$\Phi(x)\frac{\partial S_i}{\partial t} + \nabla.\varphi_i = 0 \qquad x \in \Omega, t \in J \qquad (2.1)$$

$$S_w + S_0 = 1 \qquad (2.2)$$

where φ_i is the velocity which is determined by the traditional extension of Darcy's law

$$\varphi_i = -\frac{k_{ri}(S_i)}{\mu_i} K(x) \nabla(P_i + P_g), \quad i = w,0 \qquad (2.3)$$

where μ_i is the viscosity of each phase, P_g the gravity potential, $K(x)$ and $\Phi(x)$ the absolute permeability and porosity of the material. The pressures in the two phases are related by the capillary pressure

$$P_{cap}(x,S) = P_w - P_0 = P_{cm}(x) \, p_c(S) \qquad (2.4)$$

(1) Laboratoire de Mathématiques Appliquées (URA CNRS 1204), Université de Pau et des Pays de l'Adour, avenue de l'Université, 64000 Pau, France.
(2) Équipe d'Analyse Numérique (URA CNRS 740), Université de Saint-Étienne, 23, rue Docteur Paul Michelon, 42023 Saint-Étienne Cedex 2, France.
(3) Department of Mathematics and Statistics, Utah State University, Logan, Utah 84322-3900, USA.

where $P_{cm}(x) > 0$ is the maximum of the absolute value of the capillary pressure at the point x and $p_c(S)$ is a dimensionless function such that $p_c'(S) > 0$.

By mean of a new variable called "global pressure" as in [13], [15], this system (2.1)-(2.4), may by rearranged in a form which very closely resembles to the system describing the displacement of two fully miscible fluids. This form is more suitable for mathematical purpose and will allow to get homogenization results for both miscible and immiscible fluids as in [7], [8], [9], [10], [11].

3. THEORETICAL HOMOGENIZATION OF TWO-PHASE FLOW

These type of problem has been addressed by various authors in Petroleum Engineering. Up to recently most of the works in the petroleum litterature were rather based on heuristic considerations and taking account of only one-phase flow. The main conclusion of these works is that the "effective" permeability lies between a harmonic and arithmetic average.

For our purpose, we assume that ε, the ratio between characteristic lengths of the microscopic and macroscopic scales is of order $\varepsilon \ll 1$ and then that the local porosity $\Phi^\varepsilon(x)$ and the local permeability tensor $K^\varepsilon(x)$ are rapidly oscillating functions as in Fig.1. Homogenized (effective) equations are the equations giving the effective behaviour of the fluid, i.e., from the limit of the equations when ε tends to zero. The main mathematical problem in this process is to prove that there exists such a limit (i.e. an effective equation), to define in what sense this convergence happens and finally to obtain explicitly (compute) this limit. In some similar systems the homogenized equations could be of a different type [6]. Major difficulties in this kind of problems come from the exact meaning of the limit we are considering. Actually we should use the notion of limit which gives the mean value as the "limit" of any oscillating periodic function when the period tends to zero. This kind of limit is called weak limit and does not have the property of giving the product of two weak limits equal to (weak limit) to

the weak limit of the product; for instance the mean value (weak limit) of $(\sin 2\pi x/\varepsilon)^2$ is not the square of the mean value (weak limit) of $\sin 2\pi x/\varepsilon$. To derive the homogenized equations we will not need any other hypothesis than the previous ones on the size of the heterogeneities and than the region under consideration is not intersecting the boundaries.

3.1 Homogenization without interface phenomena

In case of an immiscible displacement we assume that the capillary pressure curves $P_{cap}(x,S) = P_{cm}(x) \times p_c(S)$ do not depend on the heterogeneities. In case of a miscible displacement we assume that the dispersion tensor D depends only on the space variable x and on the concentration C, i.e. $D(x,C) = D_1(x)\, D_2(C)$. The relative permeabilities and $p_c(S)$, are assumed having the same shape all over the reservoir Ω, i.e. there is no jump on the saturation at the interfaces between heterogeneities (single type rock field).

The main theoretical results are:

i) Equations describing the homogenized flows are of the same type as the equations of the flow through the heterogeneous media. There is an effective porosity $\Phi^\sim$, effective permeabilities $K^\#$ and an effective dispersion tensor $D_1^\#$ which does not depend anymore on the microscopic scale $y = x/\varepsilon$.

ii) $\Phi^\sim$ is the arithmetic average.

iii) $K^\#$ and $D_1^\#$ are obtained by the same method as in the linear case, i.e. there exist some test functions w_ε^k solution in Ω of:

$$\begin{cases} -\nabla.(K^\varepsilon \nabla w_\varepsilon^k) = g_k & k = 1,2,\ldots,n \\ w_\varepsilon^k \,|\, \partial\Omega = 0 \end{cases} \tag{3.1}$$

where g_k has been choosen in such a way that $\nabla w_\varepsilon^k \rightarrow e_k$ (weakly in $[L^2(\Omega)]^n$).

Then $K^{\#}$ the effective permeability tensor, is given by the weak limit in $[\,L^2(\Omega)\,]^{n^2}$ of $K^{\varepsilon}\,\nabla w_{\varepsilon}^k$,

$$K^{\varepsilon}\,\nabla w_{\varepsilon}^k \to K^{\#} \qquad (3.2)$$

If moreover the repartition of the heterogeneities is uniformly periodic (Fig.2) or non uniformly periodic (Fig.3), i.e. when $K^{\varepsilon}(x) = K(x/\varepsilon)$ or $K^{\varepsilon}(x) = K(x,x/\varepsilon)$ is periodic in $y = x/\varepsilon$ then the effective permeability tensor $K^{\#}$ could be explicitly computed by:

$$K^{\#} = \text{mean value of } [K(y)\,\nabla w^k(y)] \text{ in } Y \qquad (3.3)$$

where Y is the magnified typical cell, by the change of variable $y := x/\varepsilon$ and where $w_{\varepsilon}^k(x) = w^k(y)$ is the Y-periodic solution of

$$\begin{cases} -\nabla_y.(K(y)\,\nabla_y\,w^k(y)) = 0 \\ \int_Y \nabla_y.w^k(y)\,dy = e_k \end{cases} \qquad (3.4)$$

Remark: The last results are consistent with the widely used rules of averaging when the flow is orthogonal or parallel to the stratas in a stratified porous medium. But our results allow to really compute the homogenized flow in all cases of heterogeneities such as isotropic medias, various shapes and several types of heterogeneities as long as the spatial repartition of the heterogeneities could be assumed close enough to a uniform or non uniform periodic repartition. In general cases, even with each type of porous media being isotropic, we have homogenized effective tensors which are significantly anisotropic. A numerical code [3] has been written to solve problem (3.4) via a finite element method and compute the homogenized tensor obtained in (3.3).

3.2 Homogenization with interface phenomena

In this paragraph we are investigating a model with different rock types i.e. we consider the case of a reservoir which contains several porous media with different characteristics. According to the notion of rock type, not only the absolute permeability tensor and the porosity, but also the shape of relative permeabilities and capillary pressure curves may differ in each zone. This model gives a coupled system of elliptic equation and a nonlinear degenerated diffusion-convection equation. Moreover on the interfaces the realistic transmission conditions are nonlinear and there is a jump on the saturation [14]. The essential difficulties of this problem are the discontinuities of the saturation and of the global pressure and the nonlinearity arising on the interfaces. After a change of unknown, a mixed formulation is associated to the elliptic equation and a standard weak formulation for the parabolic one. In [4] we proved that the implicit system given by time discretization has a physically meaningful and well posed solution. This formulation may be of practical use for the numerical approximation of two-phase flow in the oil industry as secondary recuperation of oil in a reservoir with various rock types.

We are now investigating the mathematical homogenization associated to this problem in order to get the effective homogeneous behavior. In view of this problem difficulties, as a first step, we have studied a linearized model [1], where saturation and pressure equations are decoupled and capillary pressure curves are linearized. Then the homogenized system is derived by using the method of multiple scales and H-convergence, both.

Under some additional regularity assumptions, there existe an "effective absolute permeability" $K^{\#}$ and an effective porosity $\tilde{\Phi}$ associated to the homogenized saturation equation :

$$\tilde{\Phi}\,\frac{\partial S}{\partial t} + \nabla.(R^{\#} + \lambda_w^{\#}(S)\,q_0^{\#}) = 0 \qquad (3.5)$$

$$R^{\#} = -K^{\#}\left(\frac{\theta_1}{c_1} + \frac{\theta_2}{c_2}\right)^{-1}\nabla S \qquad (3.6)$$

where the subscript i denotes the ith type of rock $i = 1,2$ and $c_i = \dfrac{\partial P_c^i}{\partial S}(S)$. θ_i is the proportion of the medium i in the cell Y and

$$\Phi^{\sim} = \frac{1}{|Y|}\left[\sum_{i=1}^{2} \frac{1}{c_i} \int_{Y_i} \Phi(y)dy\right]\left(\frac{\theta_1}{c_1} + \frac{\theta_2}{c_2}\right)^{-1}.$$

4. NUMERICAL SIMULATIONS

In order to test the validity of theoretical results, we have run various simulations for incompressible two-phase immiscible flow. Serveral comparisons have been done between heterogeneous and homogenized simulations for saturation and pressure contours and for production curves some of them are presented in [2]. After several tests we found the most sensitive quantity to the change of permeability would be, in our simulations, the pressure. The saturation contours is quite few sensitive and even the production curves could not discriminate accurately the right permeabilities. We should also mention that the numerical simulations have been done with a sufficiently small mesh size in the heterogeneous simulations, in order to be sure that we were numerically simulating the true heterogeneous reservoir and where not some "numerically averaging" the reservoir. Acutally we may prove [5] that taking h, the mesh size, not sufficiently small comparatively to ε, the heterogeneities size, gives a totally wrong simulation in which there is neither the heterogeneous nor the homogenized media which is simulated. In figures below, we compare the behavior of a quarter five spots with periodic heterogeneities (Fig.5) to the corresponding homogenized one. The corresponding datas are:

$K_{xx}^1 = K_{yy}^1 = 10$ md, $K_{xx}^2 = K_{yy}^2 = 1$ md.

$K_{xx}^{\#} = K_{yy}^{\#} = 6.52$ md.

Arithmetic average of $[K] = 7.75$ md.

Harmonic average of $[K] = 3.07$ md.

The pressure contours for the heterogeneous and *arithmetic average* simulations are shown in Fig.8-(a). The pressure contours for the heterogeneous and *homogenized* simulations are shown in Fig.8-(b). The pressure contours for the heterogeneous and *harmonic average* simulations are shown in Fig.8-(c).

REFERENCES

[1] Amaziane, B. and Bourgeat, A., 1988, Effective Behavior of Two-Phase Flow in Heterogeneous Reservoir, In Numerical Simulation in Oil Recovery, the IMA Volumes in Mathematics and its Application, M.F. Wheeler ed., Vol.11, Springer-Verlag, p. 1-22.

[2] Amaziane, B., Bourgeat, A. and Koebbe, J., 1990, Numerical Simulations and Homogenization of Two-Phase Flow in Heterogeneous Porous Media, To appear in J. Transport in Porous Media .

[3] Amaziane, B. and Dumont, T., 1987, Calcul de Coefficients Homogénéisés : Implémentation dans MODULEF et Résultats Numériques, Publications du Laboratoire d'Analyse Numérique, U.A. 740, Lyon Saint-Etienne, n°61.

[4] Amaziane, B., Bourgeat, A. and El Amri, H., 1990, Analysis of Two-Phase Flow in a Reservoir with Various Rock-Types, Submitted to SIAM Journal on Applied Mathematics.

[5] Amirat, Y. and Bourgeat, A., (to appear).

[6] Amirat, Y., Hamdache, K. and Ziani A., 1989, Homogénéisation d'Equations Hyperboliques du Premier Ordre, Application aux Milieux Poreux, Ann. Inst. Henri Poincaré, Analyse non linéaire, Vol.6, n°5, p. 397-417.

[7] Bourgeat, A., 1984, Homogenization Method Applied to the Behavior of a Naturally Fissured Reservoir, Mathematical Method in Energy Research, K.J. Gross ed., SIAM, p. 181-193.

[8] Bourgeat, A., 1984, Homogenized Behavior of Diphasic Flow in a Naturally Fissured Reservoir with Uniform Fractures, Computers Methods in Applied Mechanics and Engineering, n°47, p. 205-217.

[9] Bourgeat, A., 1985, Nonlinear Homogenization of Two-Phase Flow Simulation, Lecture Notes in Pure and Applied Mathematics, Marcel Dekker, New York, Vol.102, p. 207-212.

[10] Bourgeat, A., 1985, Global Behavior of Two-Phase Flows in Inhomogeneous Media, Proceedings Third International Conference on Boundary Layers, Boole Press Pub., Dublin, p. 157-160.

[11] Bourgeat, A., 1986, Homogenization of Two Phase Flows, Proceedings of the Institute on Nonlinear Analysis and Applications, Symposia in Pure Mathematics, F. Browder ed., AMS, p. 157-164.

[12] Bourgeat, A., 1988, An Effective Deal Oil Model for Two-Phase Flow in Heterogeneous Media, ASME Symposium Proceedings.

[13] Chavent, G., 1976, A New Formulation of Diphasic Incompressible Flows in Porous Media, In Lecture Notes in Mathematics, Vol.503, Springer-Verlag, p. 258-270.

[14] Chavent, G. and Jaffre, J., 1986, Mathematical Models and Finite Elements for Reservoir Simulation, Studies in Mathematics and its Applications, Vol.17, North-Holland, Amsterdam.

[15] Kruzkov, S.N.and Sukorjanskii, S.M., 1977, Boundary Value Problems for Systems of Two-Phase Porous Flow Type: Statement of the Problems Questions of Solvability Justification of Approximate Methods, Math. USSR, Sbornik, Vol.33, n°1, p. 62-80.

[16] Murat, F. and Tartar, L., 1985, Calcul des Variations et Homogénéisation, les Méthodes de l'Homogénéisation: Théorie et Applications en Physique, Coll. Direction Etudes Recherches EDF n°57, D. Bergman et al. ed., Eyrolles, Paris, p. 319-369.

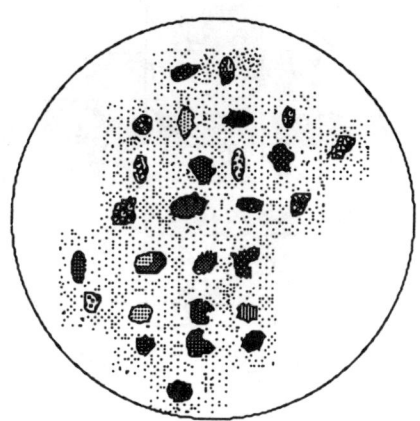

Fig.1. Example of an heterogeneous porous medium.

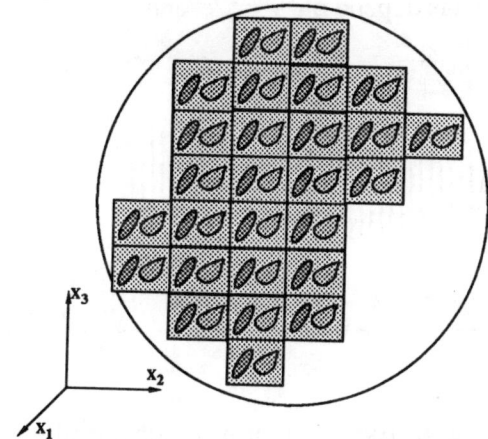

Typical cell Y

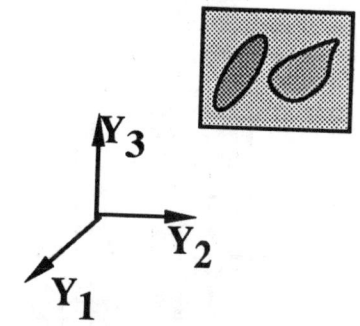

Fig.2. Periodic reparition of heterogeneities. The datas depend on $y = x/\varepsilon$ only.

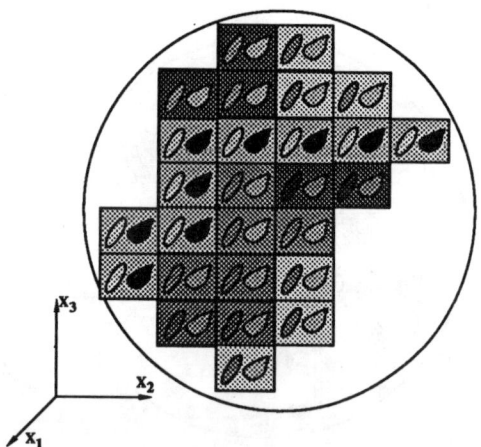

Fig.3. Non uniformly periodic repartition of heterogeneities. The datas depend on $y = x/\varepsilon$ and on x.

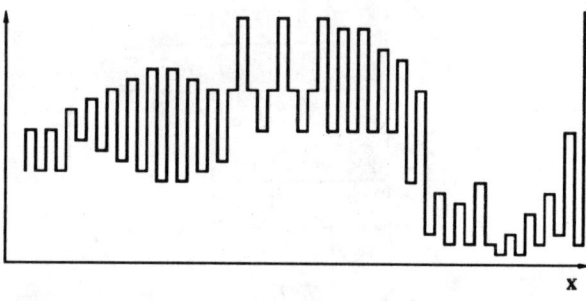

Fig.4. Example of non uniformly periodic datas.

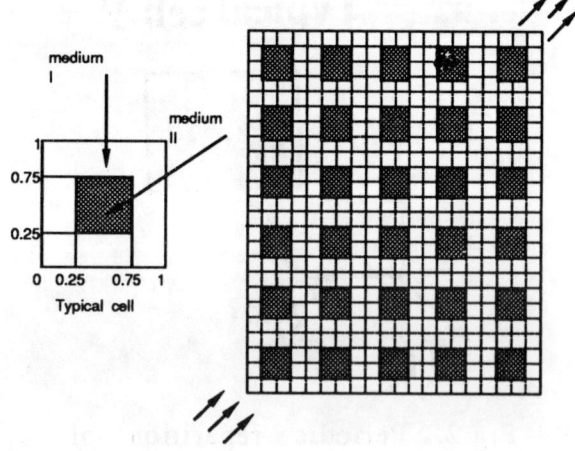

Fig.5. Heterogeneous periodic quarter five-spot.

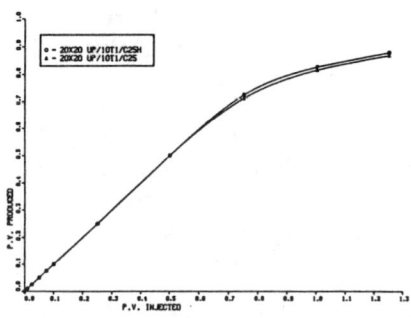

Fig.6. Production curves.
o homogenized, Δ heterogeneous.

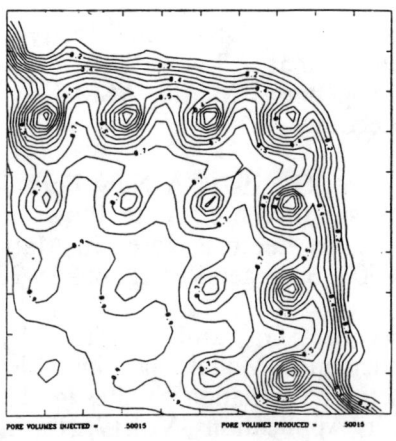

(a)

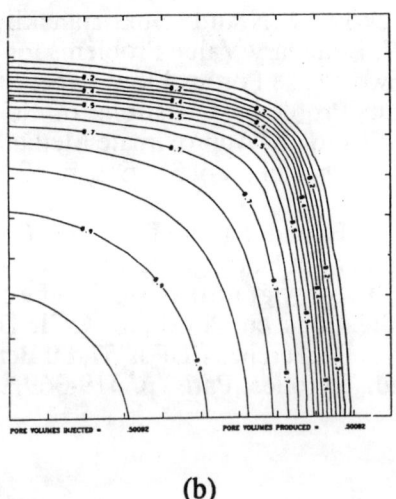

(b)

Fig.7. Comparison of the saturation contours: heterogeneous (a), homogenized (b).

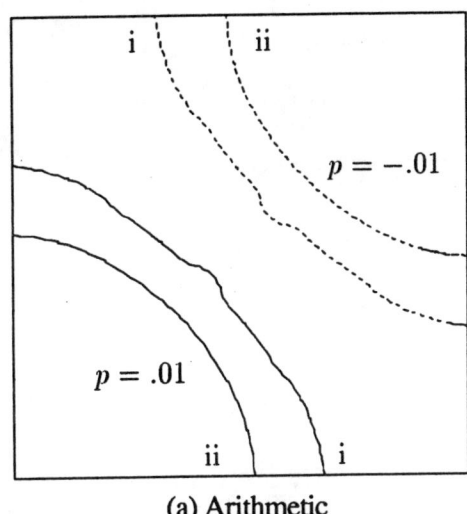

(a) Arithmetic

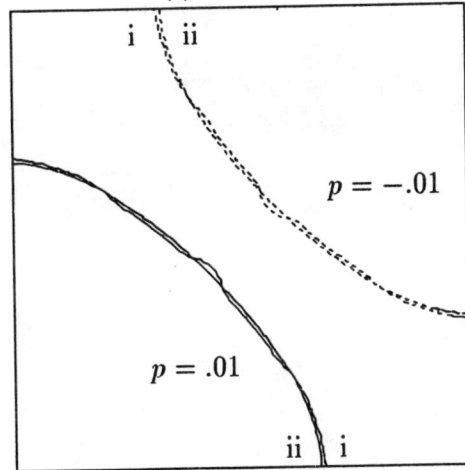

(b) Homogenized

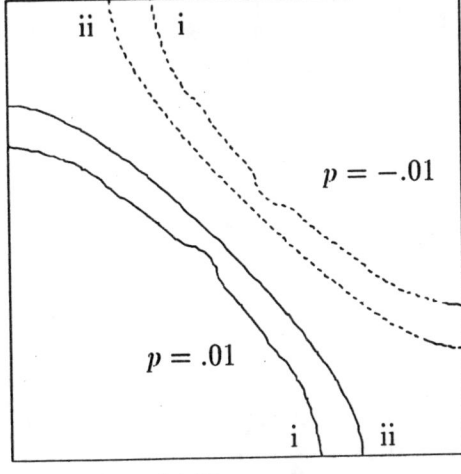

(c) Harmonic

Fig.8. Comparison of the pressure contours for various "averagings".

2nd European Conference on the Mathematics of Oil Recovery
© D. Guérillot, O. Guillon (Editors) and Éditions Technip, Paris 1990, pp. 83-90
27 rue Ginoux, 75015 Paris

Stochastic Characterization
of Grid-Block Permeabilities:
from Point Values to Block Tensors

J. J. Gomez-Hernandez and A. G. Journel[1]

ABSTRACT

Permeability is generally measured at the core scale and only at a few locations. By using geostatistical techniques, one can generate alternative high resolution images of the reservoir at the measurement scale. These images will honor both the data values at the data locations and some prior structural information as modeled, for instance, by a covariance function. However, these images must be averaged to the scale of the flow simulation grid-blocks. Two questions arise: (1) how should averaging be carried out? (2) could the grid-block values be generated directly from the core/log without the intermediary step of small scale imaging? This paper proposes avenues of answer to both questions. First, an averaging technique from point to block permeabilities accounting for the tensorial character of permeability is proposed. Second, a methodology to infer the covariances and cross-covariances between the different components of the block permeability tensor, as well as the cross-covariances between these block values and the point measurements, is proposed. This set of covariances and cross-covariances is then used to generate—directly—multiple, equiprobable spatial realizations of grid-block permeability tensors conditioned to the point measurements. The method is not constrained by small variability of the underlying point permeability distribution, nor does it require any multi-Gaussian-related hypothesis.

1 INTRODUCTION

Permeability data are seldom available at the scale required by the flow simulator and need to be averaged (upscaled) before they can be used. This paper focuses on the problem of how to assign permeability values to the flow simulator grid-blocks given that the permeability data is available at the core- scale.

In recent years stochastic modeling of reservoir parameters has been proposed in replacement of deterministic usually oversmoothed interpolation (Haldorsen, 1990). Geostatistics can be used for the

---

(1) Department of Applied Earth Sciences, Stanford University, Stanford, California 94305, USA.

characterization of the spatial variability of core-scale permeabilities (Journel and Gómez-Hernández, 1989; Gómez-Hernández, 1990). Once this characterization is completed, multiple equiprobable stochastic images of core-scale permeabilities can be generated which all honor the data at their locations.

These images contain typically hundreds of thousands of nodes and need to be coarsened (upscaled) before flow simulation can be carried out. The most common approach to the upscaling of these high-resolution images consists in obtaining an equivalent—or effective—block permeability for each grid-block as a function of the core-scale permeabilities within the block. There are several ways in which this upscaling is carried out. The simplest way amounts to identify the block permeability to some average of the internal core permeabilities, such as the geometric average. A more elaborated way identifies the horizontal and the vertical block permeabilities to two different power averages (or *p*-norms) of the core values (Deutsch, 1989). Finally, the most expensive way—and also the most accurate—requires the solution of the flow equation when no flow boundary conditions are applied to all faces of the block except for two opposite faces that are set with a constant pressure —but different for each face. The block permeability in the direction of the pressure gradient is then identified to the ratio between the total flow crossing the block and the pressure gradient given by the constant pressure boundary conditions (Warren and Price, 1961; Desbarats, 1987). This method will be referred later as the traditional pressure solver approach.

The latter technique—which is the most general and the most accurate of the three techniques mentioned—has been used successfully in the generation of grid-block values as input to flow simulators in some specific cases (Begg *et al.*, 1989). In that study, Begg *et al.* point out that such technique would fail in the case of large heterogeneities of the order of magnitude of the block size. They also point out that the block

permeabilities obtained with this technique are dependent on the boundary conditions used to solve flow within each block. In order words, block permeabilities **are not intrinsic** to the internal distribution of core values even when these core values are exhaustively known. A block permeability can not be defined which would be valid for any boundary condition. However, if the range of the core permeability covariance is finite, there will be a minimum block size beyond which the block permeability will become independent of the boundary conditions used to obtain it. This conjecture has been proven correct analytically for some simple 2d cases by Kitanidis (1990).

In this paper, a technique is presented that attempts to remove the effect of boundary conditions in the calculation of block effective permeabilities, yet capturing the tensorial character of block permeabilities with principal axes not necessarily parallel to block faces.

Then, the joint spatial variability of the core scale permeabilities and the tensor components of the grid block permeabilities is characterized by their respective covariances and cross-covariances. Once this covariance modeling is completed a joint co-simulation of all components of the block permeability tensor conditioned to the core scale data can be performed; this eliminates the costly generation of high resolution core permeability images and their upscaling, itself calling for a small flow simulation for each block (Lasseter *et al.*, 1986).

This paper considers only a two-dimensional flow problem solved by finite differences with constant block sizes, for which the maps of interface block permeabilities are required. The extension to three dimensions is straightforward, the extension to variable block sizes is feasible but tedious, the extension to block shapes other than parallelepipedic is yet unclear.

The terms macroscale and megascale will refer, respectively, to the scales at which the data is available and to the scale at which the flow equation is

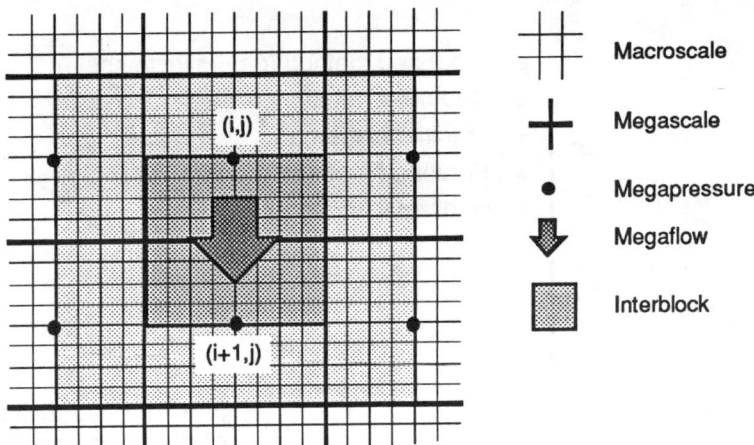

FIG. 1. Megapressures and megafluxes

numerically solved (Haldorsen, 1986). Also, because the flow equation is solved using a block centered finite difference scheme, the interface permeability rather than the block permeability will be sought. Such interface permeability may be identified to the equivalent permeability of a block centered at the interface (dark shaded area in Fig. 1).

2 A NEW UPSCALING TECHNIQUE

Figure 1 represents a heterogeneous reservoir discretized at the macroscale and which must be scaled up to the megascale. After solving the flow equation at both the macroscale and the megascale we require the following conditions to be met: (1) the megapressure ϕ_{ij} associated to block (i,j) should be equal to the storativity-weighted sum of macropressures within the block; and, (2) the megaflux $q_{i+1/2,j}$ going from block (i,j) to block $(i+1,j)$ is equal to the sum of the macrofluxes crossing the block interface.

The relationship between the megapressures and the megafluxes is given by Darcy's law. In particular for the megaflux $q_{i+1/2,j}$

$$\rho \cdot g \cdot q_{i+1/2,j} = K_{ii} \left.\frac{\partial \phi}{\partial x_i}\right|_{(i+1/2,j)} + K_{ij} \left.\frac{\partial \phi}{\partial x_j}\right|_{(i+1/2,j)} \quad (1)$$

where:

K_{ii} and K_{ij} are the two components of the interface permeability tensor in the direction orthogonal to the interface;

ρ is the fluid density;

g is the acceleration of gravity;

Note that a possible non-diagonal tensor for the interface permeability should be considered and that the megapressure gradients in both directions parallel and orthogonal to the interface have to be estimated at location $(i+1/2,j)$.

The interface permeabilities K_{ii} and K_{ij} are obtained by solving the flow equation at the macroscale as follows:

- Define an area surrounding the interface that includes a "skin" of, for example, half the block side as shown by the lighter shaded area of Fig. 1.
- Solve the flow equation at the macroscale for several boundary conditions applied to the skin outside boundary
- For each boundary condition evaluate the megaflux across the block interface as well as the megapressures of the surrounding blocks

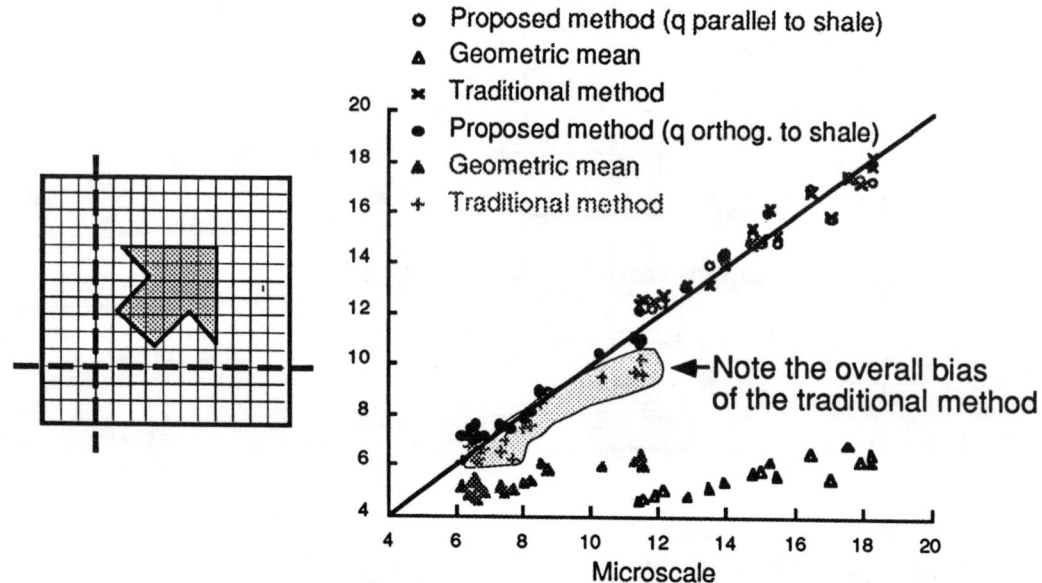

FIG. 2. *Comparison of the fluxes crossing cross-sections of the reservoir parallel and orthogonal to the shale orientation. On the left an sketch of the reservoir, showing the overall pressure gradient and two of the cross-sections for which the total fluxes were computed. On the right, the total fluxes as obtained from the flow simulation at the macroscale (abscissa axis) are plotted versus the total fluxes as obtained form the flow simulation at the megascale (ordinate axis) using the proposed upscaling method, the traditional pressure solver method and the geometric average.*

- Estimate the megapressure gradients both orthogonal and parallel to the interface
- Establish equation (1) for each boundary condition with K_{ii} and K_{ij} as unknowns and invert the resulting overdetermined system of linear equations by a least square procedure

This method is deemed superior to the methods referred in the introduction for the following reasons: (a) it provides a full permeability tensor instead of a scalar, (b) the use of a "skin" produces a flow pattern around the interface closer to the flow pattern that would be obtained from a simulation over the entire reservoir, and (c) by selecting an interface permeability tensor in a least square sense, we are retrieving a permeability tensor that performs well in average for a whole set of boundary conditions

The method proposed is tested on a synthetic reservoir with a binary distribution of permeabilities corresponding to sand and shale (pay/non-pay). The shales represent 20% of the total reservoir volume, and there is a contrast of 4 orders of magnitude between the permeability of the sand and that of the shale. The spatial variability of the shales is modeled by an anisotropic spherical covariance with ranges equal to 1/200th and of 1/20th of the reservoir extent in the vertical and horizontal directions, respectively. The reservoir was discretized into 200 × 200 cells (macroscale) to be scaled up into 20 × 20 blocks (megascale), i.e., each block is made up of 10 × 10 cells. The flow equation is solved using block centered finite differences at the macroscale with constant pressure along the four sides of the reservoir. The boundary conditions impose an overall head gradient from the bottom left corner to the upper right one so that significant megafluxes do cross all block interfaces.

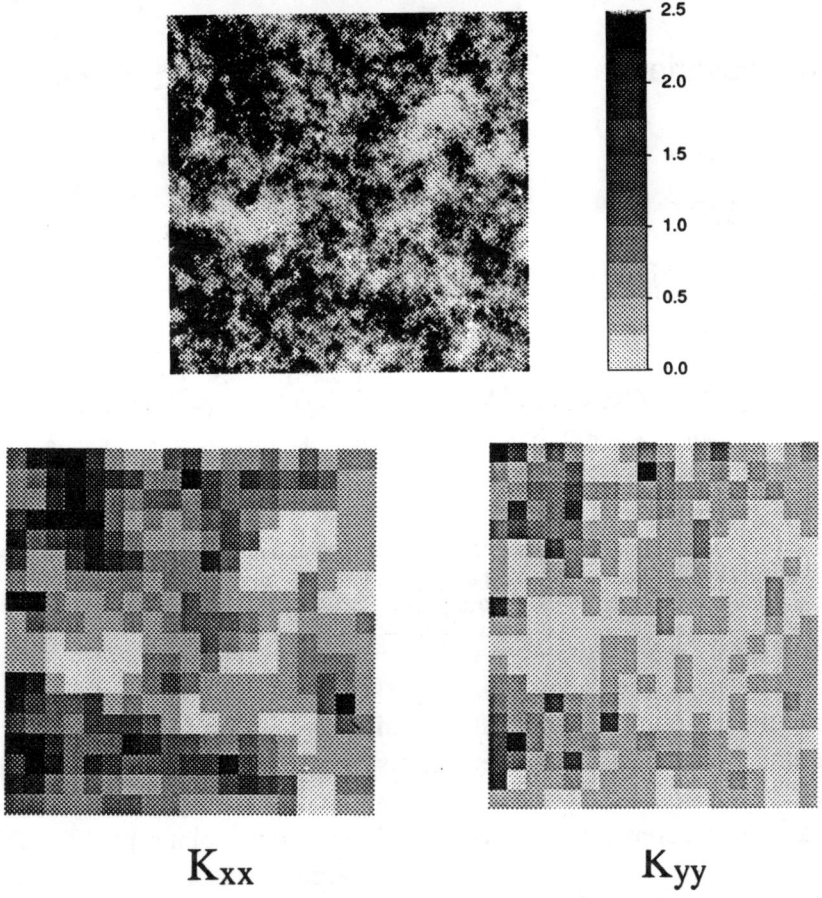

K_{xx} K_{yy}

FIG. 3. A conditional simulation of macroscale permeabilities and the resulting training images of interface permeabilities.

The flow equation is also solved at the megascale with the same above boundary conditions using three different upscaling techniques. First, the interface permeability is set equal to the geometric mean of the macroscale permeabilities contained within the block *centered* at the interface (darker shaded area in Fig. 1); second, the interface permeability is obtained using the traditional pressure solver approach applied to the block *centered* at the interface; and third, the interface permeability is obtained using the new method proposed above.

Figure 2 shows a comparison of the total fluxes crossing cross-sections parallel and orthogonal to the shales elongation, as obtained from the macroscale simulation and from the megascale simulations using the three upscaling techniques. It is clear that the geometric mean should not be used in this case; the reasons for the large underestimation of the macroscopic fluxes are two: first, the geometric mean cannot capture the anisotropy of the interface permeabilities and, second, the geometric mean gives too much weight to the shale permeabilities resulting in too low interface permeabilities. Comparing the results from the proposed method and the traditional method, both reproduce well the total fluxes parallel to the shales, but the traditional method tends to un-

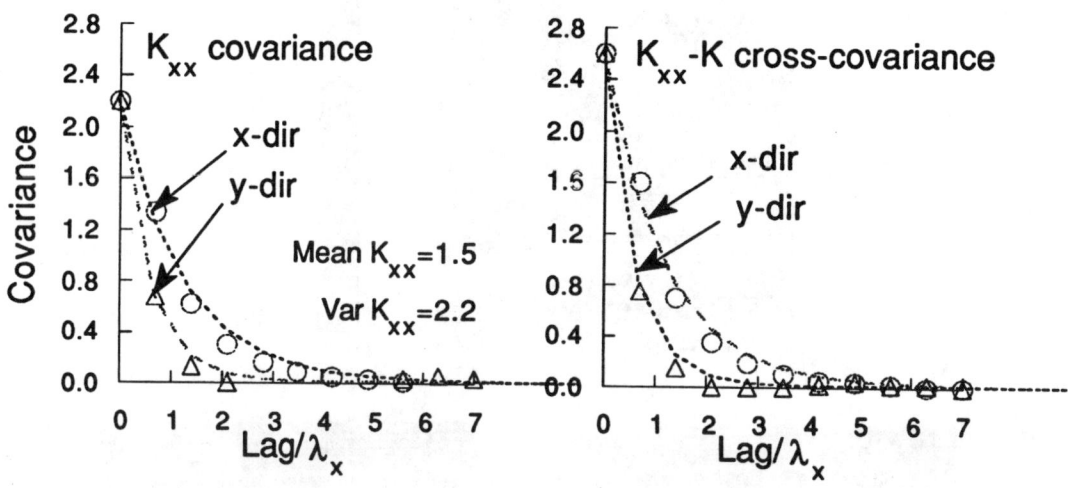

FIG 4. *Covariance of the x-component of the interface permeability tensor and cross-covariance between the x-component and the macroscale permeability. An exponential model was fitted to the inferred variograms.*

derestimate the fluxes orthogonal to the shales. This underestimation is due to the use of no flow boundary conditions in the traditional pressure solver method, which results in very low interface permeabilities for those blocks that are completely intersected by a shale.

3 STOCHASTIC CHARACTERIZATION OF INTERFACE PERMEABILITIES

In order to directly generate reservoir maps of interface permeabilities without requiring the intermediate step of generating maps of macroscale permeabilities that call for upscaling on a block by block basis, the spatial variability of the interface permeabilities must be modeled in a statistical sense. Such model can be done analytically (Rubin and Gómez-Hernández, 1990) for the specific case of isotropic multiLognormally distributed point permeability, and for blocks of different sizes embedded in an infinite reservoir. For more realistic

cases with larger variability of permeability, anisotropy and multimodality in the permeability histogram, we propose a synthetic training image approach. This approach allows the numerical inference of the joint spatial variability of macro and megascale permeabilities. It can be used to infer the expected values, variograms and cross-variograms of the components of the interface permeability tensors, and the cross-variograms between the interface values and the macroscale data.

To generate such synthetic training images the spatial variability of the macroscale permeabilities is modeled first, which amounts to the inference of a stationary model for its bivariate spatial distribution. There is no limitation to the type of model used for these macroscale permeabilities, it could be either multiGaussian-related or fully non-parametric (Journel and Alabert, 1989).

The synthetic training image approach to the characterization of interface permeabilities consists of the following steps. First generate a conditional simulation (Journel, 1974, 1989; Alabert, 1987) of macroscale permeabilities over an area at least twice as large as the correlation range of the macroscale permeability in each direction and containing at least a few hundred blocks. Using the method presented in the previous section obtain the interface permeabilities. Finally use the map of interface permeability tensors along with the map of macroscale permeabilities to infer the covariance/cross-covariances required for carrying out simulations of interface permeability tensors that are conditioned to the macroscale data.

Figure 3 shows an example of a conditional simulation of macroscale permeabilities and the two resulting maps of interface permeabilities in the horizontal and vertical directions. (In this particular case the off-diagonal terms of the interface permeability were negligible). The macroscopic permeabilities are isotropic to flow, have an anisotropic covariance and a univariate lognormal distribution; their covariance is exponential with integral scale in the x-direction, λ_x which is 10 times the cell side, and integral scale in the y-direction, $\lambda_y = \lambda_x / 2$. The megascale blocks are squares of side 0.7 λ_x. Each block consists of 7×7 cells and is expected to present some anisotropy due to the anisotropic spatial continuity of the macroscale permeabilities.

The covariances for the xx-component of the interface permeability and the cross-covariance between the xx-component and the macroscale permeability are shown in Fig. 4. Both covariances are seen to be well behaved and can be fitted by exponential models. The yy-component of the interface permeability has a coefficient of correlation of 0.98 with the xx-component and presents an anisotropy ratio $K_{yy}/K_{xx}=0.91$.

Given these models of joint spatial variability of macroscale and interface permeabilities, simulated maps of interface permeability tensors conditioned to the macroscale data can be generated directly.

For the case presented before, given the high linear correlation between the two components of the interface permeability tensor and the alignment of the principal components with the cartesian coordinates, the generation of interface permeability tensors is limited to the generation of one of the components, say K_{xx}, conditioned to the macroscale

data. The full tensor is then completed by using the anisotropy ratio computed earlier. As for the simulation of K_{xx}, the principle of sequential simulation (Journel, 1989) can be used to generate a multiGaussian field conditioned to macro- and megascale data. The algorithm proceeds as follows, first define a random path visiting once and only once all block interfaces; for each interface, draw a random realization from a model for the probability distribution of K_{xx} conditioned to the macroscale data and to all previously simulated values of K_{xx}. This conditional probability distribution function can be identified—under a multiGaussian hypothesis—to a Gaussian distribution with mean equal to the cokriging estimate of K_{xx} and variance equal to the cokriging variance. If the multiGaussian hypothesis is not deemed appropriate, the same principle of the synthetic training image can be used to infer a nonparametric statistical model for the spatial variability of the megascale permeabilities.

CONCLUSIONS

A new technique has been presented to upscale macroscale permeabilities into grid-block interface permeabilities that can be used with blocks of size of the order of magnitude of the macroscale heterogeneities. By using a least square approach based on the solution of flow for different boundary conditions, an "average" grid block permeability tensor is retrieved. This block permeability tensor reproduces better the total fluxes crossing the reservoir than more traditional techniques, especially in the direction orthogonal to the elongation of the heterogeneities.

The proposed upscaling method is then used to generate synthetic training images of grid-block permeability tensors that are used to infer a statistical model of the joint spatial variability of macroscale permeabilities and grid-block permeability tensors. This latter model is then used for the co-simulation of the components of the grid-block permeability conditioned to the macroscale data; thus avoiding the costly simulation of macroscale permeabilities and upscaling of each individual block as done in more traditional methods.

ACKNOWLEDGEMENTS. Financial support for this study was provided by the Stanford Center for Reservoir Forecasting. The authors wish to thank Yoram Rubin now at UC Berkeley, and François Alabert now with Elf Aquitaine, for the brainstorming sessions at the origin of some of the ideas developed in this paper.

REFERENCES

Alabert, F.G., 1987, Stochastic imaging of spatial distributions using hard and soft information, Ms Sc. thesis, Branner Earth Sciences Library, Stanford University, 184 p.

Begg, S.H., Carter R.R. and Dranfield, P., 1989, Assigning effective values to simulator gridblock parameters for heterogeneous reservoirs, *SPE Reservoir Engineering*, p. 455–463.

Desbarats, A.J., 1987, Numerical estimation of effective permeabilities in sand-shale formations, *Water Resour. Res., 23*(2), p. 273–286.

Deutsch, C.V., 1989, Calculating effective absolute permeability in sandstone/shale sequences, *SPE Formation Evaluation*, 343–348.

Gómez-Hernández, J.J., 1990, A case study of three-dimensional multiple indicator conditional simulation: Florida's Jay oil field, submitted for publication to *JPT*.

Haldorsen, H., 1986, Simulator parameter assignment and the problem of scale in reservoir engineering. In: L. Lake and H. Carroll (Editors), *Reservoir Characterization*, Academic Press, p. 293–340.

Haldorsen, H. and Damsleth, E., 1990, Stochastic modeling, *JPT*, p. 404–412.

Journel, A.G., 1974, Geostatistics for conditional simulation of orebodies, *Econ. Geol. 69*, (5), p. 673–687.

Journel, A.G., 1989, *Fundamentals of geostatistics in five lessons.* Short course in Geology, AGU publication, Washington, D.C., vol. 8, 40 p.

Journel, A.G. and Alabert, F.G, 1989, Non-Gaussian data expansion in the earth sciences, *Terra Nova, 1*, (2), p. 123–134.

Journel, A.G. and Gómez-Hernández J.J., 1989, Stochastic Imaging of the Wilmington Clastic Sequence, SPE paper #19857, Proc. of the 64th SPE Annual meeting, San Antonio.

Kitanidis, P.K., 1990, Effective hydraulic conductivity for gradually varying flow, *Journal of Hydrology*, in press.

Lasseter, T., Waggoner, J. and Lake, L., 1986, In: L. Lake and H. Carroll (Editors), *Reservoir Characterization*, Academic Press, p. 545–559.

Rubin, Y and Gómez-Hernández, J.J., 1990., A stochastic approach to the problem of upscaling conductivity in disordered media: theory and unconditional simulations. *Water Resources Res. 26*, (4), p. 691–702.

Warren, J.E., Price and H.S., 1961, Flow in heterogeneous porous media, *Soc. Pet. Eng. J., 12*, 14–28.

2nd European Conference on the Mathematics of Oil Recovery
© D. Guérillot, O. Guillon (Editors) and Éditions Technip, Paris 1990, pp. 91-98
27 rue Ginoux, 75015 Paris

Large-Scale Properties for Flow through a Stratified Medium: a Discussion of Various Approaches

A. Ahmadi[1], A. Labastie[2] and M. Quintard[1]

ABSTRACT

The determination of effective properties or pseudo-functions for two-phase flow through heterogeneous porous media is a problem of central importance in petroleum engineering (reservoir simulation).

Our purpose was to compare published theories by performing test case studies. The test case corresponded to the waterflooding of a two-strata heterogeneous medium. Generalized Darcy's equations were solved by using a black-oil reservoir simulator. Several simulations were performed by varying the filtration velocity and the relative importance of gravity effects versus capillary and viscous effects. Based on an extensive literature survey, two fundamental methods were essentially tested: the large-scale averaging method and an application of the fine-grid to coarse-grid method.

In the first method, the large-scale equations and the large-scale properties are obtained by averaging the Darcy's scale equations over a representative volume of the heterogeneous medium. The second method consists in performing a simulation on the whole domain with a fine grid, then obtaining the pseudo-capillary pressure and pseudo-permeabilities by averaging these results over a coarse grid. Basically in the application of these methods to a stratified medium the averaging techniques provide large-scale properties for cross-sections of the stratified medium.

The simulations showed three different fundamental types of flow corresponding to: gravity equilibrium, capillary equilibrium and dynamic flow conditions. The ability of the different methods to represent properly these various flow conditions is discussed from a comparison of the watercut curves. Criteria for choice are proposed based on dimensional analysis of the descriptive equations.

1 INTRODUCTION

Two-phase flow in heterogeneous porous media is a problem of great importance in reservoir engineering. In a numerical reservoir model, the domain is generally divided into grid blocks of 10 to 100 m. Each grid block is assumed to be homogeneous and perfectly defined by its characteristics. The properties input to the model must therefore account for the heterogeneities, as well as the fluids distribution at any scale smaller than the grid size. These properties are generally called pseudo-functions. They can be used to reduce the dimension of the

(1) Laboratoire Énergétique et Phénomènes de Transfert, École Nationale Supérieure des Arts et Métiers,
(UA CNRS 873), Esplanade des Arts et Métiers, 33405 Talence Cedex, France.
(2) Elf Aquitaine, 26, avenue des Lilas, 64018 Pau, France.

problem or to reduce the number of grid blocks in a numerical model. There is an extensive literature on this subject. The different theories can be classified as follows:

1.1 Vertical Equilibrium (VE)

This assumption corresponds to a gravity and capillary equilibrium in the direction perpendicular to the main flow. The VE approach, corresponding to a segregation of the fluids into equilibrium between hydrostatic and capillary pressure, is obviously favored by: high degree of vertical communication, large phase density difference between injected and reservoir fluids and low displacement velocity. These conditions are often satisfied in a reservoir.

The relative importance of the capillary and gravity forces leads to a classification of the VE flows as follows:

a) capillary-gravity equilibrium - The components of the capillary and gravity forces in the direction perpendicular to the main flow are of the same order of magnitude (Coats et al., 1967; Martin, 1968; Dake, 1978; Killough and Foster, 1979).

b) gravity equilibrium - The gravity forces are preponderant (Coats et al., 1967; Martin, 1968; Dake, 1978; Thomas, 1983).

c) capillary equilibrium - The fluids distribution is dominated by capillary forces (Corey and Rathjens, 1956, Yokoyama and Lake, 1981).

In each of the above cases, we can derive the saturation field in the direction perpendicular to the main flow. The dimension of the problem can thus be reduced by integrating the equations in this direction.

1.2 Dynamic Flow

The displacement process is dominated by viscous forces and transient effects. Pseudo-functions are derived from simplified solutions (Stiles, 1949; Hearn, 1971; Simon and Koederitz, 1982) or with the help of methods based on numerical simulations. The latter methods are more economically applicable to reservoirs where the 3-D distribution of fluids can be estimated reliably by studying a 2-D representative vertical cross-section.

In that case, a 2-D simulation of the flow in a representative cross-section of the reservoir is performed by using fine grids and the results are used to calculate the pseudo-functions necessary for the numerical simulation of the whole reservoir using coarse grids. The fine-grid to coarse-grid methods have been studied by different authors (Huppler, 1970; Jacks et al., 1972; Kyte and Berry, 1975; Thomas, 1983; Kortekaas, 1983; Kossack et al., 1989). These pseudo-functions may also be used to deal with the numerical dispersion effects; however, numerical problems are not within the scope of this paper.

Another approach for the calculation of the pseudo-functions is to use an averaging method: the transport equations and the effective properties at a given scale are calculated by an averaging process over the equations corresponding to a lower scale. The basis of the method is to write the local equations and the capillary pressure relation in terms of deviations to the averaged quantities. The large-scale averaging of these equations leads to terms in which the deviations remain. It is possible to obtain a closed form of these large-scale equations by representing the deviations as functions of the large-scale averages. This procedure leads to a closure problem which is very complex in the general case.

In practice, for reservoir simulations, equations similar to the local ones are used at the large scale and the properties such as the effective permeability tensors and the capillary pressure are replaced by the corresponding pseudo-functions which account for the heterogeneties. Quintard and Whitaker (1988) have shown that the large-scale equations are similar to the local-scale ones in one fundamental case called the quasi-static case, which corresponds to local capillary equilibrium. Quintard and Whitaker (1988) have found a solution to the closure problem obtained by the application of the large-scale averaging method to two-phase flow in porous media when the quasi-static assumption is satisfied. In this case, the large-scale properties (pseudo-functions) can be calculated in a relatively easy and rapid way and they are function of the large-scale saturation only. If the pressure gradients and the transient effects create significant change in the saturation over the large-scale averaging volume, then the large-scale averaging scheme gives more complex large-scale equations and closure problems featuring time-dependent, directional, dynamic behaviour (Quintard and Whitaker, 1990a and b). Specially, large-scale properties depend also on the large-scale pressure gradients, the gravity orientation and the time derivative of the large-scale saturation.

At this point it is clearly understood that the resolution of the full dynamic closure problem represents an extremely complex task. It is therefore important to determine under which condition it is possible to avoid this problem.

The purpose of this paper is to propose a first overview of this question based on numerical experiments carried out on a test problem representative of the length scale and physical properties found in classical reservoir simulations. From the numerical solution we were able to determine "pseudo-functions" by averaging the computed fields over cross-sections. From the physical properties we derived the effective parameters associated with the quasi-static averaged equations. The influence of gravity, flow rate and the capillary pressure curves on these properties have been studied. In a few particular cases the results obtained by a fine-grid to coarse-grid method have been compared to the pseudo-functions found by the large-scale averaging method restricted to the quasi-static case.

2 DIMENSIONLESS ANALYSIS

In this section, we propose an approach of the dimensional analysis based on the large-scale averaging method. The local-scale continuity and transport equations are coupled by the capillary pressure relation:

$$P_o - P_w = P_c(\varepsilon_w) \qquad (1)$$

where ε_w is the local volume-averaged volume fraction for water; P_o and P_w are the pressures of the oil and water respectively.

Introduction of the deviations to the large-scale averages as follows

$$P_\alpha = \{P_\alpha\}^\alpha + \hat{P}_\alpha \qquad (\alpha=o, w) \qquad (2)$$

leads to

$$P_c(\varepsilon_w) = \{P_o\}^o - \{P_w\}^w + \hat{P}_o - \hat{P}_w \qquad (3)$$

This relationship must be used in the capillary zone, V_c, which is the intersection of the active regions of the two phases. In this paper we will suppose that V_c is equal to V_∞ (large-scale averaging volume). If $\underline{x}$ is the centroid of V_∞ and $\underline{x}+\underline{y}$ is any current point, by using a Taylor expansion Eq. (3) gives:

$$P_c(\varepsilon_w)|_{\underline{x}+\underline{y}} = \left[\{P_o\}^o - \{P_w\}^w\right]_{\underline{x}}$$
$$+ \underline{\nabla}[\{P_o\}^o - \{P_w\}^w] \cdot \underline{y} + \left[\hat{P}_o - \hat{P}_w\right]_{\underline{x}+\underline{y}} \qquad (4)$$

The deviations can be represented as

$$\hat{P}_\alpha = (\nabla\{P_\alpha\}^\alpha - \rho_\alpha\underline{g}) \cdot \underline{b}_\alpha = \underline{\Omega}_\alpha \cdot \underline{b}_\alpha ; \quad (\alpha=o, w)(5)$$

where the vectors $\underline{b}_\alpha$ are to be calculated by resolving the closure problem (Quintard and Whitaker, 1988). Finally, Eq. (4) becomes:

$$P_c(\varepsilon_w)|_{\underline{x}+\underline{y}} = \left[\{P_o\}^o - \{P_w\}^w\right]_{\underline{x}} + \underline{\Omega}_o \cdot (\underline{y}+\underline{b}_o)$$
$$- \underline{\Omega}_w \cdot (\underline{y}+\underline{b}_w) + (\rho_o-\rho_w)\underline{g} \cdot \underline{y} \qquad (6)$$

From the inversion of Eq. (6) and large-scale averaging, we can deduct:

$$\{\varepsilon_w\}^* = \left\{P_c^{-1}\left[\left[\{P_o\}^o - \{P_w\}^w\right]_{\underline{x}} + \underline{\Omega}_o \cdot (\underline{y}+\underline{b}_o)\right.\right.$$
$$\left.\left. - \underline{\Omega}_w \cdot (\underline{y}+\underline{b}_w) + (\rho_o-\rho_w)\underline{g} \cdot \underline{y}\right]\right\} \qquad (7)$$

We use a Taylor expansion to obtain:

$$\{\varepsilon_w\}^* = \left\{P_c^{-1}\left(\{P_c\}_{\underline{x}}\right)\right\} + \left\{\frac{\partial P_c^{-1}}{\partial P_c}\bigg|_{\{P_c\}^c}\left(\underline{\Omega}_o \cdot (\underline{b}_o+\underline{y})\right.\right.$$
$$\left.\left. - \underline{\Omega}_w \cdot (\underline{b}_w+\underline{y}) + (\rho_o-\rho_w)\underline{g} \cdot \underline{y}\right)\right\} \qquad (8)$$

93

We can therefore express $\{\varepsilon_w\}^*$ as a sum of three terms as follows:

$$\{\varepsilon_w\}^* = \{\varepsilon_w\}^*_{\text{quasi-static}} + f(N_D) + g(N_g) \qquad (9)$$

The two dimensionless numbers N_D and N_g represent the possible influence of dynamic and gravity terms respectively. We have decided to adopt the following rough estimates of these two numbers:

$$N_D \simeq R_0 \max\left(\frac{V_\alpha \mu_\alpha}{K_\alpha^*}\right).\max\left(\frac{\partial \varepsilon_w}{\partial p_c}\right) \qquad (10)$$

$$N_g \simeq R_0 g(\rho_w - \rho_o).\max\left(\frac{\partial \varepsilon_w}{\partial p_c}\right) \qquad (11)$$

R_0 is the radius of the large-scale averaging volume V_∞. The calculation of these two numbers in each case studied will allow a better understanding of the phenomena.

3 TEST PROBLEM

We have chosen to study the waterflooding of a 20 m long two-strata medium. Each layer is 2.5 m thick (Fig. 1).

We study the displacement of oil by water injected at constant flow rate at the inlet face; we produce at constant pressure at the other end of the porous medium.

The two layers have significantly different petrophysical characteristics, the more permeable being at the bottom. They are both water wet; all the properties of the porous media and fluids are given in the tables 1 and 2.

phase	density (Kg/m^3)	viscosity (Pa s)
water	997	0.001
oil	836	0.002

TABLE 2 - The properties of the two phases.

The relative permeability and the capillary pressure curves for the two layers are illustrated in Figs. 2.a and 2.b.

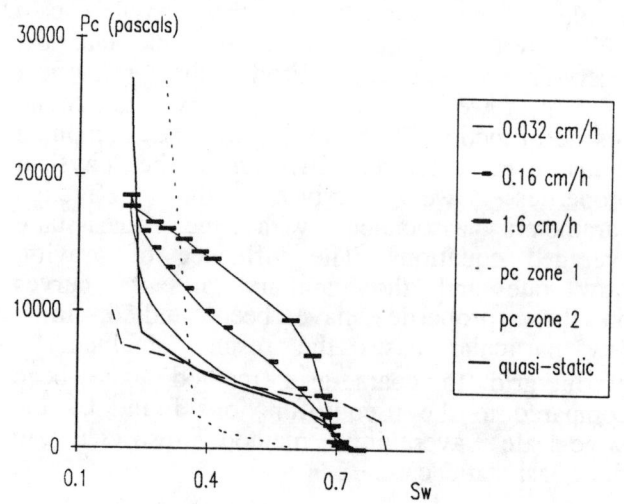

FIG. 2.a. Large-scale capillary pressure curves for different filtration velocities without gravitational effects.

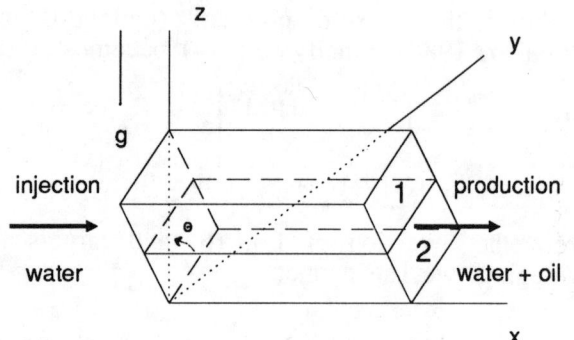

θ = 0° or 45° or 90°

K_1 =200 mD, K_2 =2000 mD

FIG. 1. The two-layered medium considered.

	oil permeability at Swi	porosity %	Swi %	Sor %	$P_c(S_w=Swi)$
layer 1	200 mD	22.5	17.8	18.5	27,000 Pa
layer 2	2000 mD	43.6	29.5	38.5	10,000 Pa

TABLE 1 - The characteristics of the two layers.

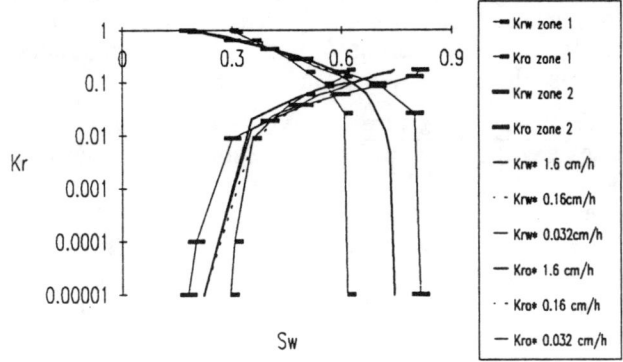

FIG. 2.b. Large-scale relative permeability curves for different filtration velocities without gravitational effects.

Two-dimensional numerical simulations have been done by the classical finite volume model SCORE, using a simultaneous solution (fully implicit) time scheme. The resulting saturation and pressure fields have been used to calculate the pseudo-functions. The local quantities are averaged over vertical cross-sections in order to obtain large-scale quantities for an equivalent homogeneous medium. Arithmetic averages over the cross-section are used for the velocity, the pressure and the capillary pressure. For large-scale saturation the average is weighted by the porosities. The large scale effective permeability is then calculated by using the generalized Darcy's equation.

4 RESULTS AND DISCUSSION

4.1 Efficiency and Intrinsic Character of the Fine-grid to Coarse-grid Method

For each case studied, the calculated pseudo-functions have been re-injected into a 1-D equivalent homogeneous medium and the resulting water-cut and oil production curves have been compared to the corresponding curves from the 2-D simulation of the heterogeneous medium using fine grids. In every case the agreement between the results is very satisfactory.

Large-scale properties have been calculated over two different cross-sections (placed at 1/3 and 2/3 of the medium), in order to test the intrinsic character of the method. It is interesting to note that for all the studied cases (including very dynamic flows: viscosity ratio of 20 or negligible capillary pressure), the method is found to be independent of the section chosen.

This justifies the use of the fine to coarse-grid pseudo-functions as reference properties for the comparison with other methods.

4.2 Influence of Different Parameters on the Pseudo-functions

. Flow rate - With the capillary pressure curves of Fig. 2.a and the gravity effects fixed, we have calculated the pseudo-functions for three different filtration velocities: 0.032 cm/h, 0.16 cm/h and 1.6 cm/h. The saturation fronts spread out obviously as the flow rate decreases. The pseudo-capillary pressure curves increase as the flow rate increases (Fig. 2.a to c) and the pseudo-relative permeabilities vary little. The dynamic number N_D is respectively equal to 0.17, 0.85 and 8.5 for the filtration velocities of 0.032, 0.16 and 1.6 cm/h.

. Gravity effects - For the three filtration velocities 0.032 cm/h, 0.16 cm/h and 1.6 cm/h, the influence of the gravity forces on the exchanges between the two layers (crossflow) has been studied by inclining the medium at the angles of 0°, 45° and 90° (Fig. 1). In our case, the gravity forces act against the capillary forces. For example, at a low flow rate (0.032 cm/h) the spontaneous imbibition due to the capillary forces decreases as the gravity effects increase.

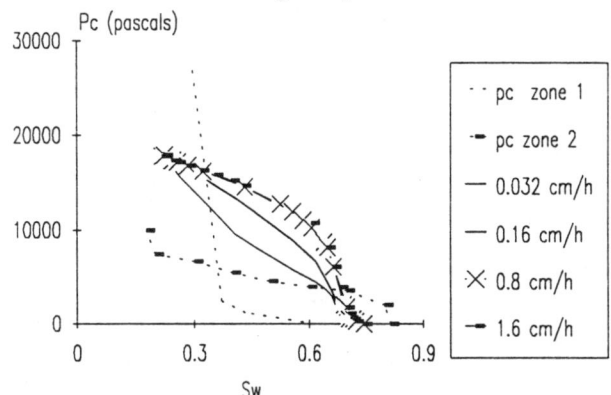

FIG. 2.c. Large-scale capillary pressure curves for different filtration velocities at maximum gravity effects (90°).

The gravitational forces seem to have little effect on the pseudo-relative permeabilities, but act on the pseudo-capillary pressure in the same way as the flow rate does. The pseudo-capillary pressure curves become higher as the gravity effects increase (Fig. 3). The gravity number N_g is equal to 0, 0.413 and 0.585 respectively for 0°, 45° and 90°.

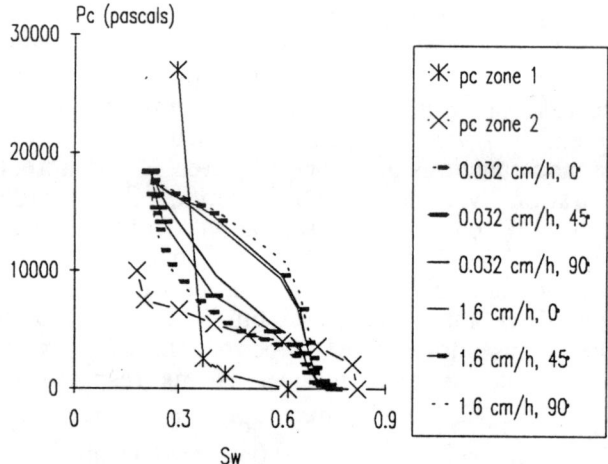

FIG. 3. Large-scale capillary pressure curves for different gravity effects at two different filtration velocities.

It is interesting to note that the range of variation of N_g is much smaller than that of N_D. Therefore, the dynamic effects have a more important role on the large-scale properties than the gravity effects. This result is coherent with the calculated large-scale properties seen above.

. Capillary pressure - We have treated two cases without capillary pressure and with a flow rate of 0.16 cm/h: 1) minimum gravity effects (0°), 2) maximum gravity effects (90°). The pseudo-relative permeabilities to oil are very close in the two cases, but the pseudo-relative permeability to water is higher in case 2.

4.3 Comparison Between the Fine-grid to Coarse-grid Method and the Large-scale Averaging Method Under the Quasi-static Assumption

We have chosen, for the time being, three cases without gravity effects for the comparison of the methods:

case	filtr. velocity	capillary pressure
1	0.16 cm/h	negligible
2	0.16 cm/h	curves of Fig. 2.a
3	0.032 cm/h	curves of Fig. 2.a

TABLE 3 - The three cases studied.

We have calculated the pseudo-functions in each case by the two methods mentioned above. In the case No. 3, there is a good

a. 0.16 cm/h, $P_c = 0$ b. 0.16 cm/h   c. 0.032 cm/h

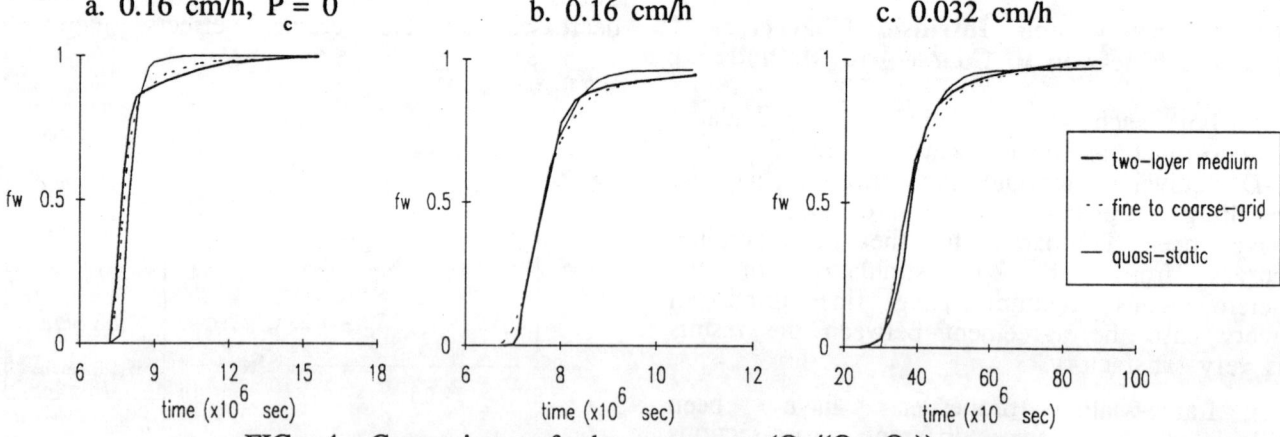

FIG. 4. Comparison of the watercut $(Q_w/(Q_w+Q_o))$ curves obtained from the simulation of the 1-D homogeneous model using the large-scale properties calculated by the two methods; without gravitational effects (0°).

agreement between the pseudo-functions calculated by the two methods; the concordance for case no. 2 is less satisfactory. For case no. 1 ($P_c=0$), the quasi-static assumption doesn't seem valid. With the same capillary pressure curves, the results obtained by the averaging method in the quasi-static case seem to be closer to the results of the fine to coarse-grid method as the flow rate decreases (Fig. 2.a).

The calculated pseudo-functions have been introduced in the corresponding homogeneous model and in each case the watercut curves have been compared (Fig. 4). In case no. 3, the results obtained by the averaging method are as satisfactory as the results of the fine to coarse-grid method from the industrial point of view. In case no. 2, the results of the two methods are comparable but there is a degradation of the quasi-static results in comparison with case no. 3. This case may correspond to the limit of the utilisation of the quasi-static assumption. For case no. 2, the results are still interesting for industrial use. For case no. 1, the fine to coarse-grid method gives a good representation of the phenomenon. On the contrary, the quasi-static assumption is no longer valid.

As the quasi-static calculation of the large-scale properties requires only solving a local problem (which is very simple in the case of the stratified systems under consideration here), it is important to know the validity domain of this approximation. We have shown above that the dimensionless analysis proposed in section 2 gives a good estimate of this validity domain.

5 CONCLUSION AND PERSPECTIVES

Several numerical simulations have given us more insight into the influence of the different parameters on the large-scale properties. Based on the large-scale volume averaging approach, new dimensionless parameters, allowing the classification of the displacements with respect to dynamic and gravity effects, have been found.

Two methods to determine pseudo-function have been studied: the fine to coarse-grid and the large-scale averaging method. The first one gives a correct solution to the problem but requires a numerical treatment of the problem over the large-scale with a fine grid. The second approach allows us to build pseudo-functions in a more general manner with computations carried out at a local scale only. However, the general closure scheme requires the resolution of several boundary value problems for different values of the input parameters (large-scale saturation, large-scale pressure gradients, etc...). The simplified version of the general closure scheme, i.e. the quasi-static case, is very attractive and simple; but further research is needed to determine more precisely under which conditions it gives a sufficiently good approximation of the required large-scale parameters.

ACKNOWLEDGEMENTS

Financial support from Elf Aquitaine Company is gratefully acknowledged.

REFERENCES

Coats, K.H., Nielsen, R.L., Terhune, M.H. and Weber, A.G., 1967, Simulation of three-dimensional, two-phase flow in oil and gas reservoirs. Society of Petroleum Engineering Journal, December, 377-388.

Corey, A.T. and Rathjens, C.H. , 1956, Effect of stratification on relative permeability. Petrol. Trans. AIME 207, 358-360.

Dake, L.P., Fundamentals of reservoir Engineering, 1978.

Hearn, C.L. , 1971, Simulation of stratified waterflooding by pseudo relative permeability curves. Journal of Petroleum Technology, July, 805-813.

Huppler, J.D. , 1970, Numerical investigation of the effects of core heterogeneities on waterflood relative permeabilities. SPE J. Forum, 10(4), 381-392.

Jacks, H.H, Smith O.J.E. and Mattax, C.C. , 1972, Modeling of three-dimensional reservoirs with two-dimensional reservoir simulator. The use of dynamic pseudo functions, 47th Fall meeting of the SPE, San Antonio, Oct. 8-11, 1972,. SPE paper n° 4071.

Killough, J.E. and Foster, H.P. Jr. , 1979, Reservoir simulation of the empire ABO field - the use of pseudos in a multilayered system. Soc. Pet. Eng. J. (Oct.), 279-291.

Kortekaas, T.F.M., 1983, Water/oil displacement characteristics in cross-bedded reservoir zones. SPE Conf., San Francisco, SPE 12112.

Kossack, C.A., Aasen, J.O. and Opdal, S.T. , 1989, Scaling-up laboratory relative permeabilities and rock heterogeneities with pseudo functions for field simulations SPE symposium, Houston, Texas. SPE 18436.

Kyte, J.R. and Berry, D.W. , 1975, New pseudos functions to control numerical dispersion.. Soc. Pet. Eng. J., (August), 269-275.

Martin, J.C. , 1968, Partial integration of equations of multiphase flow. Society of Petroleum Engineers Journal December, 370-380.

Quintard, M, and Whitaker, S. , 1988, Two-phase flow in heterogeneous porous media, the method of large-scale averaging. Transport in Porous Media 3, 357-413.

Quintard, M. and Whitaker, S., 1990a, Two-phase flow in heterogeneous porous media I: the influence of large spatial and temporal gradients, sous presse dans Transport in Porous Media.

Quintard, M. and Whitaker, S., 1990b, Two-phase flow in heterogeneous porous media II: numerical experiments for flow perpendicular to a stratified medium, sous presse dans Transport in Porous Media.

Simon, A.D. and Koederitz, L.F., 1982, An improved method for the determination of pseudo-relative permeability data for stratified systems SPE Fall Conf., New Orleans. SPE 10975.

Stiles, WM.E., 1949, Use of permeability distribution in water flood calculations. Petroleum Transactions, AIME January, 9-13.

Thomas, G.W. , 1983, An extension of pseudofunction concepts. SPE paper 12274.

Yokoyama, Y. and Lake L.W., 1981, The effects of capillary pressure on immiscible displacements in stratified porous media, 56th Annual Fall Meeting of the SPE, San Antonio, Oct. 5-7, SPE number 10109.

Numerical Schemes

2nd European Conference on the Mathematics of Oil Recovery
© D. Guérillot, O. Guillon (Editors) and Éditions Technip, Paris 1990, pp. 101-111
27 rue Ginoux, 75015 Paris

The Use of Second-Order Godunov-Type Methods for Simulating EOR Processes in Realistic Reservoir Models

K. Holing[1], J. Alvestad[1] and J. A. Trangenstein[2]

Abstract

Enhanced oil recovery using polymer or chemical flooding involves the displacement of oil by slugs with dynamic behavior critical to the response and the efficiency of the process. The numerical modelling of these processes is very difficult, since the dynamics of the slug flow lead to concentration fronts that are not self-sharpening. Conventional upstream-weighted finite differences lead to both substantial grid orientation errors and to smearing of sharp fronts.

This paper reports the formulation and performance of a second-order Godunov-type finite difference method for modelling of two-dimensional (2-D), three-component incompressible polymer floods. The scheme successfully handles realistic applications and significantly reduces both the spreading of fronts and grid orientation effects, when compared to the standard upstream-weighted finite difference schemes.

Polymer slug injection examples will be shown for both areal and cross sectional models involving various reservoir heterogeneities and realistic reservoir models. Also, a one-dimensional (1-D) example is given.

1   Introduction

The efficiency of waterflooding in moderate to high-viscosity oil reservoirs can be improved by adding polymers to the water. The polymer increases the viscosity of water, thereby reducing the mobility ratio at the front. There are several interesting physical aspects to this enhanced oil recovery technique. Because of the additional cost of the polymer, it is typically injected in slugs. While the downstream side of the slug is stable, the upstream side of the slug is significantly more viscous than the water that follows; as a result, the water injected after the slug can form viscous fingers due to instabilities at the upstream side of the slug. Therefore it is possible for water fingers to channel through the polymer bank and on to the wells. Another physical effect that complicates polymer flooding is the adsorption of polymer onto the rock matrix, thereby reducing the polymer available for mobility control.

Numerical simulation of polymer floods is used to design injection strategies that can increase the expected value of the oil production. Excessive numerical diffusion/dispersion will smear out the polymer slugs and destroy the true dynamics of the polymer slug, e.g., the forming of an oil bank ahead of the polymer slug. Additionaly the instability at the upstream side of the polymer slug is inherently a multi-dimensional effect, also requiring numerical simulation techniques with low levels of numerical diffusion/dispersion in order to capture the unstable interfaces.

On the mathematical side of the issue, polymer flooding has a very complicated chromotagraphic structure, Pope 1980[1]. For the model in this paper, there are two characteristic speeds. One is a Buckley-Leverett speed, with steep fronts (i.e., shocks in the limit of zero physical diffusion/dispersion) and continuous variations (rarefactions). The other characteristic speed is the particle velocity; this is associated with an indifferent concen-

---

(1) Statoil Postuttak, N-7004, Trondheim, Norway.
(2) Lawrence Livermore National Laboratory, PO Box 808, Livermore, California 94550-L316, USA.

tration front (due to a linear degeneracy). For arbitrary values of the concentration, there are one or two values of the water saturation at which these two speeds can coincide. Typically the characteristic directions will coincide as well, corresponding to an eigenvector deficiency.

Our purpose in this paper is to describe a numerical method that gives increased resolution of steep fronts and indifferent waves for polymer flooding, and that significantly reduces grid orientation effects in viscous fingers. We use a sequential approach to the flow equations, formulating an elliptic equation for the pressure and total fluid velocity, and a hyperbolic conservation law for the fluid components.

The outline of the paper is as follows. In Section 2, we describe the flow equations and physical assumptions. The sequential approach is described in Section 3, so that we can derive the characteristic speeds for later use in the numerical methods described in Section 4. The finite difference method for the pressure equation is described in Section 4.1, the Godunov method is outlined in Section 4.2, and the wells are discussed in Section 4.3. The numerical results are given in Section 5. Finally, the conclusions are given in Section 6.

2   Mathematical Model

We assume that the reservoir fluid flow can be described by three components, namely oil, water and polymer, forming at most two phases, namely liquid (oleic) and aqueous. We assume that the oil component forms the liquid phase and the water and the polymer components form the aqueous phase. Gravity and polymer adsorption are included. The fluid is assumed to be incompressible, and we ignore capillary pressure as well as other physical diffusive/dispersive forces.

The flow equations consist of mass conservation, Darcy's law, and the constraint that the fluid must fill the pore space. In order to describe these equations, we denote the liquid and aqueous phase densities by ρ_ℓ and ρ_a, the phase saturations by s_ℓ and s_a, c the polymer concentration, the ratio of volume of adsorbed polymer to pore volume by the function $a(c)$, the porosity by ϕ, and the phase velocities by $\vec{v}_\ell = (v_\ell^1, v_\ell^2)^T$ and $\vec{v}_a = (v_a^1, v_a^2)^T$. Further, the well flow rate is denoted by $q = q_a + q_\ell$, $q_a = f(s_a, c)q$ with $f(s_a, c)$ the gravity-free water fractional flow function, $f(s_a, c) = \lambda_a / \lambda_T$, $\lambda_T = \lambda_\ell + \lambda_a$ where λ_j, $j = \ell$, a are the phase mobilities. Then mass conservation of the components takes the form

$$\frac{\partial}{\partial t}\phi s_\ell + \nabla \cdot \vec{v}_\ell = q_\ell \qquad (1)$$

$$\frac{\partial}{\partial t}\phi s_a + \nabla \cdot \vec{v}_a = q_a \qquad (2)$$

$$\frac{\partial}{\partial t}\phi (s_a c + a) + \nabla \cdot (c\vec{v}_a) = c\, q_a \qquad (3)$$

and the pore volume constraint takes the simple form

$$s_\ell + s_a = 1. \qquad (4)$$

In order to write Darcy's law, we denote the pressure by p, the acceleration due to gravity by g, the depth by d and the rock permeability tensor by $\boldsymbol{K} = Diag\,(\kappa_1, \kappa_2)$. Then Darcy's law can be written

$$\vec{v}_j = -\boldsymbol{K} \cdot (\lambda_j \nabla p - \rho_j \lambda_j g\, \nabla d) \qquad (5)$$

for $j = \ell,\ a$.

Furthermore, we can sum the mass conservation equations for the oil and and water components (1)-(2) and use the pore volume constraint (4) to obtain the following equation for the total fluid velocity $\vec{v}_T = (v_T^1, v_T^2)^T = \vec{v}_a + \vec{v}_\ell$

$$\nabla \cdot \vec{v}_T = q. \qquad (6)$$

We will treat the equations in the forms (1)-(3), (5) and (6) in the remainder of the text.

3   Sequential Approach

The system of equations (1)-(3), (5)-(6) are of indeterminate type. In order to separate these equations into an elliptic pressure equation and a hyperbolic conservation law, we assume that the total fluid velocity is more slowly varying with time compared to the variation in time of the saturation s_a and concentration c. This allows us to take the following sequential approach to the solution of the flow equations (Trangenstein et. al. 1989[2]):

1. Solve (6) using (5), to get the pressure p and total fluid velocity $\vec{v}_T$ as functions of space $\vec{x}$ and time t, assuming that s_a and c are known functions of $\vec{x}$.

2. Rewrite the aqueous phase velocity in terms of the total fluid velocity $\vec{v}_T$

$$\vec{v}_a = f(s_a, c)\,(\vec{v}_T + \lambda_\ell[\rho_a - \rho_\ell]g\,\boldsymbol{K} \cdot \nabla d). \qquad (7)$$

3. Solve the conservation laws (2)-(3) with the aqueous phase velocity in the form (7), assuming the total fluid velocity to be a known function of $\vec{x}$. Note that the flux is a function of the conserved quantities s_a and $s_a c + a$, the total fluid velocity $\vec{v}_T$, and the gravitational term $[\rho_a - \rho_\ell]g\,\boldsymbol{K} \cdot \nabla d$.

The procedure 1.-3. is repeatedly used to increment the solution a timestep Δt forward in time; forming a time-stepping scheme of a non-standard IMPEC-type, that differs from the more standard approach by Spilette et. al. 1973[3].

It is easy to see that the pressure equation (5)-(6) is a second-order elliptic equation for p. On the other hand, the conservation law (2)-(3) and (7) is hyperbolic if and only if the matrix of derivatives of the normal flux with respect to the conserved quantities has real eigenvalues, for any choice of the normal direction. (If physical diffusive forces were present, then the conservation law would take the form of a convection-dominated diffusion equation.) Rewriting the conservation laws (2)-(3) in the following quasilinear form with respect to $w = (s, c)^T$ instead of $(s_a, s_a c + a)^T$, we obtain

$$\frac{\partial}{\partial t}(\phi w) + A_1(w)w_{x_1} + A_2(w)w_{x_2} = q_a \begin{bmatrix} 1 \\ c \end{bmatrix}. \quad (8)$$

The characteristic speeds for the j'th coordinate direction are the eigenvalues of the linearized coefficient matrix

$$A_j = \begin{bmatrix} \partial v_a^j/\partial s_a & \partial v_a^j/\partial c \\ 0 & v_a^j/(s_a + da/dc) \end{bmatrix}, \quad j = 1, 2. \quad (9)$$

Since this matrix is triangular, its eigenvalues are the diagonal entries, namely the Buckley-Leverett speed $\partial v_a^j/\partial s_a$, $j = 1, 2$ and the particle velocity speed $v_a^j/(s_a + da/dc)$, $j = 1, 2$.

The Buckley-Leverett eigenvalue can be identified as the slope of the aqueous phase velocity curve, taken as a function of s_a for fixed c. On the other hand, the particle velocity eigenvalue can be identified with the slope of a chord to the aqueous phase velocity curve from the point of zero velocity and $s_a = -da/dc$. A simple example for typical fractional flow curves shows that for each value of c there is one value of s_a (possibly two, with gravity) where the two characteristic speeds coincide. It is also easy to show that $\partial v_a^j/\partial c \neq 0, j = 1, 2$ at these points, and as a result the coincident characteristic speeds occur at an eigenvector deficiency.

In spite of these difficulties, Pope 1980[1] has described the solution of Riemann problems for this model. Johansen et. al. 1988[4] have solved the Riemann problem for generalizations to an arbitrary number of components miscible in water. This solution involves a combination of Buckley-Leverett shocks and rarefactions at constant polymer concentration, with linearly degenerate concentration fronts in case of zero polymer adsorption. (With non-zero adsorption the concentration fronts are nearly linearly degenerate). These features of the characteristics make the numerical solution of the problem difficult for several reasons. First of all, since there is so little variation in the particle-velocity speeds, the concentration fronts are not self-sharpening; numerical diffusion/dispersion in the finite difference scheme will

lead to a spreading of the concentration fronts. Secondly, in multidimensional problems, these concentration fronts will be Rayleigh-Taylor unstable, leading to viscous fingering for fronts with adverse mobility ratios (such as at the trailing edge of polymer slugs). Finally, methods that use the local characteristic structure of the flow must take care to avoid difficulties with loss of characteristic directions.

4   Numerical Method

As mentioned in the previous section, our sequential approach to solving the flow equations requires that we solve the pressure equation to obtain the pressure and total fluid velocity fields. For timesteps after the first, we use the assumption that the total fluid velocity varies slowly in time and is weakly dependent on the saturation and concentration. The pressure equation is solved only when the change in the total velocity field (measured from the time of the previous pressure solution) exceeds a given tolerance. The change is estimated using the partial derivative of Darcy's law (a posteriori estimate) to form an increment equation, introduced by Christie 1988[5]. The total fluid velocity is treated as constant in time between consequtive updates. The conservation laws (2)-(3) are advanced explicitly in time using a finite volume based method with a timestep limitation given by a Courant-Friedrich-Levy (CFL) type stability criterion. In these solutions the total fluid velocity is used in (7) for the determination of the convective fluxes of water and polymer. We will describe additional details of these two numerical methods in the following two subsections.

In 1-D calculations, we use conservative tracking of the indifferent concentration front to eliminate numerical diffusion. We track only this front because the other fronts, being self-sharpening, are well resolved by the capturing scheme. The reader is referred to Holing 1990[6] and Chern et. al. 1987[7] for details.

4.1   Pressure Equation Scheme

The pressure equation is discretized using a 9-point finite difference scheme (the stencil would be somewhat larger if the total permeability tensor were not diagonal). Following Bell et. al. 1988[8], we integrate (6) over a cell B_{ij} and apply the divergence theorem to obtain

$$\int\int_{B_{ij}} q\,dx = \int_{\partial B_{ij}} \vec{v}_T \cdot \vec{n}\,ds. \quad (10)$$

Here, $\vec{n}$ is the unit length outer normal to the relevant side of the cell, and s is the arclength along the side.

The integral along an edge, $E_{i+\frac{1}{2},j}$ say, is approximated by taking half of Simpson's rule over twice the distance

$$\int_{x_{i+\frac{1}{2},j-\frac{1}{2}}}^{x_{i+\frac{1}{2},j+\frac{1}{2}}} v_T^1 \, dx_2 \doteq$$

$$\tag{11}$$

$$\frac{1}{6}\Delta x_2 \left[(v_T^1)_{i+\frac{1}{2},j-1} + 4(v_T^1)_{i+\frac{1}{2},j} + (v_T^1)_{i+\frac{1}{2},j+1}\right] .$$

This discrete divergence operator is reflected in the total fluid velocity used in the discrete form of the conservation equation (2)-(3) which needs to be divergence free (away from the wells). Darcy's law (5) summed over the two fluid phases is discretized as follows for the x_1-direction (a similar formula can be written for the x_2-direction)

$$(v_T^1)_{i+\frac{1}{2},j} = -\left[\frac{p_{i+1,j} - p_{ij}}{\Delta x_1} - \right.$$

$$\tag{12}$$

$$\left. \frac{d_{i+1,j} - d_{ij}}{\Delta x_1} g \left(\frac{\lambda_\ell \rho_\ell + \lambda_a \rho_a}{\lambda_\ell + \lambda_a}\right)_{i+\frac{1}{2},j}\right] (\kappa_1 \lambda_T)_{i+\frac{1}{2},j}.$$

Spatially we use harmonic averaging of the mobilities, and arithmetic averaging of the densities for the gravity term to obtain the edge values. Substituting the equations (11) and (12) into (10), we obtain a linear system of equations for p which is solved using standard direct or iterative methods. Subsequently the total fluid velocity $\vec{v}_T$ is calcluated using (12). The advantage of this formulation of the pressure equation is that it has been numerically demonstrated in Bell et. al. 1988[8] to lead to an especially rotationally-invariant discretization, and thus substantially reduced grid-orientation problems.

4.2 Conservation Law Scheme

The numerical method for the solution of the hyperbolic conservation law is much more complex to describe in detail. We will summarize the most important features, and refer the reader to Colella 1990[9], Bell et. al. 1989[10], and Allen et. al. 1988[11] for details.

The numerical method is expressed in terms of the cell averages of the saturation s_a and concentration c by integrating (2)-(3) over the cell B_{ij} in space and the timestep $\Delta t = t^{n+1} - t^n$ in time, and then applying the divergence theorem

$$\int_{t^n}^{t^n+\Delta t} \int\int_{B_{ij}} q_a \begin{bmatrix} 1 \\ c \end{bmatrix} dx_1 \, dx_2 \, dt$$

$$= \int\int_{B_{ij}} \begin{bmatrix} s_a \\ s_a c + a \end{bmatrix} \phi \, \big|_{t^n}^{t^n+\Delta t} \, dx_1 \, dx_2 \tag{13}$$

$$+ \int_{t^n}^{t^n+\Delta t} \int_{\partial B_{ij}} \vec{v}_a \cdot \vec{n} \begin{bmatrix} 1 \\ c \end{bmatrix} ds \, dt.$$

This is written in discrete form as follows

$$\left(q_a \begin{bmatrix} 1 \\ c \end{bmatrix}\right)_{ij}^{n+1} \Delta x_1 \Delta x_2 \Delta t =$$

$$\left(\begin{bmatrix} s_a \\ s_a c + a \end{bmatrix}_{ij}^{n+1} - \begin{bmatrix} s_a \\ s_a c + a \end{bmatrix}_{ij}^{n}\right) \phi_{ij} \, \Delta x_1 \Delta x_2$$

$$+ \left(f_{i+\frac{1}{2},j} - f_{i-\frac{1}{2},j}\right) \Delta x_2 \Delta t + \left(f_{i,j+\frac{1}{2}} - f_{i,j-\frac{1}{2}}\right) \Delta x_1 \Delta t.$$

$$\tag{14}$$

Given values for the cell averages of the conserved quantities, all that remains to determine is numerical values for the fluxes, e.g.,

$$f_{i+\frac{1}{2},j} = \frac{1}{\Delta t} \frac{1}{\Delta x_2} \int_{t^n}^{t^n+\Delta t} \int_{x_{i+\frac{1}{2},j-\frac{1}{2}}}^{x_{i+\frac{1}{2},j+\frac{1}{2}}} v_a^1 \begin{bmatrix} 1 \\ c \end{bmatrix} dx_2 \, dt$$

$$\tag{15}$$

We will discuss the treatment of source terms due to wells in Section 4.3.

To compute the fluxes, we use the Colella extension of the corner-transport upwind scheme, Colella 1990[9]. This unsplit scheme can be recast into a predictor-corrector formalism. In the predictor, the first step is to construct monotonized slopes in the characteristic variables in each direction. (The characteristic tracing determines left and right states at the midpoints of all cell edges necessary to determine the numerical fluxes). Using 1-D characteristic information only, approximate solution states at centers of cell edges and the $t^{n+\frac{1}{2}}$-time level ((15) is approximated by the midpoint rule in space and time) are traced. This step is followed by approximations to the transverse fluxes at the stationary state in the solution to the Riemann problem defined by these states. In the corrector step, the tracing is completed by adding 2-D tracing terms giving left and right states used for the conservative flux calculation, and then an approximate flux for the corresponding Riemann problem is formed.

One of the advantages of tracing the variables used to compute the flux, $(s_a, c)^T$, rather than the conserved quantities, $(s_a, s_a c + a)^T$, is that it is easy to check that the tracing produces physical values. Near eigenvector deficiencies or strong discontinuities, it is possible that one or both of s_a and c will be either less than 0 or greater than 1; if so, then we use the cell-center value rather than the traced value. Since the wells are treated separately (see Section 4.3), the tracing is omitted all-together in these cells. Generally, when all characteristic speeds have the same sign, the flux can easily be determined as the flux at whichever of the two traced states that is the upstream state. However, for problems involving gravity and counter-current flow, we use the characteristic directions at the upstream state to construct an approximate path in (s_a, c)-space between the left and right traced states. Along each leg of this path, we construct a cubic interpolant to the characteristic speed, in order to estimate its zeros. This information allows us to calculate an approximate Engquist-Osher flux as a sum of fluxes at left or right states and points where the characteristic speeds are zero. We refer to Bell et. al. 1989[10] and Engquist et. al. 1980[12] for further details.

Since the characteristic directions can coallesce, it is necessary to take some care in constructing the numerical flux. If an eigenvector deficiency is detected at a cell edge for a Riemann problem with transonic flow (i.e., not all characteristic speeds on either side of the cell edge have the same sign and are nonzero), we use a more diffusive numerical flux, such as the Rusanov flux, Rusanov 1961[13].

The scheme requires the solution of 4 Riemann problems (i.e., 2 transverse flux calculations and 2 conservative flux calculations) for each cell at each timestep, which can be very expensive. Care has therefore been taken to efficiently implement the flux calculations. Computational experience shows that the transverse fluxes can be calculated with less accuracy than required for the conservative fluxes. (First-order evaluation of the 2-D transverse tracing terms is sufficient to obtain second-order in the combined tracing). For a more detailed discussion of the implementation, the reader is referred to Holing 1990[6].

4.3 Treatment of Wells

The wells have a great impact on the numerical solution, so special care has been taken in their treatment. Wells may limit the size of the stable timestep due to the high velocities near wells. Experience has shown that some sort of implicit treatment of wells is necessary in order to obtain good results. Explicit treatment of wells leads to unpractical small timesteps for stable solutions.

Our approach to the well modelling has been to develop a model sufficiently sophisticated to meet the main objective for this work; namely to demonstrate the potential of second-order Godunov-type methods for realistic 2-D polymer flooding. Only the well cell variables are treated implicitly, the information from the neighbouring cells is used explicitly. For each edge of a well cell, we calculate the aqueous phase flux as first-order

upstream flux. This flux calculation is implicit if the well cell is upstream for the aqueous phase, otherwise the flux is explicit.

5   Numerical results

In this section we will present numerical results from three different problems in order to illustrate the properties of the second-order Godunov-type method (SOGM). The results are compared to an analytic solution for one of the examples where this is known. In addition the presented method is compared with a method where the only difference with our new method is that single point upstream-weighted finite differences (UPSTM) are used to solve the system of hyperbolic conservation laws. UPSTM is the method that is most commonly used for these problems in the industry today.

For all the examples given we have used $\lambda_a = s_a^2/\mu_a(c)$ and $\lambda_\ell = (1 - s_a)^2/\mu_\ell$ as phase mobilities where $\mu_a(c) = \mu_a^0 (1 + a_1 c + a_2 c^2 + a_3 c^3)$ with μ_ℓ, μ_a^0, a_1, a_2, and a_3 as given constants. These phase mobilities are not representative for a particular reservoir, but their functional form is commonly used for test purposes as they reveal the main structure in the mathematical problem. They also give a gravity-free water fractional flow function with a shape which is fairly typical for realistic systems. The adsorption is set to zero in the examples shown, that is $a(c) \equiv 0$.

Generally we will discuss the solutions for the aqueous phase saturation and the polymer concentration in the aqueous phase. Additionally we will discuss the production characteristics of the aqueous phase and the polymer concentration. A uniform spatial grid (the cell sizes in the x_1- and x_2-direction may be different) is used for all the examples shown here. For all the cases studied the time-stepping of the numerical solution for the hyperbolic conservation law has been controlled automatically using a discrete stability limit estimator. Normally SOGM has been run successfully with timestep sizes restricted to a factor of 0.8-0.9 of the estimated stability limit. UPSTM generally has a lower stability limit than SOGM, and for 2-D problems with gravity UPSTM has usually been run with half the timestep sizes compared to SOGM.

For the 2-D problems we will generally show numerical solutions where the pressure equation is solved with the same frequency as the hyperbolic conservation law. However, preliminary results indicate that for most of the problems substantial savings in computational time can be made by solving the pressure equation when the total fluid velocity field has changed more than a specific limit. A conservative estimate for medium size problems is at least a halfing of the computational costs.

5.1 Example 1 (1-D example)

This example is equivalent to a no gravity Riemann problem with left and right states being $w^L = (1, .1)^T$ and $w^R = (.3, .9)^T$, respectively. Being a 1-D problem the total fluid velocity is constant in space and the boundary conditions are chosen such that the total fluid velocity is time independent as well; $1m/s$, say. Further the porosity is set equal to 1. The constants for the viscosity funtion are $\mu_a^0 = .5$, $\mu_\ell = .35$, $a_1 = 2$, and $a_2 = a_3 \equiv 0$. Hence considering the equations (2)-(3) and (7), we see that these equations form a system of hyperbolic conservation laws which may be solved directly using a 1-D version of the scheme introduced in Section 4.2.

Let $\Omega = \Omega_{1D} \times \Omega_t$ be the computational domain where $\Omega_{1D} = \{x \ (m) \in R; x \in (0,1)\}$ and $\Omega_t = \{t \ (s) \in R, t > 0\}$. The initial conditions are $w = w^R$ for all $x \in \Omega_{1D}$ and the boundary conditions at the left boundary, $x = 0$, are $w = w^L$. In the example a grid of 60 cells are used and the results at $t = .5s$ are shown in Fig. 1. The exact solution is found using the Riemann solver developed by Johansen et. al. 1989[14]. It consists from the left of a Buckley-Leverett (BL) rarefaction wave for s_a connected to a constant state followed by a contact discontinuity. This is followed by a BL wave consisting of a rarefaction wave and a BL shock. The figure shows that UPSTM has problems in resolving both the constant s_a-state and the following contact discontinuity, while SOGM resolves both the constant state and the contact discontinuity. However, the latter being an indifferent wave, is less well resolved than the self-sharpening BL shock even for SOGM. In addition, there is an overshoot in the solution at the transition between the BL-rarefaction and the constant state.

The numerical solution of this problem is vastly improved by tracking the contact discontinuity, using our HYBRID method due to Chern et. al., 1987[7]. The results of HYBRID in Fig. 2 show that it tracks the contact discontinuity perfectly. It also eliminates the overshoot at the transition between the slow BL rarefraction and the constant state. The reason for this may be explained through the fact that the front-tracking solution picks up the correct path in phase space from the initial conditions. The front capturing method uses numerical diffusion to resolve the different waves. Close to the start up time an entropy violating path in phase space is "seen" by the method but is subsequently left in favour of the entropy satisfying path. More details on this and on the formulation and implementation of the hybrid method can be found in Holing 1990[6].

Grid refinement studies have been performed to numerically demonstrate the convergence of SOGM to known analytical solutions. In order to have a non trivial flux calculation, cases including gravity effects have been studied. For such cases, we have estimated the difference in the computational expense between SOGM and UPSTM for 1-D problems to be a factor of approximately 4 pr. grid cell and timestep. The accuracy for the two methods as a function of grid size for example 1 is given in Fig. 3. The discrete L^1-norm $\|e\|_1$ used is $\max(\|e^{s_a}\|_1, \|e^c\|_1)$ where

$$\|e^{s_a}\|_1 = \Delta x \sum_{i=1}^{m} |\ s_{a_i}^{(n)} - s_a\ (x_i, t^n)\ | \qquad (16)$$

$$\|e^c\|_1 = \Delta x \sum_{i=1}^{m} |\ c_i^{(n)} - c\ (x_i, t^n)\ | \qquad (17)$$

where m is the number of cells used and $t = t^n$ is the given time. From these two above results we have estimated SOGM to be approximately 35 times more computational efficient than UPSTM for 1-D problems for 1% accuracy.

5.2 Example 2 (Areal model)

This example is a no gravity 2-D problem where we study a polymer-slug injection in a quarter of a repeated five-spot pattern. The porosity is constant and equal to 1, the absolute permeability in both directions is constant and equal to 1 Darcy. The viscosity parameters are $\mu_a^0 = 2$, $\mu_\ell = 4$, $a_1 = 15.7$, $a_2 = -32.4$, and $a_3 = 63.4$. The initial conditions are a completely oil saturated reservoir, $(s_a(t = 0) \equiv 0)$. The production/injection is at constant rates such that pure water is injected for the first 0.55 PV (pore volume) followed by polymer injection ($c = .8$) for 0.27 PV, followed by pure water injection until 5 PV have been injected. (In time units 2 years correspond to 1 PV injected). The mobility ratio at the rear of the polymer slug is 25, which means that the rear front is highly unstable. Hence severe grid orientation effects may be expected for the upstream-weighted method.

We will here show results from four different grids, coarse and fine diagonal and parallel grids. The coarse diagonal grid is 12×12 and the coarse parallel grid is 17×17. The cell sizes Δx_1 and Δx_2 are the same in both grids. The fine grids use 16 times as many cells (a refinement of 4 in each direction) as the corresponding coarse grids. First, we consider the coarse grid solutions for the polymer concentration at 1.23 PV (900 days) injected as shown in Fig. 4. The injector is located at the lower left corner of the figure and the producer at the upper right corner. Symmetry causes both the boundaries and the diagonal of the figure to be no flow boundaries, and the diagonal grid results are shown in the lower triangle while the parallel grid results are shown in the upper triangle. This figure shows that 1) UPSTM is much

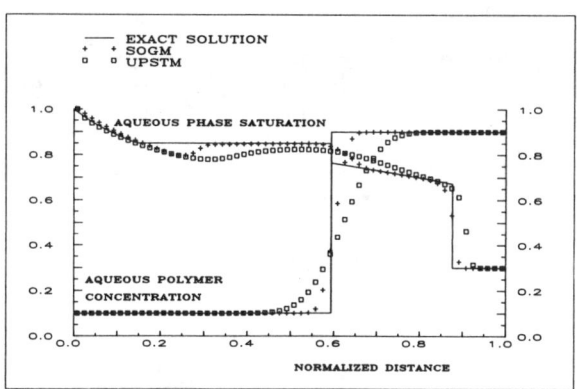

FIG. 1. Aqueous phase saturation and polymer concentration vs. cell no at $t = .5s$ for example no. 1, with 60 cells.

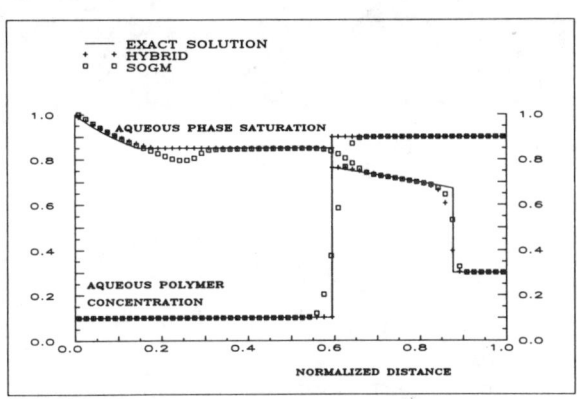

FIG. 2. Aqueous phase saturation and polymer concentration vs. cell no at $t = .5s$ for example no. 1, with 60 cells.

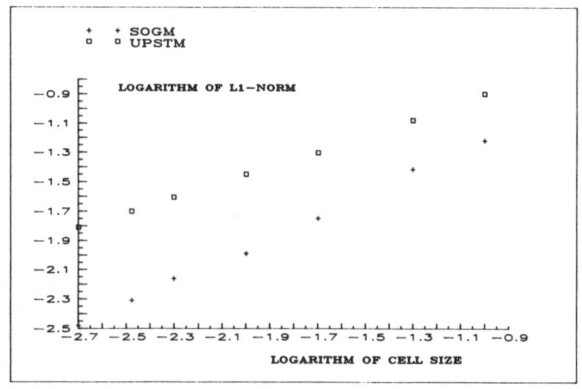

FIG. 3. Logarithm of the L^1-norm of the numerical solution vs. the logarithm of the cell size for the problem in example no. 1.

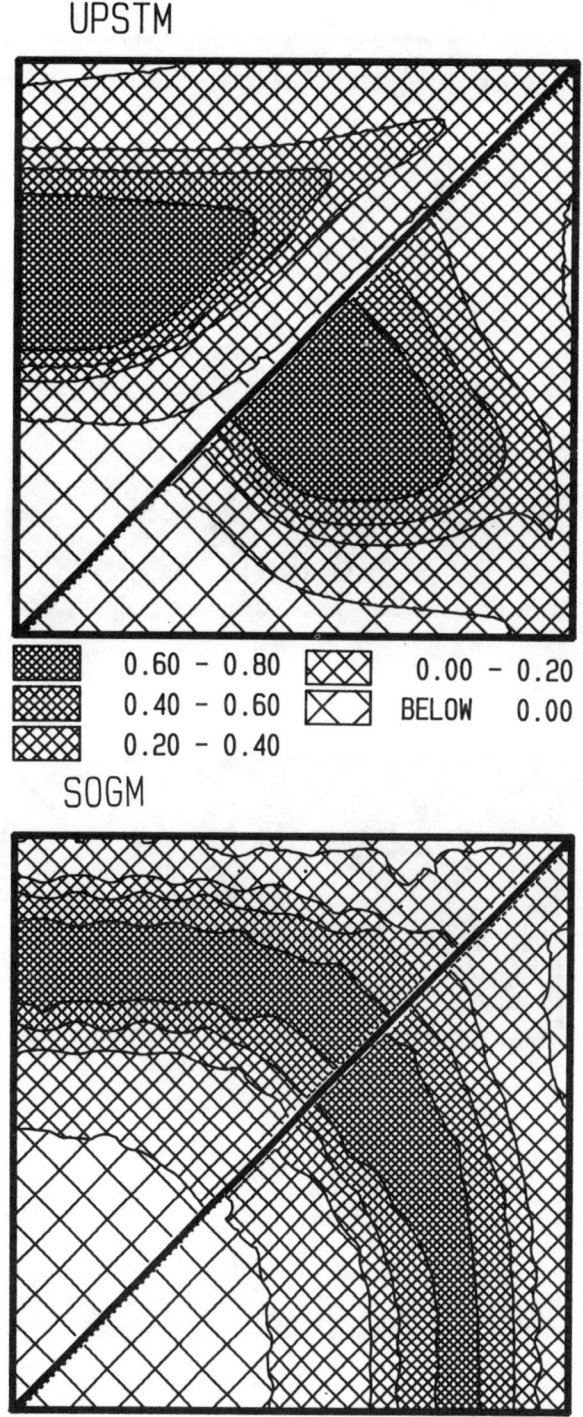

FIG. 4. Polymer concentration distribution for the coarse grid solution, parallel and diagonal symmetry element merged, example no. 2.

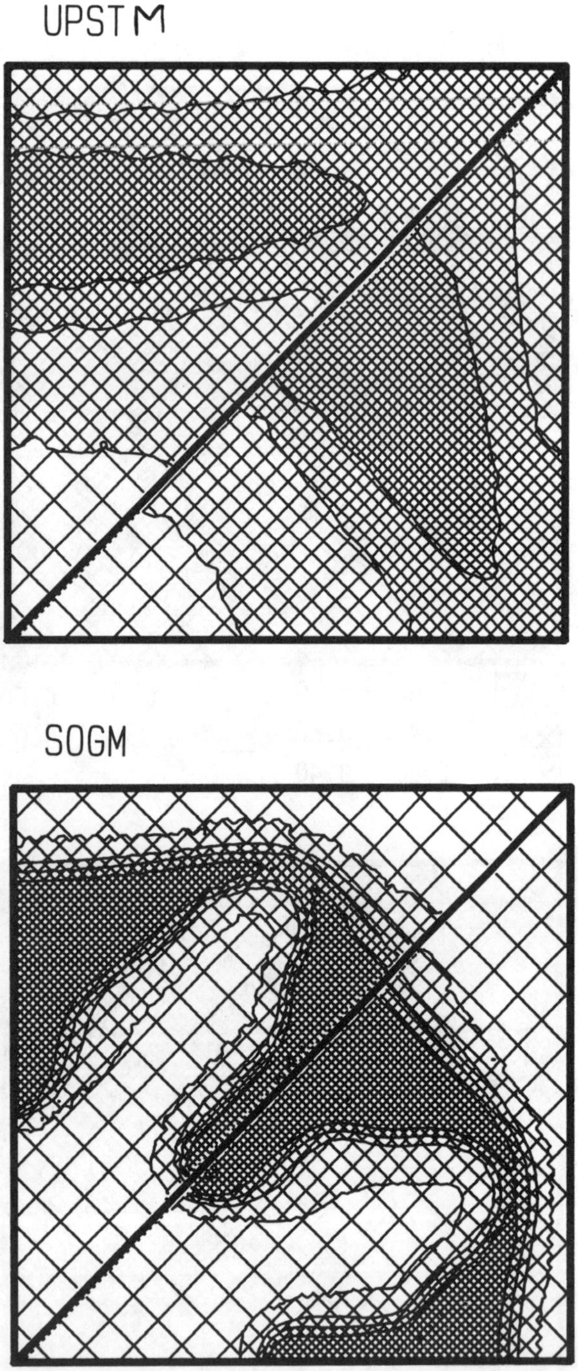

UPST M

SOGM

SOGM

▓	0.60 – 0.80
▨	0.40 – 0.60
▨	0.20 – 0.40
▨	0.00 – 0.20
▨	BELOW 0.00

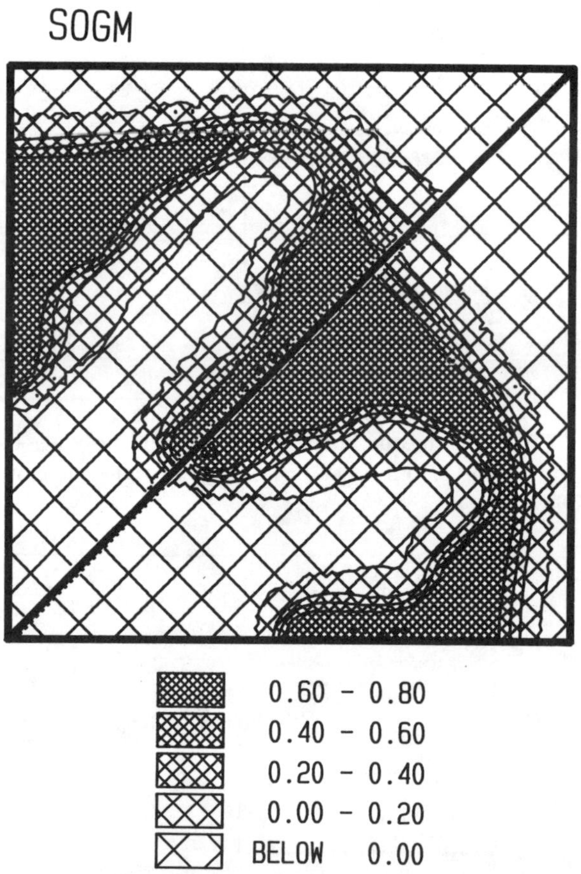

FIG. 6. Polymer concentration distribution for the fine grid solution, parallel and diagonal symmetry element merged, example no. 2 (random permeabilty distribution)

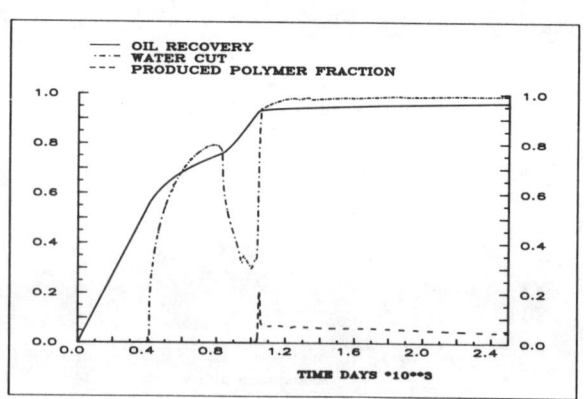

FIG. 5. Polymer concentration distribution for the fine grid solution, parallel and diagonal symmetry element merged, example no. 2.

FIG. 7. Fractional recovery, producing water cut and polymer fractions, example no. 2, fine homogeneous diagonal grid.

more diffusive than SOGM and 2) UPSTM suffers from strong grid orientation effects while SOGM shows only minor grid orientation effects. Secondly, we consider the fine grid solution for the polymer concentration as shown in Fig. 5. This figure shows that UPSTM is not able to reduce or control its grid orientation during grid refinement leading to two completely different solutions for the diagonal and the parallel grids, respectively. SOGM, however, produces nearly the same solution for both the diagonal and the parallel grids. This solution is characterised by two "fingers" of water channeling through the polymer bank one on each side of the diagonal (symmetry line). Since the rear front is miscible and unstable, viscous fingering will occur, and these are not modelled properly in this numerical solution. There is, however, a possibility that the two fingers in the solution are due to the directionality of the numerical dispersion in SOGM. That is, they may be an artifact of the method rather than a physical phenonema.

In order to test the validity of the SOGM solution the following case was simulated: 1) The absolute permeability distribution in the quarter five-spot is assumed to be stochastic with a mean of 1 Darcy and a standard deviation of 0.05 Darcy. The actual cell permeability is assigned by drawing from this distribution and subsequently smoothing the distribution. (The objective of the smoothing is to introduce some short range correlations into the permeability profile.) The injection scheme is as before, but at the end of the polymer injection a square region of 4×4 cells from the water injection corner were assigned zero polymer concentration before the pure water injection started. (This means that a square region close to the injection well was artificilly replaced with pure water at the rear end of the polymer slug. For the diagonal grid this would enhance fingering of water through the central part of the quarter five-spot). The results of this case are shown in Fig. 6 and they confirm the previous results. The dynamics of the viscous fingers should be controlled by the random variations in the permeability field, and the shape of the stable fronts preceeding the slug. Hence this suggests that two main viscous fingers will channel through the polymer bank one on each side of the symmetry line between the producer and the injector.

The symmetry for the quarter five-spot is essentially broken by the variable permeability field, but the effect is so small that we choose to present the results in the same way as for the symmetric case. From Figs. 6 and 7 we can see that SOGM predicts that the water channels through the polymer bank and therefore the polymer is produced slowly as water is flowing mainly in the already established flow channel.

5.3   Example 3 (Cross sectional model)

This final example is a polymer slug injection case in a heterogeneous and anisotropic vertical cross sectional model. The model is rectangular in shape with length $800m$, width $500m$ and thickness $100m$. It has a dip of 12° with the horizontal direction and is divided into six geological layers, each with constant thickness (Δz) and petrophysical properties as shown in Table 1.

The liquid and aqueous phase viscosities are as in the previous quarter five-spot example and the phase densities are $\rho_\ell = 780\ kg/m^3$ and $\rho_a = 1014\ kg/m^3$ for the liquid and aqueous phases, respectively. Again, the reservoir is initially completely filled with oil. There is one injection well in the model located at the lower end of the cross section and perforated in geological layer no. 4. The production well is located $712.5m$ updip and is also perforated in geological layer no. 4. The production injection rate is held constant at $3000\ m^3/day$. First pure water is injected for 1345 days (approximately the time for water break-through), then a polymer slug ($c = .8$) is injected for 1460 days, followed by pure water injection until 7500 days. Due to the unstable rear front (mobility ratio of 25) water will channel through the polymer bank in the high permeability zone in the upper part of the model.

Here we present the results of a fairly fine grid simulation with dimensions 96 × 40. From Fig. 8 we can see that SOGM predicts a much steeper decline in water cut due to the polymer injection than UPSTM. The produced polymer fraction, however, has a smaller peak and is otherwise much less than what UPSTM predicts. This is due to the fact that SOGM predicts a much more profound channeling of water through the polymer bank than UPSTM. SOGM further predicts that more polymer is bypassed due to channeling of water through the high permeability zone at the top. When the water has channelled through the polymer bank most of the water flows through this channel and only a small fraction of polymer is produced. This effect is also illustrated in the Fig. 9, which shows how the water channels through the polymer bank. From this figure we observe that the main difference between UPSTM and SOGM is the sharpness of the fronts between the polymer bank and the surroundings. UPSTM also predicts a much more significant amount of polymer around the producer than SOGM. No significant grid orientation effects are observed, as is expected since the heterogeneities dominate the fluid flow pattern. The difference in the cumulative oil production is not as large as the differences in the internal structure of the solution.

SOGM
UPSTM
WATER CUT
FRACTIONAL OIL RECOVERY
PRODUCED POLYMER FRACTION
TIME DAYS *10**3
0.0
0.2
0.4
0.6
0.8
1.0
1.0
2.0
3.0
4.0
5.0
6.0
7.0

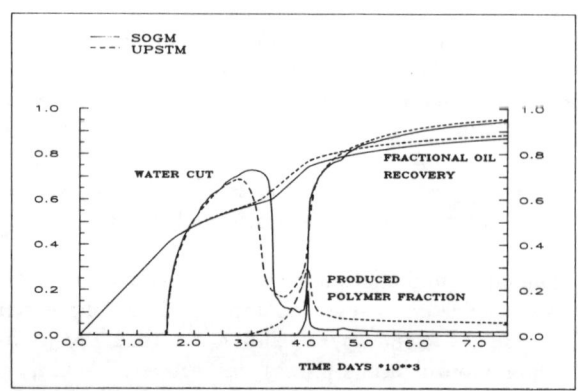

FIG. 8. Fractional recovery, producing water cut and polymer fractions for example no. 3.

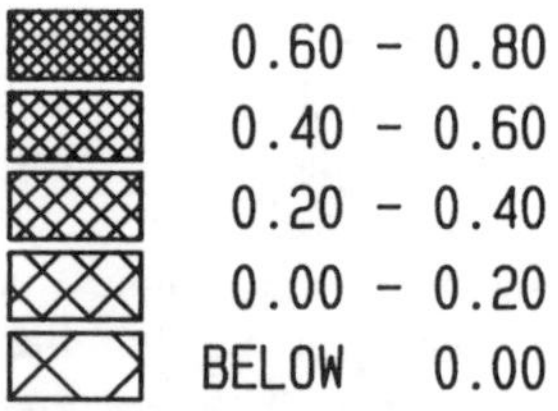

0.60 - 0.80
0.40 - 0.60
0.20 - 0.40
0.00 - 0.20
BELOW 0.00

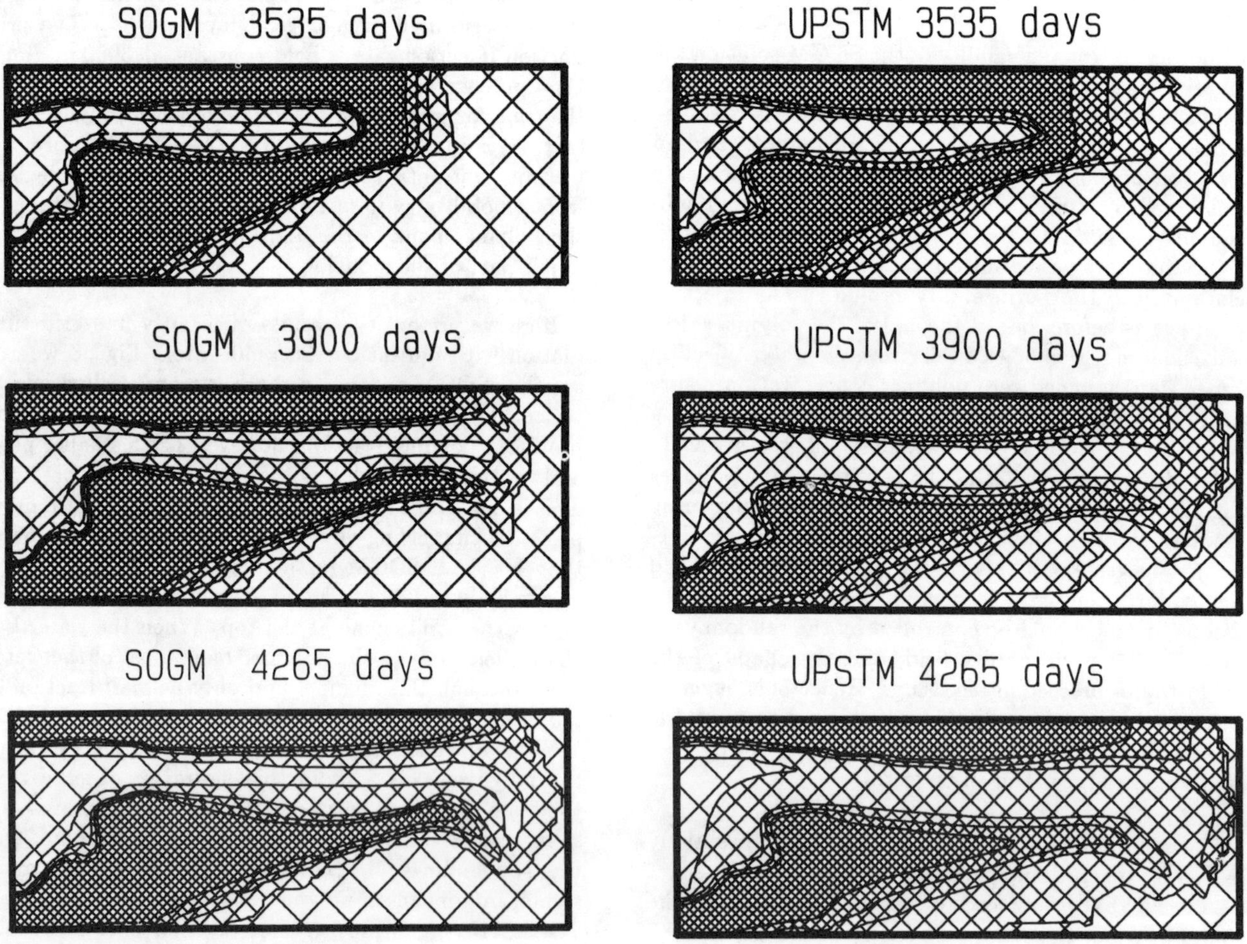

SOGM 3535 days
UPSTM 3535 days
SOGM 3900 days
UPSTM 3900 days
SOGM 4265 days
UPSTM 4265 days

FIG. 9. Polymer concentration distribution at three different times for example no. 3.

6 Conclusions

The second-order Godunov-type method presented in this paper significanly reduces both the smearing of sharp fronts and grid orientation effects when compared to standard upstream-weighted schemes. The method handles successfully both gravity effects and heterogeneity in permeability and porosity. The method has shown to be computationally significantly more efficient than standard upstream-weigthed schemes for the cases studied. However, this result needs more verification in order to extend its domain of validity.

The method seems to be promising in terms of 1) extension to more realistic reservoir geometries and rock heterogeneities, 2) development and applications to other types of chemical flooding and 3) development of more efficient schemes involving adaptive and timedependent grids. However, further research and development are needed in order to extend the applicability of the method in order to obtain a robust and competetive scheme, compared to other industry methods.

Acknowledgements

The authors would like to thank STATOIL for permission to publish this work.

References

1. Pope, G.A. (1980) *The Application of Fractional Flow Theory to Enhanced Oil Recovery* Soc. Pet. Eng. J. **20**, 191-205.

2. Trangenstein, J.A. and Bell, J.B. (1989) *Mathematical Structure of the black-oil model for petroleum reservoir simulation* SIAM J. Appl. Math. **49**, 749-783.

3. Spilette, A. G., Hillestad, J. G. and Stone, H.L. (1973) *A high-stability sequential solution approach to reservoir simulation* SPE 4542, 48th Annual Fall Meeting, Las Vegas.

4. Johansen, T. and Winther, R. (1988) *The Solution of the Riemann Problem for a Hyperbolic System of Conservation Laws Modelling Polymer Flooding* SIAM J. Math. Anal. **19**, 541-566.

5. Christie, M.A. (1988) *Application of High Resolution Simulation To Modelling Fluid Instabilities* "Mathematics in Oil Production, The Institute of Mathematics & Its Applications Conference Series", edited by Sir Sam Edwards and P.R. King, New Series Number **18**, pp. 269-284.

6. Holing, K. (1990) *Second-Order Godunov-Type Methods For One- and Two-Dimensional Polymer Flooding*, Ph.D Thesis, The Norwegian Institute of Technology, ISBN 82–7119–177–2.

7. Chern, I. and Colella, P. (1987) *A Conservative Front Tracking Method For Hyperbolic Conservation Laws* Lawrence Livermore National Laboratory, UCRL-97200 preprint (to appear in the J. Comp. Phys.)

8. Bell, J.B., Dawson, C.N. and Shubin, G.R. (1988) *An Unsplit Higher-Order Godunov Method for Scalar Conservation Laws in Multiple Dimensions* J. Comp. Phys. **74**, 1-24

9. Colella, P. (1990) *Multidimensional Upwind Methods for Hyperbolic Concentration Laws* J. Comp. Phys. **87**, 171-200.

10. Bell, J.B., Colella, P. and Trangenstein, J.A. (1989) *Higher-Order Godunov Methods for General Systems of Hyberbolic Conservations Laws* J. Comp. Phys. **82**, 362-397.

11. Allen III, M.B., Behie, G.A. and Trangenstein J.A. (1988) "Multi-Phase Flow in Porous Media: Mechanics, Mathematics and Numerics", Lecture Notes in Engineering **34**, Springer Verlag.

12. Engquist, B. and Osher, S. (1980) *Stable and Entropy Satisfying Approximations for Transonic Flow Calculations* Math. Comp. **34**, 45-75.

13. Rusanov, V.V. (1961) *Calculation of Interaction of Non-Steady Shock Waves with Obstacles* Zhur. Vychislitel'noi Mathematicheskoi Fiziki **1**, 267-279.

14. Johansen, T., Tveito A, and Winther, R. (1989) *A Riemann Solver for a Two-Phase Multicomponent Process* SIAM J. Sci. Stat. Comp. **10**, 846-879.

TABLE 1.

Geometrical and Geological Properties for the Cross Sectional Model

Layer	Δz (m)	κ_1 (D)	κ_2 (D)	ϕ
1	25.	2.6	1.1	0.33
2	15.	2.8	1.15	0.33
3	5.	1.34	0.044	0.29
4	25.	0.43	0.18	0.29
5	20.	0.18	0.08	0.27
6	10.	0.012	0.012	0.10

2nd European Conference on the Mathematics of Oil Recovery
© D. Guérillot, O. Guillon (Editors) and Éditions Technip, Paris 1990, pp. 113-120
27 rue Ginoux, 75015 Paris

Modelling Flow through Heterogeneous Porous Media with Boundary Integrals Using Higher-Order Surface Singularities

D. W. Wong[1], J. S. Archer[1] and J. M. R. Graham[2]

ABSTRACT

Numerical modelling of heterogeneous porous media has received attention at a variety of scales from pore dimensions to those of hundreds of meters. Most of the approaches to predict the behaviour of flow have focussed on enhancements to established finite difference schemes (e.g., local grid refinement, multi-grid methods). Although finite difference schemes have many advantages, in fine grids they are difficult to apply properly to curved surfaces with Neumann boundary conditions, which occur at the boundary between media of different permeability and they appear very difficult to use with multiple heterogeneities. In such circumstances boundary conditions can often be expressed using boundary integral equations, and solutions by powerful boundary integral equation methods can be applied. It may be advantageous to use these methods at a variety of scales of interest because they reduce the order of the partial differential equations of fluid motion by one, they require discretization only where the boundary conditions are expressed and are generally stable numerically because of the "smoothing" nature of integral equations in general, and Fredholm equations of the second kind in particular.

We use boundary integral equation methods with piecewise continuous surface singularities -- called "panel methods" in the aerodynamics literature -- to represent the boundary between local heterogeneous zones. The advantage of this approach is that the integral equations used can be solved analytically and hence numerical integration is never required. We adapt and expand these methods to treat permeability magnitudes ranging from zero to infinity inclusive. We introduce the use of higher order representations as an enhancement to the base flat panel method. It is also shown how the computational effort can be substantially reduced when considering numerous heterogeneous zones with the use of a technique used in electrostatics called multipole expansions. We conclude that each method has an appropriate usage to treat multiple heterogeneous zones so that computational work is minimized without lowering acceptable accuracy.

1 INTRODUCTION

Fluid transport through porous media is governed by the size, geometry and distribution of connected pore space. In such a connected pore space, at any particular scale of volumetrically averaged properties, the application of Darcy's law and mass conservation allows a solution for potential and velocity at a point. Continuous variations in

---

(1) Department of Mineral Resources Engineering, Imperial College, London SW 7 2 BP, UK.
(2) Department of Aeronautics, Imperial College, London, SW 7 2 BP, UK.

permeability can usually be treated by assuming periodic or quasi-periodic distributions but discontinuous variations -- such as permeability changing abruptly from a finite value to zero or infinity -- leads to a specification of a flux condition over a surface of arbitrary orientation. The solution of the second order partial differential equations of fluid motion which, have Neumann boundary conditions on curved surfaces, is difficult to solve using finite difference methods (Forsythe and Wasow, 1960; Peaceman, 1977). Analytical methods are of course only valid for simple geometry conditions. We demonstrate the use of integral equation methods -- expanded and adapted from "panel methods" used in aerodynamics -- for use in solving a wide range of permeability variation assemblages occuring over abrupt surfaces of arbitrary shape. The method is particularly advantageous for accurate determination of local flow acceleration because rather than solving point singularity strengths distributed over the surface using a numerical integration technique, "panel methods" use piecewise continuous singularity distributions of varying order whose solution is analytic and hence can be solved exactly.

2 PROBLEM FORMULATION

We consider a 2-dimensional steady, constant rate, incompressible displacement of one fluid by another through a uniform, isotropic porous media having isolated surfaces over which permeability varies abruptly. The mobility ratio, M, between the displacing phase and that of the displaced phase is unity. The conditions at this surface are,

$$\left(\frac{k}{\mu}\right)_1 \nabla \Phi_1 \cdot \hat{\underline{n}} = \left(\frac{k}{\mu}\right)_2 \nabla \Phi_2 \cdot \hat{\underline{n}} \qquad (1)$$

$$\nabla \Phi_1 \cdot \hat{\underline{s}} = \nabla \Phi_2 \cdot \hat{\underline{s}} \qquad (2)$$

where

$\left(\frac{k}{\mu}\right)$ = fluid mobility, effective fluid permeability / fluid viscosity

Φ_1, Φ_2 = potential in regions 1 and 2 respectively

It has been shown (see, for example, Muskat, 1946) that,

$$\widetilde{\Phi}_P = \frac{1}{2\pi} \int_{\partial S} (\hat{\underline{n}} \cdot \nabla(\Phi_1 - \Phi_2)) \Phi_S - (\Phi_1 - \Phi_2)(\hat{\underline{n}} \cdot \nabla \Phi_S) \, ds$$

$$(3)$$

where

$\Phi_P = V_\infty \cdot \underline{x} + \widetilde{\Phi}_P$ = potential at point P in flow

$\widetilde{\Phi}_P$ = potential resultant from a distribution of singularities

Φ_S = potential singularity

In order to obtain a unique solution it is necessary that $\widetilde{\Phi}_P$ must not be rotational. However, unlike $\widetilde{\Phi}_P$, the far-field potential $V_\infty \cdot \underline{x}$, may have quite general characteristics. Although we use source and sink surface singularities for Φ, vortex and doublet surface singularities can also be used (see, for example, Moran, 1984). At the surface,

$$\nabla \widetilde{\Phi} \cdot \hat{\underline{n}} = \frac{1}{2\pi} \int_{\partial S} \frac{Q}{r^2} [(x_P - x) \hat{\underline{i}} + (y_P - y) \hat{\underline{j}}] \, d\underline{s} \quad (4)$$

where Q is the source/sink strength of the ∂S surface and (x,y) are arbitrary points on the ∂S contour.

Eqn. (4) is solved by imposing the boundary condition, Eqn. (1), through an adjustment of the magnitude and direction of the far-field flow impinging on the differential surface ∂S. This is achieved through the bilinear ratio (1-M)/(1+M) such that,

$$\frac{1}{2\pi} \int_{\partial S} \frac{Q}{r^2} [(x_P - x) \hat{\underline{i}} + (y_P - y) \hat{\underline{j}}] \, d\underline{s}$$

$$= \frac{(1-M)}{(1+M)} \underline{V}_\infty \cdot \hat{\underline{n}} \qquad (5)$$

where

$$M = \left(\frac{k_2 \mu_1}{k_1 \mu_2}\right)$$

Eqn. (5) is a Fredholm equation of the second kind. The dimensionality of the problem has been reduced by one such that once the surface singularity strengths are determined, velocities and potentials everywhere else can be determined.

3 FLAT PANELS, CONSTANT SINGULARITY STRENGTH

In the "panel method" approach, nodes are placed on the surface of the region where the discontinuous variation in permeability occurs. Velocities and potentials are determined at the midpoint of the ith panel. In Fig. 1, flat panels having constant source/sink strengths Q_i are shown and we use the formulation suggested by Moran (1984). Integrating over the jth panel yields,

$$u_{ij} = \frac{1}{2\pi} \int_0^{l_j} \frac{x_s - x_s^*}{(x_s - x_s^*)^2 + y_n^2} \, dx_s^*$$

$$= -\frac{1}{4\pi} \ln \frac{r_{ij}^2 + 1}{r_{ij}^2} \qquad (6)$$

$$v_{ij} = \frac{1}{2\pi} \int_0^{l_j} \frac{y_n}{(x_s - x_s^*)^2 + y_n^2} \, dx_s^* = \frac{\varphi_{ij}}{2\pi} \qquad (7)$$

where (x_s, y_n) are the local coordinates tangential and normal to the jth panel and (u,v) are the local velocites at the ith panel caused by the jth surface singularity. Local velocities are, therefore, solved by determining source or sink strengths dependant on magnitude and direction of the right hand side of Eqn. (5). The influence of all jth panels on the ith panel yields an N^2 influence matrix A_{ij} which, when solved, yields the source strengths Q_i,

$$\sum_{i=1}^N Q_i A_{ij} = \frac{(1 - M_i)}{(1 + M_i)} \underline{V}_\infty \cdot \hat{\underline{n}}_i \qquad (8)$$

where

A_{ij} = off-diagonal logarithmic and arctangent terms from (6) & (7)

The off-diagonal terms are determined only from geometric considerations and are not dependant on the far-field velocity $\underline{V}_\infty$ or mobility ratio M_i. Eqns. (6) and (7) are determined exactly and directly so that numerical integration is never required. If point singularities are placed at nodes then velocity fluctuations may occur. Surface integration prevents such fluctuations and produces smooth, accurate velocity distributions near the permeability discontinuity surface.

The leading diagonal terms A_{ii} are determined by letting the ith midpoint occupy the midpoint of the jth panel. Since $r_{ij+1}^2 = r_{ij}^2$, $u_{ii} \to 0$. Flow on the left hand side of nodes with positive ordering in a clockwise manner have, in the limit $\varphi_{ii} \to +\pi$ whereas on the right hand side $\varphi_{ii} \to -\pi$. Off-diagonal terms are one order of magnitude less than leading diagonal terms. Once Q_i values are determined, every other point in the flow can be determined. Potential values can be calculated by Bernoulli's relation,

$$\tilde{\Phi}_i = \frac{1}{2}\rho |\underline{V}_\infty|^2 (1 - \frac{|\underline{V}|^2}{|\underline{V}_\infty|^2}) \qquad (9)$$

A large number of numerical experiments have

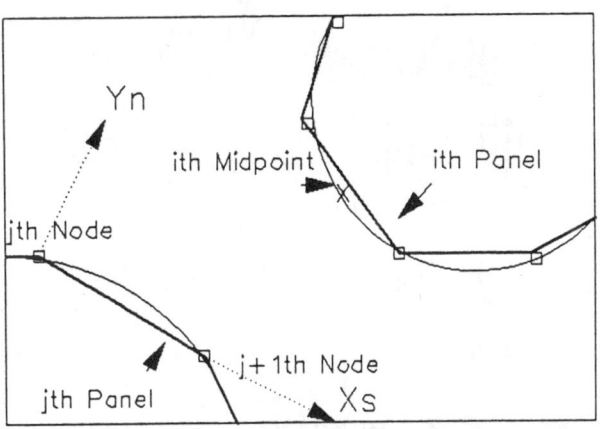

Fig. 1. Flat Panel Coverage of a Curved Permeability Discontinuity Region

demonstrated that for flow on the left hand side of convex regions -- such as would generally occur in isolated patches --differential panel angles of as much as 30° gives acceptable results ($< 5\%$). In general, these regions will be ill-defined.

Fig. 2 shows the time progression of an initially flat unit mobility displacement front through 4 elliptical heterogeneous patches. This numerical experiment -- and all numerical experiments throughout this paper -- were carried out on an Opus PC VI microcomputer having an Intel 80286 microprocessor with 25 ms hard-disk access speed and a math coprocessor. During the initialization step (INIT in Fig. 2) Q_i values are computed only for panels on a single heterogeneity and then recomputed from perturbation velocities caused by panels on other heterogeneities in order to save on computing work. In almost all cases it is overly-restrictive to solve the full N^2 case when computing the mutual perturbation of heterogeneous zones. During the execution step (EXEC in Fig. 2), the front is displaced by trapezoidal integration. Runs were periodically checked using 4th order Runga-Kutta time integration and the results were always found to be very close to the trapezoidal integration. A leap-frog technique, which has the same amount of computational effort as the trapezoidal technique, can be used if higher accuracy is desired.

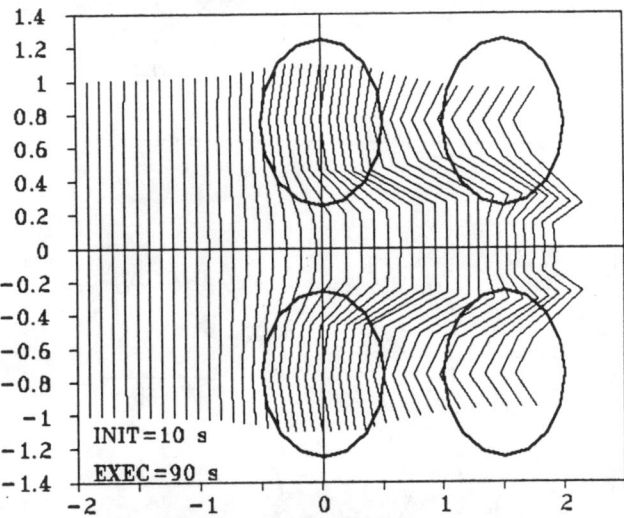

Fig. 2. Displacement Front through 4 Elliptic Heterogeneites; 16 Flat Panels per Patch; M = 1/2, 2

3.1 Multipole Expansions

If the distance from the panel to the point of interest is much greater than the characteristic size of the heterogeneous zone, the uniformly distributed source density on each panel can be approximated by a point singularity on each panel of a heterogeneous zone, as shown in Fig. 3. Although this will give less accuracy than the panel method, in many instances it is suitable for modelling heterogeneous zones far from the point of interest. The potential at an arbitrary point outside the heterogeneous patch is,

$$\widetilde{\Phi}_P = \frac{1}{2\pi} \sum_{i=1}^{N} Q_i \ln \underline{r}_{p_i} \Delta s_i \qquad (10)$$

By the law of cosines,

$$f(\underline{r}_i) = \frac{1}{2} \ln (\underline{r}^2 + \underline{r}_i^2 - 2\,\underline{r}\,\underline{r}_i \cos \theta_i) = \ln \underline{r}_{p_i} \qquad (11)$$

Expanding $f(\underline{r}_i)$ as a Taylor series about the origin,

$$f(\underline{r}_i) = f(0) + \underline{r}_i \frac{\partial f(\underline{r}_i)}{\partial \underline{r}_i}\bigg|_{\underline{r}_i=0} + \frac{\underline{r}_i^2}{2!} \frac{\partial^2 f(\underline{r}_i)}{\partial \underline{r}_i^2}\bigg|_{\underline{r}_i=0}$$
$$+ o(\epsilon^3) \qquad (12)$$

$\ln \underline{r}_p$ is now expressed as terms which are products of r_i^a and $\partial^a(\ln \underline{r})/\partial r_i^a$. Representing this in terms of an x,y coordinate axes,

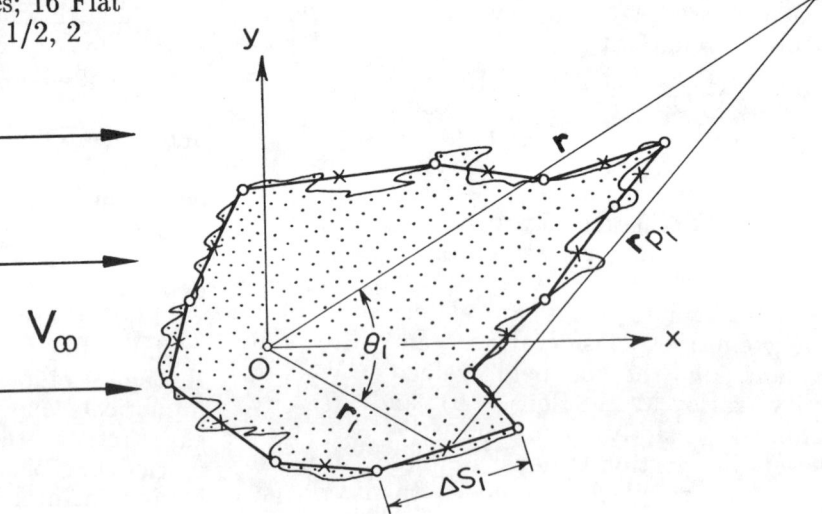

Fig. 3. Replacement of an Arbitrary Heterogeneous Region with an Equivalent Dipole

$$\tilde{\Phi}_P = \frac{1}{2\pi} \sum_{i=1}^{N} Q_i \Delta s_i \left[\ln \underline{r} \right.$$

$$+ x_i \frac{\partial}{\partial x}(\ln \underline{r}) + y_i \frac{\partial}{\partial y}(\ln \underline{r})$$

$$+ \frac{1}{2} \; x_i^2 \frac{\partial^2}{\partial x^2}(\ln \underline{r}) + x_i y_i \frac{\partial^2}{\partial x \partial y}(\ln \underline{r}) + y_i^2 \frac{\partial^2}{\partial y^2}(\ln \underline{r})$$

$$\left. + o(\epsilon^3) \; \right] \qquad (13)$$

Eqn. (13) is referred to as a multipole expansion in the electromagnetics literature (see, for example, Wangsness, 1984). It can be seen that the ln $\underline{r}$ terms and its derivatives can be moved outside each summation term since they are not functions of heterogeneity panel locations. Since the panels cannot contribute to the flow the first term, the source potential, is zero at the point P,

$$\tilde{\Phi}_S = \frac{1}{2\pi} \sum_{i=1}^{N} Q_i \ln \underline{r} \; \Delta s_i = 0 \qquad (14)$$

The remaining two expressions are the dipole and quadrapole terms. Dipole perturbation velocities at the point P are then determined by,

$$\tilde{V}_{D_x} = \frac{\mu_x}{r^2}[\; 1 - 2 \; l_x^2 \;] - 2 \frac{\mu_y}{r^2} l_x l_y \qquad (15)$$

$$\tilde{V}_{D_y} = \frac{\mu_y}{r^2}[\; 1 - 2 \; l_y^2 \;] - 2 \frac{\mu_x}{r^2} l_x l_y \qquad (16)$$

where (l_x, l_y) are direction cosines and,

$$\mu_x = \frac{1}{2\pi} \sum_{i=1}^{N} Q_i \; x_i \; \Delta s_i \qquad (17)$$

$$\mu_y = \frac{1}{2\pi} \sum_{i=1}^{N} Q_i \; y_i \; \Delta s_i \qquad (18)$$

where (μ_x, μ_y) are dipole moments.

Eqns. (15) - (18) replace a heterogeneous patch by an equivalent cylindrical one. The advantage of this approach is that once the dipole moments are determined, perturbation velocities caused by each heterogeneous zone only requires computation of eqns. (15) - (18) rather than computing the effect of each panel at each heterogeneous zone. The concept of using equivalent dipole patches to represent distributed heterogeneous patches is not in itself new (Shaw and Dawe, 1982). However, in this work, we show how the dipole moments can be computed for any arbitrary shaped patch. The use of multipole expansions may become particularly important when treating hundreds or thousands of heterogeneous zones in representations of petroleum reservoirs.

Fig. 4 shows the magnitude of the error caused by computing the dipole moment μ_x by Eqn. (17).

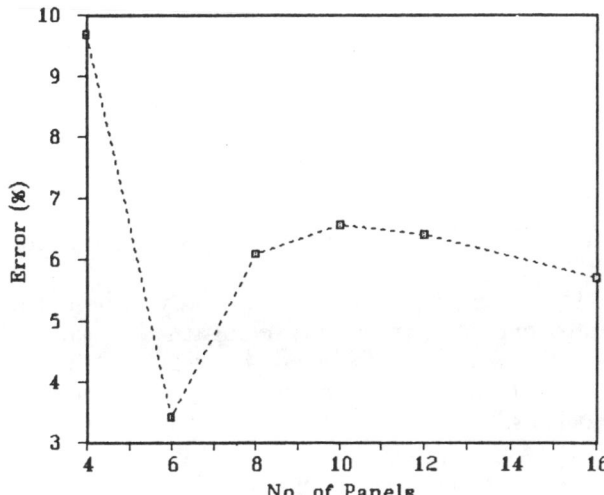

Fig.4. Error in Dipole Moment μ_x Calculation using Flat Panels for Cylindrical Heterogeneous Patch

Fig. 5 is the equivalent of Fig. 2 using dipole representations.

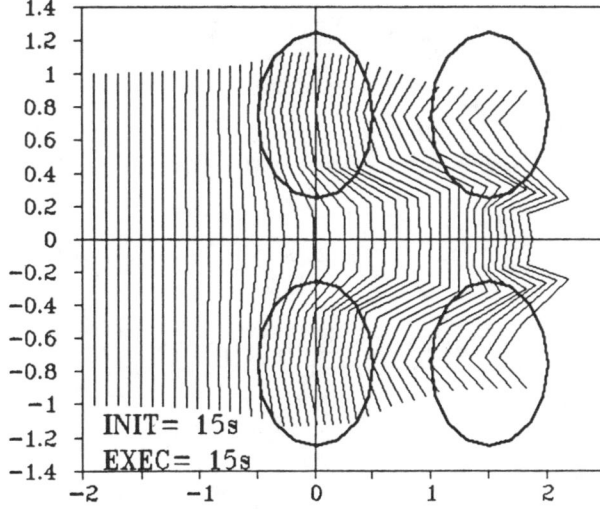

Fig.5. Displacement Front through 4 Elliptic Heterogeneities; 16 Panel Equivalent Dipoles; M = 1/2, 2

117

The close agreement in results are because dipoles accurately represent the elliptical heterogeneous patches. However, this will not be true with elongate or asymmetric patches. The slight increase in initialization in Fig. 5 compared to Fig. 2 is caused by the need to compute dipole moments. This is compensated by a six-fold reduction in the execution of the frontal displacement. Numerical experiments suggest that these equivalent heterogenous patches are acceptable when,

$$\frac{d_i}{r} < \frac{1}{5} \left| \frac{(1-M_i)}{(1+M_i)} \right| \qquad (19)$$

where d_i is the longest characteristic dimension of the heterogeneous region.

In some instances -- especially when the heterogenous region is elongate -- higher accuracy can be achieved by adding to the perturbation dipole velocities the quadrapole velocities,

$$\tilde{V}_{Qx} = \frac{\tau_{xx} \, l_x}{r^3} \left[4l_x^2 - 3 \right] + 2 \frac{\tau_{xy} \, l_y}{r^3} \left[4l_x^2 - 1 \right]$$

$$+ \frac{\tau_{yy} \, l_x}{r^3} \left[4l_y^2 - 1 \right] \qquad (20)$$

$$\tilde{V}_{Qy} = \frac{\tau_{xx} \, l_y}{r^3} \left[4l_x^2 - 1 \right] + 2 \frac{\tau_{xy} \, l_x}{r^3} \left[4l_y^2 - 1 \right]$$

$$+ \frac{\tau_{yy} \, l_y}{r^3} \left[4l_y^2 - 3 \right] \qquad (21)$$

yielding elliptical shaped equivalent heterogeneous regions. The quadrapole moments are,

$$\tau_{xx} = \frac{1}{2\pi} \sum_{i=1}^{N} Q_i \, x_i^2 \, \Delta s_i \qquad (22)$$

$$\tau_{xy} = \frac{1}{2\pi} \sum_{i=1}^{N} Q_i \, x_i y_i \, \Delta s_i \qquad (23)$$

$$\tau_{yy} = \frac{1}{2\pi} \sum_{i=1}^{N} Q_i \, y_i^2 \, \Delta s_i \qquad (24)$$

4 CURVED PANELS, VARIABLE SINGULARITY STRENGTH

In the preceeding examples, flow computed to the left of the clockwise positive progressing nodes representing panel endpoints gave good results. This was true even when panel placement was crude because there is compensatory error cancellation. Errors in source values on the upstream side of the heterogeneous zone are compensated by sink values in the downstream side such that, for symmetric shapes, exact results can be obtained at the surface midpoints (Hess and Smith, 1966). However, in concave regions the opposite is true; errors are additive and unacceptable results can occur in the region approximately one panel length away.

This problem can be accounted for by including the effect of the curvature of the surface over which the variable source density is distributed. In general, the same order of accuracy can be obtained using higher order representations as the flat panel method and the increase in computational effort will be offset by a decrease in the number of panels necessary to achieve the same accuracy. We now expand and adapt a procedure suggested by Hess (1973) to account for numerous heterogeneous zones having arbitrary mobility ratio values. The resultant integral expressions can, as before, be integrated directly and exactly.

In this method, the jth panel geometry and the effect of its variable surface singularity strength is now determined from a curve whose origin is at the jth panel midpoint (see Fig. 6). The coordinate axes -- x_s and y_n -- are tangential and normal respectively to the curved panel. The influence of the curved jth panel on the ith panel midpoint can be computed by expressing the singularity strength distribution as a power series,

$$Q_j = Q_j + \frac{dQ_j}{ds_j} s_j + \frac{d^2 Q_j}{ds_j^2} s_j^2 + o(s_j^3) \qquad (25)$$

where

Q_j = constant source/sink strength from the flat panel method

$s_j = f(x_{s_j}, y_{n_j})$ such that

$$ds_j^2 = dx_{s_j}^2 + dy_{n_j}^2$$

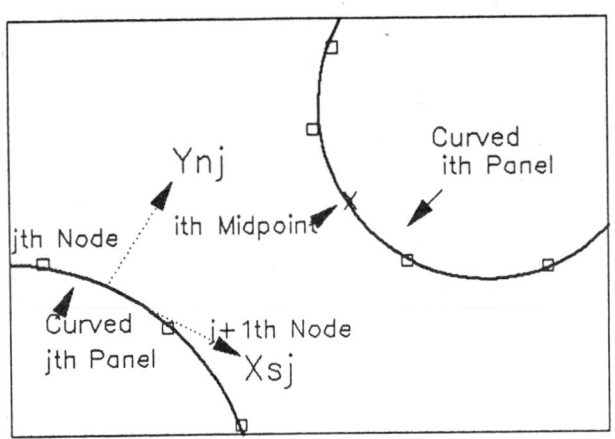

Fig. 6. Curved Panel Coverage of a Curved Permeability Discontinuity Region

By placing the y_{n_j} axis normal to the jth panel midpoint and expanding y_{n_j} as a function of x_{s_j},

$$y_{n_j} = \frac{d^2 y_{n_j}}{dx_{s_j}^2} x_{s_j}^2 + o(x_{s_j}^3) \qquad (26)$$

where

$$d^2 y_{n_j} / dx_{s_j}^2 = \text{local curvature}$$

By inserting eqns. (25) and (26) into Eqn. (5) and retaining second order terms only, we obtain integral expressions parabolic in dimension and linear in source/sink strength whose solution is analytic As before, local and tangential velocities can be calculated by,

$$u_{ij} = -\frac{1}{4\pi}\ln\frac{r_{ij}^2+1}{r_{ij}^2} - \left(\frac{dQ}{ds}\right)_j \left(\Delta y_{ij}\frac{\varphi_{ij}}{2\pi} - \frac{\Delta x_{ij}}{4\pi}\right.$$

$$\ln\frac{r_{ij}^2+1}{r_{ij}^2} + \Delta s_j\bigg) + \left(\frac{d^2 y_n}{dx_s^2}\right)_j \left(2\,\Delta x_{ij}\frac{\varphi_{ij}}{2\pi} + \right.$$

$$\left. 2\,\frac{\Delta y_{ij}}{4\pi}\ln\frac{r_{ij}^2+1}{r_{ij}^2}\right) \qquad (27)$$

$$v_{ij} = \frac{\varphi_{ij}}{2\pi} - \left(\frac{dQ}{ds}\right)_j \left(\Delta x_{ij}\frac{\varphi_{ij}}{2\pi} + \frac{\Delta y_{ij}}{4\pi}\ln\frac{r_{ij}^2+1}{r_{ij}^2}\right)$$

$$+ \left(\frac{d^2 y_n}{dx_s^2}\right)\left(2\Delta y_{ij}\frac{\varphi_{ij}}{2\pi} - 2\frac{\Delta x_{ij}}{4\pi}\ln\frac{r_{ij}^2+1}{r_{ij}^2} + \Delta s_j\right) \qquad (28)$$

Although Eqns. (27) and (28) are more complicated than Eqns. (6) and (7), once Eqns. (6) and (7) are computed, most of the remaining expressions are simple products of the ith and jth midpoint separation distances $(\Delta x_{ij}, \Delta y_{ij})$. The higher order expressions are the second and third terms in eqns. (27) and (28). Elimination of these terms would cause the solution to be identical to the flat panel method. The source singularity slope $(dQ/ds)_j$ and local curvature $(d^2 y_n/dx_s^2)_j$ must also be computed from the adjacent j-1th and j+1th panels.

5 CONCLUSIONS AND FUTURE DIRECTIONS

The "panel method" is conceptually simple to use and accuracy can be increased by a hierarchy of techniques. The most accurate method so far investigated uses higher order representations of the prescribed flux condition to compute potential and velocities in the vicinity of concave regions which arise, primarily, from the juncture of regions of several different permeabilites. Away from these regions and when only convex regions are represented, the computationally quicker flat panel method can be used with no loss of accuracy. Other heterogeneous regions can be treated by representing them as point singularities using multipole expansions.

In ongoing research at Imperial College, panel methods are being coupled to a moving point method suggested by Graham (1988) to solve general multiphase convection diffusion problems (see, for example, Ewing, 1983).

6 ACKNOWLEDGEMENT

One of the authors (DWW) is grateful to *Gulf Canada Resources Limited* and to *The British Council* for financial support to conduct this research.

7 REFERENCES

Ewing, R. E., Problems Arising in the Modeling of Processes for Hydrocarbon Recovery, in The Mathematics of Reservoir Simulation, R. E. Ewing, Ed., SIAM, 1983.

Forsythe, G. E. and Wasow, W. R., Finite-Difference Methods for Partial Differential Equations, John Wiley, 1960.

Graham, J. M. R., Computation of Viscous Separated Flow using a Particle Method, in Num. Meth. for Fluid Mech II, K. W. Morton and M. I. Baines, Ed.'s, IMA Conf. New Series No. 17, 1988.

Hess, J. L. and Smith, A. M. O., Calculation of Potential Flow about Arbitrary Bodies, Prog. Aero. Sci., 8, 1966.

Hess, J. L., Higher Order Numerical Solution of the Integral Equation for the Two-Dimensional Neumann Problem, Comp. Meth. in Appl. Mech. and Eng., 2, 1973.

Moran, J., An Introduction to Theoretical and Computational Aerodynamics, John Wiley and Sons, 1984.

Muskat, M., Flow of Homogeneous Fluids through Porous Media, J. W. Edwards, Inc., 1946.

Peaceman, D. W., Fundamentals of Reservoir Simulation, Dev. in Pet. Sci., 6, Elsevier Scientific Publishing Company, 1977.

Shaw, D. C. and Dawe, R. A., Averaging Methods for Numerical Simulations of Flow through Heterogeneous Porous Media, Trans. in Por. Media 2, 1987.

Wangsness, R. K., Electromagnetic Fields, 2nd Ed., John Wiley & Sons, 1986.

2nd European Conference on the Mathematics of Oil Recovery
© D. Guérillot, O. Guillon (Editors) and Éditions Technip, Paris 1990, pp. 121-130
27 rue Ginoux, 75015 Paris

A Finite Element Method
for Calculating Transmissibilities
in N-point Difference Equations
Using a Non-Diagonal Permeability Tensor

P. Samier[1]

ABSTRACT

Petroleum reservoir are always heterogeneous. Averaging techniques consist mostly in defining an equivalent homogeneous permeability tensor for a given heterogeneous porous medium whose absolute permeability is a space dependent function.

The equivalent permeability tensor is generally symmetric but non-diagonal: three unusual off-diagonal terms Kxy, Kxz and Kyz are to be considered in addition to the standard diagonal permeability terms in x, y and z direction.

Non-diagonal tensors arise also in a mesh whose axes are not the principal directions of the permeability tensor: an application for horizontal well simulation is briefly presented.

When predicting recovery processes, simulators use mostly a standard five-point finite volume approximation with a diagonal permeability tensor in each grid block. Introduction of a non-diagonal tensor requires higher order finite volume or finite difference schemes. According to the number of non zeros off-diagonal terms, consistent schemes may vary from a nine-point scheme to a 27-point finite volume scheme which is the most general scheme applicable to a complete tensor.

This paper describes a finite element technique used to derive transmissibility factors which are then introduced inside a finite volume black oil simulator. This method allows finite volume schemes varying from 9 to 27-point and applies to a general heterogeneous permeability system with non-uniform grid spacing.

Several examples are given in the paper, illustrating the applicability of the method.

INTRODUCTION

In this paper, we present a method to introduce non-diagonal absolute permeability tensor inside a standard block-centered finite volume simulator. The need of a non-diagonal tensor to estimate the effective permeability of a reservoir volume has already been shown by Quintard and Whitaker[1], Bamberger[2], Forges[3] ,Lake and Kassap[4], White and Horne[5],...

In a stratified reservoir with tilted layers, using a diagonal tensor requires that the x or y axis of the coordinates system is identical to the direction of bedding and therefore the tilt angles should be the same for all the layers. Whereas using a non-diagonal tensor allows to choose any direction for the x axis.

Once we obtain a non-diagonal tensor (as a result of an averaging process or rotation of the diagonal tensor in the axes of the grid system), computations of transmissibility factors should be carried out to perform the reservoir simulation. Standard 5-point scheme is not suitable and higher schemes such as the 9-point scheme for a 2D problem are needed. Derivation of transmissibilities using finite difference operator has already been made by Yanosik and Mc Cracken[6], Coats and Modine[7], Bertinger[8], Shah[9], Shiralkar and Stephenson[13] and others for diagonal tensors. All these methods are in agreement for uniform spacing and constant permeability between interblocks. Hence for non uniform spacing and heterogeneous neighbouring cells, calculations become straight forward. We find it easier to derive these transmissibilities using the finite element method for a monophasic flow and then to generalize them for multiphasic flow.

---

(1) Elf Aquitaine, 26, avenue des Lilas, 64018 Pau, France.

NEED OF COMPLETE TENSORS:

As a result of permeability averaging techniques:

Consider a parallelipedic volume containing mx.my.mz fine-scale micro grid blocks. The permeability tensor is assumed to be known and constant inside each micro-block. The aim of the scaling up is to find an absolute permeability tensor equivalent in terms of flow rates to the fine scale grid which will be discretized in the simulator as one block (fig 1).
Bamberger, Forges[3] and Guerillot[14], then Corre, Eymard[10] and Pfertzel[12] developed a powerful algorithm leading to symmetric but non-diagonal tensors.

Briefly, operations are as follows:

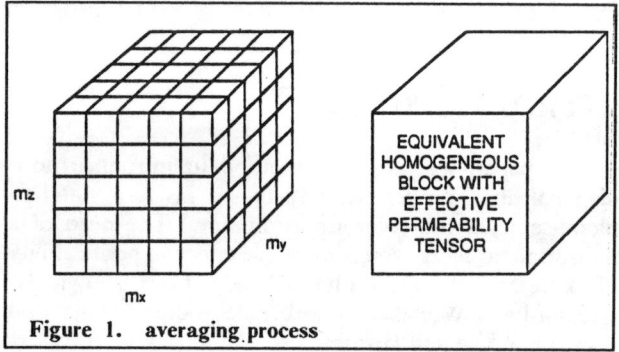

Figure 1. averaging process

1. a pressure gradient is prescribed in x direction on the 6 faces of the domain to be averaged,
2. the steady-state pressure solution is solved at each internal node by a finite element method,
3. the finite element formalism is used to obtain the Kxx component of the equivalent tensor minimizing the viscous forces energy.
4. operation is repeated in y direction so as to obtain Kxy and Kyy components,
5. operation is repeated in z direction so as to obtain Kxz, Kyz and Kzz components.

This approach is slighly similar to the Warren and Price method[11] . Main differences are the use of the finite element formalism leading to a symmetric tensor and non zero boundary conditions for the fluxes along the faces perpendicular to the pressure gradient direction giving a non-diagonal tensor.

$$\underline{K} = \begin{bmatrix} K_{xx} & K_{xy} & K_{xz} \\ K_{xy} & K_{yy} & K_{yz} \\ K_{xz} & K_{yz} & K_{zz} \end{bmatrix}$$

Non-diagonal tensors seem more physical since it implies that the principal directions of the flow do not necessarily align with the axes of the grid system.

For horizontal well modelling:

Horizontal wells technique applies to dipping reservoirs. The main advantadge is to be able to drain through several impervious layers (fig. 2).

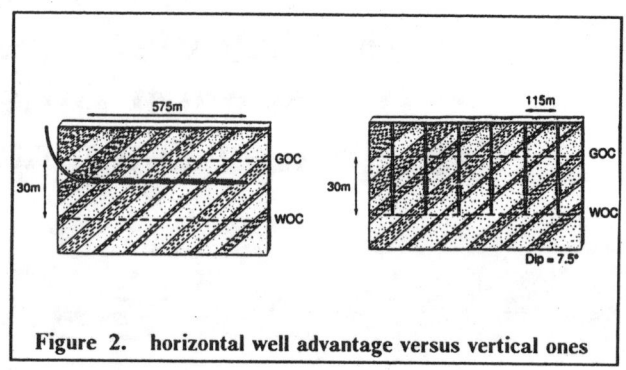

Figure 2. horizontal well advantage versus vertical ones

Hence modelling is laborious. The usual simulators adopt diagonal permeability tensors which prescribe that the mesh axes follow the same direction as the tilt of the layers. In this case, the well grid blocks communicate along a diagonal. Since the horizontal line is not an axis of the mesh, representation of horizontal contacts is unprecise. Furthermore carrying out a mesh whose well cells are communicating along a diagonal is difficult and leads to an huge number of cells (fig. 3).

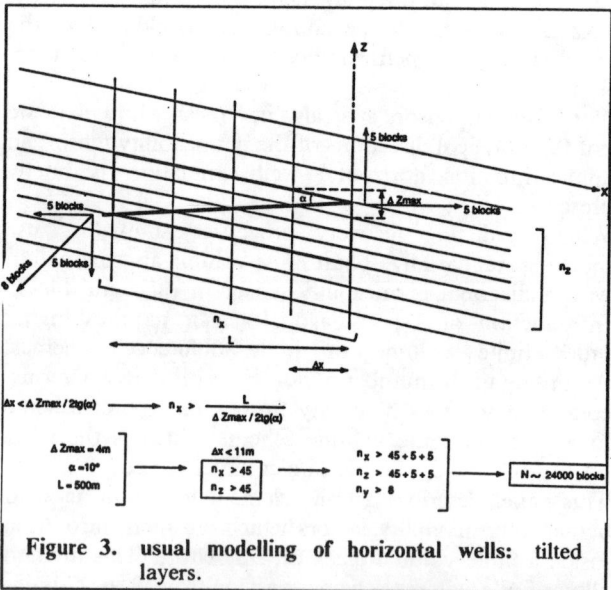

Figure 3. usual modelling of horizontal wells: tilted layers.

A much more economical technique consists of using a mesh whose x axis aligns with the horizontal well direction. Horizontal well and contacts discretization become much easier.
But since the grid axes are no more the principal directions of the flow, permeability tensor is non-diagonal and can be expressed using notations of fig. 4 as:

$$\underline{\underline{K}} = \begin{bmatrix} K_{xx} & 0 & K_{xz} \\ 0 & K_y & 0 \\ K_{xz} & 0 & K_{zz} \end{bmatrix}$$

where $K_{xx} = K_h \cos^2\theta_i + K_v \sin^2\theta_i$
$\quad K_{xz} = (K_h - K_v)\sin\theta_i\cos\theta_i$
$\quad K_{zz} = K_h \sin^2\theta_i + K_v \cos^2\theta_i$

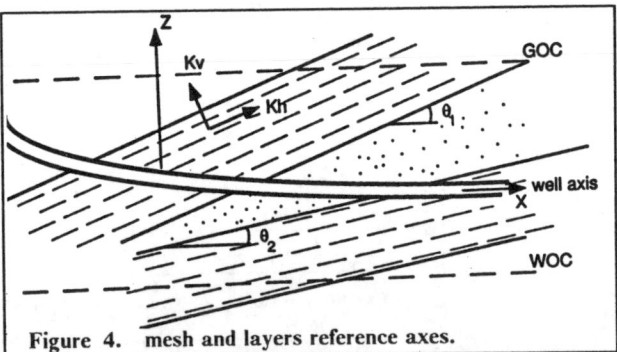

Figure 4.   mesh and layers reference axes.

Another advantage of this method is to mimic layers with different tilt angles, what is not possible with the usual modelling where the angle should be the same for all the layers.

Definition of transmissibilities:

The two previous examples show the need of introducing non diagonal tensors in the reservoir models. The point is now to see how the equations discretized by the finite volume method are affected. In discretizing the mass conservation according to the generalized Darcy's law, the permeability tensor appears only in the term $T_{i,j}$ denoted transmissibility between the cell i and the cell j. A simplified expression (assuming one component per phase and no compressibility) can be written as:

$$\sum_j T_{i,j} \frac{Kr^\varphi(i,j)}{\mu_i^\varphi}(P_j^\varphi - P_i^\varphi - \rho^\varphi g\Delta z_{i,j}) + V\frac{\Phi\Delta S^\varphi}{\Delta t} = 0$$

where:
j is the generic neighbour of the cell i (usually 6, two per each direction)
P_i^φ and S_i^φ are the pressure and saturation of phase φ in cell i
V is the volume of the cell i, μ the viscosity
ρ the density, Φ the porosity of the rock
$Kr^\varphi(i,j)$ the interblock relative permeability of fluid phase φ
$\Delta z_{i,j}$ the distance along z axis between cell i and cell j

$T_{i,j}$ is consistent with a permeability multiplied by the interblock flow surface and divided by the distance between the 2 blocks.
A method for calculating these transmissibilities consists of equating the continuous term to the discretized expression at the first order in pressure and saturation:

$$\iiint_V div(\underline{\underline{K}}\, m\, \vec{grad}P)dxdydz = \sum_j T_{ij}\, m_{ij}\,(P_j - P_i) \qquad (1)$$

where V is the grid-block volume and m the mobility of the phase.
Using a Taylor expansion for P and m, one obtains six equations (one per component of the tensor). In the particular case of an homogeneous reservoir, a uniform grid and a standard 7-point scheme, only 3 transmissibilities per grid (1 per direction due to symmetry) are unknowns. Hence this scheme (5-point in 2D, 7-point in 3D) leading to 6 different equations with only 3 unknowns is not adequate for correctly handling non-diagonal tensors. Higher order schemes with at least 6 directions in the case of a complete symmetric tensor must be used. Therefore the smallest scheme applicable to a complete tensor (Kxz, Kxy and Kxz all non zero) is a 13-point scheme.

Calculation of transmissibilities by a finite element method.

To derive these transmissibilities, the finite element formalism is prefered since it's easier for non uniform gridding and when permeability tensor differs in each element.

Theoretical formulation

For a monophasic flow:

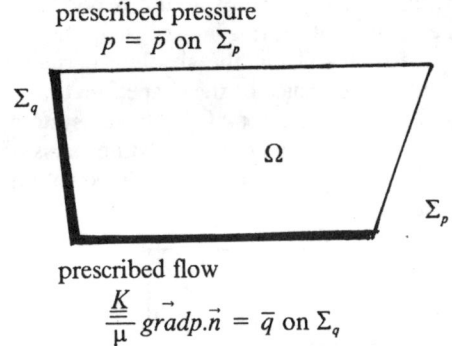

Expressing the continuity equation with the usual boundary conditions :

$$div(\frac{\underline{\underline{K}}}{\mu}\,\vec{grad}p) = 0 \text{ on } \Omega$$
$$\frac{\underline{\underline{K}}}{\mu}\,\vec{grad}p.\vec{n} = \bar{q} \text{ on } \Sigma_q \qquad p = \bar{p} \text{ on } \Sigma_p$$

gives the following weak form:

> for all function v such as v = 0 on Σ_p:
>
> $$\iiint_\Omega {}^t(\overrightarrow{\mathrm{grad}}\,v)\frac{K}{\mu}(\overrightarrow{\mathrm{grad}}\,p)d\Omega = \iint_{\Sigma_q} v\overline{q}\,d\Sigma$$
>
> with p being chosen such as : $p = \overline{p}$ on Σ_p

27-point scheme

Constraining the physical model by a 8-node parallelepipedic element and linear shape functions φ_i such as:

$$p(x,y,z) = \sum_{i=1}^{8} \varphi_i(x,y,z)P_i$$

Where:

i = element node number

P_i = unknown pressure at node i

Assuming the viscosity to be constant inside each element, the previous set of equations yields to 8 linear equations:

$$\frac{1}{\mu}[T][P] = [Q]$$

where : Q_i = pseudo-velocity at node i

T is an 8 x 8 matrix; a generic term is:
$$T_{ij} = \iiint_V {}^t(\overrightarrow{\mathrm{grad}}\varphi_i(x,y,z))K\,\overrightarrow{\mathrm{grad}}\varphi_j(x,y,z)\,dxdydz$$

The unknowns are the pressures at the 8 nodes of the element. The coefficients of this linear system depend on the geometry of each cell, on the absolute permeability in each cell and of the choice of the shape functions.

These terms $T_{i,j}$ of the matrix of the linear system are nothing more than transmissibilities relating pressure to pseudo-velocities. A 27-point scheme is obtained. Referring to fig. 5,

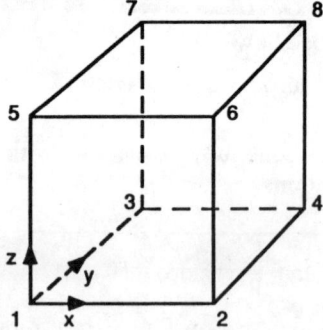

Figure 5.

the transmissibilities of the scheme are given as:

$$T_{12}^e = \frac{(2Cxx - Cyy - Czz)}{18} - \frac{Cyz}{12} = T_{78}^e$$

$$T_{13}^e = \frac{(-Cxx + 2Cyy - Czz)}{18} - \frac{Cxz}{12} = T_{68}^e$$

$$T_{14}^e = \frac{(2Cxx + 2Cyy - Czz)}{36} + \frac{Cxy}{6} = T_{58}^e$$

$$T_{15}^e = \frac{(-Cxx - Cyy + 2Czz)}{18} - \frac{Cxy}{12} = T_{48}^e$$

$$T_{16}^e = \frac{(2Cxx - Cyy + 2Czz)}{36} + \frac{Cxz}{6} = T_{38}^e$$

$$T_{17}^e = \frac{(-Cxx + 2Cyy + 2Czz)}{36} + \frac{Cyz}{6} = T_{28}^e$$

$$T_{18}^e = \frac{(Cxx + Cyy + Czz)}{36} + \frac{(Cxy + Cxz + Cyz)}{12}$$

$$T_{23}^e = \frac{(2Cxx + 2Cyy - Czz)}{36} - \frac{Cxy}{6} = T_{67}^e$$

$$T_{24}^e = \frac{(-Cxx + 2Cyy - Czz)}{18} + \frac{Cxz}{12} = T_{57}^e$$

$$T_{25}^e = \frac{(2Cxx - Cyy + 2Czz)}{36} - \frac{Cxz}{6} = T_{47}^e$$

$$T_{26}^e = \frac{(-Cxx - Cyy + 2Czz)}{18} + \frac{Cxy}{12} = T_{37}^e$$

$$T_{27}^e = \frac{(Cxx + Cyy + Czz)}{36} + \frac{(-Cxy - Cxz + Cyz)}{12}$$

$$T_{34}^e = \frac{(2Cxx - Cyy - Czz)}{18} + \frac{Cyz}{12} = T_{56}^e$$

$$T_{35}^e = \frac{(-Cxx + 2Cyy + 2Czz)}{36} - \frac{Cyz}{6} = T_{46}^e$$

$$T_{36}^e = \frac{(Cxx + Cyy + Czz)}{36} + \frac{(-Cxy + Cxz - Cyz)}{12}$$

$$T_{45}^e = \frac{(Cxx + Cyy + Czz)}{36} + \frac{(Cxy - Cxz - Cyz)}{12}$$

where:

$$Cxx = K_{xx}dy\frac{dz}{dx} \qquad Cxy = K_{xy}dz$$

$$Cyy = K_{yy}dx\frac{dz}{dy} \qquad Cxz = K_{xz}dy$$

$$Czz = K_{zz}dy\frac{dz}{dx} \qquad Cyz = K_{yz}dx$$

K refers to the absolute permeability
dx, dy, dz are the sizes of the element

Note: These transmissibilities are the contribution of an element to the global transmissibility taking into account all surrounding elements. Only T18, T27, T36 and T45 are final terms, the contributions of the 8 elements containing node 1 should be added to the the others terms. Therefore the notation T_{12}^e concerns the contribution of the element e to the final transmissibility T_{12} and we called it "elementary transmissibility".

9-point scheme:

The same method applied in the plane to a rectangular element leads to the transmissibilities of a 9-point scheme. Referring to fig. 6,

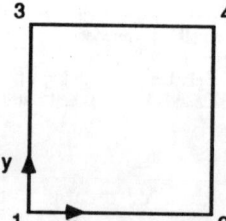

Figure 6.

the transmissibilities of the scheme are:

$$T_{12}^e = \frac{Cxx}{2} - \frac{(Cxx + Cyy)}{6} = T_{34}^e$$

$$T_{14}^e = \frac{Cxy}{2} + \frac{(Cxx + Cyy)}{6}$$

$$T_{13}^e = \frac{Cyy}{2} - \frac{(Cxx + Cyy)}{6} = T_{24}^e$$

$$T_{23}^e = \frac{-Cxy}{2} + \frac{(Cxx + Cyy)}{6}$$

Application to a finite volume simulator:

Most of finite volume simulators compute interblock transmissibilities based on 5-point scheme in 2D (7-point in 3D) and assume a diagonal tensor. To insert the transmissibilities computed previously inside a standard finite volume simulator, we consider two meshes: the simulator finite volume grid (mesh 1) and the associated finite element mesh (mesh 2) which nodes are the grid-block centers of the mesh 1 (fig 7). The aim is to derive the transmissibilities between grid-block centers of mesh 1 and then to introduce them in the simulator.

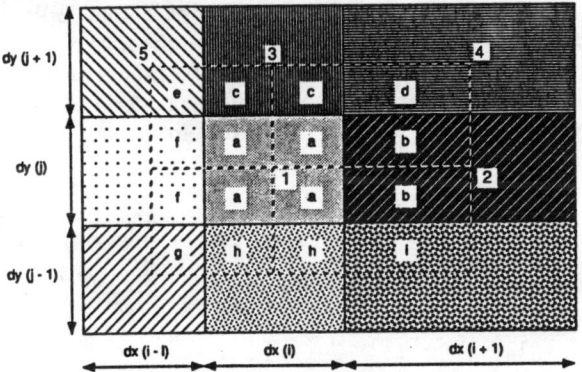

Figure 7. finite element mesh and finite volume grid overlapping: finite element mesh in dotted lines

Assuming a different tensor in each block, each element of mesh 2 is constituted of 8 volumes (4 areas in 2D) with different permeability tensors.

To calculate the transmissibilities of the element described by nodes 1 to 8 or (1 to 4 in 2D), two ways are possible:
1. use of static condensation (or internal elimination): each element of mesh 2 is decomposed in 8 micro-elements (respectively 4 elements in 2D), the complete flow-pressure system is written for each micro-element and internal nodes (where the sum of all flow around is zero) are eliminated, it is a technique to compute an equivalent permeability in the spirit of King's method[16],
2. averaging permeabilities on each element of mesh 2 instead of the finite volume grid (mesh 1) and therefore a constant non-diagonal tensor is obtained for each element (fig 8).

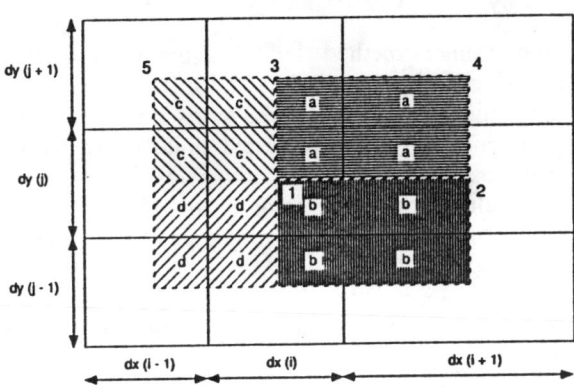

Figure 8. Finite element and finite volume overlapping meshes: finite element mesh with dotted lines

We prefer method 2. The final terms are obtained after adding the "elementary" transmissibilities of the 4 elements a, b, c, d around node 1, (respectively 8 elements in 3D) and yield with the notations of fig 8:

$$T_{12} = \frac{Cxx_a}{3} - \frac{Cyy_a}{6} + \frac{Cxx_b}{3} - \frac{Cyy_b}{6}$$

$$T_{14} = \frac{Cxy_a}{2} + \frac{Cxx_a}{6} + \frac{Cyy_a}{6}$$

$$T_{13} = \frac{Cyy_a}{3} - \frac{Cxx_a}{6} + \frac{Cyy_c}{3} - \frac{Cxx_c}{6}$$

$$T_{15} = \frac{-Cxy_c}{2} + \frac{Cxx_c}{6} + \frac{Cyy_c}{6}$$

For an homogeneous reservoir with a uniformly spaced grid and a diagonal tensor, these formulas lead to the well known 9-point scheme of Yanosik and Mc Kracken[6]. To summarize, operations are as follows:

- From the grid (mesh 1) dedicated to the simulator, the associated finite element mesh (mesh 2) which nodes are the grid-block centers of mesh 1 is built up,
- the absolute permeability equivalent tensor for each element of mesh 2 is computed by any averaging method,
- the transmissibilities between the nodes of mesh 2 are derived using the finite element technique described in the previous section and the contributions of all elements around a given node are added,
- these transmissibilities are injected inside the simulator.

Reduction of the number of points of the scheme:

This scheme is convenient in xy plane but leads to negative transmissibilities in xz or yz plane since generally reservoir mesh is distorted in z direction:

125

$$dz < < dx$$
$$dz < < dy$$ gives $$Czz > > Cxx$$
$$Czz. > > Cyy$$ and T_{12} negative

The finite element method allows negative cross diagonal terms T_{ij} for a monophasic flow and in this case gives a correct solution. However if these terms are injected in a reservoir simulator using upstream finite difference for saturations, convergence is not always possible.

This difficulty may be overcome by modifying the way of deriving the transmissibilities since there is no unique way of computing them.

Alternative methods for modifying transmissibilities may be to adopt:

1. other shapes functions,
2. non conforming elements,
3. mixed finite elements (used by Gallouët and al[15]) where the velocity becomes the variational unknown instead of the pressure.

Another track is tried, it consists of finding a necessary condition for satisfying equation (1) at the first order for each node of the finite element mesh.

For a 2D problem, developing this equation for each node and using Taylor expansions for p and m shows that for a symmetric tensor if (T12, T13, T14, T23) are elementary transmissibilities satisfying (1) over the rectangle e, then

$$T'^e_{12} = T^e_{12} + A \qquad T'^e_{13} = T^e_{13} + A$$
$$T'^e_{14} = T^e_{14} - A \qquad T'^e_{23} = T^e_{23} - A$$

are also 4 other compatible transmissibilities where A is a given constant (it's convenient to choose it as a linear function of Cxx, Cyy). A suitable choice to get rid of negative transmissibilities in a distorted mesh (high aspect ratio) is:

$$A = \frac{(C_{xx} + C_{yy})}{6} - \omega \frac{C_{xx}C_{yy}}{C_{xx} + C_{yy}} \text{ where } \omega < \frac{1}{2}$$

The harmonic mean ensures the positiveness of T12 and T13.

For 3D problem and symmetric tensors, the same method leads to 6 equations with 16 unknown transmissibilities. Transmissibilities satisfying (1) at the first order can be expressed with 10 constants A, B, C, D, E, F, G, H, I, J in the form:

$$T'^e_{12} = T^e_{12} - A - B - H$$
$$T'^e_{34} = T^e_{34} - A - B + H$$
$$T'^e_{13} = T^e_{13} - A - C - I$$
$$T'^e_{24} = T^e_{24} - A - C + I$$
$$T'^e_{14} = T^e_{14} + A - D + E$$
$$T'^e_{15} = T^e_{15} - B - C - J$$
$$T'^e_{26} = T^e_{26} - B - C + J$$
$$T'^e_{16} = T^e_{16} + B - D + F$$
$$T'^e_{17} = T^e_{17} + C - D + G$$
$$T'^e_{18} = T^e_{18} + 2D - E - F - G$$
$$T'^e_{23} = T^e_{23} + A - D - E$$
$$T'^e_{25} = T^e_{25} + B - D - F$$
$$T'^e_{27} = T^e_{27} + 2D + E + F - G$$
$$T'^e_{35} = T^e_{35} + C - D - G$$
$$T'^e_{36} = T^e_{36} + 2D + E - F + G$$
$$T'^e_{45} = T^e_{45} + 2D + E + F + G$$

Several sets of these constants help to get rid of negatives transmissibilities or to degenerate the initial 27-point scheme in schemes having a smaller number of points.

Practically, a 11-point scheme will be preferred to take into account only one cross-diagonal term Kxz for example. This scheme is derived from the 27-point scheme with the following constants:

$$A = \frac{(-5Cxx - 5Cyy + Czz)}{72} \qquad B = \frac{(Cxx + Cyy + Czz)}{72}$$
$$C = \frac{(Cxx - 5Cyy - 5Czz)}{72} \qquad D = -\frac{(Cxx + Cyy + Czz)}{72}$$
$$E = \frac{Cxy}{12} \quad F = \frac{Cxz}{12} \quad G = \frac{Cyz}{12}$$
$$H = \frac{-Cyz}{12} \quad I = \frac{-Cxz}{12} \quad J = \frac{-Cxy}{12}$$

Transmissibilities are:

$$T^e_{12} = \frac{Cxx}{6} - \frac{Czz}{12} \qquad T^e_{13} = \frac{Cyy}{4}$$
$$T^e_{15} = \frac{Czz}{6} - \frac{Cxx}{12} \qquad T^e_{16} = \frac{Cxz}{4} + \frac{Cxx}{12} + \frac{Czz}{12}$$
$$T^e_{25} = \frac{-Cxz}{4} + \frac{Cxx}{12} + \frac{Czz}{12} \quad T^e_{14} \text{ and all others} = 0$$

In the plane xz, this scheme is identical to the 9-point scheme obtained with a rectangular element.

Another case of interest is taking into account 2 cross-diagonal terms (Kxz et Kyz for example). A 15-point scheme is derived from the 27-point using the same procedure. Transmissibilities are:

$$T^e_{12} = \frac{Cxx}{4} - B \qquad T^e_{13} = \frac{Cyy}{4} - C$$

$$T^e_{15} = \frac{Czz}{4} - B - C$$

$$T^e_{16} = \frac{Cxz}{4} + B \qquad T^e_{17} = \frac{Cyz}{4} + C$$

$$T^e_{25} = \frac{-Cxz}{4} + B \qquad T^e_{35} = \frac{-Cyz}{4} + C$$

where $B = \omega_1 \dfrac{CxxCzz}{(Cxx + Czz)} \quad C = \omega_2 \dfrac{CyyCzz}{(Cyy + Czz)}$

$0 < \omega_1$ and $\omega_2 < 0.25$

This scheme is suitable for horizontal well modelling since it takes into account tilting in plane xz, reduces numerical diffusion and grid orientation effects in the plane yz perpendicular to the well axis.

In the same way, a 20-point scheme can be obtained by taking into account the 3 terms in a more economical manner.

Nevertheless the same scheme should be taken over all elements of the mesh so as to avoid material balance errors. Therefore negative transmissibilities which appear in mesh having high aspect ratio or high anisotropy can be eliminated by modifying the previous constants in the initial finite element scheme and/or, increasing the gridding in the direction where negative terms occur.

These modifications lead to a new first order correct scheme but the energy of viscous forces is no more minimum and numerical diffusion may increase. To get a stable scheme minimizing the viscous forces energy, the price should be paid in increasing the size of the grid, the usual modelling constraints are encountered.

Hence it can be noted that usual simulators adopt in 2D, for diagonal tensors, the standard 5-point scheme. This usual scheme is stable but is sensitive to grid-orientation effects, do not minimize the viscous energy and produces higher numerical diffusion than the 9-point scheme derived previously.

TESTS AND VALIDATION OF TRANSMISSIBILITIES

first validation: an heterogeneous five-spot

We use the five-spot example already published by White and Horne[5]. The corresponding mesh and the equivalent tensors obtained with Bamberger and al averaging method are indicated in fig 9. There is only one layer and 2 phases (water and oil). The coarse mesh is regular and gridding is 6 by 6. A 30 by 30 fine mesh grid is used to compare results.

The injector well is in block (1,1) while the producer is in block (30,30). Relative permeability curves are indicated in fig 10. Water viscosity is 0.5 cpo, oil viscosity varies from 1.34 to 1.43 cpo according to pressure.

The aim of this test is to validate the new method of transmissibilities computations.

4 different simulations were performed:
1. non-diagonal tensor-scaled macrosimulation, the transmissibilities are computed by the finite element method then input in a 9-point black-oil simulator "SCORE".
2. harmonically averaged macrosimulation based on the standard 5-point scheme,
3. geometrically averaged macrosimulation based on the standard 5-point scheme,
4. fine scale microsimulation based on the standard 5-point scheme.

Figure 11 compares the water contours after 100 days of injection between the coarse grid and the refined grid.

Figure 12 compares the pressure contours. Figure 13 indicates the fractional flow obtained with different methods.

It can be noticed that the non-diagonal tensor method gives the closer fine scale solution. Using geometric mean, saturations are less precise but still acceptable while the harmonic mean is for this case a very poor predictor.

Figure 12 highlites the superiority of the non-diagonal tensor- averaging technique since isobars and pressure values are almost exactly reproduced on the coarse mesh. As a comparison, the harmonic and geometric mean overestimates the pressure and leads to a wrong repartition of the pressure losses in the reservoir.

Second application: a horizontal well

This example deals with the modelling of an horizontal well in stratified geometry according to the 2 different methods mentioned previously.

This test corresponds to a typical example:
- the reservoir is drained by a horizontal well 500m long,
- oil zone is 10,5m thick and layers have a tilt angle of 10° with respect to the horizontal axis,
- for this test, the reservoir is assumed to be homogeneous with permeability values: (Kh = 1 darcy, Kv = 0.2 darcy).

This reservoir is connected to an aquifer and coning phenomeneon requires the usage of dz = 1 m in the vicinity of the well. A quarter model is used, grid sizes in x, y and z are 1000m x 500m x 210m.

The aim of this test *was to appreciate the influence of the tilt angle* on the simulation results.

For this test, 2 models were used:
1. the "conventional" model following the principal directions of the layers. Mesh is enormous (final mesh is 71 by 69 = 3963 active cells for a YZ cross-section).
2. the "9 point-scheme model" using an horizontal Y-axis but taking into account a non-diagonal permeability tensor to introduce the tilted layers (smallest final mesh to avoid negative

transmissibilities problem is 20 by 15 = 300 active cells for the same YZ cross-section).

Figure 14 exhibits the YZ cross-section of the different meshes and zooms in on the initialization of oil contacts. The horizontal well is located from y = 0 to y = 250m at z = -1813.5m. Figure 15 depicts the comparison of oil contours in the vicinity of the well. Note that the scale in Z direction is 10 times the scale in Y direction (10° looks like 60°). Model 1 requires an enormous mesh which cannot at the present time be used for 3D simulation. Initialization of oil contacts is not very precise. Transmissibility computations are correct in the direction parallel to the layers but are not correct in the vertical Z direction, therefore break-through is predicted earlier.
This test shows that the "9 point-scheme model" is adequate for horizontal well modelling in stratified geometry and is much more economical, rigorous and precise than the conventional model.

CONCLUSIONS:

Computing transmissibilities with a finite element technique, we could inject non-diagonal tensors in a reservoir simulator. This method is adequate to introduce averaging techniques leading to non-diagonal tensors in numerical simulations. It is also very convenient for modelling a horizontal well in tilted layers, and in this case can decrease the simulation costs compared to the conventional method where the mesh axes align with the dipping directions.

ACKNOWLEDGMENT

We wish to thank the people from Elf Aquitaine reservoir research department for critical comment on this study and O. De Montigny and J. Besson for their modelling experience in horizontal wells. Permission to publish was given by Elf Aquitaine.

REFERENCES

1 **Quintard M. and Whitaker S.** *"Ecoulement Monophasique en Milieu Poreux: Effet des Hétérogénéïtés Locales"*, **Journal de Mécanique théorique et appliquée vol 6 n°5, 1987.**

2 **Bamberger A..** *"Approximation des Coefficients d'Opérateurs Elliptiques"*, **rapport du Centre de Mathématiques Appliquées - Ecole Polytechnique - 1977.**

3 **Forges F.** *"Approximation d'une Equation Diffusion-Transport par des Méthodes d'Elements finis Espace-Temps"*, **Thèse de doctorat de l'Université de Paris IX - june 1980.**

4 **Lake L.W. and Kassap E.** *"An Analytical Method To Calculate the Effective Permeability Tensor of a Grid Block and Its Application in an Outcrop Study"*, **SPE 18434, presented at the SPE Symposium on Reservoir Simulation in Houston, TX, Feb 1989.**

5 **White C.D. and Horne R.N.** *"Computing Absolute Transmissibility in the Presence of Fine Scale Heterogenelty"*, **SPE 16011, presented at the 9th SPE Symposium on Reservoir Simulation in San Antonio, TX, Feb. 1987.**

6 **Yanosik J.L. and Mc Cracken T.A.** *"A Nine-Point Finite Difference Reservoir Simulation for Realistic Prediction of Adverse Mobility Ratio Displacements"*, **SPEJ, Aug. 1979,253-262.**

7 **Coats K.H. and Modine A.D.** *"A Consistent Method for Calculating Transmissibilities in Nine-Point Difference Equations"*, **SPE 12248, presented at the SPE Symposium on Reservoir Simulation in San Francisco, CA, Nov. 1983.**

8 **Bertinger W.I. and Padmanabhan L.** *"Finite Difference Solutions to Grid Orientation Problems Using IMPES"*, **SPE 12250, presented at the SPE Symposium on Reservoir Simulation in San Francisco, CA, Nov. 1983.**

9 **Shah P.C.** *"A Nine-Point Finite Difference Operator for Reduction of the Grid Orientation Effect"*, **SPE 12251, presented at the SPE Symposium on Reservoir Simulation in San Francisco, CA, Nov. 1983.**

10 **Eymard R.** *"Techniques Numériques de Simulation d'Ecoulements Polyphasiques en Milieu Poreux"*, **Thèse de doctorat de l'Université de Savoie, 1988.**

11 **Warren J.E. and Price H.S.** *"Flow in Heterogeneous Porous Media"*, **SPEJ, Sept. 1961.**

12 **Pfertzel A.** *"Sur Quelques Schémas Numériques pour la Résolution des Ecoulements Multiphasiques en Milieu Poreux"*, **Thèse de doctorat de l'Université PARIS-VI, 1987.**

13 **Shiralkar & Stephenson** *"A General Formulation for Simulating Physical Dispersion and a new Nine-point Scheme"*. **SPE 16975, presented at the SPE Annual Technical Conference in Dallas, TX, Sept. 1987.**

14 **Guérillot D, Rudkiewicz J.L., Ravenne Ch, Renard G, Galli A** *"An Integrated Model For Computer Aided Reservoir Description: From Outcrop Study to Fluid Flow Simulations."* **presented at the IOR Symposium in Budapest, April 1989.**

15 **Eymard R., Gallouët T, Joly P** *"Hybrid Finite Element Technics for Oil Recovery Simulation"* **Laboratoire Analyse Numérique, Université PARIS VI, R88015.**

16 King P.R. *"The Use of Renormalization For Calculating Effective Permeability"*, **Transport in Porous Media, 1988.**

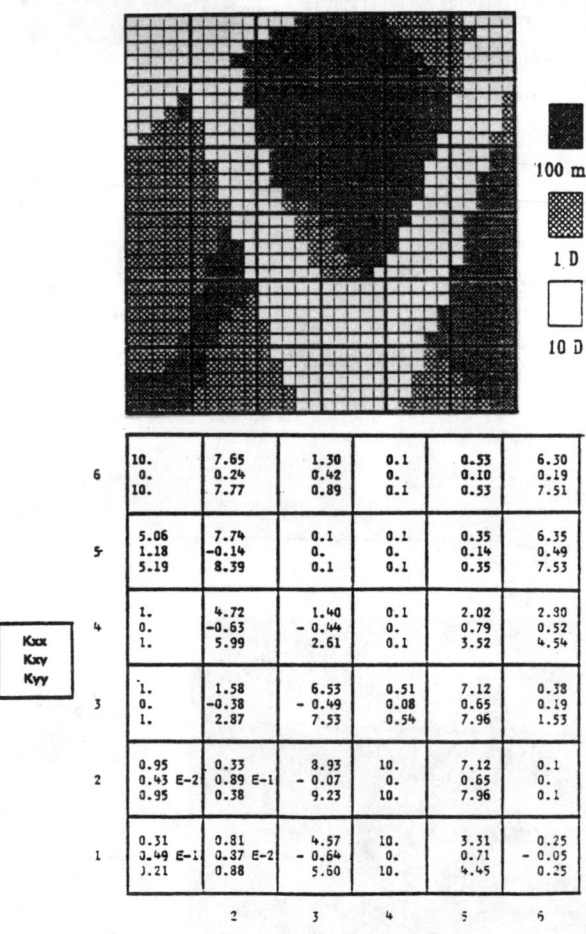

100 mD

1 D

10 D

	Kxx / Kxy / Kyy					
6	10. 0. 10.	7.65 0.24 7.77	1.30 0.42 0.89	0.1 0. 0.1	0.53 0.10 0.53	6.30 0.19 7.51
5	5.06 1.18 5.19	7.74 -0.14 8.39	0.1 0. 0.1	0.1 0. 0.1	0.35 0.14 0.35	6.35 0.49 7.53
4	1. 0. 1.	4.72 -0.63 5.99	1.40 -0.44 2.61	0.1 0. 0.1	2.02 0.79 3.52	2.30 0.52 4.54
3	1. 0. 1.	1.58 -0.38 2.87	6.53 -0.49 7.53	0.51 0.08 0.54	7.12 0.65 7.96	0.38 0.19 1.53
2	0.95 0.43 E-2 0.95	0.33 0.89 E-1 0.38	8.93 -0.07 9.23	10. 0. 10.	7.12 0.65 7.96	0.1 0. 0.1
1	0.31 0.49 E-1 0.21	0.81 0.37 E-2 0.88	4.57 -0.64 5.60	10. 0. 10.	3.31 0.71 4.45	0.25 -0.05 0.25
	2	3	4	5	6	

Figure 9. macro-blocks permeability tensors

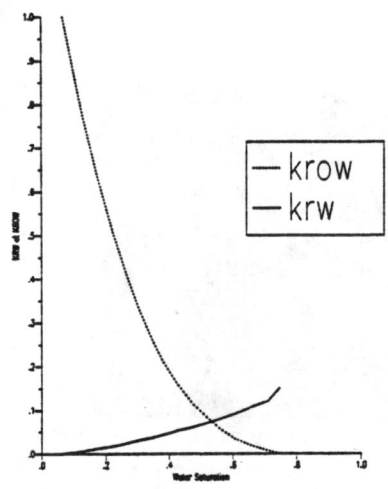

Figure 10. relative permeability curves

SATURATIONS CONTOURS AT 100 DAYS

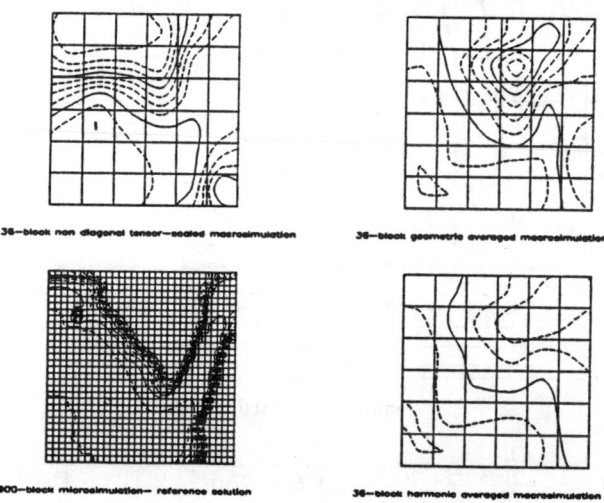

36-block non diagonal tensor-scaled macrosimulation   36-block geometric averaged macrosimulation

900-block microsimulation- reference solution   36-block harmonic averaged macrosimulation

Figure 11. water saturations contours

PRESSURE CONTOURS AT 100 DAYS

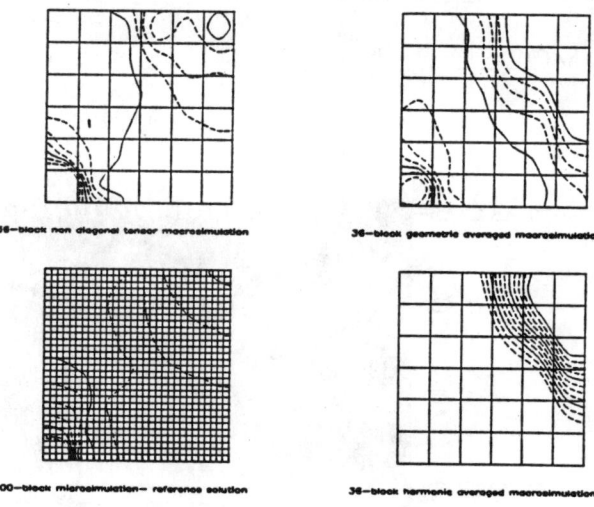

36-block non diagonal tensor macrosimulation   36-block geometric averaged macrosimulation

900-block microsimulation- reference solution   36-block harmonic averaged macrosimulation

Figure 12. pressure contours

129

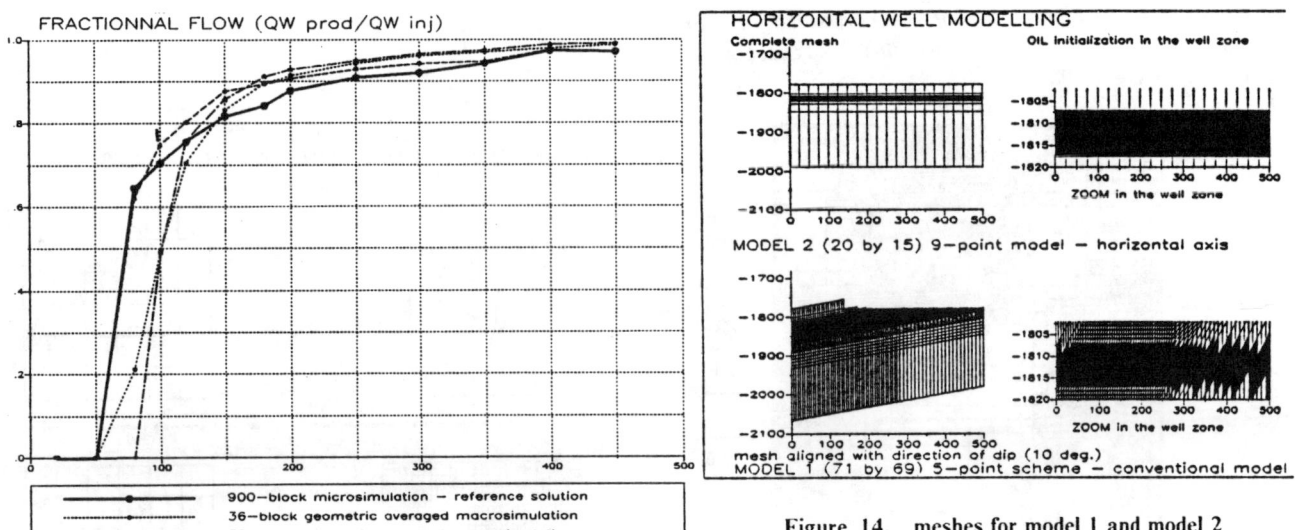

FRACTIONNAL FLOW (QW prod/QW inj)

900—block microsimulation — reference solution
36—block geometric averaged macrosimulation
36—block harmonic averaged macrosimulation
36—block non diagonal tensor macrosimulation

Figure 13.   fractional flow

HORIZONTAL WELL MODELLING

Complete mesh
OIL initialization in the well zone
ZOOM in the well zone

MODEL 2 (20 by 15) 9—point model — horizontal axis

ZOOM in the well zone
mesh aligned with direction of dip (10 deg.)
MODEL 1 (71 by 69) 5—point scheme — conventional model

Figure 14.   meshes for model 1 and model 2

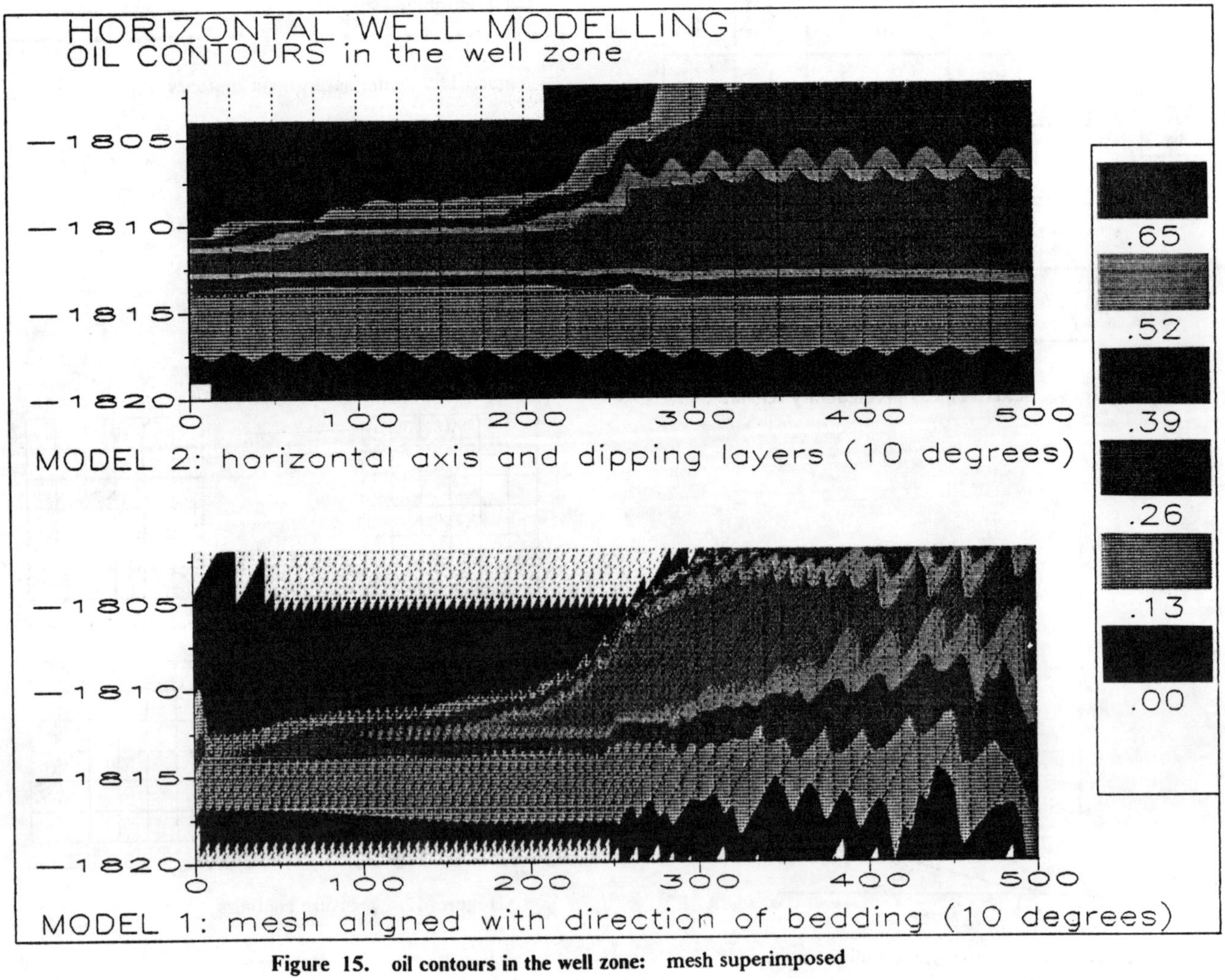

HORIZONTAL WELL MODELLING
OIL CONTOURS in the well zone

.65
.52
.39
.26
.13
.00

MODEL 2: horizontal axis and dipping layers (10 degrees)

MODEL 1: mesh aligned with direction of bedding (10 degrees)

Figure 15.   oil contours in the well zone: mesh superimposed

130

2nd European Conference on the Mathematics of Oil Recovery
© D. Guérillot, O. Guillon (Editors) and Éditions Technip, Paris 1990, pp. 131-138
27 rue Ginoux, 75015 Paris

Implicit Flux Limiting Schemes for Petroleum Reservoir Simulation

M. Blunt and B. Rubin[1]

ABSTRACT

Explicit total variation decreasing (TVD) numerical methods have been used in the past to give convergent, high order accurate solutions to hyperbolic conservation equations, such as those governing flow in oil reservoirs. To ensure stability there is a restriction on the size of time step that can be used. Many petroleum reservoir simulation problems have regions of fast flow away from sharp fronts, which means that this time step limitation makes explicit schemes less efficient than the best implicit methods.

This work extends the theory of TVD schemes to both fully implicit and partially implicit methods. We use our theoretical results to construct schemes which are stable even for very large time steps. In general these schemes are only first order accurate in time overall, but locally may achieve second order time accuracy. Results are presented for a one dimensional Buckley Leverett problem, which demonstrate that these methods are more accurate than conventional implicit algorithms and more efficient than explicit methods, where smaller time steps must be used. Results from black oil and compositional simulators are presented.

1 INTRODUCTION

Explicit flux limiting schemes for the solution of hyperbolic conservation laws have already been discussed in some detail (Sweby, 1984, 1985). For one dimensional scalar equations, it is possible to show that these schemes are total variation decreasing or TVD, which means that

the solutions are stable and will generally converge to the correct physical solution, even if the solution contains discontinuities or shocks. The schemes maintain second order accuracy in smooth regions and sharply resolve any discontinuities.

In this paper we will construct TVD schemes which are particularly appropriate for solving the system of component conservation laws in petroleum reservoir simulation. Usually the pressure field is solved implicitly, followed by an explicit update of the conservation equations, for which high order techniques can be used. For large multidimensional problems almost all the computer time is spent solving for the fluid pressures. The saturation update is comparatively fast. It is thus worthwhile to use a stable high order technique, which improves the resolution of the simulation, without appreciably increasing the total time used. The application of high order Godunov schemes in reservoir simulation has been discussed by Trangenstein et al (1986, 1989). Another second order method, flux corrected transport, has been used in multidimensional fluid flow simulations by Christie and Bond (1984). It is most efficient to solve the pressure equation only after appreciable changes in the pressure field, which may require large time steps. However, the explicit methods which have been proposed may then become unstable.

In radial flow or quasi-static situations with a front moving slowly perpendicular to a fast flow direction, such as gravity over-ride or coning, the regions of high flow drastically reduce the time step possible from explicit formulations. Thus stable implicit methods with a large time step are often used. However, implicit models are normally only first order accurate in time and space. In the regions of moderate to low flow rates, higher order techniques would greatly improve the accuracy of the models' predictions.

The ideal numerical scheme for reservoir engineer-

(1) BP Research Centre, Chertsey Road, Sunbury-on-Thames, Middlesex, TW16 7LN, UK.

ing problems would be stable for large time steps and yet resolve shock fronts accurately. This work finds such a scheme by using partially implicit methods and total variation decreasing flux limiters. The schemes are second order accurate in smooth regions, but are still stable in areas of fast flow.

2 CONSTRUCTING SCHEMES

In this section we will develop TVD schemes for a one dimensional scalar conservation equation. We begin by constructing a method which is at best second order spatially accurate and then consider temporal stability and accuracy. At the end of this section we will discuss schemes coupled with a parabolic pressure equation.

For an initial example, we will describe schemes for the solution of the Buckley-Leverett equation. This equation describes the one dimensional flow of two incompressible fluids ("water" and "oil") in a porous medium. It can be written as:

$$\frac{\partial s}{\partial t} + \frac{\partial f}{\partial x} = 0 \qquad (2.1)$$

where the flux $f(s)$ is a known function of s the saturation of the water phase: $f(s) = V f_w(s)$ where V is the total velocity of the two phase system.

2.1 Explicit Flux Limited Schemes

In this section we will briefly review explicit TVD schemes before constructing novel implicit methods. We will develop flux limiting schemes following the approach of Sweby (1984, 1985). These schemes have also been described by van Leer (1973, 1974, 1977), Sweby and Baines (1985) and Roe (1985). Other explicit higher order methods include Godunov schemes (Godunov, 1959), which have been applied to black oil models (Trangenstein, 1986) and flux corrected transport (Zalesak, 1979), which has been extended to multidimensional, multicomponent fluid flows (Christie and Bond, 1984).

First order schemes, such as the Engquist-Osher method (1984) are stable and convergent and obey the TVD criteria in the appendix. However, numerical dispersion smears out the shock fronts and so the solutions are only accurate if a large number of grid blocks are used. Most unconstrained second order schemes such as Lax-Wendroff and two point upstream weighting suffer from less numerical dispersion, but produce spurious oscillations about shock fronts.

Flux limiting methods are both accurate and stable. Thus discontinuities are resolved accurately without either excessive smearing or unphysical oscillatory instabilities. This is achieved by limiting the second order corrections to the numerical flux to obey the TVD criteria. If this is done the total variation in the solution will not increase with time: a monotonic profile (no maxima or minima) will remain monotonic without developing unstable and unphysical blips.

We write an explicit finite difference approximation to the saturation upate as follows:

$$s_i^{n+1} = s_i^n - \lambda[F_{i+1/2}^n - F_{i-1/2}^n] \qquad (2.2)$$

The superscripts and subscripts refer to the time step and the grid block respectively. $\lambda = \Delta t / \Delta x$. $F_{i+1/2}$ and $F_{i-1/2}$ represent numerical approximations to the flux, f, across the grid cell edges. If we assume that the characteristic velocity of the saturation profile, $v = df/ds$, is positive, then we find an upstream weighted approximation for F, which is accurate to second order in space (i.e; up to $O(\Delta x)^2$) and first order in time, $O(\Delta t)$:

$$F_{i+1/2}^n = F_i^n + \frac{\phi_{i+1/2}^n}{2}(F_{i+1}^n - F_i^n) \qquad (2.3)$$

F_i^n represents the flux f calculated at s_i^n. The first term in (2.3) gives single point upstream weighting. The second terms are $O(\Delta x)^2$ corrections to the numerical fluxes. ϕ, the limiter, is function of r, which is a ratio of successive flux differences:

$$r_{i+1/2}^n = \frac{F_i^n - F_{i-1}^n}{F_{i+1}^n - F_i^n} \qquad (2.4)$$

and $\phi(r_{i+1/2}^n)$ is written as $\phi_{i+1/2}^n$ in (2.3).

The results of the appendix may be used to find limits on the function $\phi(r)$ such that the scheme remains TVD. Derivations of the TVD region are given by Sweby (1984), and are also demonstrated in section 2.3 as a special case of a partially implicit method. We would also like the scheme to be second order accurate in monotonic portions of the profile. A suitable choice of function, which gives good results is the van Leer limiter:

$$\phi = \frac{r + |r|}{1 + r} \qquad (2.5)$$

It is worth pointing out that when $\phi = 0$ the scheme reverts to single point upstream, which is only first order accurate. When $\phi = r$, the scheme is the two point upstream scheme. When $\phi = 1.0$, the scheme is the midpoint scheme and when $\phi = 2.0$, the scheme uses a downstream weighted flux. All of these values of ϕ are possible when the van Leer limiter is used.

The maximum time step that can be used before the method becomes unstable (the TVD criteria are violated) is dependent on the Courant-Friedrichs-Lewy (1928), or CFL condition. If v_{max} is the largest speed (df/ds) encountered, then the CFL number is defined as λv_{max}. The van Leer limiter is stable for CFL numbers less than or equal to 1/2.

Figure 1 compares this TVD scheme with the single point upstream scheme (all explicit) for a non-linear waterflood using the following flux function:

$$f(s) = \frac{3s^2}{3s^2 + (1-s)^2} \qquad (2.6)$$

and a CFL number of 0.4. The exact solution (shown by the solid curve) has a shock front from a saturation of 0.5 to zero. The single point upwind scheme ($\phi = 0$) smears this front over several grid blocks. The TVD method decreases the numerical diffusion and resolves the discontinuity over approximately half the grid blocks required by the first order method.

This is a simplified version of the TVD schemes previously described (Sweby, 1984) since we do not have a second order time correction to the flux. Time correction will be introduced by using a partially implicit scheme. The advantage of this is that stable solutions can still be calculated, even for very large CFL numbers.

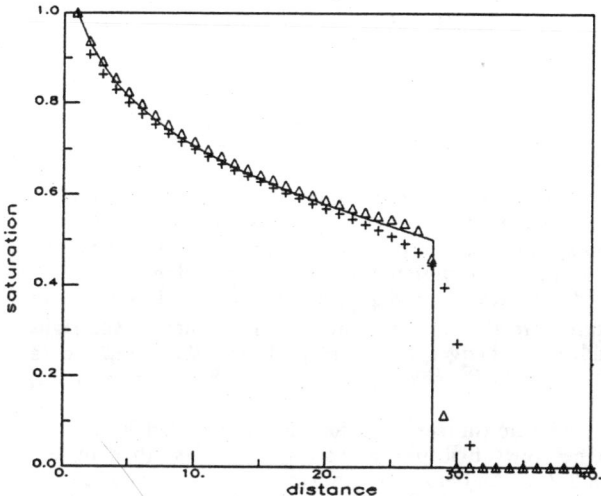

FIG. 1. Solutions with explicit schemes and a CFL number 0.4. Crosses, first order upstream; triangles, second order TVD method. The solid line is the exact solution.

2.2 Implicit Methods

Based on the explicit scheme, an implicit version of the TVD midpoint scheme will now be developed. Implicit modelling requires the calculation and use of fluxes and limiters at the unknown $n + 1th$ time level.

Following from the previous section the scheme is set up as follows:

$$s_i^{n+1} - s_i^n + \lambda[F_{i+1/2}^{n+1} - F_{i-1/2}^{n+1}] = 0 \qquad (2.7)$$

As before, if df/ds is positive, an upstream weighted approximation for F is:

$$F_{i+1/2}^{n+1} = F_i^{n+1} + \frac{\phi_{i+1/2}^{n+1}}{2}(F_{i+1}^{n+1} - F_i^{n+1}) \qquad (2.8)$$

F_i^{n+1} represents the flux f calculated at s_i^{n+1}. ϕ, the limiter, is function of r, where r is now calculated at the $n + 1th$ time level from (2.4). As above we choose the van Leer limiter, (2.5).

2.2.1 Solution by Newton-Raphson Iteration

Equation (2.7) can be solved for s_i^{n+1} by Newton-Raphson iteration (Wilkes, 1966; Froberg, 1981). Equa-

tion (2.7) may be written as: $G(s_i^{n+1}) = 0$, where:

$$G(s_i^{n+1}) = s_i^{n+1} - s_i^n + \lambda[F_{i+1/2}^{n+1} - F_{i-1/2}^{n+1}] \qquad (2.9)$$

and s_i^n is known, and the numerical fluxes, F, as functions of s_i^{n+1} may be calculated as above. The solution is obtained iteratively. For the first iteration ($k = 1$) we use $s_i^{n+1}(1) = s_i^n$. Subsequent estimates for the updated saturations are found as follows. Firstly we define the difference in saturation between two Newton iterations: $\delta s_i = s_i^{n+1}(k + 1) - s_i^{n+1}(k)$. We assume that $s_i^{n+1}(k)$ is known and we calculate δs to find a more accurate estimate of s_i^{n+1}. Then to first order in the small quantity δs:

$$G_i^{k+1} = G_i^k + \left(\frac{\partial G_i}{\partial s_j}\right)^k \delta s_j \qquad (2.10)$$

where the superscript on G labels the iteration, and the subscript labels the grid block. The value of δs_j which makes $G_i^{k+1} = 0$ obeys the expression:

$$\left(\frac{\partial G_i}{\partial s_j}\right)^k \delta s_j = -G_i^k \qquad (2.11)$$

$\partial G_i/\partial s_j$ is the Jacobian, which must be inverted to calculate δs in (2.11). For a first order scheme G_i is a function only of the saturations in the cells i and $i - 1$, and hence the Jacobian has a simple lower triangular form. However, in the more sophisticated high order schemes considered here, the Jacobian may have many non-zero elements, which makes the matrix inversion difficult. For this reason we simpify our treatment and consider only first order terms in the calculation of $\partial G_i/\partial s_j$. The quantity $-G_i^k$ on the right hand side of (2.11) is still calculated to second order, and hence we converge to the correct high order solution.

The iteration continues until the calulated δs_j, in (2.11) is smaller than some convergence criterion. In the examples we present, we calculated the maximum δs_j over all the grid blocks. If this was less than 10^{-3} the iteration was stopped. Convergence to machine accuracy usually took about another 3 iterations on simple problems. The efficiency of any implicit algorithm is limited by the number of iterations necessary to produce convergence.

2.2.2 Results

In the next section we will show that this scheme is stable for all CFL numbers.

Figures 2 and 3 show results from implicit schemes, for the flux function given by (2.6). Fig. 2 shows results using a first order implicit flux at CFL numbers of 0.4, 1 and 4, which required on average 3, 3 and 5 Newton iterations per time step respectively. Notice that the single point upstream implicit scheme is more diffusive than its explicit counterpart (Fig. 1) at the same CFL number. For CFL numbers greater than 1, outside the explicit stability limit, the shock front is very poorly resolved.

Figure 3 shows solutions using a TVD flux limiter at CFL numbers of 0.4, 1 and 4 which needed on average

3, 4 and 6 Newton iterations per time step respectively to ensure convergence. The results are superior to the first order scheme. The shock front is still more diffuse at a CFL number of 0.4 compared with an explicit method (Fig. 1), but the scheme is stable for larger timesteps.

Yee *et al* (1986a, 1986b, 1985) have proposed high order implicit schemes, but found that convergence to the correct solution by Newton iteration using the full Jacobian was very slow if a high order limited implicit flux was used. They suggested a linearized non-conservative TVD implicit scheme. The number of extra Newton iterations required for convergence with a high order method in the example above, however, was small using a first order Jacobian and was more than compensated by the improved accuracy of the solution.

In the examples below a low order Jacobian is always used, and the number of iterations needed for convergence only ever exceeds a fully first order method by one or two.

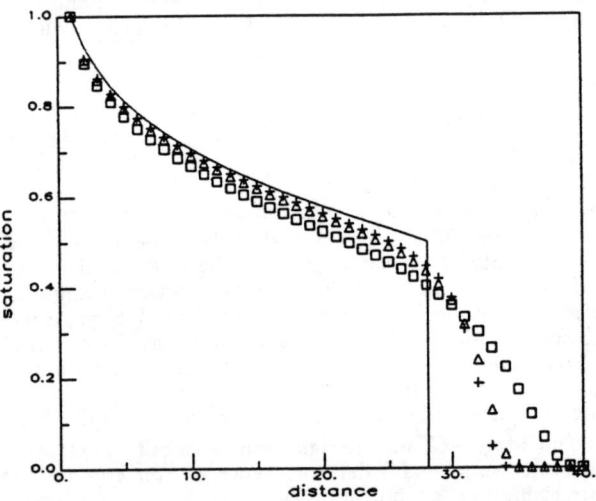

FIG. 2. Solutions with a first order implicit scheme. Squares, CFL number of 4; triangles, CFL=1; crosses, CFL=0.4.

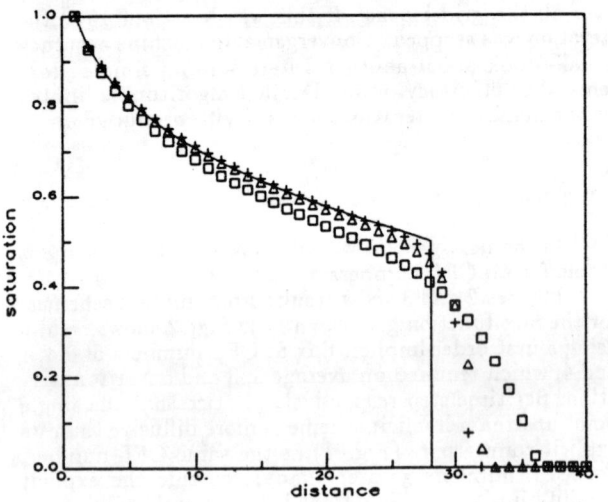

FIG. 3. Solutions with a second order implicit TVD scheme. Squares, CFL number of 4; triangles, CFL=1; crosses, CFL=0.4.

2.3 Schemes with Temporal Weighting

We will now write down a conservative difference equation for the saturation update, which is partially implicit. In the previous section we described a scheme which was first order accurate in time, but stable for large time steps. In this section we will use a partially implicit method, which gives improved time accuracy without the CFL constraint seen in totally explicit schemes. The form of the update is as follows:

$$s_i^{n+1} - s_i^n = -\lambda[(1 - \theta_{i+1/2})F_{i+1/2}^n - (1 - \theta_{i-1/2})F_{i-1/2}^n]$$
$$- \lambda[\theta_{i+1/2}F_{i+1/2}^{n+1} - \theta_{i-1/2}F_{i-1/2}^{n+1}] \qquad (2.12)$$

Yee *et al* (1986a, 1986b, 1985) have studied TVD schemes based on this equation. However, as mentioned above they solved a linearized form of this equation and only considered a fixed value of θ. Our work will be conservative, at best second order spatially accurate for both explicit and implicit fluxes, and will also allow the values of θ be different for different cell edges.

The parameters $\theta_{i+1/2}$ and $\theta_{i-1/2}$ give the degree of implicit weighting in the numerical flux across the right and left hand edges of cell i respectively. $\theta = 1$ represents a fully implicit scheme, (2.7), while $\theta = 0$ is explicit, (2.2).

The non-linear equation (2.12) is solved iteratively by the Newton-Raphson technique, as described in section 2.2.1. As before we find an upstream weighted approximation for $F_{i+1/2}^n$ using equation (2.3). $F_{i+1/2}^{n+1}$ is calculated from (2.8). The limiter ϕ is chosen to make the spatial approximation in (2.12) both accurate and TVD. The weighting θ controls the temporal accuracy of the solution: $\theta = 1/2$ gives a scheme which is second order in time.

We may write (2.12) out in full and compare it with eqn (A1) of the appendix to find the criteria for the scheme to be TVD. If we first define effective velocities or flux gradients, given by a spatial derivative;

$$v_i^n = \frac{F_i^n - F_{i-1}^n}{s_i^n - s_{i-1}^n} \qquad (2.13)$$

and a temporal derivative;

$$u_i^n = \frac{F_i^{n+1} - F_i^n}{s_i^{n+1} - s_i^n} \qquad (2.14)$$

we derive, after some algebra, the limits:

$$1 + \lambda u_i^n(\theta_{i+1/2} - \theta_{i-1/2}) \geq$$
$$\lambda v_i^n \left[(1 - \theta_{i-1/2})(1 - \frac{\phi_{i-1/2}^n}{2}) + (1 - \theta_{i+1/2})\frac{\phi_{i+1/2}^n}{2r_{i+1/2}^n}\right] \geq 0$$
$$(2.15)$$

and

$$\lambda v_i^{n+1} \left[\theta_{i-1/2}(1 - \frac{\phi_{i-1/2}^{n+1}}{2}) + \theta_{i+1/2}\frac{\phi_{i+1/2}^{n+1}}{2r_{i+1/2}^{n+1}}\right] \geq 0$$
$$(2.16)$$

Equations (2.15) and (2.16) are the major theoretical results of this paper. A scheme which conforms to the constraints above for all grid blocks i will be TVD.

2.3.1 Explicit Schemes

A partially implicit scheme will have values of θ in the region $0 \leq \theta \leq 1$. For an explicit scheme, $\theta_{i+1/2} = 0$ in all grid blocks i. Equation (2.16) reduces trivially to $0 \geq 0$, but (2.15) gives the following restriction:

$$1 \geq \lambda v_i^n \left(1 - \frac{\phi_{i-1/2}^n}{2} + \frac{\phi_{i+1/2}^n}{2r_{i+1/2}} \right) \geq 0 \qquad (2.17)$$

λv_i^n is always chosen to be positive. This expression must be obeyed for all choices $\phi_{i+1/2}$ and $\phi_{i-1/2}$. If we require:

$$2 \geq \phi(r) \geq 0 \qquad (2.18)$$

and

$$2 \geq \frac{\phi(r)}{r} \geq 0 \qquad (2.19)$$

then (2.17) is obeyed subject to the stability limit $\lambda v_i^n \leq 1/2$. If this holds everywhere, the CFL number must be less than $1/2$. We mentioned this condition in section 2.1.

2.3.2 Implicit Schemes

Here $\theta_{i+1/2} = \theta_{i-1/2} = 1$. In this case it is (2.15) which reduces to a trivial expression. From (2.16) we find:

$$\lambda v_i^{n+1} \left(1 - \frac{\phi_{i-1/2}}{2} + \frac{\phi_{i+1/2}}{2r_{i+1/2}} \right) \geq 0 \qquad (2.20)$$

If we use the same constraints on the function ϕ as in equations (2.18) and (2.19), then (2.20) is satisfied for all CFL numbers. If the limiter is TVD, there is no restriction on the size of timestep.

Before discussing the implementation of partially implicit schemes in section 3, we will briefly mention extensions of the method for variable flow directions and for fluxes derived from a pressure gradient.

2.3.3 Flow Reversal and Sonic Points

We assumed that there is an unambiguous definition of the direction of the flow, which does not change sign over a single grid block or during a time step. Imagine that we attempt to calculate the numerical fluxes $F_{i+1/2}^n$ and $F_{i+1/2}^{n+1}$. If v_i^n, v_{i+1}^n, v_i^{n+1} and v_{i+1}^{n+1} are not all of the same sign, then the upstream direction is ambiguous, and a first order Engquist-Osher (1984) flux must be used. This is defined as follows:

$$F_{i+1/2}^n = F_i^n + \int_{s_i^n}^{s_{i+1}^n} \min \left(\frac{\partial f}{\partial s}, 0 \right) ds \qquad (2.21)$$

with a similar expression for the $n+1th$ time level. Equation (2.25) reduces to an upstream weighted flux unless df/ds changes sign in the integrand (there is a sonic point). If df/ds is positive in both integrals then the upstream direction is well defined, and second order corrections to the flux may be calculated as described above. If df/ds is negative then $F_{i+1/2}^n = F_{i+1}^n$ and similarly for the $n+1th$ time level. The second order fluxes are calculated as before except that the subscript i is replaced by $i+1$ and $i-1$ by i.

If s_i and s_{i+1} straddle a sonic point then (2.21) gives: $F_{i+1/2} = f_{min}$ where f_{min} is the minimun value of f; i.e., its value where $v = df/ds = 0$. If f is known analytically, f_{min} may be calculated exactly. Where this is not the case, a quadrature rule, using only the cell centred values of f will also produce satisfactory results (Sweby, 1984).

If a sonic point is detected, then the Engquist-Osher fluxes should be used instead of the upwind values. This ensures that an unphysical static discontinuity is not produced at sonic points, (Sweby, 1984).

2.3.4 Schemes with a Pressure Gradient

We now describe cases when the flux f is derived from the gradient of a scalar field (typically the fluid pressure), where the flux function in (2.1) is given by: $f = -T(s)\frac{dP}{dx}$. The transmisibility T and the pressure P are defined at the grid cell centres. Then we write our schemes as above, but now we define the numerical fluxes by: $F_i^n = T_i^n \nabla P_i^n$, where $\nabla P_i^n = (P_{i+1}^n - P_i^n)/\Delta x$, with similar expressions at time $n+1$ and grid blocks $i+1$ and $i-1$. All the equations to update s are then just as before.

3 PRACTICAL IMPLEMENTATION

In the previous section we developed the theory of high order flux limiting schemes to include partially implicit methods. One special case, which had been studied previously, was an explicit scheme, for which we derived TVD criteria for the flux limiter ϕ. If the same limiter was used for a totally implicit formulation, then we arrived at a spatially second order accurate method which was stable for all finite CFL numbers.

We propose to use the same spatial flux limiter for our partially implicit schemes. However, this still leaves the choice of θ, the Crank-Nicholson weighting parameter undetermined. In this section we discuss several possible choices for θ, derive the stability limits for the proposed schemes and present example numerical results.

3.1 A Fixed Implicit Weighting

The analysis of the previous section allowed for a weighting θ which could change from one grid cell edge to the next. We could simplify the equations considerably by having a fixed θ, which would be determined by a global stability constraint. We will assume that θ lies in the range $1 \geq \theta \geq 0$.

We start from (2.16) with $\theta_{i+1/2} = \theta_{i-1/2} = \theta$, from which we obtain:

$$\lambda v_i^{n+1} \theta \left(1 - \frac{\phi_{i-1/2}^{n+1}}{2} + \frac{\phi_{i+1/2}^{n+1}}{2r_{i+1/2}^{n+1}} \right) \geq 0 \qquad (3.1)$$

and (2.15) gives a similar condition:

$$1 \geq \lambda v_i^n (1-\theta) \left(1 - \frac{\phi_{i-1/2}^n}{2} + \frac{\phi_{i+1/2}^n}{2r_{i+1/2}^n} \right) \geq 0 \qquad (3.2)$$

Equation (3.1) and the right hand limit of (3.2) are always obeyed for $1 \geq \theta \geq 0$. To find the global restriction on θ we consider the most constraining values of $\phi_{i-1/2}$, $\phi_{i+1/2}$ and v_i^n in (3.2). These are: $\phi_{i-1/2} = 0$, $\phi_{i+1/2}/r_{i+1/2} = 2$ and $v_i^n = v_{max}$. The left hand limit in (3.2) then becomes:

$$\lambda v_{max}(1-\theta) \leq \frac{1}{2} \qquad (3.3)$$

Equation (3.3) can be used to find the values of θ for which a scheme at a given CFL number will be stable.

If $\theta = 1/2$, the scheme is second order accurate in time, but for large CFL numbers a value of θ closer to 1 must be chosen to maintain the TVD property. A natural choice of scheme is one where θ is chosen to be as close to 1/2 as possible; i.e.,

$$\theta = \max \left[\frac{1}{2}, 1 - \frac{1}{2\lambda v_{max}} \right] \qquad (3.4)$$

Notice that this method maintains second order accuracy for CFL numbers less than 1, which allows timesteps twice as large as the explicit methods described in the previous section. However, some explicit TVD schemes which are also second order time accurate can be stable up to CFL numbers of 1 (Sweby, 1984). The advantage with this partially implicit method is that larger time steps may be automatically accommodated.

3.1.1 Results

Numerical solutions to the Buckley–Leverett equation are shown in Fig. 4 at CFL numbers of 0.4, 1 and 4. The number of Newton iterations per timestep were 3, 3 and 5 respectively, which is similar to the cases illustrated in Fig. 3. For the solutions with CFL numbers less than or equal to 1, $\theta = 1/2$ and the scheme is second order accurate in both space and time. The resolution of the shock front is comparable with that for the explicit TVD scheme shown in Fig. 1. For a CFL number of 4, $\theta = 7/8$ from (3.4). Although the shock is now smeared over several grid blocks, the solution is as accurate as a first order implicit scheme, which has a time step ten times smaller. The use of a TVD partially implicit scheme can greatly enhance the accuracy and efficiency of the saturation update.

3.2 Variable Weighted Schemes

We extend our analysis to the general case where θ is not fixed, but is only constrained by local, not global, stability criteria. This approach is of use in circumstances where the flow speed changes appreciably across the grid, and where conventional methods would unnecessarily degrade the accuracy of the solution in slow regions in order to ensure stability in portions of fast flow.

There is, unfortunately, no simple method for chosing θ to obey (2.15) and (2.16). However, it is possible to derive the following expression for the temporal weighting:

$$\theta_{i+1/2} = \max \left[\frac{1}{2}, 1 - \frac{1}{\lambda \left(u_i^n + \frac{\phi_{i+1/2}^n}{2r_{i+1/2}^n} v_i^n \right)} \right] \qquad (3.5)$$

if $u_i^n \geq v_i^n$ and:

$$\theta_{i+1/2} = \max \left[\frac{1}{2}, 1 - \frac{1 - \frac{\lambda}{2}(v_i^n - u_i^n)}{\lambda \left(u_i^n + \frac{\phi_{i+1/2}^n}{2r_{i+1/2}^n} v_i^n \right)} \right] \qquad (3.6)$$

if $v_i^n > u_i^n$, with similar expressions for $\theta_{i-1/2}$.

Although (3.5) and (3.6) may appear involved, they are easily coded in practice. All the parameters in the equations are known at the nth time step, except for the time averaged velocity u, (2.14). An estimate of u is made at each Newton iteration.

3.2.1 Results

Numerical results are shown in Fig. 5. For CFL numbers of 0.4 and 1, θ was fixed at 1/2, as in the scheme described in section 3.1. For a CFL number of 4, the value of θ was larger in regions of fast flow. However, since in this example the shock front moves with almost the maximum wave speed, the resolution of the discontinuity is similar to that for a fixed value of θ.

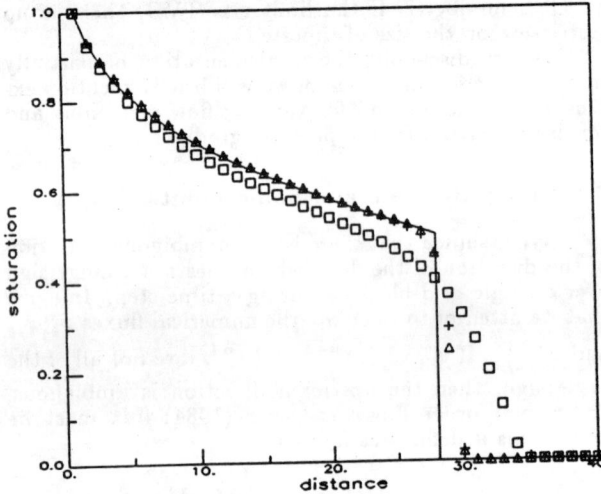

FIG. 4. Solutions with a TVD partially implicit scheme. Squares, CFL number of 4; triangles, CFL=1; crosses, CFL=0.4.

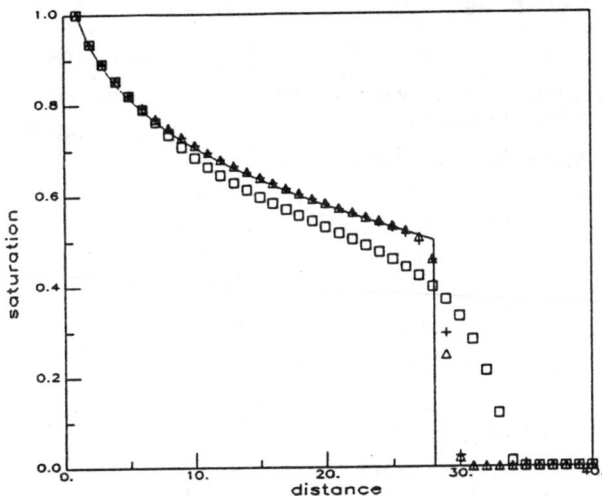

FIG. 5. Solutions with a TVD variably implicit scheme. Squares, CFL number of 4; triangles, CFL=1; crosses, CFL=0.4.

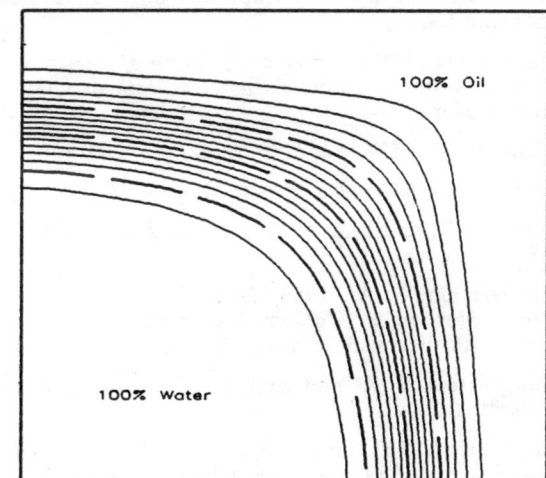

FIG. 6. Single point upstream implicit quarter five spot model. 30x30 blocks.

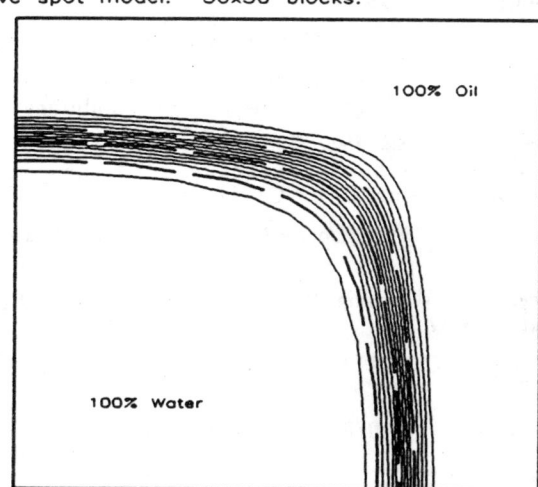

FIG. 7. TVD implicit quarter five spot model. 30x30 blocks.

3.3 Results from a Reservoir Simulator

The schemes described above, with $\theta = 1$ or $\theta = 0$ have been incorporated into black oil and compositional reservoir simulators. The pressure equation is solved implicitly with either fully implicit or fully explicit updates of the component conservation laws using the formulation of Young and Stevenson (1983). Full details of how this is done are given elsewhere (Rubin and Blunt, 1991).

We will present two sets of example results here. Figs. 6 and 7 show implicit two dimensional simulations of water displacing oil in a quarter five spot using 30×30 grid blocks and a CFL number of 2. Notice that numerical diffusion considerably smears the front for a single point upstream weighted scheme (Fig 6). To achieve the same resolution as obtained for the TVD update (Fig 7) a simulation with at least 80×80 blocks is required, which takes six times more cpu time.

Figure 8 shows the oil saturation from a 19 component compositional model. This system is described by Lee *et al* (1988). For a 125 grid block simulation, the explicit TVD method captures the features of the fine grid solution, while the single point upstream scheme not only diffuses the front, but seriously over-estimates the oil saturation at short distances. To obtain satisfactory results from this scheme requires 375 grid blocks and a computational time four times that of the TVD solution.

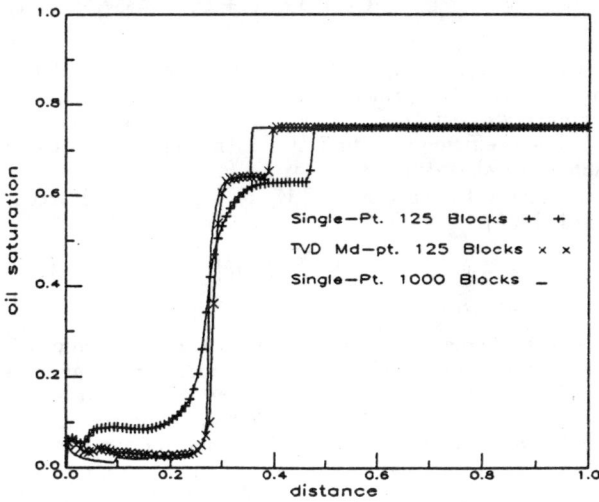

FIG. 8. Compositional model with 125 grid blocks.

4 CONCLUSIONS

We have presented several numerical schemes for the solution of conservation laws, which are stable regardless of the time step used and are at best second order spatially accurate. We developed these schemes by extending the theory of TVD flux limiters to totally and partially implicit schemes.

Compared with first order schemes these methods

offer substantially improved accuracy, particularly where the flow speeds vary across the mesh. In the examples presented we obtain the same accuracy with about one quarter to one sixth the cpu time.

Further work to apply the variable implicit scheme to a sequential reservoir simulator is required.

Acknowledgements

We are grateful to P K Sweby and J J Barley for their useful comments. We thank the British Petroleum Company plc for permission to publish this paper.

Appendix

In this appendix we define a total variation decreasing scheme and write down TVD criteria for a partially implicit numerical scheme. We can only prove these properties for a one dimensional scalar equation.

We may write a partially implicit scheme for calculating the saturation profile s as follows:

$$s_i^{n+1} - s_i^n = -C_{i-1}^n \Delta s_{i-1}^n + D_i^n \Delta s_i^n \\ -C_{i-1}^{n+1} \Delta s_{i-1}^{n+1} + D_i^{n+1} \Delta s_i^{n+1} \quad (A1)$$

where the superscripts refer to the time level and the subscripts to the grid cell in which s is evaluated, and we have written: $\Delta s_{i-1} = s_i - s_{i-1}$ and $\Delta s_i = s_{i+1} - s_i$. The coefficients, C and D, depend on the particular numerical algorithm used.

The total variation, TV, is defined as (Harten, 1983, 1984):

$$TV^n = \sum_i |\Delta s_i^n| \quad (A2)$$

A TVD scheme has $TV^{n+1} \leq TV^n$. A TVD scheme will converge to the physical solution under certain general conditions (Harten, 1983), but these do not automatically include the partially implicit methods we discuss in the text. Nevertheless, TVD schemes do possess two useful properties, which in the numerical examples we present are sufficient to give reliable and convergent solutions. The first property is that a TVD scheme will not allow unphysical oscillations to develop: a monotonic saturation profile will remain monotonic. Secondly, a TVD scheme remains bounded, which implies that it is stable.

It is possible to show (Harten, 1983, 1984) that a difference scheme written as in (A1) is TVD provided the coefficients satisfy:

$$C_i^n \geq 0 \quad D_i^n \geq 0 \\ C_i^{n+1} \geq 0 \quad D_i^{n+1} \geq 0 \\ 1 \geq C_i^n + D_i^n \geq 0 \quad (A3)$$

for all grid blocks i.

References

Christie, M. A. and Bond, D. J., 1987 *SPERE* **2** 514-522

Courant, R., Friedrichs, K. and Lewy, H., 1928 *Mathematische Annalen* **100** 32

Engquist, B. and Osher, S., 1984 *Math Comp* **34** 4575

Froberg, C-E., 1981 *Introduction to Numerical Analysis* Addison–Wesley, Reading, Massachusetts

Godunov, S. K., 1959 *Mat Sb* **47** Translation by US Dept of Commerce JPRS 7225 (1960)

Harten, A., 1983 *J Comput Phys* **49** 357

Harten, A., 1984 *SIAM J Numer Anal* **21** 1

Lee, S. T., *et al*, 1988 *SPE 18062* presented at the 63rd SPE Technical Conf, Houston, TX.

Roe, P. L., 1985 *AMS–SIAM Seminar on Numerical Methods in Fluid Mechanics*, American Math Soc, Providence

Rubin, B. and Blunt, M. J., 1991 *SPE 21222* to be presented at the 11th SPE Symposium on Res Simulation, Anaheim, CA.

Smith, G. D., 1985 *Numerical Solution of Partial Differential Equations: Finite Difference Methods* Clarendon Press, Oxford

Sweby, P. K., 1984, *SIAM J Numer Anal* **21** 995

Sweby, P. K., 1985, *Lectures in Applied Math* **22** 289

Sweby, P. K. and Baines, M. J., 1985 *J Comput Phys* **56** 135

Trangenstein, J. A., 1986, *Multiphase Flow in Porous Media: Mechanics, Mathematics and Numerics* Lecture Notes, IBM Scientific Center, Bergen, Norway

Trangenstein, J. A. and Bell, J. B., 1989 *SIAM J App Math* **49** 1

Van Leer, B., 1973 *Towards the Ultimate Finite Difference Scheme, I* Lecture Notes in Physics **18** Springer, Berlin

Van Leer, B., 1974 *J Comput Phys* **14** 361

Van Leer, B., 1977 *J Comput Phys* **23** 1

Wilkes, M. V., 1966 *A Short Introduction to Numerical Analysis* Cambridge University Press, Cambridge

Yee, H. C., 1986a *Proc Seminar on Computational Aerodynamics* AAA Special Publication ed M Hafez, University of California, Davis, California

Yee, H. C., 1986b *Computers and Maths with Applications* **12A** 413

Yee, H. C., Warming, R. F. and Harten, A., 1985 *J Comput Phys* **57** 327

Young, C. C. and Stephenson, R. E., 1983 *SPEJ* **23** 727-742

Zalesak, S. T., 1979 *J Comput Phys* **31** 355

2nd European Conference on the Mathematics of Oil Recovery
© D. Guérillot, O. Guillon (Editors) and Éditions Technip, Paris 1990, pp. 139-146
27 rue Ginoux, 75015 Paris

The Use of Boundary Element Method in Front Tracking for Composite Reservoirs

J. Kikani[1] and R. N. Horne[2]

ABSTRACT

This paper demonstrates a new approach using the Boundary Element Method (BEM) to solve for pressure transient behavior in composite and sectionally homogeneous reservoirs.

A boundary element solution is proposed in Laplace space to a piecewise homogeneous reservoir with arbitrary geometry of each region. Any number of such regions with different rock and fluid properties can be included in the solution procedure. This formulation can solve fluid injection problems which show composite behavior (as in steam injection and CO_2 flooding). In addition, impermeable barriers of any shape and orientation as well as large pressure support sources (aquifers) can be included.

At the same time as pressure transients are calculated, fluid velocity is determined at the phase front. Injection fluid front velocity is therefore calculated as part of the solution and the fronts can be tracked quite accurately. Example calculations for radial and rectangular composite reservoirs are presented.

BEM solutions are more accurate than other numerical methods such as finite element and finite difference techniques because the method retains the free space Green's function of the governing diff-

erential operator as a global weighting function in the integral equation.

1. INTRODUCTION

Long term performance of a reservoir is intimately associated with the breakthrough of unwanted fluids. These unwanted fluids could be the cap gas, bottom water or injected fluids. In order to predict the fluid breakthrough times one needs to keep track of the movement of fluid interfaces. This is a fairly difficult task in three phase, three dimensional situations. A host of common and special techniques have been suggested to track the fluid fronts in the reservoir. Most of these techniques are approximate. Analytical techniques, on the other hand are limited in their ability to consider complex situations.

A front tracking problem is in essence a moving boundary problem. One way to handle this is the transformation of the problem to a moving coordinate system. The origin of such a coordinate system is located at the fluid interface and moves in time with the velocity of the front. This of course, requires one to know the velocity of the interface which is usually unknown. In the moving coordinate system the injection phase fronts are station-

---

(1) ECL Petroleum Technologies, 6408 Englewood, Colorado 80110, USA.
(2) Department of Petroleum Engineering, Stanford University, Stanford, California 94345-2225, USA.

ary in time and can be used to solve pressure transient or other problems. *Ramey*, (1970), showed that the assumption of stationary boundary for well testing type applications is reasonable under most injection/production situations.

A recent technique called the boundary element method is used to approximately track fluid fronts in a piston type displacement problem. The problem for single phase slightly compressible fluid is formulated in terms of pressure equation. Compatibility conditions at the interface accounts for mobility and storativity ratio differences between the injected and in-situ fluids. These equations including interface conditions are cast into an integral equation formulation.

Transformation to a local coordinate system reduces the dimensionality of the problem and necessitates evaluation of the unknowns at boundary nodes only. Thus, the procedure is called the boundary element method. It precludes the requirement for discretizing the entire problem domain and is quite conducive to representing irregular boundaries. In addition, the basis functions used in the numerical integration procedures are the fundamental solution to the governing equation. This representation improves the accuracy of the method significantly over other numerical techniques.

As part of the solution procedure for composite or sectionally homogeneous reservoirs, both the pressure and the pressure derivatives (spatial) at the interface are obtained as part of the boundary solution. Thus, without any additional computational effort, the front velocities are obtained. The front is then moved in an Eulerian time stepping scheme. The approximation comes from the fact that the displacement of the front to the new position is not completely current with the front velocities. Although not significant, the errors are nevertheless still present. This adds an additional constraint on time stepping schemes.

1.1 BOUNDARY ELEMENT METHOD

Developed in the realm of potential theory, this method is now being used in a variety of engineering applications. This technique is based on surface discretization instead of domain discretization used in finite difference and finite element schemes.

In addition, the formulation using this technique is based on the weak formulation of governing equation. This is similar to the finite element and collocation type schemes. The weak formulation has smoothing properties which improve the characteristics of the solution. One difference from the finite element method is the basis or weighting functions used in the integral scheme. The weighting functions in a finite element type formulation has a local basis or has compact support whereas the boundary element method uses global basis functions. These global basis functions are the fundamental solution or the Green's function of the governing equation. Use of the fundamental property of the governing equation improves the accuracy of the solution. In addition, this provides good convergence property for the integral equation. Being a surface method, the errors due to grid orientation are eliminated and increased flexibility in handling complex reservoir configurations and production schemes is obtained. An extensive reference list is provided in *Kikani* (1989). *Ligget and Liu* (1983) and *Pina* (1984) gave the formulation and solution procedure for both real and Laplace space versions of unsteady state flow of single phase fluid in aquifers and transient heat conduction problem respectively. *Kikani and Horne* (1988) presented a comparison of real and Laplace space solutions for the transient problem. Extension of this technique to nonlinear problems has been difficult. The way to get around nonlinearities is to delegate all the nonlinearities to the right hand side of the governing equation and use them as body force terms. This requires a domain integral to be performed, for which the problem domain has to be discretized. The charm of the boundary procedure is then reduced.

1.2 SECTIONALLY HOMOGENEOUS RESERVOIRS

Large scale features in reservoirs such as finite length faults, permeability barriers or different properties in front of and behind a flood zone represent zonal heterogeneities. In certain situations it is important to capture flow details due to the presence of such heterogeneities. From the pressure transient testing point of view the above problems are divided in two categories. The first category consists of wells producing from reservoirs containing internal boundaries such as constant pressure 'holes' or permeability barriers. The second category consists of fluid injection problems which create mobility and storativity contrasts in the reservoir. Composite reservoir models are used extensively to analyze well test in water injection, steam injection, in-situ combustion and miscible injection projects.

These problems have received limited treatment because of the difficulty of handling them analytically. Multiple boundaries of finite dimensions have not been solved analytically because of the difficulty in posing the boundary conditions in an otherwise homogeneous flow field. *Hantush and Jacob* (1966) considered an eccentric well within a bounded aquifer with a leaky caprock. *Britto and Grader* (1988) presented type curves for a well producing external to a circular or elliptic no-flow or constant pressure boundary. The solution presented is not amenable to multiple boundaries. As for the composite reservoir models, only radial and elliptic systems have been solved analytically. In reality, the injection fronts are never truly radial for several reasons. Background drift caused by injection and production wells nearby can affect the geometry of the propagating front and so can outer boundary effects.

Conceptually, both the above categories of problems belong to one group. This group consists of internal boundaries (sub-regions). The internal boundaries could be stationary as in finite length faults or permeability barriers, or they can be mobile as in injection or phase fronts. Thus, these two categories of problems can be treated in a generic fashion.

2.   PROBLEM FORMULATION

Figure 1 shows a schematic of a composite reservoir. The phase front can take any arbitrary shape depending on the flow field. The governing equation and compatibility conditions are discussed by *Kikani and Horne* (1989). The parameters governing the system are the mobility ratio (M), storativity ratio (F_s), and the distance to the discontinuity (R_D). The integral equations for the inner and outer regions are given by

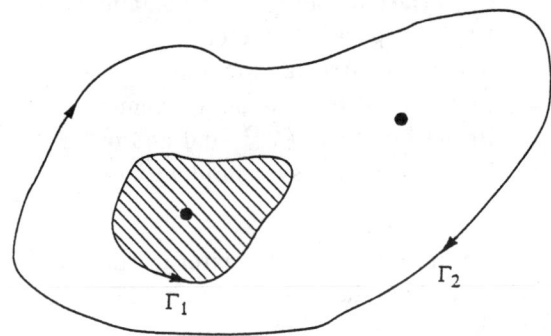

FIG. 1. Schematic of a composite reservoir*

$$\theta_1 \bar{p}_{D1} = \int_{\Gamma_1} \left(\bar{G}_1 \frac{\partial \bar{p}_{D1}}{\partial n} - \bar{p}_{D1} \frac{\partial \bar{G}_1}{\partial n} \right) dS \quad (1)$$
$$+ \frac{1}{s} \sum_{i=1}^{n_{w1}} \bar{G}_{1i} Q_{D1i}$$

$$\theta_2 \bar{p}_{D2} = \int_{\Gamma_1 + \Gamma_2} \left(\bar{G}_2 \frac{\partial \bar{p}_{D2}}{\partial n} - \bar{p}_{D2} \frac{\partial \bar{G}_2}{\partial n} \right) dS \quad (2)$$
$$+ \frac{1}{s} \sum_{i=1}^{n_{w2}} \bar{G}_{2i} Q_{D2i}$$

where,

$$\bar{G}_1(x_D, y_D, \xi, \zeta; s) = -K_0(|r_D|\sqrt{s}) \quad (3)$$

$$\bar{G}_2(x_D, y_D, \xi, \zeta; s) = -K_0(|r_D|\sqrt{s\frac{M}{F_s}}) \quad (4)$$

and

$$r_D{}^2 = (x_D - \xi)^2 + (y_D - \zeta)^2 \quad (5)$$

θ_i are the internal angles between two adjacent boundary elements and are given by

$$\theta = 2\pi \qquad if \ (x_D, y_D) \ \epsilon \ \Omega \quad (6)$$

$$\theta = \theta \qquad if \ (x_D, y_D) \ \epsilon \ \Gamma_1 \quad (7)$$

$$\theta = 2\pi - \theta \qquad if \ (x_D, y_D) \ \epsilon \ \Gamma_2 \quad (8)$$

where,

141

θ_i =   subtended angle between elements

Γ_i = i^{th} surface contour of the domain Ω

s =   Laplace space parameter

G_i =   free space Green's function

n =   unit normal to a boundary element

$K_0(x)$=   Bessel function of 2nd kind and order 0

(ξ, ζ) =   local coordinate system identifiers

Equation 8 implies that the exterior angles between adjacent boundary elements should be considered for interior sub-regions.

The second term in Eqs. 2 and 3 arise due to the presence of line sources or sinks in the domiain.

3. SOLUTION PROCEDURE

For an arbitrarily shaped reservoir the outer boundary is divided into N_2 nodes which represent the geometry sufficiently. An initial position and shape of the inner boundary or the phase front is assumed. This is divided into N_1 number of nodes. Linear interpolation functions for both pressure and fluxes are used in between the nodes. The choice has been made based on accuracy and computational effort. The interpolation function is defined as

$$p_D = \frac{\left[(p_D)_{j+1} - (p_D)_j\right]\xi + \left[\xi_{j+1}(p_D)_j - \xi_j(p_D)_{j+1}\right]}{\xi_{j+1} - \xi_j}$$

(9)

where ξ is the local coordinate varying along the element and the inequality, $(\xi_j < \xi < \xi_{j+1})$ holds. Similar interpolation function is defined for the fluxes.

The integral equations 2 and 3 suggest that if the pressure and its normal derivative at all points along the boundaries are known, one can calculate the pressure at any internal point of the reservoir, be it in the inner sub-region or the outer (in-situ fluid) region. In actuality, both the pressure and its normal derivative are not known at all points. On the external boundary, one piece of information – either the pressure or the flux is known from the boundary conditions. At the injection front, on the other hand, nothing is known. Thus, the integral equations are first used to evaluate these nodal unknowns. Equations 2 and 3 have to be solved separately and

coupled through the interface conditions. Figure 2 shows the partition of the boundaries into two regions. Region I consists of the fluid interface and the domain internal to it. Region II contains both the external boundary and the internal surface (interface). In other words, region II is the annular region between the interface and the boundary.

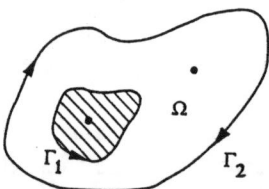

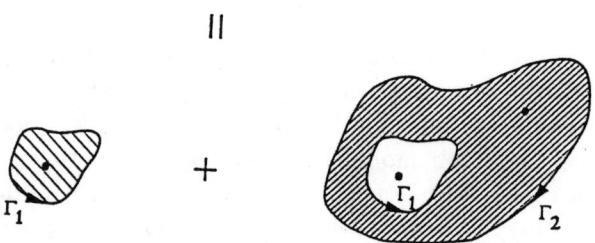

FIG. 2. Partition of reservoir boundary in two regions*

Discretized equations are written for both regions. The inner region is a simply connected surface and N_1 equations in $2N_1$ unknowns are obtained. This is because at the interface no information is known. For the annular region N_1+N_2 equations in N_2+2N_1 unknowns are obtained. This region is a multiply connected surface in the complex analysis sense. The two regions are coupled through $2N_1$ compatibility conditions at the interface. This fully specifies the problem. The matrix equation is solved to give the interface unknowns. Using the auxiliary equation 2 again, pressures at any interior location (well pressure or interference pressure) can be calculated. The solution thus obtained is still in Laplace space and is inverted to the real space by the use of Stehfest algorithm.

3.1 VELOCITY CALCULATIONS

At the interface, both the pressure and its normal

derivatives are known as part of the boundary solution, thus the interface velocities can be calculated from the following equations

$$v^2 = v_x^2 + v_y^2 \qquad (10)$$

$$\alpha = \tan^{-1}\frac{v_y}{v_x} \qquad (11)$$

The velocities are calculated using Eq. 12

$$v_x = -\frac{k}{\mu}\frac{\partial p_D}{\partial n}\frac{\partial n}{\partial x_D} \qquad (12)$$

where,

v_x, v_y = front velocity in x and y directions resp.
$x_{t+\Delta t}$ = x-coord. of the node at time $t + \Delta t$
α = direction vector of the interface node
n = direction normal to a boundary element

An identical equation can be written for pressure gradient in the y-direction. Once the velocities are calculated, the location of the front at time $t + \Delta t$ is given by

$$x_{t+\Delta t} = x_t + v_x\Delta t \qquad (13)$$

$$y_{t+\Delta t} = y_t + v_y\Delta t \qquad (14)$$

This explicit movement of the interface places restrictions on the size of the time steps that can be taken. A simple error analysis for logarithmic time stepping scheme is provided in the appendix.

3.2 COMPUTATIONAL EFFORT

Representation of irregular geometry requires a substantial number of nodes. The computational workload increases as the cubic power of the number of nodes because the matrices obtained are full. The reason for obtaining a full matrix is the global basis function in BEM compared to the local dependence of solution in other numerical techniques. Of course, for the same accuracy, a fewer number of nodes are required with the boundary element method. The biggest effort however, is in the inversion of the Laplace space solution. For solution at one time step, 6 or 8 such matrix inversions are required. In addition, the constraint of time stepping with front tracking restricts one to take 20 - 50 points per log cycle of time, depending on the error tolerance. More research is required in reducing this effort to make the method competitive in terms of computer time.

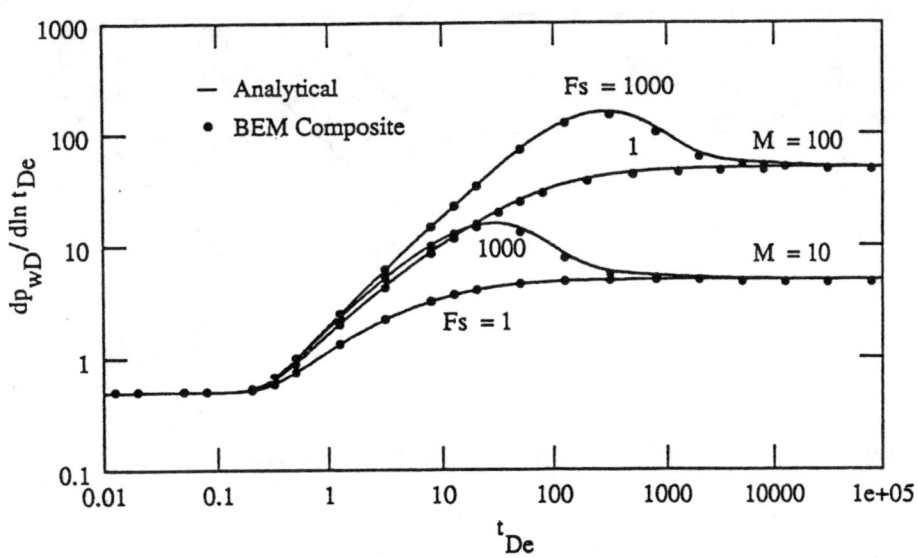

FIG. 3. Effect of mobility and storativity ratios on the derivative response of a radial composite reservoir*

4. RESULTS AND DISCUSSION

The boundary element solution for both internal boundaries and composite reservoirs have been verified against known analytical solutions. The accuracy of the method has been established. The pressure derivative behavior of a composite reservoir with a stationary front is shown in Fig. 3. The semilog pressure derivative is considered because it is more sensitive to errors. The effect of mobility and storativity ratios are shown on a log-log plot. The dimensionless time is based on the distance to the phase front, i.e.,

$$t_{De} = t_D / R_D^2 \qquad (15)$$

The solid dots are the solutions generated with the BEM and the solid lines are the analytical solutions. The first semilog straight line appears as a flat line with a value 0.5. After a long transition depending on the mobility and storativity ratio, a second semilog straight line occurs corresponding to the outer region. The slope of this straight line is given by a value of M/2. The effects of finite drainage region is shown in Fig. 4. The mobility ratio considered is 10 and the storativity ratio is 1000. The ratio R_{eD}/R_D represents the size of the outer boundary compared to the injection front. For R_{eD}/R_D values of less than 100, the semilog straight line corresponding to the outer region will not be seen. The comparison with the analytical solution is quite good.

Figure 5 is an example of boundary element solution for a non-circular drainage region that is difficult to treat with analytical methods. For some thermal injection wells intercepted by a vertical fracture, the displacement front moves in a direction normal to the plane of the fracture at early times. The swept region at early times can be idealized as a low width to length ratio rectangle. The effect of the width to length ratio of a rectangular composite reservoir on the pressure transient behavior is shown in Fig. 5. The dimensionless time is based on the length of the inner zone. The mobility and storativity ratios used are 200 and 16.67 respectively. It is evident that higher the aspect ratio of the rectangular front longer is the transition zone before the outer region is felt. The rectangular front gets more elliptic in shape as it propagates further in the reservoir and finally becomes pseudoradial. These distinct changes in the shape of the front influences the pressure transient behavior and the assumption of stationary front of a particular shape can cause significant errors.

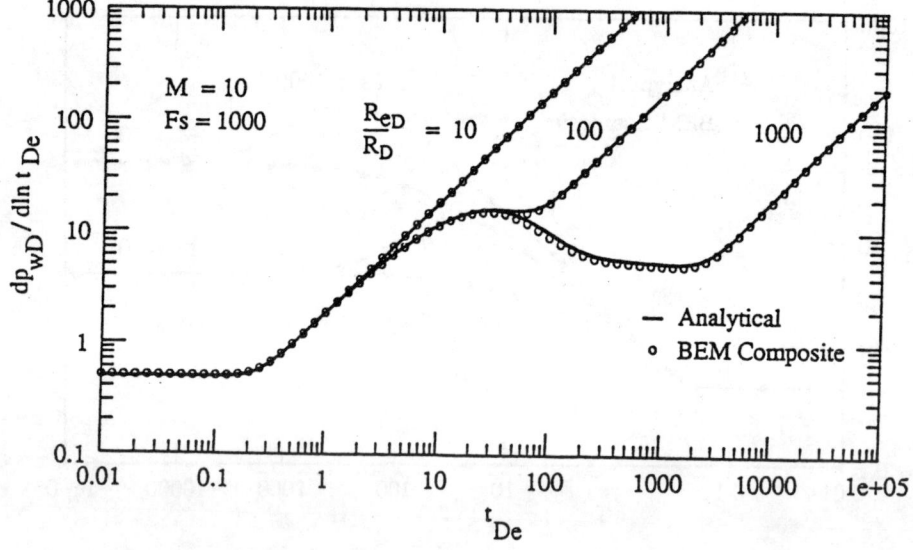

FIG. 4. Pressure derivative response of a bounded radial composite reservoir

5. REFERENCES

Britto, P.R. and Grader, A.S., 1988, The Effect of Size, Shape, and Orientation of an Impermeable Region on Transient Pressure Testing, SPE Formation Evaluation, p. 595-606.

Kikani, J., 1989, Use of Boundary Element Method for Streamline Generation and Pressure Transient Testing, Ph.D Dissertation, Stanford University, 216 p.

Kikani, J. and Horne, R.N., 1988, Pressure Transient Analysis of Arbitrary Shaped Reservoirs by the Boundary Element Method, SPE paper #18159 presented at the Annual Mtg. Houston.

Kikani J. and Horne, R.N., 1989, Modeling Pressure Transient Behavior of Sectionally Heterogeneous Reservoirs by the Boundary Element Method, SPE paper #19778 presented at the Annual Mtg. San Antonio.

Liggett, J.A. and Liu, P.L-F., 1983, The Boundary Integral Equation Method for Porous Media Flow George Allen and Unwin, London.

Pina, H.L.G., 1984, Time Dependent Potential Problems. In: C.A. Brebbia (Editor), Boundary Element Methods in Computer Aided Engineering Martinus Nijhoff Publishers, Dodrecht.

Ramey, H.J., 1970, Approximate Solutions for Unsteady State Liquid Flow in Composite Reservoirs, J. of Canadian Petroleum Technology, p. 32-37.

6. APPENDIX

Approximate Error Analysis in Front Tracking.

The velocity at a point i (on the front) at a time t_j is given by $\vec{v}_i^j$. This velocity ($\vec{v}_i^j$) is used to move the front to another location for the next time step. For a radial system

$$v \propto \frac{1}{R_D} \qquad (16)$$

where, R_D is the radial distance.

If it is assumed that the similarity variable holds for this situation, then

$$\frac{R_D^2}{t} = c \qquad (17)$$

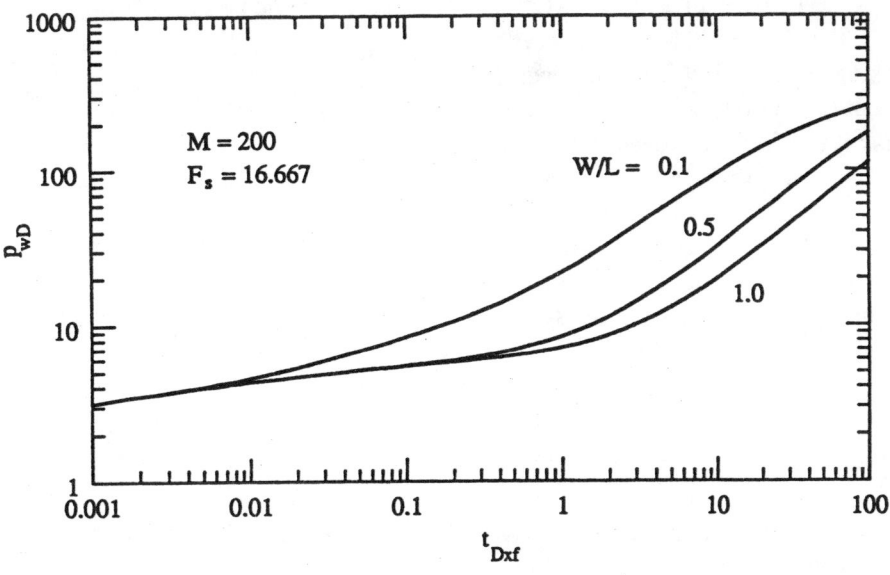

FIG. 5. Effect of aspect ratio on the pressure response of a rectangular composite reservoir

where, c is a constant

$$\Rightarrow R_D \propto c\sqrt{t} \tag{18}$$

$$\Rightarrow \vec{v}_i^j \propto \frac{c_1}{\sqrt{t}} \tag{19}$$

By the same token,

$$v(t + \Delta t) = \vec{v}_i^{\Delta j} \propto \frac{c_1}{\sqrt{t + \Delta t}} \tag{20}$$

The difference between the velocities at the time t and $t + \Delta t$ is given by

$$\Delta \vec{v}_i^j = -\frac{1}{2}\frac{\Delta t}{t\sqrt{t}} + O\left[\left(\frac{\Delta t}{t}\right)^2\right] \tag{21}$$

Thus,

$$\frac{\Delta \vec{v}_i^j}{\vec{v}_i^j} = -\frac{1}{2}\frac{\Delta t}{t} + HOTS \tag{22}$$

Now, for logarithmic time stepping, the incremental time step is given by the following relation

$$\frac{\Delta t}{t} = \left(10^{\frac{1}{N}} - 1\right) \tag{23}$$

where, N is the number of points per log cycle of time. The percentage change in velocity over a logarithmic time step can be evaluated using Eqs. 22 and 23 as

$$\frac{\Delta \vec{v}_i^j}{\vec{v}_i^j} = -\frac{1}{2}\left(10^{\frac{1}{N}} - 1\right) \tag{24}$$

Numerically speaking, if 20 point/log cycle were taken, then the percentage change in velocity will be 6.1. This implies that if the front is moved with the velocity calculated from the previous time step, it will be overestimated by 6.1 %. 50 points/log cycle will reduce this to about 2 %, although the computational load will increase proportionately. The error tolerance will determine the number of time steps per log cycle.

Reservoir Simulations

2nd European Conference on the Mathematics of Oil Recovery
© D. Guérillot, O. Guillon (Editors) and Éditions Technip, Paris 1990, pp. 149-156
27 rue Ginoux, 75015 Paris

Control Volume Method to Model Fluid Flow on 2D Irregular Meshing

I. Faille[1]

INTRODUCTION

Modelling fluid flow in porous media requires the approximation of an elliptic partial differential operator such as $\operatorname{div} \frac{\bar{\bar{K}}}{\mu} \overrightarrow{\operatorname{grad}} p$, which appears when substituting Darcy's law in mass conservation equation. When the medium is heterogeneous, the permeability tensor $\bar{\bar{K}}$ is discontinuous and depends on the lithologic nature of the medium. In order to take this dependence well into account, the grid used to discretize the set of equations is such that cell boundaries are aligned with geological discontinuities. In the particular case of 2D basin modelling [1], discussed here, the grid used follows the stratigraphic layers and is made of irregular trapeziums: FIG 1.

The aim of this study is to **discretize** $\operatorname{div} \frac{\bar{\bar{K}}}{\mu} \overrightarrow{\operatorname{grad}} p$ **on such an irregular meshing, using a cell centered control volume method.** The approximated solution is piecewise constant and defined by its values in each cell: the value in one cell approximates the exact solution at the isobarycentre of the cell. The discrete problem is obtained by integrating the mass conservation equation over each grid block. Difficulties in applying this method to basin modelling come from:

1. **the irregularity of the meshing**: convergence properties of cell centered method are usually stated for uniform grids, or irregular rectangular grids.

2. **the anisotropy and discontinuity of the permeability tensor.**

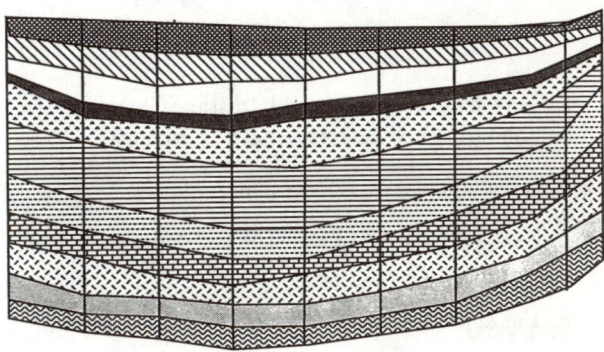

Figure 1: grid used to represent a sedimentary basin, with different sedimentary layers.

Recently, several papers have appeared in the litterature, dealing with control volume methods on irregular meshing. These methods are point centered methods: an initial grid allows us to define the points where the solution is to be approximated and a secondary one allows us to define the control volumes where mass balances are to be written. In [2], control volumes are built in order to have a simple form of the discrete mass balance equation. In [3],[4] an approximated solution is defined on the entire domain (finite element type approximation) and mass

---

(1) Institut Français du Pétrole, 1 et 4, avenue de Bois-Préau, 92506 Rueil Malmaison, France.

balance is written over control volumes surrounding each node. These methods handle well with irregular meshing due to complex domain geometry. However, when the irregularity of the meshing comes from the description of heterogeneities, it seems difficult to use these methods since the control volumes can not be chosen.

The present paper deals with **cell-centered control volume method** and is organized as follows: in the first part, we focus on the irregularity of the grid by looking at a more simple problem: approximation of Laplace operator. A cell centered, conservative scheme with a consistent approximation of the fluxes is built. Some of its mathematical properties (weak consistency) are discussed. In the second part, the scheme is generalized to $\text{div} \, \overline{\overline{K}} \, \overrightarrow{\text{grad}} P$: the discontinuity of $\overline{\overline{K}}$ is taken into account by writing the continuity of the fluxes on each cell boundary. Lastly, some numerical experiments are presented.

notations:

We introduce the following notations:

- h the grid size i.e. a scalar such that there exists two constants c and d satisfying the inequality:

$$c\,h \leq l \leq d\,h$$

for any length l of the grid (distance between two neighbouring cell centers, length of an edge).

- V_k a cell and ω_k its isobarycentre.

- $\overline{p}$ the exact solution of the problem under consideration.

- p_h the approximated piecewise constant solution.

- $\overline{P}_h$ the vector whose components are the values of $\overline{p}$ at each cell center.

- P_h the vector whose components are the values of p_h in each cell.

1 APPROXIMATION OF LAPLACE OPERATOR

We consider the following problem:

$$\begin{cases} -\Delta p = f \text{ in } \Omega \\ p = p_b \text{ on } \Gamma \end{cases} \quad (1)$$

where

- Ω is an open bounded subset of R^2,

- Γ its boundary.

Integrating the equation over a control volume V_k (i.e over a mesh) leads to:

$$\int_{V_k} -\Delta p \, d\omega = \int_{V_k} f \, d\omega \quad (2)$$

which can be rewritten, using Green formula and dividing by S_k, the surface of the control volume:

$$\frac{1}{S_k} \sum_{\delta \text{ edge of } V_k} \int_\delta -\overrightarrow{\text{grad}} p . \vec{n}_k \, d\sigma = f_k \quad (3)$$

where

- f_k denotes an approximation of $\frac{1}{S_k} \int_{V_k} f \, d\omega$

- $\vec{n}_k$ the outward normal to the cell boundary i.e. on the edge δ, $\vec{n}_k = \pm \vec{n}_\delta$, $\vec{n}_\delta$ is the normal to edge δ orientated in an arbitrary way.

Finally, $\int_\delta \overrightarrow{\text{grad}} p . \vec{n}_k \, d\sigma$ is approximated by a linear combination of p_h values in the neighbouring cells of δ. Therefore, equation (3) reduces to a linear relation between p_h values and the discrete problem can be written as a linear system whose unknowns are the values of p_h:

$$A_h \, P_h \, = \, B_h$$

The control volume method is defined as soon as the approximation of $\int_\delta \overrightarrow{\text{grad}} p . \vec{n}_k \, d\sigma$ is specified. The choice of a particular approximation is guided by the properties of cell centered finite volume methods on 2D irregular rectangular meshing. We build a scheme satisfying the two criteria which appear as being important for the quality of a finite volume scheme and we present some of the mathematical properties of the proposed scheme.

1.1 Guide

Usually, in finite difference methods, the approximation is chosen in order to get a consistent scheme: the truncation error T_h (obtained by substituting the values of the approximate solution P_h by the values of the exact solution $\overline{P}_h$ in the discrete equation) converges to zero as the grid size is reduced: $T_h = A_h \overline{P}_h - B_h$ is such that

$$||T_h||_\infty \longrightarrow 0 \text{ when } h \longrightarrow 0$$

Different studies [5],[6] about cell-centered control volume methods applied to irregular rectangular grids, have proven that **consistency is not a necessary condition for a scheme to be convergent.** Looking at these results in more details, two characteristics appear as being important for the quality of the scheme:

- **a unique approximation of $\int_\delta \overrightarrow{\text{grad}} P . \vec{n}_\delta \, d\sigma$** i.e. the same expression (except for the sign) is used when writing mass balance over the two meshes adjacent to edge δ. The scheme then obtained is said to be **conservative** because the approximated solution satisfies mass balance over any group of cells and so over the whole domain.

- a consistent approximation of $\int_\delta \overrightarrow{\text{grad}}p.\vec{n}_\delta$ i.e.

$$L_\delta = l_\delta[\overrightarrow{\text{grad}}p.\vec{n}(\omega_\delta) + \alpha_\delta] \qquad (4)$$

where l_δ denotes the length of edge δ, ω_δ its center and α_δ a scalar converging to zero when the grid size is reduced.

1.2 The scheme

To build a scheme satisfying the two conditions listed above, we have to specify how a consistent approximation of the flux on each edge is obtained.

Let us consider an edge δ: FIG 2. **The basic idea is to approximate P at two points H,B situated on both side of δ and on the straight line Δ_δ,** going through ω_δ in the direction orthogonal to δ. An expression for L_δ is then given by:

$$L_\delta = l_\delta \frac{P_H - P_B}{d(H,B)} \qquad (5)$$

where

- P_H, P_B are approximations of p at points H and B;

- d(H,B) is the distance between H and B.

In order to get a simple expression for P_H, point H is chosen on the straight line joining the centers of two adjacent cells. Hence a linear interpolation between the values of p_h in these two cells gives an expression for P_H. For instance, in the case of FIG 2, H is the intersection between line Δ_δ and line (M, M') and

$$P_H = \alpha P_M + (1 - \alpha)P_{M'}$$

where

-
$$\alpha = \frac{d(H,M')}{d(M,M')} \qquad (6)$$

- $P_M, P_{M'}$ are the values of p_h in the two cells M and M'.

B and P_B are obtained in a similar way. Hence, the expression of the flux on an edge is a linear combination between four of the six values of p_h in its neighbouring cells, the choice of the cells is determined by the direction of the edge. When equation (3) is written for a particular cell, we get a relation between at most nine values of p_h (the value in the considered cell and the values in the eight neighbouring cells) so **the scheme is a nine points one.**

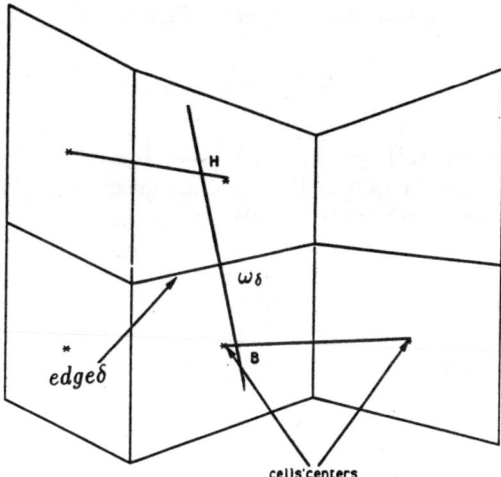

Figure 2: points used in the approximation of the flux on edge δ.

1.3 Mathematical properties

1.3.1 stability

Under the assumption that the grid is not too much distorted, the matrix A_h is irreducibly diagonally dominant with positive real entries on the diagonal and negative real entries outside the diagonal. Therefore, the discrete solution satisfies a discrete maximum principle and the inverse of A_h is boundered independently of h.

1.3.2 consistency-weak consistency

Let us examine the truncation error. As P_H and P_B are second order approximations of p at points H and B and as ω_δ is not the middle of [H,B], L_δ is a first order approximation of the flux :

$$L_\delta = l_\delta(\overrightarrow{\text{grad}}p.\vec{n}_\delta(\omega_\delta) + \alpha_\delta)$$

with $\alpha_\delta = O(h)$.
The truncation error in the cell V_k can then be written:

$$(T_h)_k = \frac{1}{S_k} \left(\sum_{\delta \text{ edge of } V_k} \vec{n}_\delta.\vec{n}_k l_\delta[(\overrightarrow{\text{grad}}p.\vec{n}_\delta)(\omega_\delta) + \alpha_\delta] \right) - f_k$$

As α is only a O(h) term, $\|T_h\|_\infty$ doesn't reduce to zero with h. **The scheme is not consistent.** However, **the truncation error converges towards zero in a weaker sense:**

for any infinitely smooth function Φ with compact support in Ω, the truncation error T_h satisfies:

$$\int_\Omega \Phi T_h d\omega \longrightarrow 0 \text{ when } h \longrightarrow 0. \qquad (7)$$

We call this property "weak-consistency" of the scheme. Details of the proof are given in Appendix A, but let us see more precisely the properties of the truncation error. It is the sum of two terms:

- a first one, associated with the finite volume formulation and the use of Green formula, which has a good behaviour as it is a O(h) term:

$$(T_h^0)_k = \frac{1}{S_k} \sum_{\delta \text{ edge of } V_k} (\vec{n}_\delta . \vec{n}_k) l_\delta (\overrightarrow{\text{grad}p}.\vec{n})(\omega_\delta) - f_k \qquad (8)$$

- a second one associated with the approximation of the flux on each edge.

$$(T_h)_k = \frac{1}{S_k} \sum_{\delta \text{ edge of } V_k} (\vec{n}_\delta . \vec{n}_k) l_\delta \alpha_\delta \qquad (9)$$

Because of the conservative character of the scheme and the consistant approximation of the fluxes, this term converges towards zero in the weak sense given above (7).

2 APPROXIMATION OF div $\bar{\bar{K}} \overrightarrow{\text{grad}p}$

The problem under consideration now is:

$$\begin{cases} -\text{div } \bar{\bar{K}} \overrightarrow{\text{grad}p} = f \text{ in } \Omega \\ p = p_b \text{ on boundaries} \end{cases} \qquad (10)$$

where p represents the pressure.
Following the method presented in the first part, the construction of a scheme is reduced to the approximation of the flux on each edge:

$$L_\delta = \int_\delta \bar{\bar{K}} \overrightarrow{\text{grad}p}.\vec{n}_\delta d\sigma$$

As the medium is heterogeneous, $\bar{\bar{K}}$ can have different values in the two cells adjacent to δ. **This discontinuity implies a discontinuity of $\overrightarrow{\text{grad}p}$ on the edge, the continuous quantity is $\bar{\bar{K}} \overrightarrow{\text{grad}p}.\vec{n}_\delta$.** Thus, we choose an approximation of $\bar{\bar{K}} \overrightarrow{\text{grad}p}.\vec{n}_\delta$ at the middle ω_δ of edge δ, which uses this continuity:

1. consider P^* an approximation of p at ω_δ

2. use this value to approximate $\bar{\bar{K}} \overrightarrow{\text{grad}p}.\vec{n}_\delta$ in the two cells adjacent to δ:
Let us consider, for instance, a vertical edge δ. If $\bar{\bar{K}}_R$ is the permeability tensor in the right cell and $\bar{\bar{K}}_R^T$ its transpose, we have the following equality:

$$\bar{\bar{K}}_R \overrightarrow{\text{grad}p}.\vec{n}_\delta = \overrightarrow{\text{grad}p}.(\bar{\bar{K}}_R^T \vec{n}_\delta)$$

Then an expression for L_δ^R, the approximation of L_δ on the right side of δ, is

$$L_\delta^R = l_\delta K_R \frac{P_R - P^*}{d(\omega_\delta, R)} \qquad (11)$$

where

- R is a point located on the straight line Δ_R, going through ω_δ and in the direction $(\bar{\bar{K}}_R^T \vec{n}_\delta)$

- $K_R = \| \bar{\bar{K}}_R^T \vec{n}_\delta \|_2$

Like in the case of Laplace operator, R is chosen on the straight line joining the centers of two adjacent cells (FIG 3). P_R is a linear approximation of p at point R.
In a similar way, an expression for L_δ^L (left side) is obtained.

3. write the continuity of $\bar{\bar{K}} \overrightarrow{\text{grad}p}.\vec{n}_\delta$ on the edge, to get an expression for P^* i.e

$$L_\delta^R = L_\delta^L$$
$$\left[\frac{K_R}{d(\omega_\delta, R)} + \frac{K_L}{d(\omega_\delta, L)} \right] P^* = \frac{K_R}{d(\omega_\delta, R)} P_R + \frac{K_L}{d(\omega_\delta, L)} P_L$$

4. substitute this expression of P^* in one of the two expressions of the flux to get the final expression:

$$L_\delta = l_\delta \frac{K_R K_L d(\omega_\delta, L) d(\omega_\delta, R)}{d(\omega_\delta, L) K_R + d(\omega_\delta, R) K_L} (P_R - P_L) \qquad (12)$$

The linear approximation of P at point R is specified in Appendix B.

When the cells are rectangular and the main permeability directions parallel to the coordinate axes, Δ_L and Δ_R are simply the line orthogonal to the edge δ. The scheme then reduces to the expression of the usual five point scheme with harmonic averaging of the permeabilities.

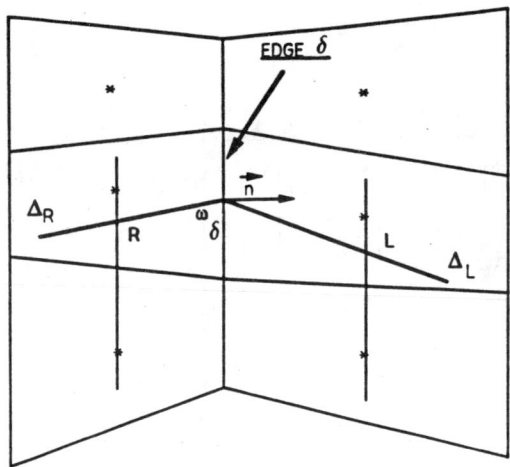

Figure 3: points used in the approximation of the flux on the edge δ when the mediun is anisotropic and heterogeneous

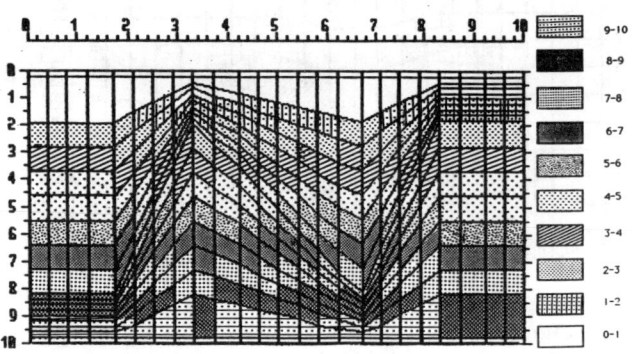

Figure 4: solution, piecewise constant, for test1: Kershaw's test

3 NUMERICAL RESULTS

We first give results concerning the approximation of Laplace operator and then concerning the approximation of div $\overline{\overline{K}} \overrightarrow{grad}p$.

3.1 for Laplace problem

Three series of results are presented:

- <u>test 1</u>: FIG 4
  The particularity of the first test, called Kershaw's test [8], lies in the grid used, as it is very distorted. The problem solved is Laplace problem (1) on the domain of FIG 4 with

  - f=0
  - p=10 on the upper boundary, p=0 on the bottom boundary
  - no flux on lateral boundaries.

The exact solution is $\overline{p}(x, y) = y$. FIG 4 gives the piecewise constant approximated solution: each cell is coloured in accordance with the value of the approximated function in that cell. FIG 5 gives the isovalue curves of the computed solution: these curves are not altered by the distortion of the grid as they would have been with a five points scheme.

- <u>test 2</u>: FIG6
  We test the scheme in the case of a more complicated solution. Laplace problem is solved with:

  - f=0

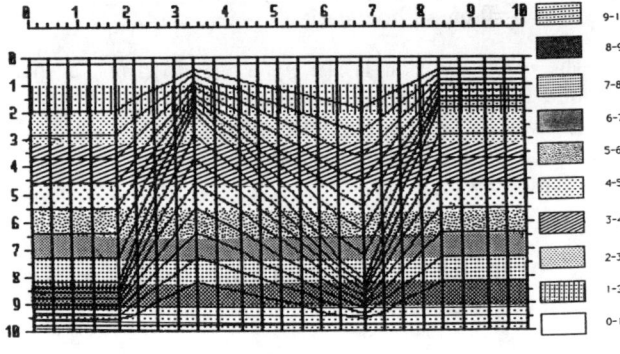

Figure 5: isovalue curves for test1: Kershaw's test

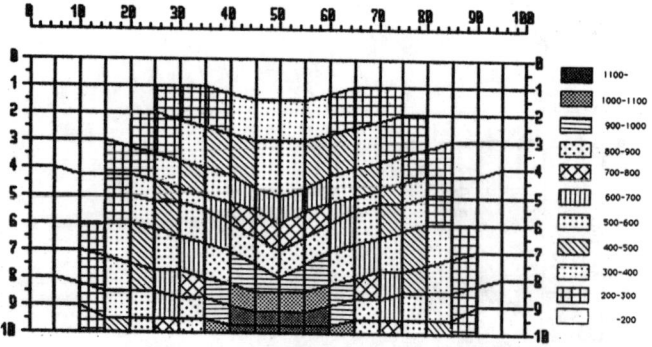

Figure 6: solution, piecewise constant, for test 2

- $P(x, 10) = \frac{2500}{\sqrt{2*\pi}}(1 - cos(2\frac{\pi x}{100})) + 100$ on the bottom boundary

- P=100 on the three other boundaries.

FIG 6 shows the computed solution. Since the exact solution is known [7], errors have been calculated: the maximum absolute error is 0.7 and the maximum relative error is 0.18 %.

- <u>test 3</u>: convergence test, FIG7 and FIG8
Test 3 illustrates some of the convergence properties of the scheme. We give the solution computed with a particular grid and with the grid obtained once the edges of the initial grid have been divided by two. The problem solved is

- $f = -2[x(x-10) + y(y-10)]$

- p=0 on all boundaries.

the exact solution is $\overline{p}(x,y) = x(x-10)y(y-10)$ and FIG give the relative errors in each cell for both grids: FIG7 and FIG8. The errors are much smaller for the second grid, in fact the relative error is almost divided by four.

3.2 for a discontinuous and anisotropic permeability

The permeability tensor has main directions along and across the stratigraphy $(\vec{s}, \vec{a})$: FIG 9. FIG 10 gives the solution obtained for the following boundary conditions and values of permeability :

- P=0 on the top boundary,

- P=100 on the bottom boundary,

- no flow on the lateral boundaries,

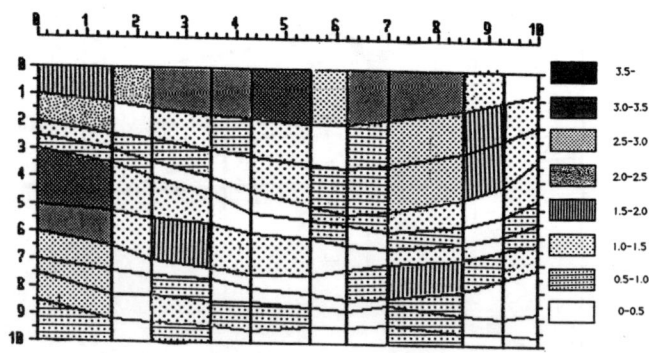

Figure 7: test 3: relative error for the coarse grid, as a percentage of the exact solution

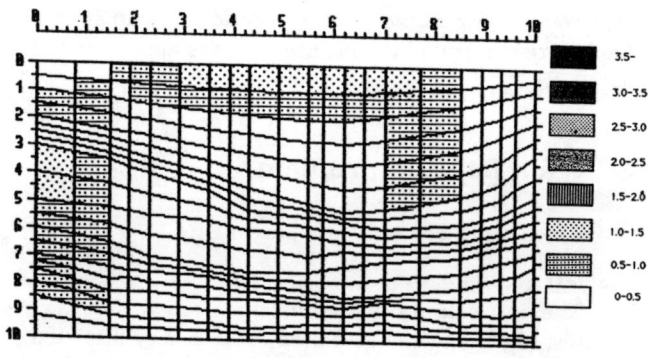

Figure 8: test 3: relative error for the refined grid

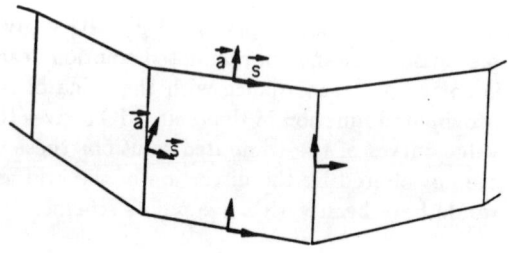

Figure 9: along and across stratigraphy directions in a layer

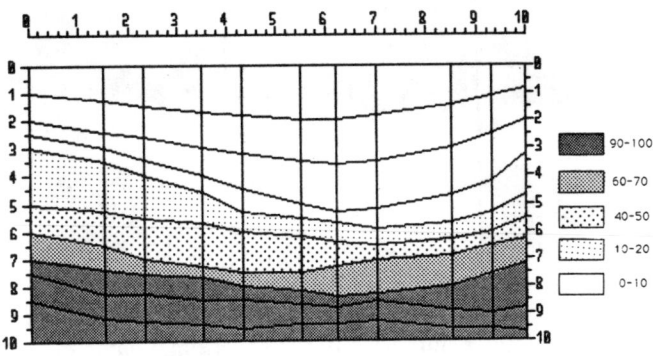

Figure 10: solution for an anisotropic and discontinuous permeability tensor

- $K_s = 1$, $K_a = 10^{-2}$ in the first four rows and the last three ones,

- $K_s = 1$, $K_a = 10^{-4}$ in the three other rows.

where K_s and K_a denote the values of permeability tensor in the two directions $\vec{s}, \vec{a}$.

The numerical solution reflects well the main features of the permeability tensor:

- its anisotropy: the isovalue curves follow the stratigraphic layers as K_s is bigger than K_a.

- its discontinuity: the gradient of the solution is more important in the three rows where K_a is lower.

The scheme has also been compared to results obtained with a bilinear finite element approach; in all the cases the two methods are comparable and we observe the good behaviour of the cell centered method.

References

[1] Ungerer P.,Burrus J.,Doligez B.,Chenet P.Y.,Bessis F. Basin evaluation by integrated two dimensional modelling of heat transfer, fluid flow, hydrocarbon generation and migration. AAPG Bull V14,309-335 March 1990

[2] Heinemann Z.E. and Brand C. Modeling reservoir geometry with irregular grids. SPE 18412. Feb 1989.

[3] Rozon B. A generalized finite volume discretization method for reservoir simulation. SPE 1814. Feb 1989.

[4] Forsyth P.A. A control finite element method for local mesh refinement. SPE 18415. Feb 1989.

[5] Manteuffel T.A. and White A.B. The numerical solution of second order boundary value problem on non uniform meshes. Math. Comp. 47 511-536 1986

[6] Forsyth P.A.Jr and Sammon P.H. Quadratic convergence for cell centered grids. Applied Num. Math. 377-394 North-Holland 1988

[7] Carslaw H.S and Jaeger J.C. Conduction of heat in solids. second edition Clarendon Press Oxford.

[8] Kershaw D.S. Differencing of the diffusion Equation in Lagrangian Hydrodynamic Codes J. Comp. Physics, vol 39, pp375-395 1987

Appendix A

We want to show that for any infinitely smooth function Φ, with compact support in Ω,

$$\int_\Omega \Phi T_h \, d\omega \longrightarrow 0 \text{ when } h \longrightarrow 0 \qquad (13)$$

As written in part one, the truncation error is divided into two parts T_h^0 and T_h^1 given by (8),(9). As T_h^0 is a O(h) term, it satisfies:

$$\int_\Omega \Phi T_h^0 \, d\omega \longrightarrow 0 \text{ when } h \longrightarrow 0 \qquad (14)$$

Proof of this property for T_h^1 is not as straight forward and includes the following steps:

1. there exists a subsequence of (T_h^1) which converges to a limit, called T. Indeed $\|T_h^1\|_\infty$ is boundered independently of h, so there exists a subsequence of (T_h^1) which converges in L_∞ weak * towards T.

2. this limit satisfies:

$$\int_\Omega \Phi T \, d\omega = 0 \qquad (15)$$

so

$$\int_\Omega \Phi T_h^1 d\omega \longrightarrow \int_\Omega \Phi T d\omega = 0 \text{ when } h \longrightarrow 0$$

To prove (15), we introduce the piecewise constant function Φ_h associated with Φ:

$$\Phi_h(\omega) = \Phi(\omega_k) = \Phi_k \text{ for } \omega \text{ in } V_k$$

On one hand:

$$T_h^1 \xrightarrow{L_\infty weak*} T \text{ and } \Phi_h \xrightarrow{L_1} \Phi$$

$$\text{so } \int_\Omega \Phi_h T_h^1 d\omega \longrightarrow \int_\Omega \Phi T d\omega$$

and on the other hand:

$$\int_\Omega \Phi_h T_h^1 d\omega \longrightarrow 0$$

Indeed,

$$\int_\Omega \Phi_h T_h^1 d\omega = \sum_k S_k \Phi_k (T_h^1)_k$$
$$= \sum_k \Phi_k \sum_{\delta \text{ edge of } V_k} \vec{n}_\delta . \vec{n}_k l_\delta \alpha_\delta \quad (16)$$

If V_k and $V_{k'}$ are the two adjacent cells to the edge δ: $\vec{n}_\delta . \vec{n}_k = -\vec{n}_\delta . \vec{n}_{k'}$ and integration by parts of (16) leads to

$$\int_\Omega \Phi_h T_h^1 d\omega = \sum_\delta \vec{n}_\delta . \vec{n}_k l_\delta \alpha_\delta (\Phi_k - \Phi_{k'}) \quad (17)$$

Finally, as $l_\delta, \alpha_\delta, \Phi_k - \Phi_{k'}$ are all O(h) terms, (17) is

$$\int_\Omega \Phi_h T_h^1 d\omega = \sum_\delta O(h^3)$$

and therefore converges towards zero when h reduces to zero.

We can notice that it's the conservative character of the scheme which enables the integration by parts: α_δ appears with opposite signs in the truncation errors of the two cells adjacent to δ.

Appendix B

We detail here the linear aproximation chosen for P_R when the permeability tensor is not continuous (second part).

Point R is on the line D joigning two centres of cells M and M': FIG11. $\text{grad } p$ is not continuous on this line as $\overline{\overline{K}}$ can take different values in the two cells V_M and $V_{M'}$. Although it seems difficult to take this discontinuity well into account, we suggest the following approximation:

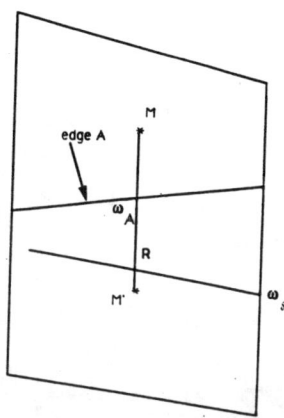

Figure 11: points used in the approximation of P_R

1. we introduce $\tilde{p}$ the value of p on the edge A.

2. we approximate this value by writing the continuity of $\text{grad } p.\vec{n}_A$ on the edge A like in steps 2 and 3 of the approximation of $div \overline{\overline{K}} \text{ grad } p$.

3. we choose for P_R a linear approximation between P_M and $\tilde{p}$ if R is on $[M, \omega_A]$ and between $P_{M'}$ and $\tilde{p}$ otherwise.

2nd European Conference on the Mathematics of Oil Recovery
© D. Guérillot, O. Guillon (Editors) and Éditions Technip, Paris 1990, pp. 157-163
27 rue Ginoux, 75015 Paris

Heterogeneous Porous Media and Domain Decomposition Methods

M. S. Espedal, R. Hansen, P. Langlo[1]
O. Sævareid[1] and R. E. Ewing[2]

Abstract

The physics governing flow in porous media are characterized by localized phenomena taking place on scales that are small compared to the overall size of the reservoir. To obtain proper resolution, local refinement, both in space and time, seems to be promising. The approach is especially attractive within the framework of domain-decomposition, where the global solution is constructed from local computations on sub-domains where the resolution and even the entire solution strategy can be specially designed to match local properties of the model. The strategy is well suited for parallel computer architectures as well as integration into existing large scale simulators.

In the present paper we will focus on heterogeneous rock properties. Core-samples, well-tests and production data provide information on widely different scales. To be useful in simulations, all the data have to be brought to the scale of the discrete blocks of the simulator. For this purpose, a stochastic approach based on assuming a multivariate lognormal distribution of permeability data will be investigated in the context of local grid refinement.

1   INTRODUCTION

Although a porous media intrinsically is deterministic, it contains a complex mixture of properties at many scales. Partly because of the complexity and partly because of incomplete information of the reservoir, stochastic techniques are used to describe and model different features of the media [1,2,3,4,5].

In this paper we will study a heterogeneous, two-dimensional reservoir where the spatial permeability distribution is modeled as a random function. Except for very idealized cases, such models has to be solved numerically. The numerical calculation of flow processes require that we are able to assign permeability and dispersion values to blocks of different sizes in the model. We have to relate and model processes which occur at smaller scales to a manageable computational scale.

Our main objective in this paper is to test recently developed numerical schemes [6] on such heterogeneous models for reservoir flow. The algorithms are based on operator splitting and domain decomposition methodes and give an excellent tool for local refinement of the computational grid. This means that we have to give appropriate permeability/dispersion values to blocks of variable size in the numerical model.

2   MODEL

We shall consider a model describing immiscible flow where gravity and compressibility are neglected. As noted, we will assume that the permeability of the reservoir, $\bar{K}(\mathbf{x})$ exhibit random fluctuations which can be modeled as a random function with a multivariable lognormal distribution of the

(1) Department of Applied Mathematics, University of Bergen, Allegaten 55, 5007 Bergen, Norway.
(2) Department of Mathematics, University of Wyoming, PO Box 3036, University Station Laramie, Wyoming 82071, USA.

157

point permeabilities; $\tilde{Y}(\mathbf{x}) = \ln \tilde{K}(\mathbf{x})$. The stochastic behavior of the permeability will of course generate fluctuations in both the pressure and the saturations of the fluids and we will assume that these also are normally distributed.

Both in hydrology [2] and in reservoir flow problems [1] stochastic models of this kind have been studied.

Recently Rubin and Gomez-Hernandes [7] have developed an approach for the upscaling of the permeability of a stochastic model for hydrological flow based on the lognormal distribution. The procedure is based on a perturbation expansion of the fluctuations δY where $\delta Y = \tilde{Y} - \langle \tilde{Y} \rangle$, $\tilde{Y} = \ln \tilde{K}$ and $\langle \cdot \rangle$ denotes the expected value.

We will extend these results to our model for immiscible reservoir flow.

Let Ω be the flow domain with a characteristic dimension R, $(R = 1)$. In the numerical framework, Ω is partitioned into a large number of discrete elements with characteristic dimension l. We assume that $\tilde{K}$ is given as a constant on each of these elements. The block averaged expected value of $\tilde{K}$, is given by [7]:

$$K = K_G \cdot F(C_Y(\mathbf{x}, \mathbf{x}'), L) \qquad (1)$$

where K_G is the geometrical mean of the block permeability, L is the characteristic dimension of the block where the averaging is performed, and $C_Y(\mathbf{x}, \mathbf{x}') = \langle \delta Y, \delta Y' \rangle$ is the covariance of $\tilde{Y} = \ln \tilde{K}$. Note that $\lim_{L \to \infty} K = K_G$ and $\lim_{L \to 0} K = K_A$, where K_A is arithmetic mean.

We will adopt an isotropic exponential covariance given by:

$$C_Y(\mathbf{x}, \mathbf{x}') = \sigma^2 \exp(-|\mathbf{x} - \mathbf{x}'|/I) \qquad (2)$$

where σ^2 is the variance of $\tilde{Y}$ and I is the correlation length. We may note that the function $F(C_Y(\mathbf{x}, \mathbf{x}'), L) \to 1$ as $\sigma^2 \to 0$.

Then the model is characterized by the following quantities:

Characteristic lengths:
l, L, R and I $\qquad (3)$
and the variance σ^2.

Following the procedure given by Rubin and Gomez-Hernandez, we get the following partial differential equation [8] for the block averaged quantities, where we have kept terms to lovest order in the fluctuations δY.

$$\nabla \cdot \mathbf{v} = q_1(\mathbf{x}, t), \qquad (4)$$

$$\mathbf{v} = -\mathbf{A}(\mathbf{x}) \cdot \nabla p, \qquad (5)$$

$$\phi \frac{\partial}{\partial t} S + \nabla \cdot (f(S)\mathbf{v}) +$$
$$-\epsilon \nabla \cdot (\mathbf{D}_1(S, \mathbf{x}) \cdot \nabla S) + \qquad (6)$$
$$-\alpha \nabla \cdot (\mathbf{D}_2(S, \mathbf{x}) \cdot \nabla S) = q_2(\mathbf{x}, t),$$

where $\mathbf{v}$ is the total Darcy velocity, p the total fluid pressure, S denotes the saturation of water and ϵ and α are parameters scaling the diffusion/dispersion terms. We restrict ourselves to two space dimensions and assume the absolute permeability tensor to have the form

$$\mathbf{K}(\mathbf{x}) = k_x(\mathbf{x})\mathbf{ii} + k_y(\mathbf{x})\mathbf{jj}, \qquad (7)$$

where k_x, k_y is given by eqs.(1)-(2).

Let λ_i, $i = w, o$ denote the mobility of water and oil respectively. We then have

$$\mathbf{A}(S, \mathbf{x}) = \mathbf{K}(\mathbf{x})(\lambda_w + \lambda_o), \qquad (8)$$

$$f(S) = \frac{\lambda_w}{\lambda_w + \lambda_o}, \qquad (9)$$

$$\mathbf{D}_1(S, \mathbf{x}) = \mathbf{K}(\mathbf{x})\frac{\lambda_w \lambda_o}{\lambda_w + \lambda_o}\frac{dp_c}{dS}$$
$$\mathbf{D}_2(S, \mathbf{x}) = (\frac{d}{ds}\bar{f}(S))^2 |\mathbf{v}|(\mathbf{ii} + \mathbf{jj}) \qquad (10)$$

where p_c is the capillary pressure. The mobilities and the capillary pressure are assumed to be known functions of the water saturation, and $\bar{f}(S)$ is the convex hull of $f(S)$.

In deriving the equations, we have assumed that $\lambda_w + \lambda_o$ is a slowly varying function of S and that the longitudinal and transverse dispersivity coefficients can be set equal to α.

3 SOLUTION PROCEDURE.

We shall adopt a sequential solution strategy to handle the system (4)-(6). With the assumed weak saturation dependence in (4)-(5), we may solve for pressure and velocity at a given time with acceptable accuracy. The velocity is then assumed known when we calculate the saturation at the next timelevel from (6).

The solution procedure is presented in several papers [9-17], so we will only give the main steps here.

The pressure/velocity equation (4)-(5) are solved using a Galerkin formulation. An preconditioned iterative procedure, based on domain decomposition is used to solve the algebraic equation. A typical grid for this calculation is given in Fig.1.

The saturation equation (6) is solved using an operator-splitting technique [10]. The hyperbolic part, with a modified fractional flow function, $\bar{f}(S)$, is solved in the first step [9]. This solution is used to approximate the time derivative along the approximate characteristics in the second step. Here too, an iterative procedure based on domain decomposition is used to solve the resulting equations. The main features of the solution procedure are:

- Almost no grid orientation effects are present.
- Numerical diffusion is small compared to the physical diffusion in the model.
- We may use long global timesteps with acceptable accuracy.
- Fixed and adaptive local grid refinement capabilities are included.
- The algorithms have good parallel properties.

This solution procedure is tested earlier on a simplified heterogeneous model [6] with good results. Here, we will extend these calculation to the heterogeneous model presented in section 2.

Figure 1: Typical composite mesh for pressure and velocity computations.

4 COMPUTATIONAL RESULTS.

The computational domain Ω is the unit rectangle with an injector in the lower left corner and a producer in the upper left. We chose the boundary conditions to be

$$\mathbf{v} \cdot \mathbf{n} = 0, \quad \mathbf{x} \in \partial\Omega, \tag{11}$$

$$\nabla S \cdot \mathbf{n} = 0, \quad \mathbf{x} \in \partial\Omega, \tag{12}$$

where $\mathbf{n}$ denotes the unit normal out of Ω. In each of the realizations of $\tilde{K}$, we will solve the deterministic problem based on the gridblocks of dimension l. In these calculation the dispersion is zero ($\alpha = 0$) and the permeability $K = \tilde{K}$. We will compare the numerical solution based on the averaged quantities with these high resolution solutions. The mass balance error of all the runs are within 1 percent.

4.1 Variance $\sigma^2 = 0.25$

1. High resolution solution. The permeability is generated on blocks given by a 180×180 grid, which define the scale l. We have chosen the correlation length $I = O(l)$. The compuational results are shown in Fig.2a where capillary forces are neglected. In Fig.2b capillary forces are included, $\epsilon = 10^{-2}$. Based on mathematical theory and earlier investigation [16] the finegrid solution reported here accurately reflect the true solution of the differential equations. Therefore it should give a good base for comparance with averaged solutions.

2. Upscaled permeability solution. Since σ^2 is small, we may approximate the upscaled permeability by the block geometric mean K_G [7]. As in the first example $I = O(l)$ and the size of the coarse grid where the permeability is averaged is $L = 20l$. This means that $l \sim I \ll L \ll 1$. The computational results based on this approximation is given in Fig.3a and Fig.3b. In Fig.3a only dispersion is included with $\alpha = 2\sigma^2 \cdot 10^{-2}$. In Fig.3b capillary forces also are included with $\epsilon = 10^{-2}$. A direct comparison of Fig.2a,b and Fig.3a,b shows a very good agreement. This means that the convective dispersion equation is capable of representing the effects of heterogenities in the model.

3. Models witch correlation $I \gg l$. We choose the correlation length $I = 10l$ which give a porous media with channeling as shown in Fig.6. Again we make a high resolution calculation as shown in Fig.4a (capillary forces neglected). For the calculation based on averaged quantities we choose $L = I$ (Fig.4b) and $L = 2I$ (Fig.4c). Theoretical hydrology [7] predict that the average of the permeability should be close to the arithmetic mean in this case. Experimentally we have verified that this average indeed gives the best result. As expected 18×18 coarse grid solution resolves most of the dynamics with the longest correlation lengths. The coarser grid 9×9, gives a more smeared result. The figure also shows the local refinement used around the saturation front.

Fig. 4d is based on the same data as Fig 4b, except that dispersion is included with $\alpha = 2\sigma^2 \cdot 10^{-2}$.

4.2 Variance $\sigma^2 = 1.00$

1. **High resolution, 180×180 grid solution.** We have chosen $I = O(l)$ and again we get an accurate solution given in Fig. 5a and 5b at two different timesteps.

2. **Upscaled permeability/dispersion.** The solution is based on $L = 20l$ and the dispersion coefficient $\alpha = \sigma^2 \cdot 10^{-2} (\epsilon = 0)$. Although we get a fairly good average representation of the high resolution

solution, an anisotropic dispersion tensor may be needed in this case [4,18]. In the diagonal region, where we have the largest velocity, the dispersion is too small. Fig.5c and 5d give the results for the same time steps as in Fig.5a,b.

5 Conclusion.

The examples shows that the numerical shceme based on operator splitting and domain decomposition methods is well suited for the solution of heterogeneous models. We have based the averages on theoretical results obtained in the hydrological literature, which seems to give good results, even for fairly long correlation lengths. Further, we may note that the nonlinearities limits the width of the saturation front for dispersion models in the same way as for models with capillary diffusion. In the present study the dispersion tensor is isotropic, but we will extend the model to unisotropic cases in future work.

Also, we will continue the research on scale dependent averages of permeability and dispersion for stochastic models. Both the saturation dependence in the dispersion tensor and the effect of the saturation fluctuations in the pressure/velocity equation is important topics for further work.

Futher, we will seek to extend the models to cases where the correlation lenght may vary over the computational area.

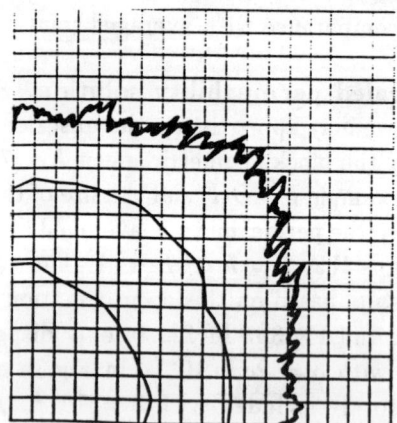

Figure 2: a) 180×180 high resolution, $I = l$, $\sigma^2 = 0.25$, $\epsilon = \alpha = 0$.

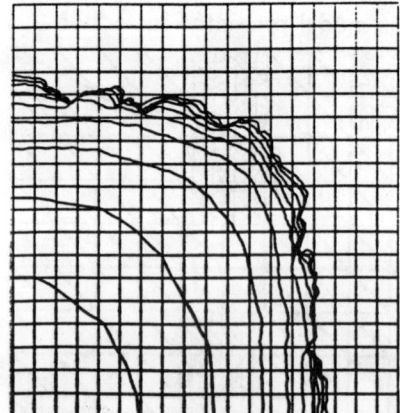

Figure 2: b) Like 2a with capillary forces included, $\epsilon = 10^{-2}$.

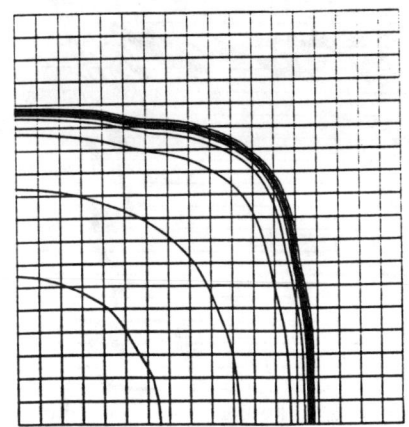

Figure 3: a) Like 2a with geometric averaged 9×9 blocks and dispersion, $\alpha = 2\sigma^2 10^{-2}$.

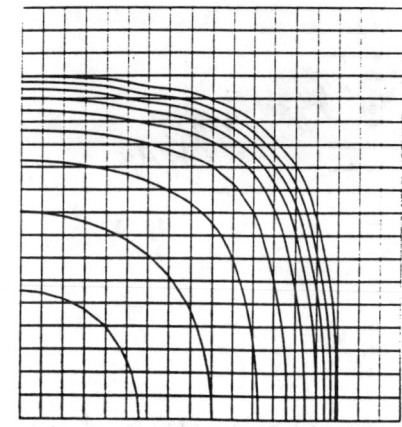

Figure 3: b) Like 3a with capillary forces included, $\epsilon = 10^{-2}$.

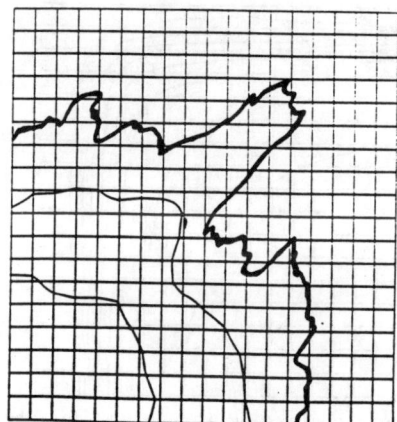

Figure 4: a) 180×180 high resolution, $I = 10l$, $\sigma^2 = 0.25$, $\epsilon = \alpha = 0$.

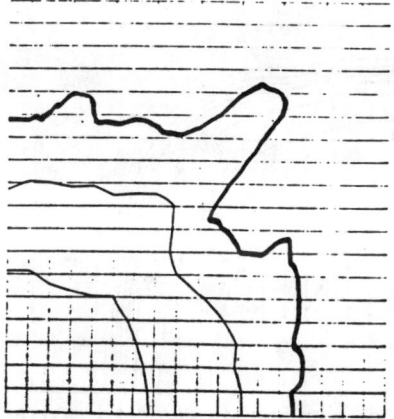

Figure 4: b) Like 4a with arithmetric averaged 18×18 blocks.

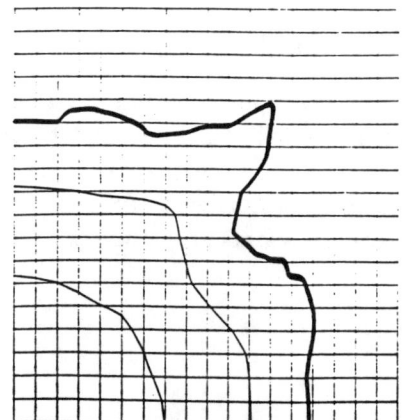

Figure 4: c) Like 4a with arithmetric averaged 9×9 bloks.

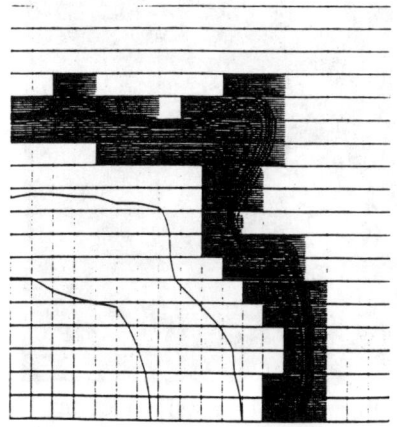

Figure 4: d) Like 4b with dispersion included, $\alpha = 2\sigma^2 10^{-2}$.

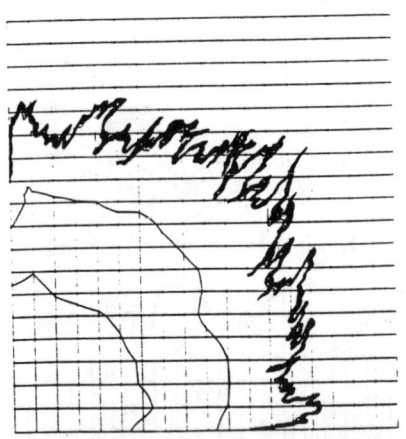

Figure 5: a) 180×180 high resolution, $I = l$, $\sigma^2 = 1.0$, $\epsilon = \alpha = 0$.

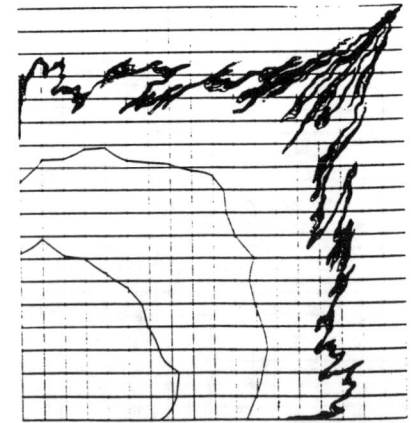

Figure 5: b) Like 5a at a later time step.

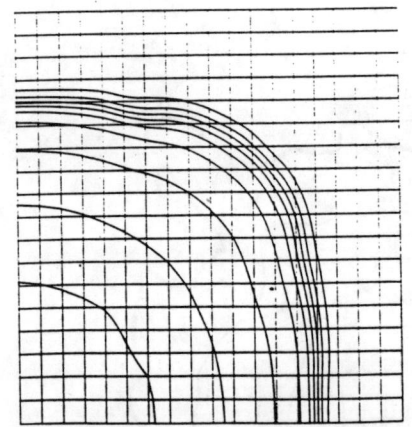

Figure 5: c) Like 5a with geometric averaged 9×9 blocks and dispersion, $\alpha = 2\sigma^2 \cdot 10^{-2}$.

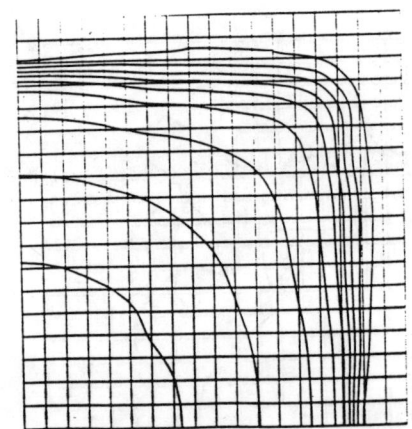

Figure 5: d) Like 5c at a later time step.

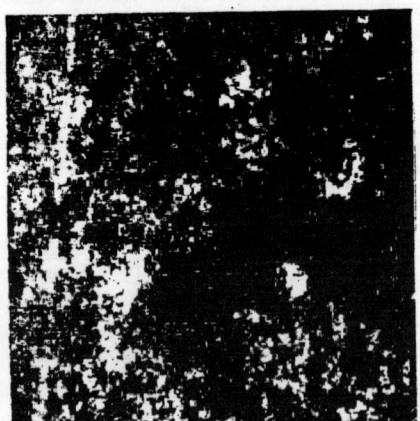

Figure 6: Permeability density plot for Fig.4a.

Acknowledgments

This research was supported in part by the Norwegian Research Council of Science and Humanities (NAVF), and VISTA, a research cooperation between the Norwegian Academy of Science and Letters and Den norske stats oljeselskap a.s (Statoil).

References

[1] T.J. Lasseter, J.R. Waggoner and L.W. Lake, Reservoir heterogeneities and their influence on ultimate recovery, in Reservoir Characterization, Eds. L.W. Lake and H.B. Carroll, *Academic Press, 1986.*

[2] G. Dagan, Flow and Transport in Porous Formations, *Springer-Verlag, Berlin-Heidelberg, 1989.*

[3] H. Haldorsen, Simulator parameter assignment and the problem of scale in reservoir engineering, in Reservoir Characterization, Eds. L.W. Lake and H.B. Carroll, *Academic Press, 1986.*

[4] R.E. Ewing, T.F. Russell and L.C. Young, An Anistropic Coarse-Grid Dispersion Model of Heterogeneity and Viscous Fingering in Five-Spot Miscible Displacement That Matches Experiments and Fine-Grid Simulations, SPE 18441, *Proceedings Tenth SPE Symposium on Reservoir Simulation,* Housten, Texas, February 6-8, 1989, 447-446; and *SPE Res. Eng.,* (to appear).

[5] H. Haldorsen and E. Damseth, Stochastie modeling, *J. Pet. Tech., Vol. 42, 1990, p 404-413.*

[6] M.S. Espedal, R.E. Ewing, T.F. Russell, and O. Sævareid, Reservoir Simulation using mixed methods, a modified method for characteristics, and local grid refinement, *Proceedings of Joint IMA/SPE European Conferance on the Mathematics of Oil Recovery,* Robinson College, Cambridge University, July 25-27, 1989

[7] Y. Rubin and J.J. Gomez-Hernandez, A Stochastic approach to the problem of upscaling of conductivity in disordered media: Theory and Unconditional Numerical Simulations. by To be published in Water Resources Res.

[8] P. Langlo, M.S. Espedal, *Report Dep. of Math. Univ. of Bergen,* To be published.

[9] J. Douglas, Jr. and T.F. Russell, Numerical methods for convectiondominated diffusion problems based on combining the method of characteristics with finite element or finite dif-ference procedures,*SIAM J. Numer. Anal., 19 (1982), 871-885.*

[10] M.S. Espedal and R.E. Ewing, Characteristic Petrov-Galerkin subdomain methods for two-phase immiscible flow, *Comp. Meth. Appl. Mech. and Eng., 64 (1987), 113-135.*

[11] R.E. Ewing, Domain decomposition techniques for efficient adaptive local grid refinement, Domain Decomposition Methods (T.F. Chan, R. Glowinski, J. Periaux, O.B. Widlund, eds.), *SIAM, Philadelphia, Pennsylvania, 1989,* 192-206.

[12] R.E. Ewing, Adaptive grid refinements for transient flow problems, Adaptive Methods for Partial Differential Equations (J.E. Flaherty, P.J. Paslow, M.S. Shephard, and J.D. Vasilakis, eds.), *SIAM, Philadelphia, Pennsylvania, Chapter 14, 1989, 194-205.*

[13] R.E. Ewing, M.S. Espedal, J.A. Puckett, and R.S. Schmidt, Simulation techniques for multiphase and multicomponent flows, *Communications in Applied Numerical Methods, 4 (1988), 335-342.*

[14] H.K. Dahle, Adaptive characteristic operator-splitting techniques for convection-dominated diffusion problems in one and two space dimensions, *Rep. No. 85,* Dep. of Applied Math. University of Bergen, 1988.

[15] H.K. Dahle, M.S. Espedal, and R.E. Ewing, Characteristic Petrov-Galerkin subdomain methods for convection diffusion problems, *IMA Volume 11* Numerical Simulation in oil Recovery (M.F. Wheeler, ed.), *Springer-Verlag, Berlin, 1988, 77-78.*

[16] H.K. Dahle, M.S. Espedal, R.E Ewing and O. Sævareid, Characteristic adaptive sub - domain methods for reservoir flow problems, Numerical Methods for Partial Differential Equations, (to appear).

[17] H.K. Dahle, M.S. Espedal and O. Sævareid, Domain Decomposition for Reservoir Flow Problems, Proceedings 1989 Conferance on Domain Decomposition Methods, *SIAM Publications,* (to appear).

[18] Y. Rubin, Stochastic Modeling of Macrodispersion in Heterogeneous Poros Media. *Water Resources Res. Vol 26, 133-141.*

2nd European Conference on the Mathematics of Oil Recovery
© D. Guérillot, O. Guillon (Editors) and Éditions Technip, Paris 1990, pp. 165-168
27 rue Ginoux, 75015 Paris

Parallel Simulation of Petroleum Reservoirs

J. Larsen[1,2] and N. Bech[1,3]

ABSTRACT

The parallelisation of an oil reservoir simulator is described. The target machine architecture is an array of transputers. The reservoir simulator is described. Since the most time consuming part of the reservoir simulation is the solution of the resultant systems of equations the parallelisation of this is discussed. The scaling properties and performance of a key routine are reported.

1 INTRODUCTION

As part of a major ESPRIT II funded project, "Operating Systems and Programming Environments for Parallel Computers - Supernode II" the parallelisation of a commercial oil reservoir simulator is being carried out. The target architecture is an array of transputers, more specifically the Supernode architecture with a reconfigurable network of transputer processing elements.

The oil reservoir simulator application is one of four applications that are implemented in order to assess the systems software and tools that is developed within the project. The reservoir simulator application will integrate with the development of a parallel library and the development of a parallelising Fortran compiler.

The most compute intensive part of the application is the solution of the resulting systems of equations. In the current version of COSI three different equation solvers are applied, viz. a band solver from Linpack (Dongarra et al. 1977), a sparse solver, SESYS, using Gaussian elimination with drop of small elements and iterative refinement, and finally a Krylov subspace method, ORTHOMIN.

As part of the Supernode II project the NAG library is being implemented on the Supernode architecture. This includes a band solver which will be used by the parallel version of COSI for smaller systems. The parallelisation is concentrated on the sparse matrix solver with drop of small elements. But instead of iterative refinement we use Krylov subspace methods to accelerate the convergence.

2 COSI

COSI (Compositional Simulation) is a 3-dimensional, 3-phase, fully compositional, isothermal reservoir simulation model. The simulator contains a dual permeability option for the description of naturally fractures in reservoirs.

The area of application covers natural depletion, gas and water injection for pressure maintenance, cycling of gas condensate reservoirs and other miscible recovery methods.

The formulation uses a Soave-Redlich-Kwong of Peng-Robinson equation of state description of phase equilibrium (other equations of states can easily be implemented). An extended black-oil description of the phase behaviour is avail-

This work is supported by the Commission of the European Communities under ESPRIT contract P2528.

(1) Danish PARSIM Consortium, c/o Math-Tech ApS, Rosenstandsvej 4C, DK-2920 Charlottenlund, Denmark.
(2) Math-Tech ApS, Rosenstandsvej 4 C, DK-2920 Charlottenlund, Denmark (*Permanent Affiliation*).
(3) Risø National Laboratory, Postbox 44, DK-4000 Roskilde, Denmark (*Permanent affiliation*).

able as an option.

Dual permeability reservoirs are modelled by the continuum approach where the fracture system and matrix blocks are considered as two overlapping continua. The description of matrix-fracture exchange takes into account gravitational effects and saturation distributions within individual matrix blocks.

The basic mass conservation equation includes a diffusion term which accounts for mass transport driven by concentration gradients.

Space integration is performed by the integral finite difference technique which avoids reference to a specific coordinate system. The method encompasses standard Cartesian and cylindrical coordinates. In addition, it is possible to specify difference grids formed by an arbitrary combination of different grid cell types such as Cartesian, cylindrical and/or irregular grid cells. This means, among other things, that flow patterns around wells can be described more accurate in a field scale simulation than is usually possible. More over, grid orientation effects can be alleviated by application of irregular, curvilinear grid cells.

Additional advantages of the integral finite difference (IFD) method are:

- IFD includes the nine-point scheme where diagonally aligned adjacent grid cells are included in the difference operator (reference is mode to the 2-D case here).

- IFD includes the seven point scheme where the grid cells have a hexagonal section.

- IFD handles reservoirs characterized by a non-diagonal permeability tensor.

- Dead or non-active grid cells do not exist.

- Local grid refinement can be restricted to the location where it is needed.

The time integration method is fully implicit. The time step size is adjusted automatically to maintain a specified accuracy.

The resulting non-linear systems of algebraic equations are solved by the Newton-Raphson method.

The system of linear algebraic equations is solved optionally by a direct band solver (Dongarra et al., 1979) or iteratively by a sparse matrix solver (Houbak, 1985) or an algorithm based on incomplete factorisation and ORTHOMIN acceleration (Vinsome, 1976).

COSI has been coded in standard Fortran77. During the development of the computer code it has been considered a major objective to design a flexible program in which new features compatible with the overall scope are easily included. The code is well suited to serve as a basic framework for the development of special purpose simulators.

Consider the multi-phase flow of N_c components in a double porous rock. The mass conservation equation for each component and in each rock medium together with the Darcy's law, relating the flow velocity of each phase in each rock medium with pressure, are the fundamental equations on which COSI is based (for details see Bech, 1984).

The numerical solution of the system of equations is based upon the integral finite difference (IFD) method (see e.g. Preuss and Bodvarsson, 1983). The reservoir is divided into subregions, referred to alternatively as volumes, elements or grid cells, of arbitrary number, N_v, and shape.

The mass conservation equations are integrated over the volume of each cell and the resulting equation is transformed by means of Gauss' divergence theorem, leading to a system of ordinary differential equations. By solving this system implicitly there results a system of non-linear, algebraic equations

$$\mathbf{F}(\mathbf{y}) = \mathbf{0}. \tag{1}$$

The number of equations and unknowns are $N = (1+N_c)N_m N_v$ where N_m is the number of rock media.

The system is solved by means of the Newton-Raphson technique, i.e.

$$\mathbf{A}\mathbf{x} = \mathbf{b}, \tag{2}$$

where $A = \mathbf{F}'(\mathbf{y}_0)$ is the Jacobian, $\mathbf{y}_0$ is an approximation to the solution, $\mathbf{b} = -\mathbf{F}(\mathbf{y}_0)$ and $\mathbf{x} = \mathbf{y} - \mathbf{y}_0$. Only two to three iterations are necessary for convergence. The matrix A is sparse and non-symmetric.

3 THE TEST ARCHITECHTURE

The transputer T800 is a single chip processor featuring a floating point arithmetic with peak performance in excess of 1.5 Mflops (million floating point operations per second) and four on chip links each with a communication band width of 10 Mbits per second.

The test configuration is shown in Figure 1. The host transputer has four Mbytes of memory and the workers has 1 Mbyte of memory each.

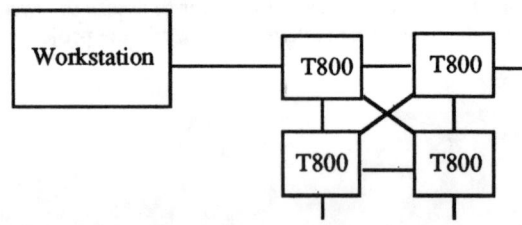

Figure 1. Test configuration.

Software allows for arbitrary splitting of the code between the host and the transputer array. Using the interface

Client Stub Source Writer (CSSW) from Cresco Data part of COSI runs on the host work station and the compute intensive part of COSI runs on the transputer array.

4 THE ALGORITHM

Since the band solver will be provided by another partner in the Supernode project we shall only describe the two sparse techniques, SESYS, and ORTHOMIN.

4.1 SESYS

The sparse linear solver, SESYS, uses Gaussian elimination with drop of small elements during the elimination procedure to create an approximate solution. In the existing code the solution is iteratively refined. We have tested the parallelisation of the elimination procedure on a processor ring (see Fig. 2).

The rows of the matrix are distributed to each processor in the ring in an interleaved fashion. For P processors, row i is stored in the memory local to processor i mod P. Partial pivoting is performed by interchanging row numbers rather than interchanging rows. The elimination row is broadcasted to all processors and the elimination is then performed in parallel. For a banded structure this will work efficiently provided the half band width is larger than the number of processors.

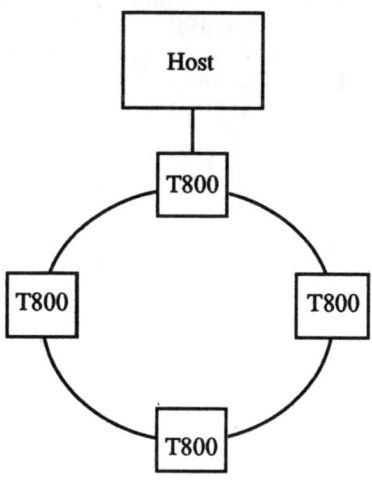

Figure 2. Processor ring

Care must be taken when the performance of a particular algorithm is measured. For a fixed size problem, the performance gained by adding more processors will saturate; but by allowing the problem size to grow with the number of processors an almost linear speed up as measured e.g. by the number of floating point operations (flop), can be obtained (cf. Gustafson et al., 1988).

The performance of good parallel algorithm will scale almost linearly with the number of processors. In addition a good parallel algorithm should utilise the processing power efficiently such that e.g. the ratio between actual flop rate and peak flop rate is as large as possible. A solution procedure with good scaling properties might be better on a parallel computer than the best serial one.

4.2 ORTHOMIN

The ORTOHOMIN method belongs to the class of Krylov subspace method where the solution is projected onto a Krylov subspace. The full projection method with a dimension that equals the order of the system of equations is a direct method; but its extensive use stems from the fact that a subspace of low dimension yields a good approximation to the solution. Different choices of subspace base vectors and different choices of expansions coefficient yield different methods (Saad, 1989). We give here the algorithm for ORTHOMIN (Vinsome, 1976).

We are seeking a solution to equation (1) with an initial guess x_0. We define $r_0 = b - Ax_0$, and $v_1 = r_0$. Then for $j = 1, 2, \ldots, m$ we have

$$x_j = x_{j-1} + \alpha_j v_j, \tag{3}$$

$$r_j = r_{j-1} - \alpha_j Av_j, \tag{4}$$

where

$$\alpha_j = (r_{j-1}, Av_j)/(Av_j, v_j), \tag{5}$$

$$v_{j+1} = r_j - \sum_{i=1}^{j} \beta_{ij} v_j, \tag{6}$$

in which

$$\beta_{ij} = (r_j, Av_i)/(Av_i, v_i), \tag{7}$$

with a stop criteria based on the norm of r_j being small.

We have reproduced the algorithm here in order to demonstrate that only matrix-vector multiplication and dot products enter the algorithms. It is thus possible to implement the algorithm using level one and level two BLAS (Basic Linear Algebra Subroutines).

Apart from ORTHOMIN we will implement GMRES (Saad, 1986) and as preconditioners we use inter alia SESYS.

4 TEST RESULTS

We report here a test of the scaling properties of the elimination procedure in SESYS. We have used one, two and three processors in a ring structure with the host transputer. The results are listed in Table 1.

TABLE 1
Test results for parallel elimination procedure

P	N	CPU-time (sec)	Mflops
1	256	51	0.22
2	256	28	0.40
3	256	13	0.62
3	512	133	0.67

From Table 1 it is readily seen that the performance of the algorithm scales almost linearly with the number of processors. It is also seen that by increasing the order N of the system of equations better utilisation of the processors is obtained. In the implementation we have used a Fortran compiler from 3L. By using BLAS routines coded in assembler it is possible to substantially increase the flop rates. In counting the number of floating operation we have counted both multiplications and additions.

5 CONCLUDING REMARKS

From an ongoing development of a parallel reservoir simulator test results for a key algorithm in a parallelised elimination routine are reported. Good scaling performance is obtained.

REFERENCES

Bech, N., 1984, The EFP Basic Reservoir Simulation Model - Proposal for Mathematical Formulation and Numerical solution Approach. Energy Research Program of the Danish Ministry of Energy, Report no. 7, RISØ-M-2425.

Dongarra, JJ., Moler, C.B., Bunch, J.R. and Stewart, G.W., 1979, LINPACK, Users Guide, SIAM Philadelphia.

Gustafson, J.L., Montry, G.R. and Benner, R.E., 1988, Development of parallel methods for a 1024-processor hybercube. SIAM J. Sci. Stat. Comput., 9, p. 609-638.

Houbak, N. (1985). SESYS - A Sparse Matrix Linear Equation Solver, Users Guide. Energy Research Program of the Danish Ministry of Energy, Report no. 12, RISØ-M-2527.

Saad, Y., 1986, GMRES: A generalized minimal residual algorithm for solving nonsymmetric linear systems. SIAM J. Sci. Stat. Comput., 7, p. 856-869.

Saad, Y., 1989, Krylov subspace methods on supercomputers. SIAM J. Sci. Stat. Comput., 10, p. 1200-1232.

Vinsome, P.K.W., 1976, ORTHOMIN, at Iterative Method for Solving Sparse Sets of Simultaneous Linear Equations, SPE 5729, SPE-Aime Fourth Symposium on Numerical Simulation of Reservoir Performance, Los Angeles, Feb. 19-20, 1976.

2nd European Conference on the Mathematics of Oil Recovery
© D. Guérillot, O. Guillon (Editors) and Éditions Technip, Paris 1990, pp. 169-176
27 rue Ginoux, 75015 Paris

Comprehensive Mathematical Modeling
of Horizontal Wells

M. R. Islam[1]

ABSTRACT

Recently, most technological advances in the petroleum industry have been in the areas of horizontal wells. Horizontal wells have been reported to produce up to 10 times more oil than that produced by vertical wells. Even though recent advances in horizontal well technology have led to a large drop in drilling and operation costs, very little has been done to advance mathematical modeling of a horizontal well in order to achieve more realistic prediction capabilities. This paper describes a reservoir simulator capable of dealing with important features of a horizontal wells, including wellbore dynamics and near wellbore radial flow.

In this paper, Darcy's law is used together with conservation of mass to model fluid flow in the reservoir. Radial flow (radial geometry) is assumed near the wellbore and linear flow (cartesian geometry) is used farther away from the wellbore. These equations are coupled with a rigorous wellbore model. The wellbore model constists of time dependent, three-phase, one-dimensional, mass and momentum balance equations. The simulator has the capability of solving reservoir and wellbore equations with different degree of implicitness. It also allows the selection of turbulent or laminar flow in the wellbore. The reservoir simulator is applied to an isothermal problem in a hypothetical reservoir in order to demonstrate some of the features which are necessary for modeling a horizontal well.

1. INTRODUCTION

Horizontal well technology has been one of the most studied topics in petroleum industry in recent years. The increasing interest for using horizontal wells is evident from the

---

(1) Department of Petroleum Engineering, University of Alaska, Fairbanks, Alaska 99775, USA.

number of studies reported in recent years (Joshi, 1988; Islam and George, 1989; Butler, 1988; Doan *et al*, 1990). Due to more extensive contact with the reservoir, horizontal wells allow fluid flow at lower fluid velocities still providing total flows which are economic. The applicability of horizontal wells has been tested in thermal applications, such as steam flooding (Proctor *et al*, 1987), steam-assisted gravity drainage (Joshi, 1986; Ong and Butler, 1989), and electromagnetic heating (Islam and Chakma, 1990a); in reservoirs containing bottom water (Proctor *et al.*, 1987; Islam and Chakma, 1990a) and many other areas. However, the most advancement in horizontal well technology has been in the areas of drilling and drainhole operations. The most neglected area of horizontal well technology is the reservoir simulation. Most papers published in this area deals with simplistic representation of the horizontal well (Rial, 1984; Gussis, 1985). These models neglect radial flow near the wellbore and, more importantly, the multiphase fluid dynamics in the wellbore. Only recently, Stone *et al.* (1989) reported the first paper dealing with wellbore dynamics in a horizontal well. In modeling gravity drainage in a bitumen reservoir, they used Darcy's law for describing fluid flow in the reservoir and momentum, mass and energy balance of emulsion (oil and water) and gas in the wellbore. However, they used productivity index concept for representing radial flow near the wellbore. More recently, Collins *et al.* (1990) reported a simplistic model for simulating horizontal well flow in a field. They used Darcy's law in a dual porosity representation of a horizontal well. This model unnecessarily introduces simplicity and does not deal with wellbore dynamics with any rigorousness. Recently, Jelen (1990) used a new technique for solving fluid flow

equations in different domains independently. This method was particularly useful for modeling a horizontal well in which several domains exhibit very different types of fluid flow. However, Jelen did not report any rigorous treatment of the wellbore itself.

This paper takes similar approach as Stone *et al.* but introduces hybrid grid system in order to describe radial nature of fluid flow near the wellbore. The simulator is a compositional, isothermal model. However, example runs presented in this paper are conducted in a black-oil (three-phase, three-component) mode.

2. MATHEMATICAL FORMULATION

2.1. Reservoir Flow Equations

The model is based on compositional, three-phase, hybrid grid formulation. The reservoir flow near wellbore is described by a radial geometry whereas the flow farther away from the wellbore is described by a cartesian geometry. The continuity equation for the i-th component, for cartesian flow farther away from the wellbore, is given by:

$$\nabla(\rho_i v_i) = -\frac{\partial C_i}{\partial t} \qquad (1)$$

The same for radial flow near the wellbore is given by:

$$\frac{1}{r}\frac{\partial}{\partial r}(\rho_i v_i) = -\frac{\partial C_i}{\partial t} \qquad (2)$$

Darcy velocities, for a given phase, p, is given for farther away from the wellbore by:

$$\bar{v}_p = - \frac{\check{K} \, K_{rp}}{\mu_p} (\nabla p_p - \gamma_p \nabla z) \quad (3)$$

The same for near wellbore is given by:

$$\bar{v}_p = - \frac{\check{K} \, K_{rp}}{\mu_p} (\frac{\partial p}{\partial r} - g(\gamma_p, \theta)) \quad (4)$$

In the above formulation, γ_p is the acceleration due to gravity multiplied by density of the phase, p. Equations (3) and (4), along with Equations (1) and (2), constitute the flow equations in the porous region of the reservoir.

2.2. Wellbore Flow Equations

Islam (1990) showed through experimental results that fluid flow in a horizontal well under realistic reservoir conditions is stratified. Also, Islam and Chakma(1990b) reported that, in heavy oil reservoirs, gas and oil flow simultaneously with gas bubbles being entrained by the liquid mass. Following this representation, we consider two phases in the wellbore - the oil phase (containing gas bubbles) and the water phase, the overall flow being segregated. The two-fluid equations for conservation of mass momentum and energy are derived by Ishii (1975) for a general three-dimensional two-phase flow. One-dimensional forms can be obtained by integrating across the duct area. Neglecting viscous interactions, and

pressure changes at the interface due to surface tension, the equations for mass and momentum conservation for phase j are given by:

$$\frac{\partial}{\partial t}(\alpha_j \bar{\rho}_j) + \frac{\partial}{\partial x}(\alpha_j \rho_j \bar{U}_j) = 0 \quad (5)$$

$$\alpha_j \bar{\rho}_j \frac{\partial \bar{U}_j}{\partial t} + \alpha_j \rho_j \bar{U}_j \frac{\partial \bar{U}_j}{\partial x} + \alpha_j \frac{\partial \bar{P}_j}{\partial x}$$

$$-\alpha_j q_j - (\bar{P}^* - \bar{P}_j)\frac{\partial \alpha_j}{\partial x} = 0 \quad (6)$$

Here x is the spatial coordinate along the duct axis and t is time, ρ_j, P_j, and U_j denote respectively the density, pressure and velocity of phase j and P^* is the interface pressure. The over-bars refer to phase average quantities defined by:

$$\bar{a}_j = A_j^{-1} \int_{A_j} a_j \, dA$$

$\alpha_j = A_j/A$ is the fraction of the duct area occupied by phase j. Note that in the above formulation averages of all products have been taken as equal to products of averages and all flow variables represent mean values over some short time interval.

2.3. Boundary Conditions

Boundary conditions between cell blocks of radial and cartesian geometry are defined following Pedrosa and Aziz (1986). Therefore, the productivity index concept is eliminated from this study. Between radial and cartesian systems, the following boundary conditions were applied:

$$\frac{\partial p}{\partial x}\Big|_{\Gamma_c} = \frac{\partial p}{\partial r}\Big|_{\Gamma_r} \qquad (7)$$

where Γ_c and Γ_r represent the boundaries of the cartesian and radial domain, respectively. Also, the following boundary condition has to be realized:

$$p\Big|_{\Gamma_c} = p\Big|_{\Gamma_r} \qquad (8)$$

2.4. Reservoir Fluid Data

In order to demonstrate the capability of the numerical simulator, primary depletion test was performed in the presence of a bottom-water zone. Such reservoirs are not economically produced with vertical wells and horizontal wells appear to be attractive alternatives. Figure 1 shows the reservoir geometry used for numerical simulation. The horizontal well is placed in the middle of the oil zone. The water zone contains an upper layer of transition zone and a lower layer 100% water-saturated zone. The horizontal well extends to 500 m. Other reservoir characteristics and fluid properties are given in Table 1. The water-oil relative permeability and capillary pressure data were taken from Islam and Chakma (1990c). As discussed earlier, it was assumed that oil and gas flow as a mixture and, therefore, were treated as one phase. This assumption has been experimentally validated for heavy oil reservoirs by Islam and Chakma (1990b).

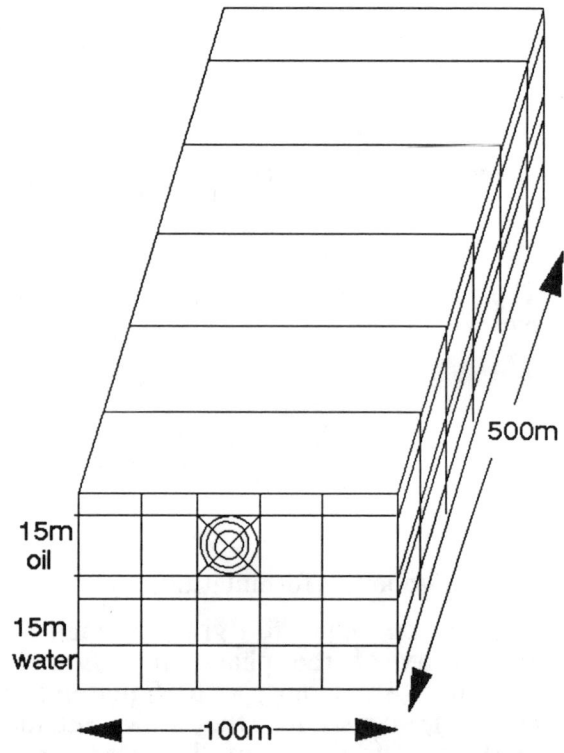

FIG.1 Reservoir Geometry and Wellbore

TABLE 1:

Reservoir and Fluid Properties

Initial Oil Saturation	80%
Initial Water Saturation	20%
Reservoir Pressure (kPa)	5,500
Porosity	30%
Permeability (μm^2)	4
Oil Viscosity (mPa.s)	1,500
Water Viscosity (mPa.s)	1.0
Oil Density (g/ml)	0.9
Water Density (g/ml)	1.0
Wellbore diameter (cm)	8.9

3. SOLUTION METHODS AND ACCURACY

Since the pressure dependent coefficients in this paper have been treated implicitly, the resulting system of algebraic equations is non-linear. This system is solved by means of the Newton-Raphson procedure. The system of linearized equations generate matrices of sparse, banded structure. Direct method was used to solve these equations.

The wellbore equations are solved by an algorithm, called SIMPLE, outlined by Patankar (1980). Such algorithm has been used commercially by Rosten and Spalding (1986). In this technique, the relevant variable ϕ_p is solved at finite-difference node P, using a finite-volume form of the conservation equation

$$a_p \phi_p = \Sigma a_i \phi_i + S_\phi \qquad (9)$$

where ϕ_p is the value of ϕ at node P, at the center of its control volume, ϕ_i the value of ϕ at the relevant neighboring point at the center of its control volume (including the 'old' time value ϕ), a_p the coefficient for node P which, in absence of sources and boundary conditions, is the sum of the a_i's, and a_i the coefficients for relevant neighbouring nodes. The source term in the finite-difference equation is linearized for stability and the convection contribution is upwinded. The simulator also has an option for introducing turbulent flow regime as outlined by Stones *et al.* (1989). However, in this paper we limit our discussion to laminar flow which is likely to prevail in heavy oil reservoirs.

In order to validate the numerical simulator, the problems of SPE Benchmark II and V were solved. Results were well within the range provided by the benchmarks. These problems, however, validated only vertical well problems. The horizontal well prediction could not be validated due to the lack of any benchmark.

4. RESULTS AND DISCUSSION

Using the reservoir and fluid data as given in Table 1, numerical simulation runs were conducted for i) primary depletion, ii) gas injection and iii) water injection. All these cases are relevant to heavy oil reservoirs which is the main focus of this paper for horizontal well applications. For this case, the vertical well was placed in the center of the reservoir so that the drainage took place with a radius of 50 m. A no-flow boundary condition was applied to the 50 radius periphery. Numerical simulation runs for primary depletion were conducted with a rate constraint of 10 m^3/day for vertical wells and 25 m^3/day for horizontal wells. The reason for different rates was that vertical wells were immediately invaded by the water zone if the flow rate was high. One of the main reasons for implementing a horizontal well is to be able to produce at higher rates still delaying the water breakthrough. Figure 2 compares watercut of vertical and horizontal wells. Note that the water breakthrough took place almost instantaneously for the vertical well. Water breakthrough takes place after 100 days in the horizontal well (case b). Note that if pressure drop along the wellbore is not considered, numerical results show considerable delay in water breakthrough (case c). Such an optimistic result arises from the fact that this model estimates uniform water front rising from the bottom water. However, pressure drop along the wellbore deforms this front and water breakthrough takes place somewhere close to the point where the horizontal well turns to the vertical portion.

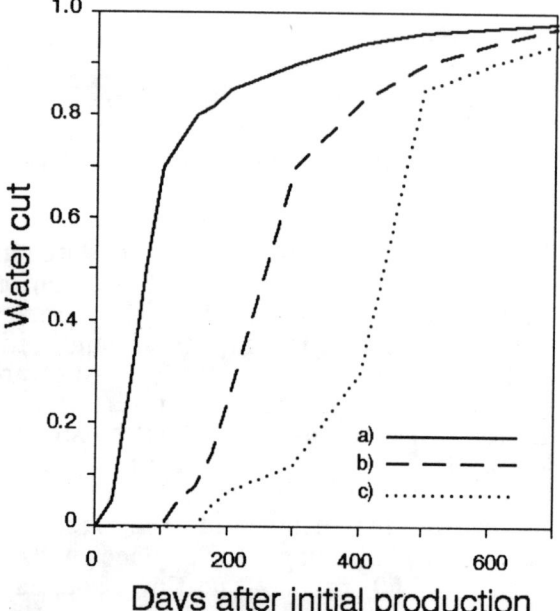

FIG. 2 Comparison of Water Cuts in Vertical and Horizontal Wells
(a: vertical well, b: horizontal well, c: horizontal well (without Δp in the wellbore)

These numerical runs show how horizontal wells can be used to control bottom water encroachment. Figure 3 compares the saturation fronts in vertical and horizontal wells. Results are shown after 150 days of oil production. In both cases, the saturation front was taken at the base of the vertical run (just after deviation point for the horizontal well). Note that the growth of the water- saturation front is controlled

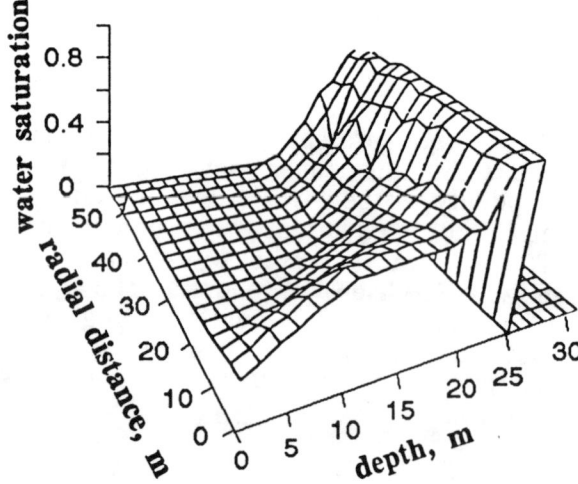

(a) Vertical well after 150 days

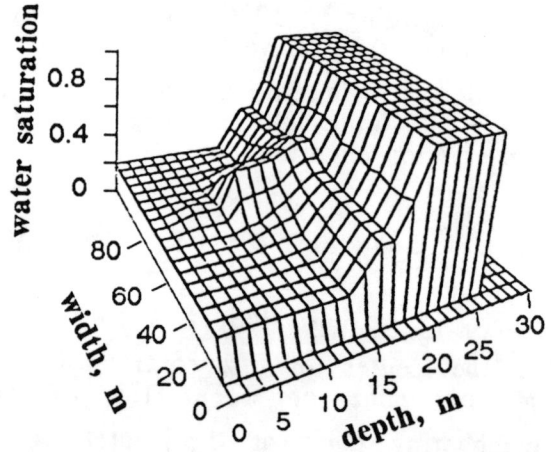

(b) Horizontal well after 150 days

FIG. 3 Comparison of Water Saturation Front in Vertical and Horizontal Wells

remarkably for the horizontal well case even though the fluid withdrawal rate was much higher for the horizontal well case.

Figure 4 compares the oil production rates of the vertical and horizontal well cases. Obviously, the oil production rate is much higher in the vertical case than in the horizontal case. The improvement by the horizontal well is further accentuated by the delayed water breakthrough. Note that the elimination of pressure drop in the wellbore leads to overestimation of the oil production rate.

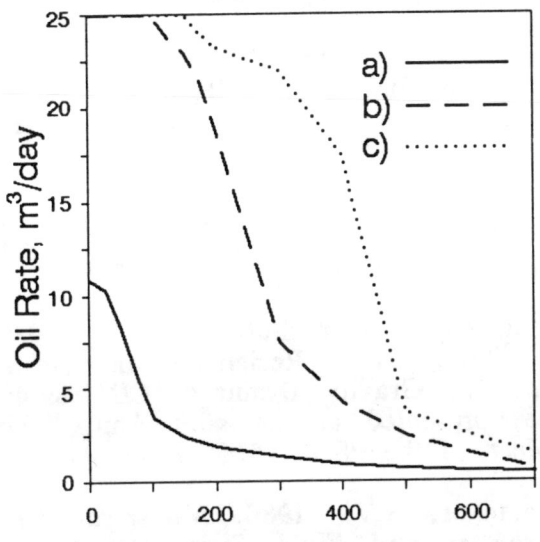

FIG. 4. Comparison of Oil Rates in Vertical and Horizontal Wells (a:vertical well; b: horizontal well; c: horizontal well without Δp in the wellbore)

The second series of numerical runs was conducted to model water injection. For this case, a combination of two horizontal wells (500 m long, horizontal placement and separated by 100 m). For this case a uniform reservoir of 30 m thickness and 100 m width was assumed. The injection well was placed in the bottom of the reservoir in order to have gravity stabilization of the water front. This justified the use of same relative permeability curve as in the primary case which was essentially applicable to a stable and stabilized front. The same system was used for gas injection. However, for this case, the injection was carried out through the well situated at the top of the reservoir. Also, pressure

drop along the wellbore was neglected for the injection wells since only single phase flow was involved. Relative permeability curves for the gas-oil system was taken from Islam and Chakma (1990b). Results of these runs are shown in Figure 5. All these runs were conducted with a injection rate constraint of 50m^3/day.

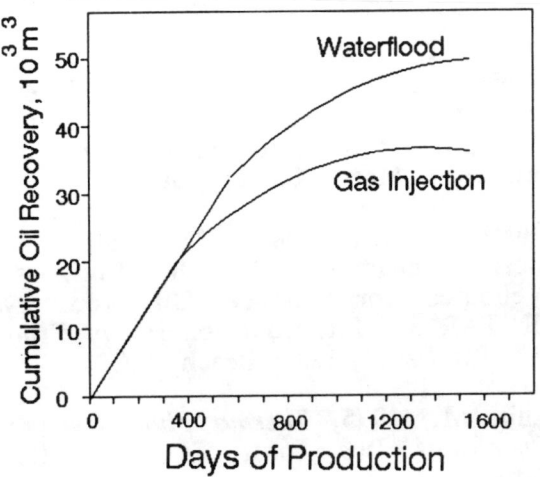

FIG. 5 Recovery Performance of Waterflood and Gas Injection

5. CONCLUSIONS

A complete mathematical formulation for modeling horizontal wells is presented. The model consists of hybrid-grid formulation along with rigorous treatment of the wellbore dynamics. Examples of numerical results are provided for a selected case of reservoir simulation.

6. NOMENCLATURE

C	concentration
$\bar{\bar{K}}$	absolute permeability tensor
K_{ri}	relative permeability to phase i
p	pressure
q	flow rate
S_i	saturation of the phase i
t	time
U	velocity in the wellbore

7. REFERENCES

Butler, R.M., 1989, The Potential for Horizontal Wells for Petroleum Production, J.Can.Pet.Tech.,May-June, 39.

Collins, D. Nghiem, L., Sharma, R., Li, Y., and Jha, K., 1990, Field-Scale Simulation of Horizontal Wells, paper CIM/SPE 90-121, presented at the CIM/SPE Int. Tech. Meet., Calgary, AB.

Doan, Q., Farouq Ali, S.M., and George, A.E., 1990, Scaling Criteria and Model Experiments for Horizontal Wells, paper presented at the CIM/SPE Int. Tech. Meet., Calgary, June.

Gussis, G.L., 1985, "Simulation of Steam Injection Through Horizontal Wellbores for Viscous Oil Recovery, UNITAR 3rd Int. Conf. on Heavy Crude and Tar Sands, Long Beach, CA.

Ishii, M., 1975, *Thermo Fluid Dynamic Theory of Two Phase Flow*, Eyrolles, Paris.

Islam, M.R., 1990, Comprehensive Physical and Numerical Modeling of a Horizontal Well, SPE 20627, presented at the SPE Ann. Conf. and Exhib.

Islam, M.R. and George, A.E., Sand Control in Horizontal Wells in Heavy Oil Reservoirs, SPE 18789, presented at the SPE Calif. Reg. Meet.

Islam, M.R. and Chakma, A., 1990a, A New Recovery Technique for Heavy Oil Reservoirs With Bottom Water, SPE/DOE 20258, paper presented at the SPE/DOE Seventh Symp. on EOR, Tulsa, OK.

Islam, M.R. and Chakma, A., 1990b, Mechanics of Bubble Flow in Heavy Oil Reservoirs, SPE 20070, presented at the SPE Calif. Reg. Meet.

Islam, M.R. and Chakma, A., 1990c, Mathematical Modeling of Enhanced Oil Recovery by Alkali Solutions in the Presence of Cosurfactant and Polymer, J. Pet. Eng. Sci., in press.

Jelen, J., Multidomain Method in Reservoir Simulation, paper presented at the 1990 Summer Computer Simulation Conf., July, Calgary, AB.

Joshi, S.D., 1986, A Laboratory Study of Thermal Oil Recovery Using Horizontal Wells, paper SPE/DOE 14916, SPE/DOE Fifth Symp. on EOR.

Joshi, S.D., 1988, Augmentation of Well Productivity With Slant and Horizontal Wells, JPT, June, 729 p.

Ong, T.S. and Butler, R.M., 1989, Wellbore Flow Resistance in Steam-Assisted Gravity Drainage, CIM 89-40-58, presented at the 40th Ann. Tech. Meet. of the PS of CIM, Banff, AB.

Patankar, S.V., 1980, *Numerical Heat Transfer and Fluid Flow*, Hemisphere, Washington, D.C.

Pedrosa, O.A. and Aziz, K., 1986, Use of Hybrid Grid in Reservoir Simulation, SPEJ, Nov., 611 p.

Proctor, M.L., George, A.E., Farouq Ali, S.M., 1987, Steam Injection Strategies for Thin, Bottom Water Reservoirs, SPE 16338, SPE Calif. Reg. Meet.

Rial, R.M., 1984, 3D Thermal Simulation Using A Horizontal Wellbore for Steamflooding, paper SPE 13076, Ann. Conf. and Exhib., Dallas, TX.

Rosten, H.I. and Spalding, D.B., 1986, *Beginner's Guide and User Manual*, CHAM TR/100, London, U.K.

Stone, T.W., Edmunds, N.R., Kristoff, B.J., 1989, A Comprehensive Wellbore-/Reservoir Simulator, SPE 18419, presented at the SPE Symp. Reservoir Simulation, Houston, TX.

Analytical Methods

2nd European Conference on the Mathematics of Oil Recovery
© D. Guérillot, O. Guillon (Editors) and Éditions Technip, Paris 1990, pp. 179-186
27 rue Ginoux, 75015 Paris

Curvilinear Grid Generation Techniques

C. L. Farmer and D. E. Heath[1]

ABSTRACT

Algebraic and numerical techniques for constructing curvilinear simulation grids are reviewed. Attention is given to the control of smoothness and orthogonality while deforming grids around geological structures, well patterns, faults and reservoir boundaries. A new method for constructing grids using global minimisation is outlined.

1 INTRODUCTION

The chance of finding an analytical solution, or the accuracy of a numerical solution, to the partial differential equations of reservoir simulation, depends upon the choice of an appropriate coordinate system.

In this paper we will consider the construction of grids for numerical simulation. Although the flow equations are solved numerically, analytical techniques are often used to construct the coordinate system which defines the grid by joining selected points in the coordinate system by straight lines. We will review some of the literature on analytical and numerical grid generation which, in our view, has particular promise for reservoir simulation applications.

Comprehensive reviews of grid generation have been provided by Thompson, Warsi and Mastin (1982, 1985). Brief reviews have been given by Baker (1989), Eiseman (1982, 1985) and Thompson (1984). These reviews are aimed primarily at the aerospace industry but are of general relevance to any problem requiring a grid. A review specific to oil reservoir simulation has been provided by Heinemann and Brand (1989). There is an extensive literature on grid generation as represented by the conference proceedings edited by Häuser and Taylor (1986) and Sengupta et al. (1988).

The general problem in reservoir grid generation is to construct grids which are as smooth and orthogonal as possible and such that the density and orientation of grid lines is controlled by some combination of :

(a)   reservoir boundaries
(b)   geological features such as faults and layering
(c)   well positions
(d)   flow paths
(e)   fluid contact or flood front orientation
        and position.

This problem can be regarded as one in the theory of the approximation of functions (defined using analytical formulae or interpolation on a fine grid). Thus a best approximation is constructed on a coarse grid in which the corner points of the grid cells, in addition to the function values associated with the cells, are variables to be computed.

---

(1) AEA Petroleum Services Winfrith, Dorchester, Dorset, DT2 8DH, UK.

2 ALGEBRAIC METHODS

In this section we describe some aspects of grid construction based on interpolation formulae. These techniques are called, in the literature, "algebraic". Many of the contributions to this approach have been made by Eiseman (1979,1985,1988). Algebraic methods are fast, but orthogonality, smoothness, and control are difficult to automate.

2.1 Interpolating Points

Let $\{x_i : i=1, N\}$ be the 2-D or 3-D coordinates of a set of N distinct points. Let $\{\kappa_i : i=1, N\}$ be a set of basis functions defined on the interval [0,1] satisfying the conditions

$$\kappa_i(\xi_j) = \delta_{ij} \tag{1}$$

with $0 = \xi_0 < \xi_1 < ... < \xi_i < ... < \xi_N = 1$.

Then the formula

$$x(\xi) = \sum_{i=1}^{N} x_i \, \kappa_i(\xi) \tag{2}$$

defines a continuous curve. If the functions κ_i are differentiable then the curve will be differentiable. Typical forms of κ_i are Lagrange interpolation or piecewise linear interpolation functions.

Hermite interpolation can be used to incorporate derivative information into point interpolation but the subsequent curves are sensitive to the gradients and have a tendency to produce unwanted oscillations (Faux and Pratt, 1979).

2.1.1 Multi-surface interpolation

The "multi-surface" method of point interpolation uses the formula

$$x(\xi) = x_0 + \sum_{i=1}^{N-1} \frac{G_i(\xi)}{G_i(1)} (z_{i+1} - z_i) \tag{3}$$

where

$$G_i(\xi) = \int_0^{\xi} \kappa_i(\zeta) \, d\zeta \tag{4}$$

The points $\{z_i : i=1, N\}$ are control points. If these points are equispaced in ξ-space and κ_i are piecewise linear basis functions then

$$x(\xi_i) = \frac{1}{2}(z_i + z_{i+1}) \tag{5}$$

and

$$\dot{x}(\xi_i) = \frac{(z_{i+1} - z_i)}{\delta\xi} \tag{6}$$

where $\delta\xi = 1/(N-1)$.

Figure 1 illustrates multi-surface interpolation over the interval [i,i + 2].

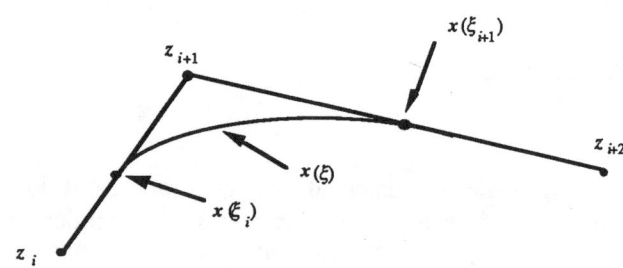

FIG. 1. Multi-surface interpolation

For further details see Thompson et al. (1985) and Eiseman (1985, 1988). This form of interpolation is qualitatively similar to that of B-splines and Bézier curves as used in computer graphics applications (see, for example, Watt, 1989).

2.2 Algebraic grids

A vector function $x(\xi,\eta)$ of two parameters, ξ and η satisfying $0 \le \xi \le 1$, $0 \le \eta \le 1$, defines a mapping between a rectangle in ξ-η space and a region in x-y space.

2.2.1 Transfinite interpolation

An algebraic grid is defined by the image grid in x-y space of a uniform rectangular grid in ξ-η space. Given four continuous curves we wish to construct an interpolation surface the boundaries of which coincide with these curves, as in Fig. 2.

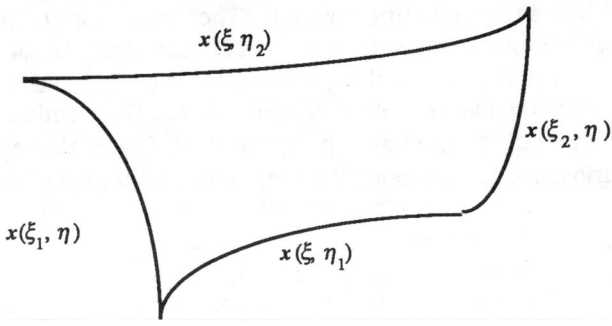

FIG. 2. Transfinite interpolation

An interpolation formula which achieves this feature is called "transfinite" because it is exact at all points on the boundary. Such a formula is given by the "Boolean sum"

$$
\begin{aligned}
x(\xi, \eta) = &\sum_{i=1}^{2} \kappa_i(\xi)\, x(\xi_i, \eta) \\
&+ \sum_{j=1}^{2} \kappa_j(\eta)\, x(\xi, \eta_j) \\
&- \sum_{i,j=1}^{2} \kappa_i(\xi)\, \kappa_j(\eta)\, x(\xi_i, \eta_j)
\end{aligned} \qquad (7)
$$

The generalisation of Eq. 7 to 3-D and a detailed description can be found in Thompson et al. (1985). In transfinite interpolation the boundary curves may be defined by any convenient technique, but in our applications this is usually achieved using an interpolation formula for curves. A grid can be defined by selecting a grid of points $\{\xi_i, \eta_j\}$ in ξ–η space and computing the coordinates $x(\xi_i, \eta_j)$. Note that these mesh points need not coincide with those of any 1-D mesh used to define the boundary curves. Equation 7 also generalises to any number of surfaces to be interpolated.

Complex "multi-block" configurations can be built by joining together a set of simple coordinate systems defined on individual blocks as discussed by Weatherill and Forsey (1984). Although we have used coordinates with a rectangular topology, we can generalise to any convenient connectivity pattern of the grid blocks. So, for example, in some blocks we could use cylindrical coordinates to more accurately represent the flow patterns near wells. This has been considered in detail by Pedrosa and Aziz (1985) and Gottardi and Vignati (1990).

2.2.2   Internal Control

We may wish to concentrate gridlines in a particular sub-region, which can be done by introducing extra, interior, boundary surfaces. This can be difficult and time consuming and so Eiseman (1988) has suggested the use of a coarse mesh of control points to define the internal boundaries using the multi-surface interpolation.

3   NUMERICAL METHODS

Many numerical approaches to grid generation are based upon the use of partial differential equations. Such methods, while able to produce smooth, near orthogonal grids, are complicated to use when internal controls are needed. Further, this approach is not a natural technique for the solution of the adaptive approximation of functions problem described earlier. Thompson et al. (1985) give a detailed review of partial differential equation methods. These methods have been applied in the petroleum literature by, for example, Goslinga (1983) and Lambeth and Dawe (1987). Frequent use has been made of conformal methods using the solution of a single-phase flow problem in a uniform medium, for example, Sonier and Chaumet (1974).

Another possibility is to use variational methods, such as that of Brackbill and Saltzman (1982), which is becoming a popular technique in computational fluid dynamics (see, for example, Hawkins and Kightley, 1989).

The variational method proceeds by formulating a functional measure of smoothness and orthogonality subject to a distributed weight function. Where the weight function is large, the cells will tend to be small. A partial differential equation is then deduced by finding the Euler equations for the extremum of the grid functional. These equations are then solved using an iterative technique. The method is elegant but requires some knowledge of tensor analysis and the Euler equations are very hard to deduce in 3-D. Further, it is difficult to devise weight functions which enable us to solve the approximation of functions problem.

One could try to avoid the need to derive the Euler equations by direct discretisation of the grid functional. However, as noted by Kennon and Dulikravich (1985), this leads to decoupling problems if central differences are used. That is, the equation centred at the point (x_{ij}, y_{ij}) is independent from x_{ij} and y_{ij}.

3.1 Local minimisation techniques

In this section we consider curvilinear grid generation techniques based on the construction of an objective function which, by definition, is a function of the grid point positions and so does not require discretisation. For simplicity of exposition we will consider the 2-D case, but the techniques of this section generalise easily to 3-D.

Let $\{x_{ij} : i=1,N_x : j=1,N_y\}$ denote the coordinates of the points in a curvilinear mesh. We assume that the pairs of points $\{(x_{ij}, x_{i+1,j}) : i=1,N_x - 1 : j=1,N_y\}$ are joined by straight lines to form one family of coordinate curves and $\{(x_{ij}, x_{i,j+1}) : i=1,N_x : j=1,N_y - 1\}$ are joined to form the other.

Define the vectors

$$r_{i+1,j} = x_{i+1,j} - x_{ij} , \quad i=1,N_x-1 , \; j=1,N_y \qquad (8)$$

$$r_{i-1,j} = x_{i-1,j} - x_{ij} , \quad i=2,N_x \quad , \; j=1,N_y \qquad (9)$$

$$r_{i,j+1} = x_{i,j+1} - x_{ij} , \quad i=1,N_x \quad , \; j=1,N_y - 1 \quad (10)$$

$$r_{i,j-1} = x_{i,j-1} - x_{ij} , \quad i=1,N_x \quad , \; j=2,N_y \qquad (11)$$

Let a_{ij} denote the area of the cell whose corners are $(x_{ij}, x_{i+1,j}, x_{i+1,j+1}, x_{i,j+1})$. Kennon and Dulikravich (1985) define the objective function

$$f = \sum_{i=2}^{N_x-1} \sum_{j=2}^{N_y-1} (\alpha\, s_{ij} + (1 - \alpha)\, b_{ij}) \quad (12)$$

where $0 \le \alpha \le 1$; $s_{i,j}$ is a local smoothness measure defined by

$$\begin{aligned}
s_{ij} &= (a_{ij} - a_{i-1,j})^2 \\
&+ (a_{ij} - a_{i,j-1})^2 \\
&+ (a_{i-1,j} - a_{i-1,j-1})^2 \\
&+ (a_{i,j-1} - a_{i-1,j-1})^2
\end{aligned} \qquad (13)$$

and b_{ij} is a measure of local orthogonality defined by

$$\begin{aligned}
b_{ij} &= (r_{i+1,j} \cdot r_{i,j+1})^2 + (r_{i,j-1} \cdot r_{i+1,j})^2 \\
&+ (r_{i-1,j} \cdot r_{i,j-1})^2 + (r_{i,j+1} \cdot r_{i-1,j})^2 \quad (14)
\end{aligned}$$

Kennon and Dulikravich then prescribe the positions of the boundary points and minimise f by using the Fletcher-Reeves (1964) conjugate gradient algorithm with an exact line search. The line search is straightforward since f is quartic in the line parameter and so the minimisation along a given direction is found by examining the roots of a cubic polynomial. These authors view their method as a post-processor for smoothing grids, and for removing inverted cells containing points with a negative Jacobian. However, the technique appears to have potential as a grid generation method as discussed by Carcaillet et al. (1986).

The idea of using direct minimisation as a grid generation method was pursued by Kumar and Kumar (1988) who introduced an objective function which is quadratic in the grid coordinates. However, Kumar and Kumar do not pose an optimisation problem to be solved by an iterative method; instead, they introduce a "local" grid function which is minimised sequentially, cell by cell. Nevertheless we can formulate the Kumar and Kumar approach as a local minimisation (in the optimisation theory sense) using an alternating variables method where the variables are the vector coordinates of each grid point. As a local measure of smoothness and orthogonality Kumar and Kumar use

$$\begin{aligned}
S_{ij} &= \lambda_{i+\frac{1}{2},j} \, |\, r_{i+1,j} \,|^2 + \lambda_{i-\frac{1}{2},j} \, |\, r_{i-1,j} \,|^2 \\
&+ \lambda_{i,j+\frac{1}{2}} \, |\, r_{i,j+1} \,|^2 + \lambda_{i,j-\frac{1}{2}} \, |\, r_{i,j-1} \,|^2
\end{aligned}$$

$$(15)$$

$$\begin{aligned}
B_{ij} &= [(x_{i+1,j} - x_{i-1,j}) \cdot (x_{i,j+1} - x_{ij})]^2 \\
&+ [(x_{i+1,j} - x_{i-1,j}) \cdot (x_{i,j-1} - x_{ij})]^2 \\
&+ [(x_{i,j+1} - x_{i,j-1}) \cdot (x_{i+1,j} - x_{ij})]^2 \\
&+ [(x_{i,j+1} - x_{i,j-1}) \cdot (x_{i-1,j} - x_{ij})]^2
\end{aligned}$$

$$(16)$$

They then minimise

$$F = \sum_{i,j} (\alpha\, S_{ij} + (1 - \alpha)\, B_{ij}) \qquad (17)$$

by successively fixing all variables except S_{ij} and B_{ij} in one cell. They then choose x_{ij} to minimise

$$\alpha\, S_{ij} + (1 - \alpha)\, B_{ij} \qquad (18)$$

This involves the solution of a set of simultaneous equations in two unknowns. We have found that the Kumar and Kumar approach provides a good smoothing or post-processing algorithm if convergence is not required, but is very slow if tight convergence is needed.

3.2   Global minimisation methods

We now outline some of our own work in which weight functions are introduced so that the optimisation involves a discontinuous objective function to be minimised with some combinatorial aspects to the problem. We are interested in the following two problems;

(a)   The scattered data set problem

Given a set of scattered data points

$\{z^m : m=1,M\}$ find a grid $\{x_{ij} : i=1,N_x : j=1,N_y\}$ such that

$$G = f + \sum_m v (z^m - c^m)^2 + \sum_{i,j} \mu N_{ij}(1 - N_{ij}) \tag{19}$$

is a minimum subject to the constraint that each grid cell is convex. In Eq. (19), f is the Kennon and Dulikravich (1985) objective function of Eq. (12), c^m is the position of the centre of the cell occupied by point m, N_{ij} is the number of scattered data points located in cell (i,j), and v and μ are positive weights. The idea behind this optimisation problem is to construct a grid such that there is not more than one scattered point in each grid cell and that each scattered point is at the centre of a cell. Applications of this might be to designing a grid to conform to a dense well pattern, a set of sample points for permeability measurements, or as part of the moving point to fixed mesh interpolation in the moving point method (Farmer, 1987).

(b)   The function approximation problem

Let $w(x)$ be a bounded function. Over a cell (i,j) we define the "variance" of w to be

$$v_{ij} = \int_{cell} [w(x) - w_{av}]^2 \tag{20}$$

where

$$w_{av} = \frac{\int_{cell} w(x)\ dx}{\int_{cell} dx} \tag{21}$$

We wish to construct a grid which minimises

$$V = f + \gamma \sum_{i,j} v_{ij} \tag{22}$$

subject to the constraint that each grid cell is convex. In equation (22) f is the Kennon and Dulikravich objective function and γ is a positive weight. Our interest in this problem stems from our research on the use of structured random noise to model heterogeneous reservoirs (Farmer,1990). We conjecture that a grid which follows high and low permeability streaks and other geological structures will not only reduce memory requirements but also help the convergence of linear solvers.

We have found that the simulated annealing algorithm of Kirkpatrick et al.. (1983) provides a robust approach to the solution of problems such as those stated above.

Figure 3 shows a smooth, near orthogonal, grid which has been constructed to follow the scattered data points indicated by the bold dots.

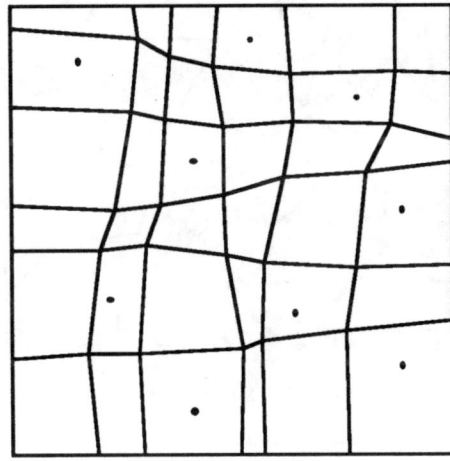

FIG. 3.   The scattered data set problem

Figure 4 displays a function defined on a 30x30 grid. A key relating the shading patterns to numerical values is given in Fig. 10. In Fig. 5 we represent the pattern on a rectangular, 10x10 grid, where the coarse grid values are arithmetic averages of the fine grid values within a coarse grid cell. Figure 6 shows a 10x10 curvilinear grid which has a low degree of orthogonality and smoothness and accurately captures the pattern of Fig. 4 as shown in Fig. 7. In Figs. 8 and 9 we increase the orthogonality and smoothness, yet still capture the main features of the underlying pattern. A detailed description and evaluation of this algorithm is given by Farmer, Heath and Moody (1991).

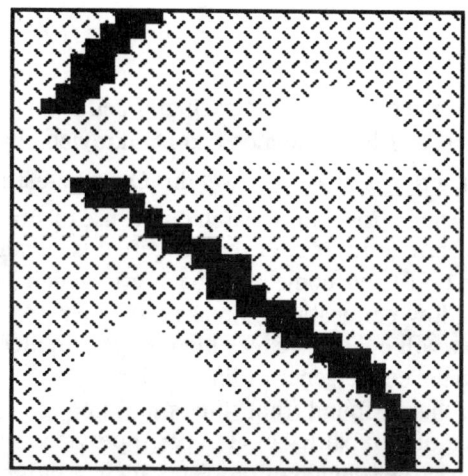

FIG. 4. 30x30 Test pattern

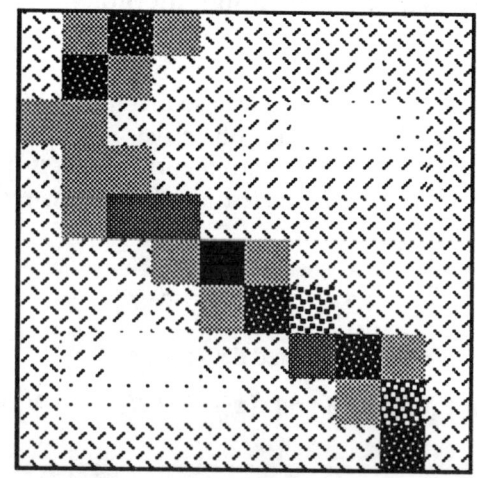

FIG. 5. 10x10 Averaged pattern

FIG. 6. 10x10 grid - low smoothness and orthogonality

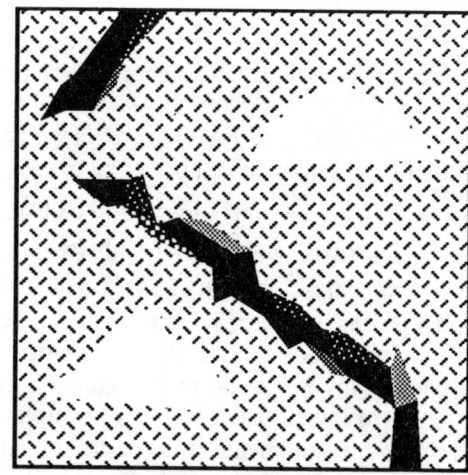

FIG. 7. 10x10 pattern corresponding to Fig. 6

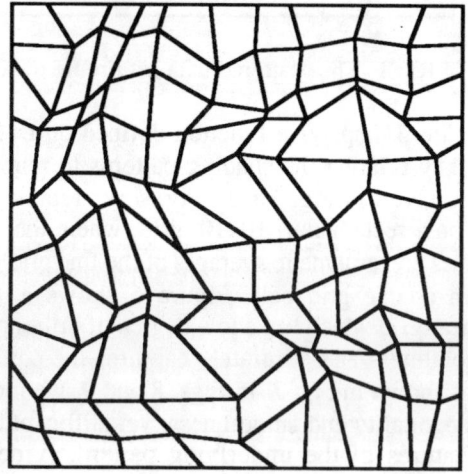

FIG. 8. 10x10 grid - high smoothness and orthogonality

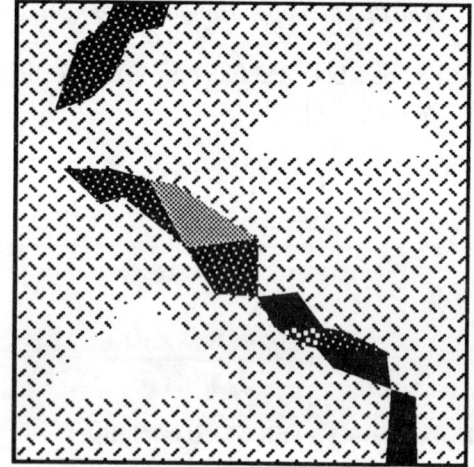

FIG. 9. 10x10 pattern corresponding to Fig. 8

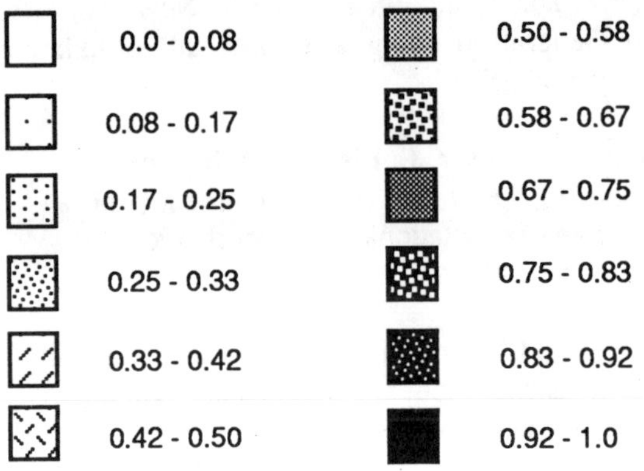

☐	0.0 - 0.08	▦	0.50 - 0.58
⊡	0.08 - 0.17	▦	0.58 - 0.67
▦	0.17 - 0.25	▦	0.67 - 0.75
▦	0.25 - 0.33	▦	0.75 - 0.83
▨	0.33 - 0.42	▨	0.83 - 0.92
▧	0.42 - 0.50	■	0.92 - 1.0

FIG. 10. Key for Figs 4-9.

4   CONCLUSIONS

In this paper we have reviewed the problems and techniques of curvilinear grid generation. Robust techniques are now available for constructing grids which conform to

      (a) boundary shapes
      (b) internal boundaries
      (c) scattered data points
      (d) spatially variable functions.

To make use of these developments we need to develop reservoir simulators capable of accurately and systematically handling the off-diagonal terms which arise from using non-orthogonal grids.

ACKNOWLEDGEMENT

This research was funded by the AEA Technology Corporate Research Programme.

REFERENCES

Baker, T.J., 1989, Developments and Trends in Three-Dimensional Mesh Generation, Applied Numerical Mathematics, **5**, 275-304.

Brackbill, J.U., and Saltzman, J.S., 1982, Adaptive Zoning for Singular Problems in Two Dimensions, J.Computational Physics, **46**, 342-368.

Carcaillet, R., Kennon, S.R., and Dulikravich, G.S., 1986, Generation of Optimum Three-Dimensional Grids. In: D. Rues and W. Kordulla (Editors), Proceedings of the 6th GAMM-Conference on Numerical Methods in Fluid Dynamics, Viewig.

Eiseman,P.R.,1979, A Multi-Surface Method of Coordinate Generation, J.Computational Physics, **33**, 118-150.

Eiseman, P.R.,1982, Orthogonal Grid Generation, Applied Mathematics and Computation, **10/11**, 193-233.

Eiseman, P.R.,1985, Grid Generation for Fluid Mechanics Computations, Annual Reviews of Fluid Mechanics, **17**,487-522.

Eiseman, P.R., 1988, A Control Point Form of Algebraic Grid Generation, International Journal for Numerical Methods in Fluids, **8**, 1165-1181.

Farmer, C.L., 1987, Moving Point Techniques. In: J. Bear and M.Y. Corapcioglu (Editors), Advances in Transport Phenomena in Porous Media, Martinus Nijhoff Publishers, 953-1004.

Farmer, C.L., Numerical Rocks, 1990. In: F.J. Fayers and P.R. King (Editors), Proceedings of the 1st European Conference on the Mathematics of Oil Recovery, Oxford University Press.

Farmer, C.L., Heath, D.E., and Moody, R.O., 1991, A Global Optimisation Approach to Grid Generation, SPE 21236, Reservoir Simulation Symposium, Anaheim, Feb. 17-20, 1991.

Faux, I.D., and Pratt, M.J., 1979, Computational Geometry for Design and Manufacture, Ellis Horwood.

Ferguson, W.I., and Wadsley, A.W.,1986, The Construction of Curvilinear Co-ordinate Grid Systems for Reservoir Simulation, SPE 15857, SPE European Petroleum Conference, London, Oct. 20-22, 1986.

Fletcher, R., and Reeves, C.M., 1964, Function Minimisation by Conjugate Gradients, Computer Journal,**7**, 149-154.

Goldthorpe, W.H., and Chow, Y.S., 1985, Unconventional Modelling of Faulted Reservoirs: A Case Study, SPE 13526, Reservoir Simulation Symposium, Dallas, Feb. 10-13,1985.

Goslinga, J., 1983, The Use of Adaptive Grids in Numerical Simulations, SPE 12266, Reservoir Simulation Symposium, San Francisco, Nov. 15-18, 1983.

Gottardi, G., and Vignati, L., 1990, Hybrid Grid Black Oil Reservoir Simulator, Journal of Petroleum Science and Engineering, **3**, 345-360.

Häuser, J., and Taylor, C. (Editors), 1986, Numerical Grid Generation in Computational Fluid Dynamics, Proceedings of the International Conference held at Landshut, West Germany, July 14-17, 1986. Pineridge Press, 1986.

Hawkins, I.R., and Kightley J.R., 1989, Grid Adaption for Turbulent Flow Problems. In: P.M. Gresho, C. Taylor and R.L. Sani (Editors), Numerical methods in Laminar and Turbulent Flow, Pineridge Press.

Heinemann, Z.E., and Brand, C.W., 1989, Gridding Techniques in Reservoir Simulation. In: Proceedings of the First and Second International Forum on Reservoir Simulation, Alpbach, Sept.12-16,1988 and Sept. 4-8,1989, 339-425, Published by Paul Steiner.

Kennon, S.R., and Dulikravich, G.S., 1985, A Posteriori Optimisation of Computational Grids. AIAA Paper 85-0483.

Kirkpatrick, S., Gelatt, Jr., C.D., and Vecchi, M.P., 1983, Optimisation by Simulated Annealing, Science, **220** (4598), 671-680.

Kumar, A., and Kumar, N.S., 1988, A New Approach to Grid Generation Based on Local Optimisation. In: Sengupta et al. 1988, 177-184.

Lambeth, N., and Dawe, R.A., 1987, Boundary and Crossflow Behaviour During Displacement in Nodal Systems, SPE 16972, 62nd Annual Technical Conference and Exhibition, Dallas, Sept. 27-30, 1987.

Pedrosa, O.A., and Aziz, K., 1985, Use of Hybrid Grid in Reservoir Simulation, SPE 13507, Reservoir Simulation Symposium, Dallas, Feb. 10-13,1985.

Roach, P.J., and Steinberg, S., 1985, A New Approach to Grid Generation using a Variational Formulation, AIAA Paper 85-1527.

Sengupta, S., Häuser, J., Eiseman, P.R., and Thompson, J.F. (Editors), 1988, Numerical Grid Generation in Computational Fluid Mechanics '88, Pineridge Press, 1988.

Sonier, F., and Chaumet, P., 1974, A Fully Implicit Three-Dimensional Model in Curvilinear Coordinates, SPEJ, August 1974, 361-370.

Thompson, J.F., Thames, F.C., and Mastin, C.W., 1974, Automatic Numerical Generation of Body-Fitted Curvilinear Coordinate System for Field Containing Any Number of Arbitrary Two-Dimensional Bodies. J.Computational Physics, **15**, 299-319.

Thompson, J.F., Warsi, Z.U.A., and Mastin, C.W., 1982, Boundary-Fitted Coordinate Systems for Numerical Solution of Partial Differential Equations - A Review, J.Computational Physics, **47**, 1-108.

Thompson, J.F., and Warsi, Z.U.A., 1983, Three-Dimensional Grid Generation from Elliptic Systems, AIAA Paper 83-1905.

Thompson, J.F., 1984, Grid Generation Techniques in Computational Fluid Dynamics, AIAA Journal, **22**(11), 1505-1523.

Thompson, J.F., Warsi, Z.U.A., and Mastin, C.W., 1985, Numerical Grid Generation, North-Holland.

Thompson, J.F., 1988, Some Current Trends in Numerical Grid Generation. In: K.W. Morton and M.J. Baines (Editors). Numerical Methods for Fluid Dynamics III, Oxford University Press, 1988, 87-100.

Wadsley, W.A., 1980, Modelling Reservoir Geometry with Non-Rectangular Coordinate Grids, SPE 9369, 55th Annual Technical Conference and Exhibition, Dallas, Sept. 21-24, 1980.

Watt, A.,1989, Fundamentals of Three-Dimensional Graphics, Addison-Wesley.

Weatherill, N.P., and Forsey, C.R., 1984, Grid Generation and Flow Calculations for Complex Aircraft Geometries Using a Multi-Block Scheme, AIAA Paper 84-1665.

2nd European Conference on the Mathematics of Oil Recovery
© D. Guérillot, O. Guillon (Editors) and Éditions Technip, Paris 1990, pp. 187-195
27 rue Ginoux, 75015 Paris

An Analytical Investigation by the Method of Characteristics of Gravity Stabilised Gas Injection

R. W. S. Foulser[1]

ABSTRACT

This paper considers the gravity stabilized displacement of oil by gas. The displaced oil leaves a region of lower saturation in which oil drains under the influence of gravity and viscous forces. Component exchange between the oil and gas causes compositional changes in the gas invaded region. This results in more viscous oil near the injection point because of extraction of lighter oil components by the gas. However, near the displacement front the oil becomes less dense and less viscous due to the absorption of the previously extracted lighter components. The phase behaviour is simply explained using a ternary representation.

Equations describing this process in one dimension are described. Their hyperbolic character is discussed and they are analysed using the method of characteristics. As a result it is shown that the process is not one in which the oil drains into the oil ahead of the displacement front but should be thought of as oil being left behind. A significant consequence of this is that the oil produced has the original composition uncontaminated by any gas components. Additionally it is shown that the fractional recovery depends mainly on the oil properties just above the oil bank. These are easily derived from phase behaviour calculations, and are largely independent of oil viscosification effects occurring in the upper regions of the oil column.

1 INTRODUCTION

Oil production by gravity stabilised nitrogen or hydrocarbon gas injection appears to have potential to recover additional oil from North Sea reservoirs either in secondary or tertiary displacements. The secondary recovery process in a massive reservoir is envisaged as the injection of gas in the crest of the reservoir with production from the oil leg below, see Figure 1. A sharp interface exists between the gas and oil. In the tertiary process the injection of gas in the crest of the reservoir displaces oil and water downward with the development of an oil bank from which the oil is produced, see Figure 2. To recover the oil as early as possible, and thus improve the economics, oil needs to be

---

(1) Division of Petroleum Reservoir Technology, Winfrith Technology Center, Dorchester, Dorset DT2 8DH, UK.

produced from wells whose completions are moved down as the oil bank moves down and a good strategy is to try to maintain an oil bank of constant thickness.

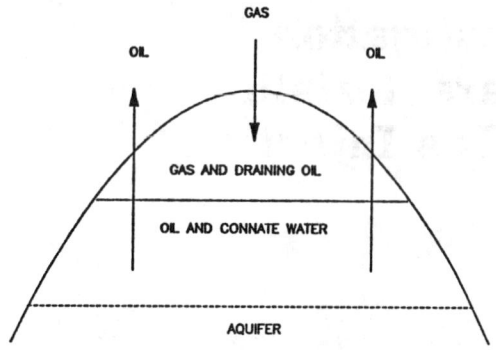

FIGURE 1 : SCHEMATIC OF SECONDARY GAS FLOOD

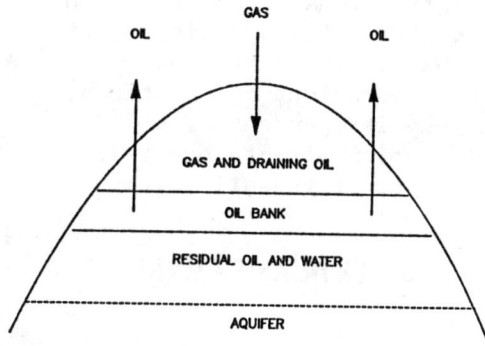

FIGURE 2 : SCHEMATIC OF TERTIARY GAS FLOOD

In both secondary and tertiary floods there is an exchange of components between the oil and gas. At the top of the reservoir, the gas extracts components from the oil but at the bottom the gas donates lighter components to the oil. Generally speaking the compositional effects are beneficial to oil recovery [1].

A fundamental question regarding the process is the oil production rate which may be achieved. Oil drainage in the reservoir determines an absolute limit. This paper concerns the development of a one dimensional model aimed at giving an appreciation of the interconnections between oil recovery, recovery rates and compositional changes.

The method of characteristics used here became prominent in the analysis of EOR processes with the paper of Helfferich [2]. This fractional flow approach is used in Buckley Leverett theory [3] and was effectively generalised by Pope to analyse EOR processes [4].

Dumore used a method of characteristics to analyse condensing and vapourising gas drives in secondary flooding. Composition and velocity effects were considered in the model [5]. This model, extended to include gravity terms, forms the basis of the work reported here.

2 DEVELOPMENT OF MODEL

2.1 Phase Behaviour

Phase behaviour in gas systems is discussed by Stalkup [6] so here only the briefest discussion is included.

Injected gas may vapourise some reservoir oil and reservoir oil may absorb some gas. Some mixtures may be single phase, however, other mixtures may remain gas and liquid phases with compositions which differ from the original fluids before mixing. The phase behaviour relationships can be represented approximately on a triangular diagram by grouping the components into three pseudo-component. The pseudo-components considered in this paper are N2, C1 and C2+ (C2 and all higher molecular weight components) but the analysis is general. The type of behaviour considered is shown in Figure 3. Tie lines in the two phase region join phases that can exist in equilibrium. Mixtures outside the two phase region are single phase, eg N2 gas at the apex and the reservoir oil whose composition is N2=0.0, C1=0.13, C2+=0.87. The representation of mixture compositions and phase behaviour in this manner can only be approximate since the individual components within the pseudo-component group will have different volatilities and not be distributed in the same proportions between phases.

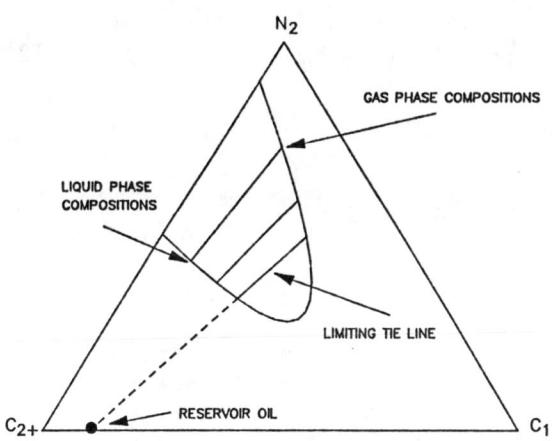

FIGURE 3 : PHASE BEHAVIOUR REPRESENTATION FOR
A THREE COMPONENT SYSTEM

It is assumed here that the temperature and pressure is fixed. If an overall composition is single phase then the fluid molar density m, density ρ, and viscosity μ are fixed. If the overall composition, $\underline{x}$ (vector of mole fractions), lies in a two phase region, then it must lie on a tie line which identifies the compositions of the two phases into which it splits $\underline{x}_o$ and $\underline{x}_g$. The phase molar densities m_o and m_g, densities ρ_o and ρ_g, and viscosities μ_o and μ_g are thus fixed. The phase volume fractions, S_o^* and S_g^*, are fixed since:

$$\underline{x} = \frac{\underline{x}_o S_o^* m_o + \underline{x}_g S_g^* m_g}{S_o^* m_o + S_g^* m_g} \qquad (1)$$

where:

$$S_o^* + S_g^* = 1$$

When an aqueous phase is present these volume fractions are defined by:

$$S_\alpha^* = \frac{S_\alpha}{1 - S_a} \qquad \alpha = o, g$$

where: S_a is the aqueous phase volume fraction

In an N_c component system an overall composition, $\underline{x}$, is determined by N_c-1 concentrations. The fact that the physical properties and phase saturations depend only on N_c-1 overall fluid concentrations will be used in the analysis below.

In a gas displacement the gas phase is more mobile than the oil phase so the process can be envisaged as one in which the gas undergoes 'multiple contacts'; that is the gas contacts some reservoir oil, comes into equilibrium with it, then moves forward to contact fresh oil. The forward moving gas thus changes its composition and induces compositional changes in the oil left behind. Consideration of this process for the system represented by Figure 3 shows that changes in the composition of the gas will continue until it reaches the composition on the tie line whose extension passes through the reservoir oil. It is assumed here that the tie line is unique so that the limiting tie line can be identified independently of the details of the gas/liquid mobilities. (It is interesting to ask whether there is a unique limiting tie line in a multi-component system).

2.2 Multiphase Fractional Flow

The flow of aqueous, oleic and gaseous phases is assumed to be determined by Darcy's law:

$$U_\alpha = - \frac{K_r k_{r\alpha}}{\mu_\alpha} \left(\frac{\partial P_\alpha}{\partial z_\alpha} - \rho_\alpha g \right) \qquad \alpha = a, o, g \quad (2)$$

An overall fluid flow rate, U, may be defined by:

$$U(z,t) = U_a + U_o + U_g \qquad (3)$$

and if capillary pressure is neglected

$$P_a = P_o = P_g$$

so that:

189

$$\frac{\partial P_a}{\partial z} = \frac{\partial P_o}{\partial z} = \frac{\partial P_g}{\partial z} \qquad (4)$$

then equation (2) may be rearranged in fractional flow form:

$$U_a = \frac{\lambda_a}{\lambda}\left(U + Kg(\lambda_o(\rho_a-\rho_o) + \lambda_g(\rho_a-\rho_g))\right) \qquad (5)$$

$$U_o = \frac{\lambda_o}{\lambda}\left(U + Kg(\lambda_g(\rho_o-\rho_g) + \lambda_a(\rho_o-\rho_a))\right) \qquad (6)$$

$$U_g = \frac{\lambda_g}{\lambda}\left(U + Kg(\lambda_a(\rho_g-\rho_a) + \lambda_o(\rho_g-\rho_o))\right) \qquad (7)$$

where:

$$\lambda_\alpha = \frac{k_{r\alpha}}{\mu_\alpha} \qquad \alpha=a,o,g \qquad (8)$$

and:

$$\lambda = \lambda_a + \lambda_o + \lambda_g \qquad (9)$$

2.3 Multiphase Conservation Law

Conservation of component i results in:

$$\frac{\partial}{\partial t}(S_a m_a x_a^i + S_o m_o x_o^i + S_g m_g x_g^i) +$$

$$\frac{1}{\phi}\frac{\partial}{\partial z}(U_a m_a x_a^i + U_o m_o x_o^i + U_g m_g x_g^i) = 0 \qquad (10)$$

Substitution of equations (5), (6) and (7) into (10) results in:

$$\frac{\partial F_i}{\partial t} + \frac{\partial G_i}{\partial z} = 0 \qquad i=1,N_c \qquad (11)$$

where:

$$F_i = \sum_\alpha S_\alpha m_\alpha x_\alpha^i \qquad (12)$$

and

$$G_i = \frac{1}{\phi}\sum_\alpha \frac{\lambda_\alpha}{\lambda} m_\alpha x_\alpha^i \left(U + K g \sum_\beta \lambda_\beta(\rho_\alpha-\rho_\beta)\right) \qquad (13)$$

As discussed above, fixing N_c-1 concentrations fixes the phase physical properties, saturations and also the relative permeabilities since these are assumed to be functions of saturation. Equation (11) represents N_c equations in N_c-1 independent overall concentrations and the unknown velocity U. It may be conveniently expressed in matrix notation as:

$$J_F \frac{\partial y}{\partial t} + J_G \frac{\partial y}{\partial z} = 0 \qquad (14)$$

where:

$$y = (x^1, \ldots\ldots x^{N_c-1}, U)$$

$$J_F = \left(\frac{\partial F_i}{\partial y_1} \ldots\ldots \frac{\partial F_i}{\partial y_{N_c-1}}, 0\right)$$

$$J_G = \left(\frac{\partial G_i}{\partial y_1} \ldots\ldots \frac{\partial G_i}{\partial y_{N_c}}\right)$$

This is a set of equations of the hyperbolic type if certain conditions, which are discussed in the next section, are satisfied.

2.4 Multiphase Characteristic Velocities

A 'composition' y, moves at a constant velocity V, if it satisfies the convection equation:

$$\frac{\partial \underline{y}}{\partial t} + V \frac{\partial \underline{y}}{\partial z} = 0$$

Premultiplying by the matrix J_G and using equation (14) to eliminate the spatial variation results in:

$$(J_G - V J_F) \frac{\partial \underline{y}}{\partial t} = 0 \qquad (15)$$

This homogeneous equation set can only have a non trivial solution if:

$$\det (J_G - V J_F) = 0 \qquad (16)$$

This is a polynomial in V of degree N_c-1. The system of equations is of hyperbolic type if all the possible values of V are real. Thus given a 'composition' $\underline{y}$, characteristic velocities V may be found, together with the corresponding eigenvectors indicating the directions of what Helffrich called paths [2]. The paths are determined by the integration of equation (15) using continuously varying eigenvalues. The paths represent possible continuous compositional changes in a flood. Which paths are realised in practice depends on the initial and injection compositions. Note that in the application discussed here the overall flowrate U varies because of changes in the overall fluid composition.

2.5 Non-Hyberbolic Regions

Fayers has examined solutions to equation (16) for three phase flow in the special case of constant flow U, constant phase composition and neglecting gravity terms [7]. He has shown that for some relative permeability formulations (based on Stone's expressions [8], [9]) there may be small regions where the velocity is not real and so equations (14) are non-hyperbolic.

A similar study has ben carried out by Shearer and Trongenstein for different three phase relative permeability formulations [10]. They too found small non-hyperbolic regions. They believe such regions to be consequences of non-physical relative permeabilities produced by their formulations. Further, numerical studies have shown that if the initial data is posed in the hyperbolic region then the computed solution tends to avoid the non-hyperbolic regions [11]. Marchesin and Medenis have shown that with Corey relative permeabilities real characteristic velocities are produced even when gravity is included [12, 13].

In this paper we shall assume that either real characteristic velocities are always produced, or that non-hyperbolic regions are avoided by the flow behaviour.

2.6 Shocks

By considering the accumulation and flow of each component in a region around a shock moving with velocity V, the following jump condition is produced:

$$G_i^- - G_i^+ = V (F_i^- - F_i^+) \qquad i=1 \dots N_c \qquad (17)$$

where the + and - refer to the upstream and downstream directions either side of the shock.

Following Dumore [5], equation (17) may be written:

$$\sum_\alpha P_\alpha^+ x_\alpha^{i+} = \sum_\alpha P_\alpha^- x_\alpha^{i-} \qquad i=1 \dots N_c \qquad (18)$$

where

$$P^+ = \left[\frac{\lambda_\alpha^+}{\lambda^+} \left\{ \frac{K\,g}{\phi} \sum_{\beta\alpha} (\rho_\beta^+ - \rho^+)\, \lambda_\beta^+ + \frac{U^+}{\phi} \right\} - VS_\alpha^+ \right] m_\alpha^+$$

and similarly for P_α^-.

Summing over components gives the constraint:

$$\sum_\alpha P_\alpha^+ = \sum_\alpha P_\alpha^- \qquad (19)$$

Equation (18) shows that the phase compositions across a shock lie in a $N_p^+ + N_p^-$ - 1 dimensional subspace of the N_c dimensional compositional space where

N_p^+ is the number of phases present upstream of the shock and N_p^- is the number of phases downstream. Equation (19) adds a further constraint so that in general the phase compositions across a shock lie in an $N_p^+ + N_p^-$ -2 dimensional subspace. When $N_p^+ = N_p^- = 2$, as in Dumore's analysis, the compositions across a shock lie in a plane.

In the event that there are two phases upstream (oil and gas) and one downstream (or vice versa) the phase compositions must be collinear. This corresponds to the situation discussed in the section on phase behaviour where the forward contact procedure eventually terminated at the tie line collinear with the contacted oil composition. Similarly if a single phase gas is injected to produce an immiscible displacement of oil then the composition of the produced phases will lie on the tie line whose extension passes through the injected gas composition.

2.7 Component Concentration Changes

If there are three phases (oil, gas and water) upstream, two phases downstream (oil and water) and the water component (component N_c) resides exclusively in the water phase; then the water equation separates out from equation (18) to give:

$$P_w^+ = P_w^- \qquad (20)$$

$$P_o^+ x_o^{i+} + P_g^+ x_g^{i+} = P_o^- x_o^{i-} \qquad i=1, N_c-1 \qquad (21)$$

and

$$P_o^+ + P_g^+ = P_o^- \qquad (22)$$

Hence the equilibrium oil and gas compositions at the displacement front must be collinear with the composition of the oil being displaced.

3 INTERPRETATION

3.1 Composition Distribution

As discussed by Helffrich [2], the composition distribution in a flood may be determined by constructing the compositional route from the compositional paths and shock front conditions. The distribution will divide up into a number of regions where the compositions vary smoothly and compositional jumps occurring at the shock fronts. For continuous gas injection at a constant rate from time zero the compositions and shock fronts propagate at a constant rate so the distribution upstream of a front simply stretches with time. This behaviour is familiar in the Buckley Leverett analysis of waterflooding [3].

3.2 Implications

The linear growth of the compositional distribution above the oil displacement front means that the quantity of oil components in this region grows linearly with time. This material must be accumulating as a result of <u>oil being left behind at the displacement front</u>. Thus the oil recovery process is not one in which oil drains down into the oil below.

The first implication of this is that the viscosification of the oil that occurs near to the gas injection point can have no direct effect on the oil recovery.

A second implication is that the displacement performance is directly dependent on the compositions, physical properties and flow characteristics of the fluids around the displacement front, ie the original oil and the limiting tie line oil and gas.

A third implication of this is that the recovery efficiency for all components of oil must be the same since oil with the original composition is left behind. In secondary mode the efficiency will be given by:

$$E_s = \frac{U_o^-}{\phi V S_{oi}} \qquad (23)$$

where S_{oi} is the original oil saturation and V is the displacement front velocity. In tertiary mode, given that the oil is extracted at a rate to maintain a constant oil bank thickness, the efficiency will be given by:

$$E_t = \frac{U_o^-}{\phi V S_{or}} \qquad (24)$$

where S_{or} is the residual oil saturation to waterflooding.

The remainder of the paper will make these implications clearer by analysing what is happening around the displacement front.

3.3 Secondary Displacement

At the front equations (18) and (19) imply:

$$P_o^+ x_o^{i+} + P_g^+ x_g^{i+} = P_o^- x_o^{i-} \qquad i=1, N_c \qquad (25)$$

$$P_o^+ + P_g^+ = P_o^- \qquad (26)$$

where the original composition $\underline{x}_o^-$ is known as are the tie line oil and gas compositions $\underline{x}_o^+$, $\underline{x}_g^+$.

P_o^- is a function of the oil saturation S_o^-, the shock front velocity V, and the Darcy velocity U^-. However $S_o^- = S_{oi}$, and if the oil is extracted at a constant rate U_o^-, so that $U^- = U_o^-$, then P_o^- is a function only of V.

P_o^+ and P_g^+ are functions of S_o^+, U^+ and V. However, S_o^+ and U^+ are related

by having to lie on the compositional path tracing back to the inlet 'compositional' boundary condition. Thus P_o^+ and P_g^+ can be thought of as functions of S_o^+ and V only. Under these circumstances equations (25) and (26) provide the two equations necessary for S_o^+ and V to be found.

The displacement efficiency can then be found from equation (23). Note that details of the compositional variations occurring in the drainage region enter into the solution of the front variables only through the relationship between S_o^+ and U.

If composition changes do not affect the Darcy velocity, $U = U^+ = U^- = $ constant, equations (25) and (26) can be applied directly to find S_o^+ and V. Consequently in this case the recovery efficiency is entirely independent of compositional changes in the drainage region.

3.4 Tertiary Displacement

In this case the conditions at the front are described by equations (20), (21) and (22), with the following functional dependencies:

$$P_w^+ = P_w^+(S_o^+, S_w^+, U^+, V)$$

$$P_o^+ = P_o^+(S_o^+, S_w^+, U^+, V)$$

$$P_g^+ = P_g^+(S_o^+, S_w^+, U^+, V)$$

$$P_w^- = P_w^-(S_o^-, U^-, V)$$

$$P_g^- = P_g^-(S_o^-, U^-, V)$$

U^+, S_o^+ and S_w^+ are all related by the compositional path tracing back to the boundary condition and may be parametrized by say S_o^+. If the total fluid extraction rate (water + oil), U^-, is fixed, then there are three unknowns, S_o^+, V and S_o^- which may be found by solution of equations (20) to (22). The displacement efficiency can then be found by using equation (6) to determine U_o^- and then applying equation (24). Note that compositional variations occurring in the drainage region enter the efficiency through the relationship between U^+, S_o^+ and S_w^+.

If compositional changes do not affect the Darcy velocity, $U = U^+ = U^- =$ constant. Also, the water saturation S_w^+ is related to the front velocity V and the oil saturation S_o^+ by the requirement that the characteristic velocity for S_w^+ is V. This leaves three unknowns, S_o^+, S_o^- and V which may be found by solution of equations (20), (21) and (22). Consequently, in this case, the recovery efficiency is again independent of the compositional changes in the drainage region.

4 CONCLUSIONS

Compositional effects in the gravity stabilised displacement of oil by gas can be analysed using the method of characteristics.

The analysis shows that the process should be thought of in terms of oil being left behind by the displacement front, not oil draining into it.

The displacement efficiency depends mainly on the original oil and limiting tie line oil and gas compositions and flow properties.

The details of the compositional distribution in the drainage region appears to have only a small effect in as much as the compositional changes alter the overall flowrate in the drainage region above the displacement.

ACKNOWLEDGEMENT

The author would like to thank Dr S G Goodyear and Mr L J Roberts for reviewing this paper. The work has been funded by the UK Department of Energy who have kindly given permission for the paper to be published.

REFERENCES

[1] Ypma, J.G.J. Compositional Effects in Gravity-Dominated Nitrogen Displacement. SPE 14416 64th Annual Technical Conference and Exhibition. Las Vegas 1985.

[2] Helffrich, F.G. General Theory of Multicomponent, Multiphase Displacement in Porous Media. SPE 8372 54th Annual Fall Technical Conference an Exhibition. Las Vegas 1979.

[3] Buckley, S.E. and Leverett, M.C. Mechanism of Fluid Displacement in Sands. Trans AIME, v146, p107, 1942.

[4] Pope, G.A. The Application of Fractional Flow Theory to Enhanced Oil Recovery.

[5] Dumore, J.M., Hagoort, J. and Risseeuw, A.S. An Analytical Model for One Dimensional, Three Component Condensing and Vaporising Gas Drives. SPEJ April 1984.

[6] Stalkup, F.I. Miscible Displacement. SPE Monograph No 8. SPE 1983.

[7] Fayers, F.J. Extension of Stone's Method 1 and Conditions for Real Characteristics in Three-Phase Flow. SPE Reservoir Engineering. November 1989.

[8] Stone, H.L. Probability Model for Estimating Three Phase Relative Permeabilities. J. Pet. Tech. February 1970.

[9] Stone, H.L. Estimation of Three Phase Relative Permeability and Residual Oil Data. J. Cdn. Pet. Tech. Oct-Dec 1973.

[10] Shearer, M. and Trangenstein, J.A. Loss of Real Characteristics for Models of Three-Phase Flow in a Porous Medium. Transport in Porous Media 4: 499-525, 1989.

[11] Bell, J.B., Trangenstein, J.A. and Shubin, G.R. Conservation Laws of Mixed type Describing three phase flow in a porous media. SIAM J Appl Math, 46, 1986.

[12] Marchesin, D. and Medeiros, H.B. A Note on Gravitational Effects in Multiphase Flow. Catholic University of Rio de Janeiro, May 1988.

[13] Corey, A.T., Rathjens, C.H., Henderson, J.H. and Wyllie, M.R.J. Three Phase Relative Permeability. Pet Trans AIME, 207, 1956.

2nd European Conference on the Mathematics of Oil Recovery
© D. Guérillot, O. Guillon (Editors) and Éditions Technip, Paris 1990, pp. 197-204
27 rue Ginoux, 75015 Paris

Dispersive Mixing in Unstable Displacement

L. J. T. M. Kempers[1]

ABSTRACT

The stable displacement of miscible fluids through a porous medium that has many small-scale permeability variations exhibits dispersive mixing between the fluids. The unstable displacement exhibits two flow regimes: one in which the displacement is dominated by viscous fingers and one in which the displacement is dominated by dispersive mixing due to the permeability variations. In the viscous-finger-dominated regime the mixing zone expands linearly in time; in the dispersive-mixing-dominated regime the mixing zone expands as the square root of time. We have estimated the condition of transition between the two flow regimes. This condition has been tested by monitoring the development of the mixing zone in detailed numerical simulations of our own and by evaluating simulations reported by Araktingi and Orr, by Crump and by Moissis, et al. In addition, we found that when the unstable displacement is dominated by dispersive mixing, the expansion of the dispersive-mixing zone can be calculated according to the analytical model proposed by Kempers (1989). This model should be considered as a better alternative to the conventional Koval or Todd and Longstaff models for displacements with moderately mobility ratios.

1 INTRODUCTION

When a fluid in a porous medium is displaced by a less viscous fluid, the displacement front may be unstable: a small perturbation of the displacement front grows. The resulting large perturbations, which are called viscous fingers, are illustrated in Fig. 1a. This figure was obtained from a numerical simulation of an unstable displacement front between two miscible fluids.

The finger pattern of Fig. 1a was generated by the small disturbances of the displacement front resulting from spatial variations in the permeability. In this case the spatial permeability variations were very small: the standard deviation σ was only 5% of the average log-permeability (corresponding to a geometrical standard deviation $S = e^{\sigma}$ of 1.05). In fact, the porous medium was practically uniform. This is illustrated by the stable displacement seen in the same grid when displacing and displaced fluids are given the same viscosity (Fig. 1b).

An unstable displacement in a non-uniform porous medium that nonetheless has a statistically homogeneous spatial permeability distribution is shown in Fig. 1c. In this case the geometrical standard deviation S is 10. The pattern of Fig. 1c clearly shows large fingerlike disturbances of the displacement front. The fingers of this pattern are severely affected by the large permeability contrasts. This is shown by comparing Fig. 1c to a simulation of a stable displacement in the same grid (Fig. 1d). Unlike the stable displacement in the uniform medium (Fig. 1b), the stable displacement in the

---

(1) Koninklijke/Shell Exploratie en Produktie Laboratorium, PO Box 60, 2288 GD, Rijswijk (ZH), the Netherlands.

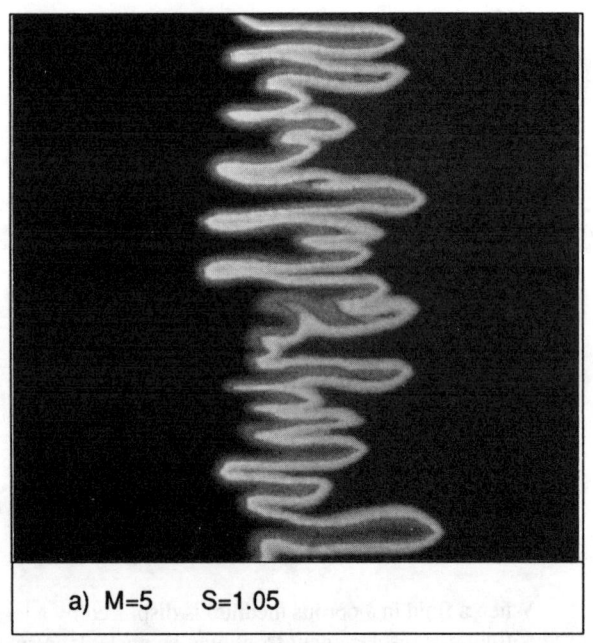

a) M=5   S=1.05

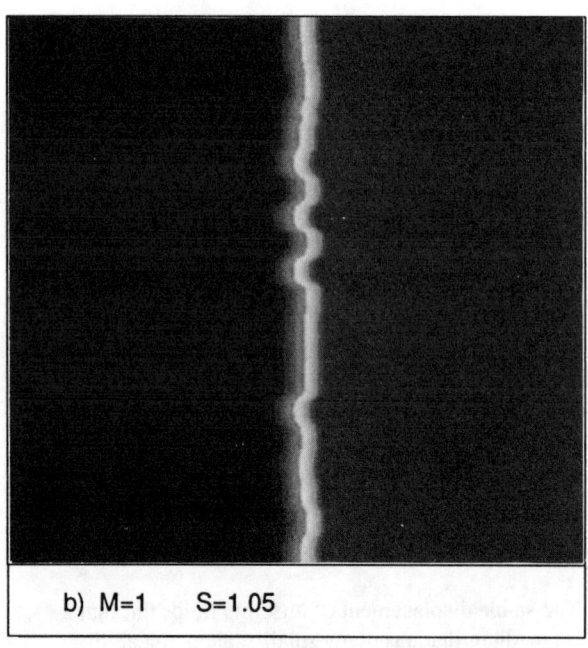

b) M=1   S=1.05

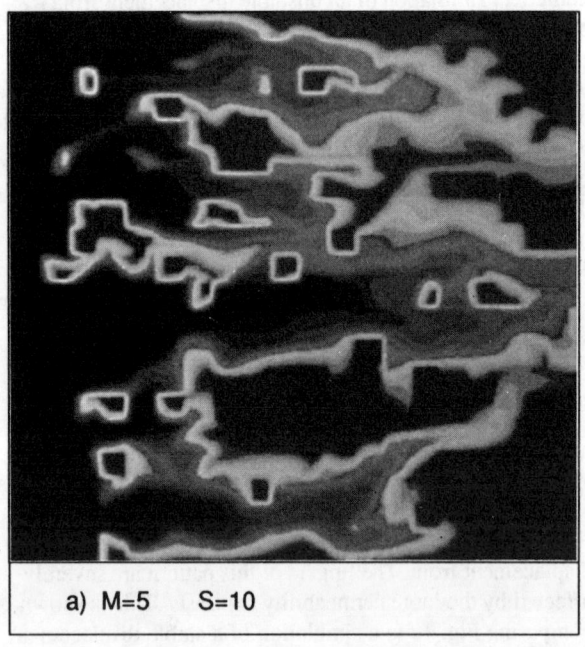

a) M=5   S=10

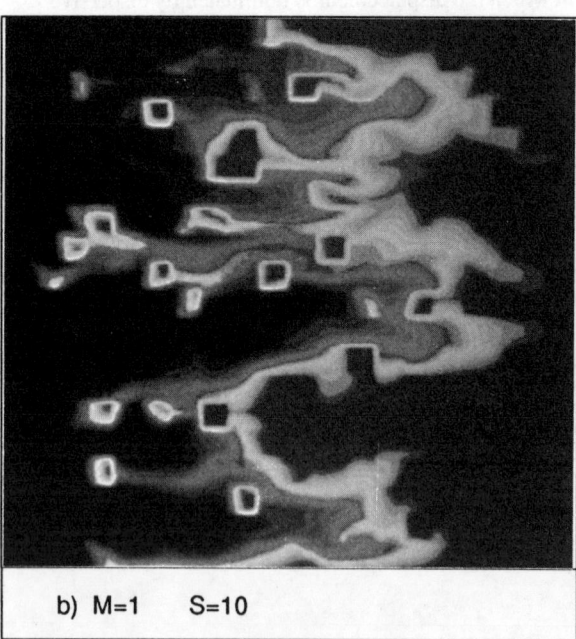

b) M=1   S=10

FIG. 1 Effect of small-scale heterogeneity on unstable displacement.

non-uniform medium still has large fingerlike disturbances, although they are somewhat shorter than the fingers in the unstable displacement. Furthermore, the fingers in the stable displacement follow the same paths as in the unstable displacement.

In this paper we quantify the condition at which an unstable displacement in a non-uniform medium becomes dominated by small-scale permeability variations; this condition should be checked before the fingering model of Koval or Todd and Longstaff is applied. The work described here also assists in the design of slugs in miscible flooding or polymer flooding, because it shows that in very non-uniform media the effects of viscous fingers may be negligible.

2 CONDITION FOR VISCOUS FINGERING AS PRESENTED IN THE LITERATURE

The most recent and complete treatment of the condition for unstable displacement that takes the stabilising effect of dispersion into account is the one of Coskuner and Bentsen (1989). Their instability condition for a system with length L and transverse dimensions B and H is:

$$\frac{U \dfrac{d\mu}{dC} - kg \dfrac{d\rho}{dC} \sin\gamma}{\bar{\mu} D} \frac{\partial \bar{C}}{\partial x} \frac{L^2}{\Omega}$$

$$\cdot \left[\left(\frac{1}{\Omega} + \frac{D'}{D} \right) \left(\frac{1}{\Omega} + 1 \right) \right]^{-1} > \pi^2 , \qquad (1)$$

with $\Omega = \dfrac{L^2(B^2 + H^2)}{B^2 H^2}$ (for a two-dimensional system in

which $H = 0$, $\Omega = \dfrac{L^2}{B^2}$) and

where U = displacement velocity
  μ = viscosity of mixture
  ρ = density of mixture
  C = injectant concentration
  k = permeability
  g = gravitational acceleration
  γ = dip angle
  ϕ = porosity

D = the longitudinal dispersion
D' = the transverse dispersion.

The term $\partial \bar{C} /\partial x$ is accordingly the average concentration gradient and $\bar{\mu}$ is the average viscosity. The treatment of Coskuner and Bentsen covers the more specialised treatments of Perrine (1961) and Peters, et al. (1984).

Condition (1) includes the well-known instability conditions that the mobility ratio M, defined as the ratio of the viscosity of the displaced fluid to that of the displacing fluid, is larger than 1 and that the displacement velocity exceeds the critical velocity.

We define the (un)stable regime as the set of conditions that do (not) violate condition (1).

3 FLOW REGIMES IN THE UNSTABLE REGIME

3.1 Estimation of transition

If a displacement is affected more by small-scale permeability variations than by viscous fingers, some viscous fingers may be present during the displacement, but they are so small that they cannot be discriminated from distortions of the displacement front caused by the permeability variations.

This is not a new idea. Young (1986), for instance, used the idea to defend his match of simulations of unstable, miscible displacement with the solution of the convection-dispersion equation. Araktingi and Orr (1986) also used it to explain their simulations of unstable, miscible displacement in heterogeneous media. They discovered that at heterogeneity index HI > 0.3 flow is dominated by the permeability field and viscous fingers are not present. (HI is defined as

$$HI = \sigma^2 \frac{\lambda}{L} , \qquad (2)$$

where σ = standard deviation of the log-normal
      permeability distribution
  λ = correlation length of the spatial permeability
      variation.)

Because Araktingi and Orr conducted simulations only with M = 20, however, their results cannot be generalised.

Below we present an estimation, based on this idea, of the condition at which a displacement front in the unstable regime is affected more by small-scale permeability variations than by viscous fingers. (The estimation is restricted to cases in which gravity can be neglected.) When the condition applies, the displacement front exhibits a dispersive-mixing zone. For that reason we refer to such a displacement as being **dispersive-mixing dominated**. When the condition does not apply, that is, when the displacement front is affected more by viscous fingers than by small-scale permeability variations, we call the displacement **viscous-finger dominated**.

The expression we use for the length of the mixing zone due to viscous fingers in a homogeneous, uniform medium is given by Koval (1962):

$$X_0 - X_{100} = (E - \frac{1}{E}) X, \qquad (3)$$

where X_C is the distance travelled by a plane with concentration C averaged over a transverse cross section, X is the average distance travelled and the Koval E-factor is defined by

$$E = (0.22 \, M^{0.25} + 0.78)^4 . \qquad (4)$$

The length of the mixing zone due to small scale permeability variations in a stable displacement without viscous effects (so M = 1) can be estimated with a commonly used expression for the mixing-zone length (Perkins and Johnston, 1963), with the dispersivity D/U given by the theory of Gelhar and Axness (1983):

$$X_{10} - X_{90} = 3.62 \sqrt{\lambda \, \sigma^2 \, X} . \qquad (5)$$

A comparison of Eqs. (3) and (5) suggests an estimate of the condition at which the displacement is dispersive-mixing dominated, namely

$$3.62 \sqrt{\frac{\lambda}{X} \sigma^2} > E - \frac{1}{E} . \qquad (6)$$

The point at which the transition from dispersive-mixing dominated to viscous-finger dominated displacement can then be calculated by substituting Eq. (2) in Eq. (6):

$$\frac{X_{tr}}{L} = HI \left(\frac{3.62}{E - \frac{1}{E}} \right)^2 . \qquad (7)$$

At X smaller than X_{tr}, the displacement is dispersive mixing dominated; at larger X it is viscous-finger dominated.

3.2 Discussion

Transition (7) indicates that a displacement tends to be dispersive-mixing dominated instead of viscous-finger dominated:
* at a large correlation length λ (provided λ<<L, to satisfy the condition of dispersive behaviour);
* at a small X;
* at a large standard deviation σ of the permeability variation;
* at a small mobility ratio M > 1.

Transition (7) clearly shows that when the heterogeneity index exceeds a certain value, the displacement is dominated by small-scale permeability variations. The condition thus explains the observation of Araktingi and Orr (at mobility ratio of 20) that when the heterogeneity index HI is larger than 0.3, the displacement is dominated by the permeability field. In fact, it predicts that, at M = 20, the transition from viscous-finger dominated to dispersive-mixing dominated occurs exactly at HI = 0.3.

Transition (7) also implies that a displacement is dispersive-mixing dominated in the beginning (at small X) and becomes viscous-finger dominated later (at larger X), as was observed by Moissis, et al. (1988).

Crump (1988) demonstrated that the mixing parameter model of Todd and Longstaff fails at low mobility ratios when the small-scale heterogeneity of the medium has the most influence on the distortion of the interface. This can be better understood in light of the results presented here: if the effects of unfavourable mobility ratio on the mixing-zone length are smaller than the effects of non-uniformity permeability, the displacement can be considered as dispersive-mixing dominated. The mixing parameter model of Todd and Longstaff is then no longer applicable and should be replaced by a dispersion model.

4 LENGTH OF DISPERSIVE-MIXING ZONE

If a displacement is dispersive-mixing dominated instead of viscous-finger dominated, then the increase of the mixing-zone length with time can be better described with a model that provides for an increase of the mixing-zone length with the square root of time than one that provides for an increase that is proportional to time, such as the one of Todd and Longstaff. We show below a way to calculate the length of a dispersive-mixing zone so that it is consistent with the dependence on the square root of time.

In an earlier paper (Kempers, 1989) we quantified the length of a dispersive-mixing zone that expands with the square root of time. In that paper no restriction was made in the assumptions or in the derivation of the random-walk model that limited the analysis to the stable regime, except for the assumption of the presence of a dispersive-mixing zone. This assumption, though, was then successfully tested for the stable regime.

In a linear displacement the injectant concentration, averaged over a transverse cross-section, is given by

$$\overline{C} = \frac{1}{2} \, \mathrm{erfc} \left(\frac{\frac{x}{L} - I}{2\sqrt{I}} \, \sqrt{Pe} \right), \qquad (8)$$

where x = longitudinal coordinate
       L = length of porous medium
       I = dimensionless time, defined as Ut/L where t is time
      Pe = Peclet number: $Pe = UL/D$
      D = dispersion coefficient.

This dispersion coefficient is further defined by

$$D/U = \lambda \, \sigma^2 \, \alpha \, . \qquad (9)$$

The dispersivity ratio α takes into account the effect of the fluid properties on the mixing-zone length. It is equal to 1 by definition when the fluid properties have no effect. The model presented in Kempers (1989) quantifies α. In Fig. 2 we have plotted α as a function of mobility ratio for both the stable and the unstable regimes.

5 VALIDATION

The transition in the unstable regime between dispersive-mixing dominated and viscous-finger dominated displacement, given by Condition (7), and the use of the random-walk model for dispersive-mixing zones in the unstable, dispersive-mixing dominated regime were validated with a series of numerical simulations. Transition (7) was also tested against the simulation runs of Araktingi and Orr (1988), Moissis, et al. (1988) and Crump (1988).

5.1 Set-up of simulations

We carried out a series of 16 simulation runs of unstable, miscible displacement in a medium with irregular, small-scale permeability variations. The purposes of the simulations

were:
1. to validate Transition (7),
2. to show that if (7) correctly predicts dispersive mixing, the length of the mixing zone can be calculated with the random-walk model of Kempers (1989).

The simulations did not take gravity into account and were run in the unstable regime, so the mobility ratio was larger than 1, the displacement velocity exceeded the critical velocity and the numerical dispersion did not suppress the fingers. The grid consisted of 80 by 80 blocks. Squares of 4 by 4 blocks had a uniform permeability. There was no correlation between permeabilities of adjacent squares. The distribution of the permeability values was log-normal. The mixing-zone development in the runs was monitored by calculating Pe at each 0.01 pore volume injected (see Kempers, 1989).

To establish the dispersive character of the mixing zone (and the negligible presence of viscous fingers), we required the same behaviour from the mixing zone expansion as we have observed in the stable regime (in Kempers (1989)). This implies that we required a constant course of the Peclet number with time (variations within 10%). We also required a sufficiently high correlation coefficient R^2 (above 0.99) between Eq. (8) and the mixing zone. (In case the reference run with M=1 had a Peclet variation of more than 10% (but always below 15%), we demanded that the Peclet variation was not more than 15%. In case the reference run with M = 1 had a correlation coefficient below 0.99, we demanded that the correlation coefficient of the simulation run with M > 1 was not more than 0.01 less than the correlation coefficient of the reference run.)

In the simulation series we varied, among other things, S and M. The lowest S value was 1.05, the highest 10; M varied between 1.5 and 100. We also varied the aspect ratio L/B. The aspect ratio is equal to the ratio between the longitudinal correlation length of the permeability and the transverse correlation length.

For the validation of Condition (7), we also considered the simulation runs of Araktingi and Orr, Crump and Moissis, et al. Because neither Araktingi and Orr nor Crump conducted their runs with the aim of investigating dispersive mixing in the unstable regime, we had to discriminate the runs exhibiting a dispersive-mixing zone from the runs exhibiting viscous fingers. We assumed that a simulation run of Araktingi and Orr had a dispersive-mixing zone if they evaluated the run as being dominated by the permeability field; we only took those runs into consideration that did not have an extremely large correlation length of the permeability ($\lambda/L \leq 0.7$). A simulation run of Crump was judged to have a dispersive-mixing zone if Crump found a mixing parameter that was dependent on the mobility ratio.

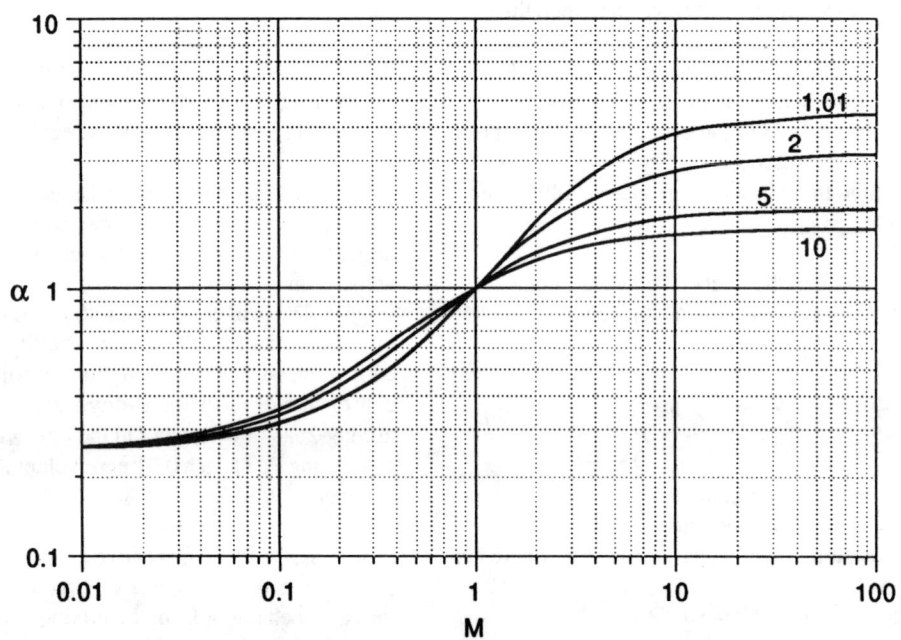

FIG. 2 Dispersivity ratio α as a function of mobility ratio M
(parameter: geometrical standard deviation S)

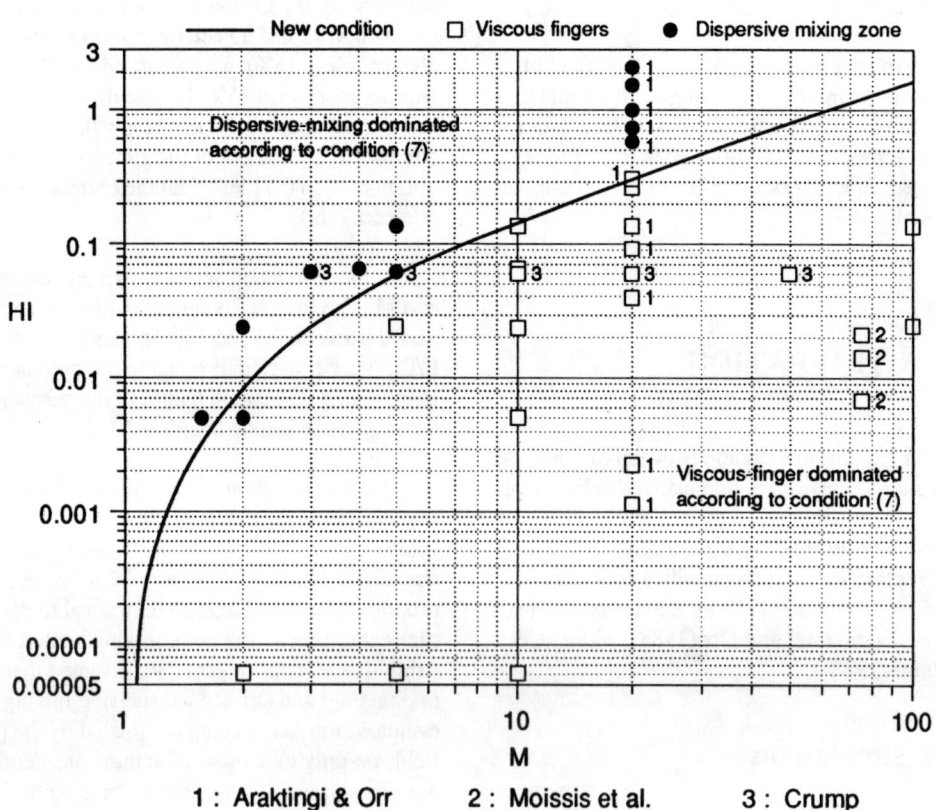

1 : Araktingi & Orr   2 : Moissis et al.   3 : Crump

FIG. 3 Validation of new condition (7)
Simulation runs are evaluated at X=L

5.2 Results

5.2.1 Transition (7)

All our runs, as well as those of Araktingi and Orr, Crump and Moissis, et al., have been summarised in Fig. 3, in which the heterogeneity index HI is plotted against M. The figure shows a very good agreement between the simulation results and Transition (7).

5.2.2 Random-walk model

All runs that exhibited a dispersive-mixing zone are listed in the table below. Their α values (α_s) are systematically higher than those predicted by the random-walk model (α_m).

TABLE
Runs that exhibit a dispersive-mixing zone

S	M	$\frac{L}{B}$	R^2	α_s	α_m	$\frac{\Delta\alpha}{\alpha}$	$\frac{X_{tr}}{L}$
2.60	2	27.2	0.99	1.58	1.48	6%	2.69
1.57	1.5	6.8	0.996	1.63	1.38	15%	1.93
2.60	2	1.0	0.992	1.80	1.48	18%	2.69
5	4	6.8	0.98	2.10	1.61	23%	1.67
2.60	2	6.8	0.99	2.03	1.48	27%	2.69
1.57	2	6.8	0.992	2.55	1.72	33%	0.61
10	5	1.0	0.96	2.24	1.50	33%	2.40
1.57	2	1.0	0.994	3.40	1.71	50%	0.61

This systematic underestimation is attributed to the violation of the condition that the longitudinal correlation length of the permeability be much larger than the transversal correlation length (Kempers, 1989). The table shows that the highest deviation occurs in the run with L/B=1 for which Transition (7) predicts viscous fingers for X/L > 0.61 but in which a dispersive-mixing zone is nonetheless observed.

6 CONCLUSIONS

1. There are two flow regimes in unstable displacement: a viscous finger dominated regime and a dispersive-mixing dominated regime due to small-scale permeability variations. In the first regime the mixing zone expands linearly with time; in the second regime the mixing zone expands with the square root of time. The transition between the flow regimes is given by Condition (7).

2. The transition between the flow regimes, Condition (7), has been validated by a series of numerical simulations of displacement in the unstable regime and by simulation rungs of Araktingi and Orr, Crump and Moissis, et al.

3. The length of the dispersive-mixing zone in the unstable regime can be calculated with the random-walk model of Kempers (1989), as expressed in Eqs. (8) and (9) and Fig. 2.

ACKNOWLEDGEMENT

The author is indebted to Gokhan Coskuner, Jacques Hagoort, Dick Ligthelm, John van Wunnik for many useful discussions and to the management of Shell Internationale Research Maatschappij BV for permission to publish this paper.

REFERENCES

Araktingi, U.G. and Orr Jr., F.M., 1988, Viscous fingering in heterogeneous porous media, SPE 18095.

Crump, J.G., 1988, Detailed simulations of the effects of process parameters on adverse mobility ratio displacements, SPE/DOE 17337.

Coskuner, G. and Bentsen, R.G., 1989, Effect of length on unstable miscible displacements, J. Can. Pet. Techn. 28 No. 4, pp. 34-44.

Moissis, D.E., Miller, C.A. and Wheeler, M.F., 1988, A parametric study of viscous fingering in miscible displacement by numerical simulation, in: M.F. Wheeler, ed., Numerical simulation in oil recovery, Springer, New York.

Gelhar, L.W. and Axness, C.L., 1983, Three-dimensional stochastic analysis of macrodispersion in aquifers, Water Resources Research, Vol. 19, No. 1, pp. 161-180.

Kempers, L.J.T.M., 1989, Effect of fluid properties on convective dispersion: comparison of analytical model with numerical simulations, 1st European Conference on the

Mathematics of Oil Recovery, Cambridge.

Koval, E.J., 1962, A method for predicting the performance of unstable miscible displacement in heterogeneous media, Soc. Pet. Eng. J. 1962, pp. 145-154.

Perkins, T.K. and Johnston, O.C., 1963, A review of diffusion and dispersion in porous media, Soc. Petr. Eng. J., pp. 70-84.

Perrine, R.L., 1961, The development of stability theory for miscible liquid-liquid displacements, Trans. AIME, 222, pp. 17-25.

Peters, E.J., Broman, W.H. and Broman, J.A., 1984, A stability theory for miscible displacement, SPE 13167.

Todd, M.R. and Longstaff, W.F., 1972, The development, testing and application of a numerical simulator for predicting miscible flood performance, J. Pet. Techn. 24, pp. 874-882.

Young, L.C., 1986, The use of dispersion relationships to model adverse mobility ratio miscible displacements, SPE 14899.

2nd European Conference on the Mathematics of Oil Recovery
© D. Guérillot, O. Guillon (Editors) and Éditions Technip, Paris 1990, pp. 205-210
27 rue Ginoux, 75015 Paris

Fluid Flow in Porous Media and Related Rock Mechanics Problems

M. Boutéca and J. P. Sarda[1]

ABSTRACT

The equations used to describe fluid flow in reservoirs are usually worked out without taking rock deformations directly into account. Indeed these deformations are included by correcting the compressibility of the fluid. On the basis of coupled equations for rock deformations and fluid flow, this paper analyzes the rock mechanics aspects of the diffusivity equation used by reservoir engineering specialists. The paper emphasizes the corresponding restrictive assumptions. It then goes on to explain the effective computing of rock deformations and fluid flow by iteration between a mechanical model of the deformable medium and a flow model.

1 INTRODUCTION

In the mechanics of deformable bodies, the equilibrium of a solid can be calculated if the boundary conditions are known as they are given in terms of stress and/or displacement, and if the body forces (gravity) are known. The history of loading must also be added if irreversible deformations are liable to appear. For rocks, a new force comes into play, i.e. pressure. The pressure field itself can be determined only if the pressure or flow boundary conditions are known together with the source terms (fluid injection or production). Likewise, the deformation of the rock (matrix and pores) changes the fluid flow conditions. This is obviously a coupled problem. That has been studied by Biot [1,2] for single-phase flow. Biot ended up with a system of differential equations in which fluid flow is described by a diffusivity equation in which the variable is the relative displacement of the fluid in relation to the solid (more precisely, the divergence of this relative displacement).

Hydraulics specialists also describe fluid flow by a diffusivity equation. However, the variable is no longer the divergence of a displacement but the pressure itself. Initially, equations are worked out without taking the deformation of the rock into account. This deformation is subsequently included by way of the compressibility coefficient modified to take part of the deformation of the solid into account. This paper describes the assumptions for going from Biot's diffusivity equation to the pressure diffusivity equation.

The single-phase production phase in reservoirs is often of brief duration, and specialists in fluid mechanics have had to develop models capable of computing the pressure field for multiphase flow. the complexity of the problem is such that it is illusory to try to couple rock deformation and multiphase flow in the same model. This paper describes the joint com-

---

(1) Institut Français du pétrole, 1 et 4, avenue de Bois-Préau, 92506 Rueil-Malmaison, France.

puting of rock deformation and multiphase flow by means of an iterative process.

2 EQUATIONS WORKED OUT BY M. A. BIOT

For a detailed description of Biot's theory, one should consult the article by Laurent and Quettier [4] and refer to the articles by Biot [1,2] and Coussy [3]. Let us assume a homogeneous isotropic medium subjected to a stress system σ_{ij} to infinity. The resulting displacement field is called u_i. Let us assume a fluid circulating in the porous medium. The displacement vector of the fluid is called U_i. For a given volume element, the relative displacement w_i of the fluid in relation to the solid is given by:

$$w_i = \Phi(U_i - u_i) \qquad (1)$$

The variation in the fluid content ζ is given by:

$$\zeta = -\nabla \bullet w \qquad (2)$$

Assuming the isotropy of the material and elastic behavior, Biot worked out the following equations:

$$\mu\nabla^2 u_i + (\mu + \lambda_u)\nabla e - \alpha M \nabla \zeta = 0 \qquad (3)$$

$$\dot{\zeta} - \frac{k.M_u}{\eta}\nabla^2 \zeta = 0 \qquad (4)$$

in which

$$\lambda_u = \lambda_d + \alpha^2 M \qquad M_u = \frac{2\mu + \lambda_d}{2\mu + \lambda_u}M \quad (5)$$

and the behavior laws are written:

$$\sigma_{ij} = 2\mu\varepsilon_{ij} + (\lambda_d + \alpha^2 M)tr(\varepsilon_{ij})\delta_{ij} - \alpha M\zeta\delta_{ij} \quad (6)$$

$$\Delta P_{pore} = \alpha M tr(\varepsilon_{ij}) - M\zeta \qquad (7)$$

the second behavior law, can also be written in the form:

$$\frac{\Delta P_{pore}}{M} = \alpha e - \zeta \qquad (8)$$

Comments on Biot's Equations

In the absence of any fluid, the equilibrium equations (3) can be reduced to:

$$\mu\nabla^2 u_i + (\mu + \lambda_u)\nabla e = 0 \qquad (9)$$

in which we find the equilibrium equations conventionally used in elastic solid mechanics.

Likewise, in Eq. (4) we recognize a diffusivity equation having a form that is familiar to hydraulics specialists. However, the variable is not the pressure but the variation in the fluid content, which includes a volumetric deformation term for the solid.

3 GOING FROM BIOT'S EQUATIONS TO THE PRESSURE DIFFUSIVITY EQUATION

To express the diffusivity equation in the form with which hydraulics specialists are familiar, we have to substitute the pressure for the variation in the fluid content ζ. For this we use Eq. (8).

In an homogeneous body the elastic parameter M is given by:

$$\frac{1}{M} = \frac{\alpha - \Phi}{K_s} + \frac{\Phi}{K_{fl}} = [\alpha - \Phi]c_s + \Phi \cdot c_{fl} \quad (10)$$

in which:

$$c_s = \frac{1}{K_s} = \frac{1}{V_s}\left[\frac{\partial V_s}{\partial P}\right]_{P_{pore} - P_{conf}} \qquad (11)$$

where V_s represents the volume of the skeleton, i.e. the set made up by the grains, cement and occluded porosity.

The value of Biot's coefficient α varies between:

$$\Phi \leq \alpha \leq 1 \qquad (12)$$

The compressibility of the pore fluid c_{fl} is much greater than the compressibility of the skeleton c_s. It thus seems initially allowable to approximate coefficient M by:

$$\frac{1}{M} = [\alpha - \Phi]c_s + \Phi \cdot c_{fl} \simeq \Phi \cdot c_{fl} \qquad (13)$$

Now let us return to the behavior law (8). The volumetric deformation e can be written:

$$e = -\frac{\partial V_b}{V_b} \qquad (14)$$

and the behavior law (8) becomes:

$$\Phi \cdot c_{fl} \cdot \Delta P_{pore} = -\alpha \frac{\partial V_b}{V_b} - \zeta \qquad (15)$$

3.1 Incompressible Medium

The behavior law (15) becomes:

$$\Phi \cdot c_{fl} \cdot \Delta P_{pore} = -\zeta \qquad (16)$$

Furthermore we have:

$$M \ll \lambda_d \Longrightarrow \lambda_u \equiv \lambda_d \Longrightarrow M_u \equiv M \qquad (17)$$

By incorporating (16) in (4) and by taking (17) into account, we obtain:

$$\nabla^2 P_{pore} = \frac{1}{K} \dot{P}_{pore} \qquad K = \frac{k}{\Phi \eta c_{fl}} \qquad (18)$$

3.2 Compressible Medium

The diffusivity equation used by hydraulics specialists implicitly contains the following assumptions:

$$M \ll \lambda_d \Longrightarrow \lambda_u \equiv \lambda_d \Longrightarrow M_u \equiv M \qquad (19)$$

$$\alpha \simeq 1 \qquad (20)$$

$$e = -\frac{\partial V_b}{V_b} \simeq -\frac{\partial V_p}{V_b} = -\Phi \frac{\partial V_p}{V_p} = -\Phi c_{mp} \cdot \partial P_{pore} \qquad (21)$$

in which c_{mp} is the matrix compressibility defined in reservoir engineering practice by:

$$c_{mp} = -\frac{1}{V_p} \left[\frac{\partial V_p}{\partial P_{pore}} \right]_T \qquad (22)$$

The behavior law (8) becomes:

$$\Phi \cdot c_{fl} \cdot \Delta P_{pore} = -\Phi \cdot c_{mp} \cdot \Delta P_{pore} - \zeta \qquad (23)$$

The equivalent compressibility c_{eq} can then be defined as:

$$c_{eq} = c_{fl} + c_{mp} \qquad (24)$$

and the behavior law (8) can be written as:

$$\Phi \cdot c_{eq} \cdot \Delta P_{pore} = -\zeta \qquad (25)$$

The diffusivity equation is written:

$$\nabla^2 P_{pore} = \frac{1}{K} \dot{P}_{pore} \qquad K = \frac{k}{\Phi \eta c_{eq}} \qquad (26)$$

3.3 Comments

The assumptions made by specialists in fluid mechanics concerning the deformation of the rock are thus of two sorts:

- The deformation of the skeleton is neglected. This assumption is used several times
  - in Eq. (13),
  - in Eq. (17), where equating λ_d to λ_u implies that the deformation of the rock is negligible compared to the deformation of the fluid,
  - in Eq. (21), while assuming $\partial V_b \simeq \partial V_p$

- The decrease in pore pressure is implicitly interpreted as an equivalent increase in confining stress, or an equivalent increase of the normal components of the stress tensor.

3.4 The Limits of the Empirical Treatment of Rock Mechanics Aspects

In reservoir engineering, the evolution of rock properties is defined as a function of pore pressure. A pressure change δP_{pore} induces an effective stress change δP_{pore}, which in turn results in a change in permeability, compressibility and pore volume. Based on the oedometric test for which detailed calculating is given in the Appendix, we show that a reduction in pore pressure does not necessarily correspond to an isotropic increase in effective stress. The oedometric test is a vertical loading test of a test specimen in which the lateral deformations are blocked. In a first approximation, this test simulates the compaction of a reservoir. The detailed calculating is given in the Appendix. The main result is (42):

$$\frac{\delta\sigma'_r}{\delta\sigma'_z} = \frac{\nu}{1-\nu}$$

Taking Eq. (34) into account we obtain:

$$\delta\sigma_r - \alpha\delta P_{pore} = \frac{\nu}{1-\nu}\left[\delta\sigma_z - \alpha\delta P_{pore}\right] \qquad (27)$$

Since the vertical total stress σ_z does not vary in time, we obtain:

$$\delta\sigma_r = \alpha\frac{1-2\nu}{1-\nu}\delta P_{pore} \qquad (28)$$

Hence when pressure decreases, the total stress σ_r decreases.

From Eq. (27) the effective stress change can be written as:

$$\delta\sigma'_r = -\alpha\frac{\nu}{1-\nu}\delta P_{pore} \qquad (29)$$

When pore pressure decreases, the effective stress σ'_r thus increases, but it increases in appreciably smaller proportion ($\nu/(1-\nu) \in [1/4, 1/2]$) than required by the assumption of the isotropic reduction of the rock volume.

4 PROPOSAL FOR THE COUPLING OF RESERVOIR AND ROCK-MECHANICS MODELS

Let us go back to Eq. (4) and introduce (8). We obtain:

$$\frac{\dot{P}_{pore}}{M} - \alpha\dot{e} - \frac{kM_u}{\eta M}\nabla^2 P_{pore} + \frac{\alpha k M_u}{\eta}\nabla^2 e = 0 \quad (30)$$

Namely:

$$\left[\frac{\eta}{kM_u}\dot{P}_{pore} - \nabla^2 P_{pore}\right] - \alpha M\left[\frac{\eta}{kM_u}\dot{e} - \nabla^2 e\right] = 0 \qquad (31)$$

Likewise, let us return to Eq. (3) and introduce (8). We obtain:

$$\mu\nabla^2 u_i + (\mu + \lambda_u)\nabla e - \alpha^2 M\nabla e + \alpha\nabla P_{pore} = 0 \quad (32)$$

Namely:

$$\left[\mu\nabla^2 u_i + \left(\mu + \lambda_u - \alpha^2 M\right)\nabla e\right] + [\alpha\nabla P_{pore}] = 0 \qquad (33)$$

A direct solution of the coupled problem seems possible for single-phase flow without making use of reservoir-engineering software. However, it is entirely illusory to try to deal with multiphase flow without the help of specialists in fluid mechanics. An examination of Eqs. (31) and (33) suggests a coupling method between rock-mechanics models and fluid-mechanics models. The solution might be as follows:

1. Initializing the pressure field solving Eq. (31) with $\alpha = 0$. The reservoir model is used without any modification. The value of compressibility is modified to become: $c = 1/(\Phi M_u)$

2. The pressure map thus generated is used to solve Eq. (33). Since term $\alpha\nabla P_{pore}$ is known, it is introduced in the right hand side of the equation. We can thus use a conventional code of rock mechanics in which the term ($\alpha\nabla P_{pore}$) intervenes as a body force.

3. Solving all of Eq. (31). Since deformation e was calculated in the second stage, the multiplying term of α in (31) is known. We can thus use a conventional reservoir engineering code in which the multiplying term of α intervenes as a source term.

4. Iteration of the second and third stages with a convergence criterion.

CONCLUSION

Starting from coupled equations for rock deformation and fluid flow we have shown the assumptions required for working out the diffusivity equation used by hydraulics specialists. These assumptions can be summed up as follows:

- Deformation of the skeleton (or matrix) is neglected.

- Deformation of the rock depends solely on how

pressure evolves., i.e. the mechanical problem of how stresses evolve during pressure decrease is assumed to be solved by an equal increase of the normal components of the effective stress tensor at each point. This simplification overlooks the fact that the mechanical problem, just like the hydraulic problem, requires the solving of a system of partial derivative equations concerning the displacements of the rock. In practice, an isotropic compression is automatically substituted for a more complex reality. This can have important consequences, especially on the compaction of the rock, on its permeability and on the estimating of borehole stresses for wells drilled after production. It is thus important, if possible, to handle the flow problem and the rock mechanics problem with the same degree of severity.

Whether the problem is dealt with from the standpoint of rock mechanics or that of fluid mechanics, it is illusory to try to deal rigorously with the deformation of rocks and multiphase flow (with possible phase change) within the same model. The method of solving that we propose, based on an iterative process, should enable these difficulties to be overcome.

REFERENCES

[1] Biot M.A. (1941) - "General Theory of Three Dimensional Consolidation", J. Appl. Phys., 12, 155-164

[2] Biot M.A. (1962) - "Mechanics of Deformation and Acoustic Propagation in Porous Media", J. Appl. Phys., 33, 1482-1498

[3] Coussy O. (1989) - "Thermomechanics of Saturated Porous Solids in Finite Deformation", European Journal of Mechanics, A/Solids, vol. 8, n. 1, 1-14

[4] Laurent J.,Quettier L. (1989) - "Comportement des milieux poreux consolidés dans le domaine élastique", Rock at Great Depth,Vol. 2, pp. 915-921, Balkema

NOMENCLATURE

c_{fl} compressibility of the fluid (counted positively)

c_s compressibility of the skeleton (grains + cement + occluded porosity) (counted positively)

$e = \nabla \bullet u = tr(\varepsilon_{ij}) = $ volumetric deformation (counted positively for a compression)

k permeability

K_{fl} incompressibility modulus of the fluid

K_s incompressibility modulus of the skeleton

M poroelastic coefficient

M_u undrained poroelastic coefficient

P_{pore} pore pressure

P_{conf} confining pressure

u displacement of the rock (positive for a compression)

U displacement of the fluid (same sign as u)

w relative displacement of the fluid in relation to the solid

α Biot's coefficient

$\varepsilon_{ij} = \frac{1}{2}[u_{i,j} + u_{j,i}]$

λ_d Lamé's coefficient for drained conditions

λ_u Lamé's coefficient for undrained conditions

μ shear modulus

ν Poisson's ratio (drained)

Φ open (or connected) porosity

σ_{ij} stress (positive for a compression)

σ_r radial stress

σ_r' effective radial stress

σ_θ orthoradial stress

σ_θ' effective orthoradial stress

σ_z vertical stress

σ_z' effective vertical stress

$\tau_{r\theta}, \tau_{rz}, \tau_{z\theta}$ shear stresses in cylindrical coordinates

ζ variation of fluid content (positive for expelled fluid)

η viscosity of the fluid

APPENDIX A: OEDOMETRIC TEST

A.1 Boundary Conditions - Assumptions

We are working in radial symmetry. If u_1, u_2 and u_3 are the displacement components in the x, y and z directions respectively, we have:

- $u_1 = u_2$
- $\frac{\partial}{\partial \theta} = 0$

We assume zero gravity forces:

- $\vec{g} = \vec{0}$
- $\frac{\partial P_{pore}}{\partial z} = 0$

We will not deal with the transient phase, and so we will write:

- $\frac{\partial u_1}{\partial z} = \frac{\partial u_2}{\partial z} = 0$

We will assume a uniform deformation field and a uniform pressure field:

- $\frac{\partial u_3}{\partial r} = 0$
- $\frac{\partial P_{pore}}{\partial r} = 0$

The "boundary condition" is given by:

- $\frac{\partial u_1}{\partial r} = \frac{\partial u_2}{\partial r} = 0$

A.2 Stress-strain relations

Biot showed that the solving of the coupled elasticity problem involved substituting for the total stress σ a fictive stress σ' defined by:

$$\sigma = \sigma' + \alpha P_{pore} \qquad (34)$$

The equilibrium equations are written using total stresses, but only the effective stress is involved in the energy. The relations between stresses and strains are thus relations between effective stresses

and strains. For cylindrical coordinates, these relations can be written:

$$\delta\sigma'_r = \lambda_d \Delta + 2\mu\varepsilon_r \qquad (35)$$

$$\delta\sigma'_\theta = \lambda_d \Delta + 2\mu\varepsilon_\theta \qquad (36)$$

$$\delta\sigma'_z = \lambda_d \Delta + 2\mu\varepsilon_z \qquad (37)$$

$$\varepsilon_r = \frac{\partial u_1}{\partial r} \quad \varepsilon_\theta = \frac{1}{r}\left[u_1 + \frac{\partial u_2}{\partial \theta}\right] \quad \varepsilon_z = \frac{\partial u_3}{\partial z} \qquad (38)$$

$$\Delta = \varepsilon_r + \varepsilon_\theta + \varepsilon_z \qquad (39)$$

A.3 Solving the Problem

By taking the assumptions and boundary conditions into account, we have:

$$\Delta = \frac{u_1}{r} + \varepsilon_z \qquad (40)$$

Deformation ε_r is zero at all points. This means that displacement $u_1 = u_2$ is zero at all points. The volumetric deformation is then reduced to the vertical deformation ε_z. The relations between stresses and strains (35-37) become:

$$\delta\sigma'_r = \lambda_d \varepsilon_z \quad \delta\sigma'_\theta = \lambda_d \varepsilon_z \quad \delta\sigma'_z = (\lambda_d + 2\mu)\,\varepsilon_z \qquad (41)$$

The effective stress changes are then linked by the following equation:

$$\frac{\delta\sigma'_r}{\delta\sigma'_z} = \frac{\lambda_d}{\lambda_d + 2\mu} = \frac{\nu}{1 - \nu} \qquad (42)$$

2nd European Conference on the Mathematics of Oil Recovery
© D. Guérillot, O. Guillon (Editors) and Éditions Technip, Paris 1990, pp. 211-218
27 rue Ginoux, 75015 Paris

Composition Paths in Binary CO_2-C_{10} Displacements: Effects of Reservoir Heterogeneity and Crossflow on Displacements with Limited Solubility

K. K. Pande[1,*] and F. M. Orr, Jr[1]

ABSTRACT

Material balance equations are formulated for the flow of two-phase, two-component mixtures in a porous medium consisting of two layers with differing permeabilities. Effects of viscous crossflow between the layers are modeled under the assumption that enough crossflow has taken place that fluids in the two layers are in vertical pressure equilibrium. The resulting set of coupled hyperbolic partial differential equations is solved using the method of characteristics.

Example solutions are reported for displacements of decane by CO_2. Three layer permeability ratios are considered, 1.5, 3.0, and 10.0, and the solutions are compared with the corresponding solutions without fluid crossflow.

1. INTRODUCTION

Computation of oil recovery from petroleum reservoirs requires modeling multiphase, multicomponent flow in heterogeneous porous media. The governing equations are usually solved by numerical methods because the equations are too complex to be solved analytically. Analytical solutions can be developed, however, for special limiting cases that can be used to verify the numerical solutions. Analytical solutions are also valuable in obtaining a mechanistic understanding of the recovery process.

The method of characteristics (MOC) has been applied by several investigators to describe multicomponent, multiphase flow in porous media. The classical description of immiscible displacement of oil by water in a homogeneous, one-dimensional porous medium developed by Buckley and Leverett (1942) can also be solved by the MOC as shown by Scheidegger (1957). Solutions to this problem consist of a region of spreading waves that emanates from the inlet injection condition and results in continuous variations in saturation up to the upstream side of an intermediate discontinuity to the downstream initial condition. Helfferich (1981) extended the Buckley-Leverett displacement theory to incorporate phase behavior effects in multiphase displacements. Helfferich's theory was subsequently extended by Dumore et al. (1984) to include the effects of volume change on mixing. Dumore et al. applied this theory to describe three-component, two-phase flow in $C_1 - C_4 - C_{10}$ systems. Monroe et al. (1986) extended the model of Dumore et al. to four component, two-phase systems for quaternary CO_2 - hydrocarbon systems.

All of the models discussed so far are restricted to the description of one-dimensional flow. Zapata and Lake (1981) developed a model to incorporate two-dimensional flow effects caused by viscous crossflow in the direction transverse to flow and applied it to immiscible water-oil displacements in a linear, two-layer flow system. Zapata and Lake used the vertical equilibrium assumption (Coats et al. 1967, 1971, Martin 1968) to achieve vertical pressure equilibrium between the layers and obtain a pseudo one-dimensional problem. The work presented here extends the work of Zapata and

(1) Department of Petroleum Engineering, Stanford University, Stanford, California 94305-220, USA.
(*) Now with Chevron Oil Field Research Company, PO Box 446, La Habra, California 90633-0446, USA.

Lake by including the effects of phase behavior in two-phase, two-component flow in a two-layer system.

2. MATHEMATICAL MODEL

Material balance equations for a linear, two layer flow system are derived with the assumptions: (1) the system is in vertical equilibrium, (2) phases are in local chemical equilibrium, (3) partial molar volumes of components are constant, (4) all fluids are incompressible, and (5) there are no gravitational, viscous fingering, dispersion, or capillary pressure effects. The layers are coupled at the injector and producer. Hence the injection pressure and the production pressure is the same in each layer, and the total pressure drop across the flow system is also the same in each layer.

$$R_{\phi h_1} \frac{\partial C_i^1}{\partial t_D} + \frac{\partial (F_i^1 q_{1D})}{\partial x_D} + q_{D_{xf_i}} = 0 \; , \; i = 1, n_c \quad (1)$$

$$R_{\phi h_2} \frac{\partial C_i^2}{\partial t_D} + \frac{\partial (F_i^2 q_{2D})}{\partial x_D} - q_{D_{xf_i}} = 0 \; , \; i = 1, n_c \quad (2)$$

where the amount of component i crossflowing is

$$q_{D_{xf_i}} = F_i^k \frac{\partial q_{1D}}{\partial x_D} \; , \qquad i = 1, n_c \quad (3)$$

The value of F_i^k in Eq. (3) depends on the direction of crossflow. If $\partial q_{1D}/\partial x_D < 0$, then crossflow is from Layer 1 to Layer 2, and $F_i^k = F_i^1$. If $\partial q_{1D}/\partial x_D > 0$, crossflow is from Layer 2 to Layer 1, and $F_i^k = F_i^2$.

The overall volume fraction and overall fractional flow of component i are given by

$$C_i = \sum_{j=1}^{n_p} c_{ij} S_j \; , \qquad F_i = \sum_{j=1}^{n_p} c_{ij} f_j \; , \quad i = 1, n_c$$

The dimensionless variables are defined as

$$t_D = \frac{q_t t}{\overline{\phi} H W L}, \qquad R_{\phi h_k} = \frac{\phi_k h_k}{\overline{\phi} H}, \qquad q_{kD} = \frac{q_k}{q_t}$$

and the following definitions apply,

$$\overline{\phi} = \frac{\sum\limits_{k=1}^{n_l} \phi_k h_k}{\sum\limits_{k=1}^{n_l} h_k}, \qquad H = \sum_{k=1}^{n_l} h_k, \qquad q_t = \sum_{k=1}^{n_l} q_k$$

$$q_k = W k_k h_k \lambda_{rt}^k \frac{dp}{dx_D}\Big|_k , \; \lambda_{rt}^k = \sum_{j=1}^{n_p} \lambda_{rj}^k = \sum_{j=1}^{n_p} \frac{k_{rj}^k}{\mu_j^k}$$

The initial condition is a uniform initial composition distribution in the flow system. The inlet boundary condition is that fluid of constant composition is injected at constant flow rate.

The differential material balance equations only apply in regions where the composition varies continuously. Integral material balances describe discontinuous variations in composition. The shock velocity of component i in layer k, Λ_i^k, is obtained from the integral material balances:

$$\Lambda_i^1 = \frac{(F_i^1 q_{1D})^{II} - (F_i^1 q_{1D})^I - F_{xf_i}^k (q_{kD}^{II} - q_{kD}^I)}{R_{\phi h_1} (C_i^{1\,II} - C_i^{1\,I})}, \; i = 1, n_c \quad (4)$$

$$\Lambda_i^2 = \frac{(F_i^2 q_{2D})^{II} - (F_i^2 q_{2D})^I + F_{xf_i}^k (q_{kD}^{II} - q_{kD}^I)}{R_{\phi h_2} (C_i^{2\,II} - C_i^{2\,I})}, \; i = 1, n_c \quad (5)$$

where for crossflow from Layer 1 to Layer 2, $F_{xf_i}^k = F_{xf_i}^1$ and $q_{kD} = q_{1D}$, and for crossflow from Layer 2 to Layer 1, $F_{xf_i}^k = -F_{xf_i}^2$ and $q_{kD} = q_{2D}$. The superscripts I and II refer to opposite sides of the shock.

The overall fractional flow of component i in the crossflowing fluid is evaluated using the vertical equilibrium condition (Zapata 1981) and is assumed to be given by

$$F_{xf_i}^k = \frac{F_i^{k\,I} \lambda_{rt}^{k\,I} + F_i^{k\,II} \lambda_{rt}^{k\,II}}{\lambda_{rt}^{k\,I} + \lambda_{rt}^{k\,II}} \; , \quad i = 1, n_c \quad (6)$$

Vertical equilibrium requires that the vertical pressure gradient be zero everywhere. Therefore, at any point in the flow system the pressure in Layer 1 is equal to the pressure in Layer 2 ($p^1 = p^2$). Hence, the horizontal pressure gradient in Layer 1 is equal to that in Layer 2, and q_{1D} is given by

$$q_{1D} = \frac{k_1 h_1 \lambda_{rt}^1}{k_1 h_1 \lambda_{rt}^1 + k_2 h_2 \lambda_{rt}^2} \quad (7)$$

3. METHOD OF CHARACTERISTICS SOLUTION

Solution of Eqs. (1) and (2) by the method of characteristics (MOC) requires calculation of the velocity at which a given overall composition moves through the porous medium. The eigenvalue problem is:

$$\lambda \frac{\partial C_i^1}{\partial x_D} = \frac{q_{1D}}{R_{\phi h_1}} \sum_{m=1}^{n_c-1} \frac{\partial F_i^1}{\partial C_m^1} \frac{\partial C_m^1}{\partial x_D} + A\,C \; , \; i = 1, n_c \quad (8)$$

$$\lambda \frac{\partial C_i^2}{\partial x_D} = \frac{q_{2D}}{R_{\phi h_2}} \sum_{m=1}^{n_c-1} \frac{\partial F_i^2}{\partial C_m^2} \frac{\partial C_m^2}{\partial x_D} + B\,C \; , \; i = 1, n_c \quad (9)$$

where

$$C = (F_i^1 - F_i^2) \sum_{m=1}^{n_c-1} \left[\frac{\partial q_{1D}}{\partial C_m^1} \frac{\partial C_m^1}{\partial x_D} + \frac{\partial q_{1D}}{\partial C_m^2} \frac{\partial C_m^2}{\partial x_D} \right]$$

The system of equations in matrix form for a binary system is

$$\begin{bmatrix} a_{11} & a_{12} \\ a_{21} & a_{22} \end{bmatrix} \begin{bmatrix} \dfrac{\partial C_1^1}{\partial x_D} \\ \dfrac{\partial C_1^2}{\partial x_D} \end{bmatrix} = \lambda \begin{bmatrix} \dfrac{\partial C_1^1}{\partial x_D} \\ \dfrac{\partial C_1^2}{\partial x_D} \end{bmatrix} \quad (10)$$

where the elements of the matrix are defined as:

$$a_{11} = \frac{q_{1D}}{R_{\phi h_1}} \frac{\partial F_1^1}{\partial C_1^1} + A D, \qquad a_{12} = A E$$

$$a_{21} = B D, \qquad a_{22} = \frac{q_{2D}}{R_{\phi h_2}} \frac{\partial F_1^2}{\partial C_1^2} + B E$$

where

$$D = (F_1^1 - F_1^2) \frac{\partial q_{1D}}{\partial C_1^1}, \qquad E = (F_1^1 - F_1^2) \frac{\partial q_{1D}}{\partial C_1^2}$$

The values of A and B in Eqs. (8) to (10) depend on the direction of fluid crossflow. For crossflow from Layer 1 to Layer 2, $A = 0$ and $B = 1/R_{\phi h_2}$. For crossflow from Layer 2 to Layer 1, $A = 1/R_{\phi h_1}$ and $B = 0$.

Construction of a solution from the inlet injection condition to the outlet initial condition requires that continuous variations and shocks be combined in a way that also satisfies a constraint on wave velocities. The velocity constraint requires the wave velocity to decrease monotonically from the downstream initial condition to the upstream injection condition.

4. RESULTS

MOC solutions for displacements in low $(k_1/k_2 = 1.5)$, medium $(k_1/k_2 = 3.0)$, and high $(k_1/k_2 = 10.0)$ permeability contrast layered systems, with no porosity contrast $(\phi_1/\phi_2 = 1)$ and no thickness contrast $(h_1/h_2 = 1)$ are described herein. Layer 1 is the higher permeability layer. The injection gas composition is pure CO_2, and the initial oil composition is pure decane. In the following discussion, CO_2 is referred to as component 1 and C_{10} is referred to as component 2. Results are reported in units of overall volume fractions.

According to the Peng-Robinson equation of state (1976), at 1600 psia and 160°F the liquid phase contains 64.02 vol % CO_2 and the vapor phase 98.22 vol % CO_2. Phase viscosities, μ_j, were calculated with the Lohrenz-Bray-Clark correlation (1964) to be $\mu_g = 0.028$ cp and $\mu_o = 0.125$ cp. The end-point viscosity ratio for pure CO_2 displacing pure decane is 10.45. Relative permeabilities of the fluid phases are given by

$$k_{rg} = k_{rg}^o \left[\frac{S_g}{1 - S_{wr} - S_{or}} \right]^{n_g}, \quad k_{rl} = k_{rl}^o \left[\frac{1 - S_g - S_{or} - S_{wr}}{1 - S_{wr} - S_{or}} \right]^{n_l}$$

where $k_{rg}^o = 1.0$, $k_{rl}^o = 1.0$, $S_{or} = 0.1$, $S_{wr} = 0.0$, $n_g = 2.0$, and $n_l = 2.0$.

The phase diagram for the two layer, CO_2 - C_{10} system is shown in Fig. 1. Nine different regions are shown on the phase diagram based on the phases present in each layer in that region. There are four composition paths through any point in the phase diagram. Two paths, a fast path and a slow path, are associated with each direction of crossflow. Wave velocities along these paths are discontinuous at phase boundaries. Therefore, a phase transition in either layer is always accompanied by a discontinuous variation in composition (shock). Velocity variations along paths that originate from the initial and injection compositions violate the velocity constraint. Hence there is a shock from the initial composition (leading shock) and another shock from the injection composition (trailing shock). Based on analysis of the driving forces for viscous crossflow, Pande and Orr (1988, 1990) argue that fluid crossflows from the fast layer to the slow layer at the leading shock and in the reverse direction at the trailing shock. Both the leading and trailing shocks are intermediate discontinuities because the shock velocity equals the wave velocity on one side of the shock (Jeffrey 1976). The leading shock is an upstream intermediate discontinuity (UID), and the trailing shock is a downstream intermediate discontinuity (DID) (Pande 1989). The wave velocity on the upstream side of the leading shock is given by the fast eigenvalue for crossflow from Layer 1 to Layer 2, while the wave velocity on the downstream side of the trailing shock is given by the slow eigenvalue for crossflow from Layer 2 to Layer 1:

$$\lambda = \frac{q_{kD}}{R_{\phi h_k}} \frac{dF_1^k}{dC_1^k} \quad (11)$$

where $k=1$ for the leading shock and $k=2$ for the trailing shock. The coherence condition, which states that the wave velocity must be the same for each component (Helfferich 1981), requires that $\Lambda_1^1 = \Lambda_1^2$ and because the shock velocity must equal the wave velocity of the composition that is the limit of a continuous variation:

$$\Lambda_1^1 = \Lambda_1^2 = \lambda \quad (12)$$

Equation (12) was solved for the composition on the upstream side of the leading shock and the composition on the downstream side of the trailing shock by Newton-Raphson iteration. Mapping of those compositions on a phase diagram such as that shown in Fig. 1 shows that there is a phase transition from single-phase liquid flow to two-phase flow in Layer 1 across the leading shock and a phase transition from single-phase gas flow to two-phase flow in Layer 2 across the trailing shock. The solution between the leading and trailing shocks must include at least one more phase transition shock because it is impossible to connect the leading and trailing shock compositions without traversing a phase boundary. Any shocks that lie between the leading and trailing shocks are referred to as intermediate shocks. The equations that govern these shocks depend on the direction of crossflow across the shock, the type of discontinuity, and any path constraints that apply to the upstream and downstream compositions. Because it is difficult to determine these solution features a priori, the differential material balance equations were solved by finite differences (FD) and the type of discontinuity was deduced from the features of the FD composition profiles.

The solution route for low permeability contrast (LPC) is shown in Fig. 1 and is described as follows: there is a leading shock (LS A → B), a leading region of continuous variation (LRCV B → C), a leading intermediate shock (LIS C → D), an intermediate RCV (IRCV D → E), a trailing intermediate shock (TIS E → F), a trailing RCV (TRCV F → G), and a trailing shock (TS G → H).

The composition route is time invariant, but each point on the route has an associated velocity. Those velocities are used to compute composition, saturation, and flowrate profiles at a given time as shown in Figs. 2 and 3. Examination of Fig. 2 reveals that there is a leading zone of constant state (LZCS C → C) and a trailing ZCS (TZCS F → F). These zones of constant state are obtained because the LIS and TIS are intermediate discontinuities, and thus, are the limit of a continuous variation only on one side of the shock. The LIS is an UID and the TIS is a DID. The direction of crossflow across these shocks can be deduced from the flowrate profile in Fig. 3. Fluid crossflows from the slow layer to the fast layer across the LIS and in the reverse direction across the TIS. The flowrate profile indicates that there is no crossflow between the layers over a large portion of the displacement, and that a large amount of crossflow occurs at shock fronts. In addition, more crossflow is observed around the leading and trailing shocks than around the intermediate shocks.

The saturation profiles show a phase transition across each shock. A phase transition from single-phase liquid flow to two-phase flow occurs in Layer 2 across the LIS and a transition from two-phase flow to single-phase gas flow occurs in Layer 1 across the TIS. The composition on the downstream side of the LIS is constrained to lie on the fast path defined by the upstream composition (B) of the LS. This path constraint arises because the wave velocity along most of the slow path is lower than the velocity of the TS. Hence, a path switch from the

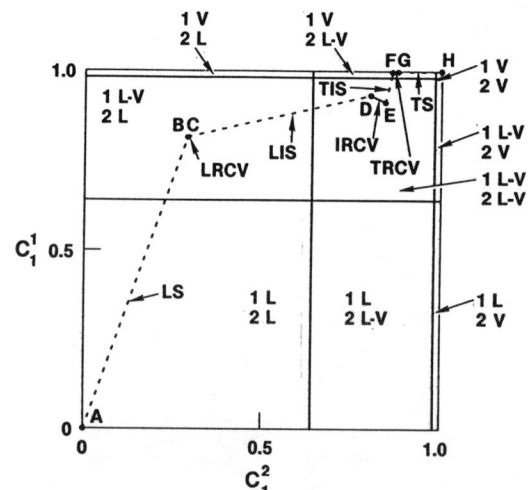

Figure 1. Solution Route for $k_1/k_2 = 1.5$

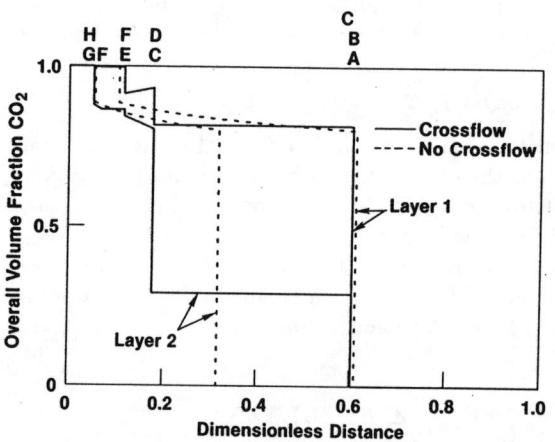

Figure 2. Composition Profiles for $k_1/k_2 = 1.5$

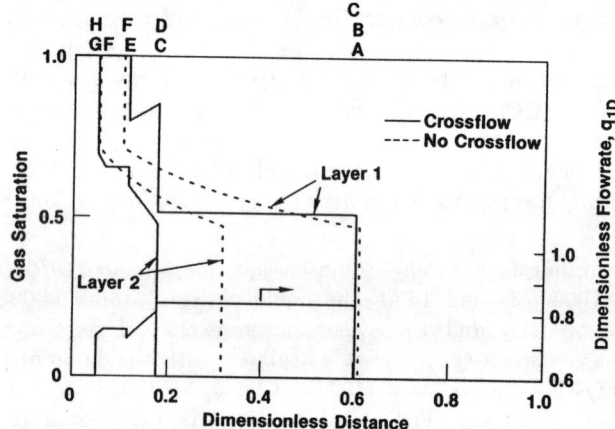

Figure 3. Saturation and Flowrate Profiles for $k_1/k_2 = 1.5$

214

fast path to the slow path is not allowed in the region between the LS and LIS. Velocity constraints also preclude path switches between the TS and the TIS. Hence, the composition on the upstream side of the TIS must lie on the slow path defined by the downstream composition (G) of the TS. The LIS and TIS are coupled because the composition on the downstream side of the TIS must be on the same path as the composition on the upstream side of the LIS. These two compositions must lie on the same path because there are no allowable path switches in the region of the phase diagram where there is two-phase flow in both layers simultaneously. The procedure used to solve these two shocks simultaneously is described by Pande (1989).

In order to illustrate how viscous crossflow impacts displacements, we also report composition and saturation profiles for displacements without crossflow. Displacement performance without crossflow was computed by performing a layer by layer integration of the MOC solution for uniform flow (Pande 1989). As Figs. 2 and 3 show, the LS and TS in the displacement with crossflow move slightly slower, whereas the TIS moves slightly faster, than the corresponding shocks in the displacement without crossflow. The LIS, however, travels much slower than the LS in Layer 2 in the displacement without crossflow. That happens because the flowrate in Layer 2 decreases as fluid crossflows into Layer 1. As a result, two-phase flow occurs in both layers simultaneously over a much smaller portion of the displacement when there is viscous crossflow.

The solution route for medium permeability contrast (MPC) is shown in Fig. 4. The features of this solution are similar to those for LPC, except that the leading intermediate and trailing intermediate shock fronts have coalesced into a single shock front, referred to as the intermediate shock (IS C → D). As indicated by the composition profile in Fig. 5, this shock is a UID. The path constraints on the upstream and downstream sides of this shock are identical to those for the downstream side of the LIS and the upstream side of the TIS in the LPC case. The flowrate profile in Fig. 6 indicates that fluid crossflows from the slow layer to the fast layer across this shock. In addition, the saturation profiles show that a phase transition occurs in both layers across this shock. In Layer 1, there is a phase transition from two-phase flow to single phase gas flow, while in Layer 2 there is a transition from single-phase liquid flow to two-phase flow. As a result, there is no region in this displacement where two-phase flow occurs in both layers simultaneously. In contrast to the LPC case, there is a large RCV between the LS and the ZCS. Fluid crossflows from the fast layer to the slow layer in this region, but the amount of crossflow is small because the slope, dq_{1D}/dx_D, is small in this region. Unlike the LPC case, which had no crossflow over a large portion of the

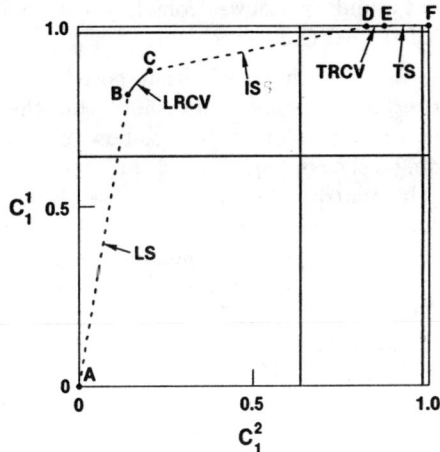

Figure 4. Solution Route for $k_1/k_2 = 3.0$

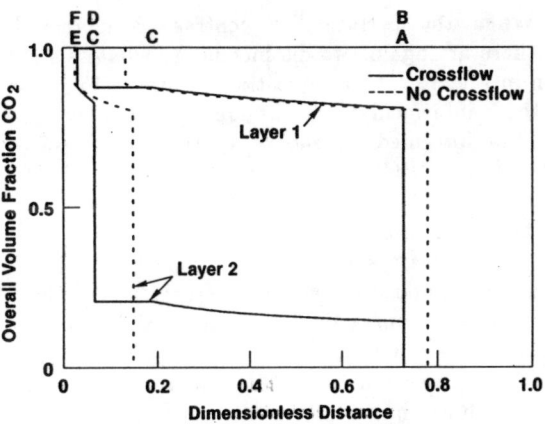

Figure 5. Composition Profiles for $k_1/k_2 = 3.0$

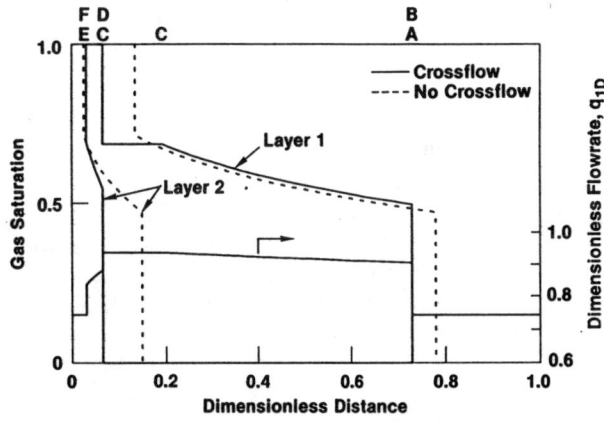

Figure 6. Saturation and Flowrate Profiles for $k_1/k_2 = 3.0$

215

displacement, fluid crossflows from Layer 1 to Layer 2 over most of the MPC displacement.

Comparison of the shock front positions with and without crossflow in Figs. 5 and 6 shows that the LS and TS in the displacement with crossflow are near the corresponding shocks in the displacement without crossflow. In contrast to the LPC case, however, the trailing shock in the displacement with crossflow is slightly ahead of the corresponding shock without crossflow. Also, with the increase in permeability contrast from 1.5 to 3.0, the difference in positions between the leading shocks for the displacement with and without crossflow is larger. The intermediate shock in the displacement with crossflow is located well behind the leading shock in Layer 2, and the trailing shock in Layer 1 in the displacement without crossflow. The reason for that behavior is that the flowrate in Layer 2 decreases substantially as fluid crossflows into Layer 1 behind the intermediate shock. Recall that a similar situation occurred for LPC.

When the permeability contrast is increased to 10.0, there are again two distinct intermediate shocks as shown in Fig. 7. The composition profiles in Fig. 8 show that these intermediate shocks are separated by a ZCS, while the intermediate shocks in the LPC case were separated by a RCV. Therefore, the LIS is a DID and the TIS is an UID. As in the LPC case, the LIS and TIS must be solved simultaneously, because the composition on the downstream side of the TIS must lie on the same path as the composition on the upstream side of the LIS. In fact, those compositions are identical because both compositions are not the limit of a continuous variation, but rather correspond to a ZCS. The saturation profiles in Fig. 9 show that a phase transition from two-phase flow to single-phase gas flow occurs in Layer 1 across the LIS, and a transition from single-phase liquid flow to two-phase flow occurs in Layer 2 across the TIS. Thus, single-phase gas flow occurs in Layer 1 simultaneously with single-phase liquid flow in Layer 2 in the region between the LIS and the TIS. Two-phase flow occurs in Layer 2 over a small region because the TS and the TIS are located close together. The flowrate profile in Fig. 9 shows that fluid crossflows from Layer 1 to Layer 2 across the LIS and in the reverse direction across the TIS. A larger amount of crossflow occurs across the TIS than the LIS as indicated by the larger change in q_{1D} across that shock. As in the MPC case, fluid crossflow from Layer 1 to Layer 2 occurs over a large portion of the displacement due to the large region upstream of the leading shock where the composition varies continuously. As before, the amount of crossflow in that region is small.

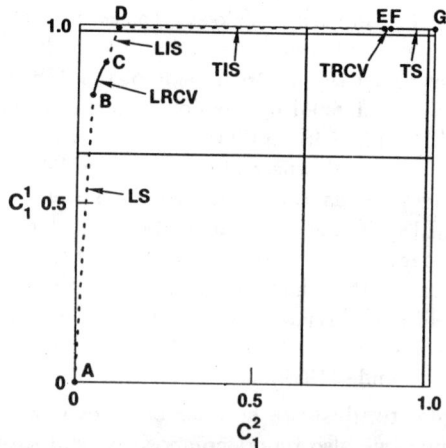

Figure 7. Solution Route for k_1/k_2 = 3.0

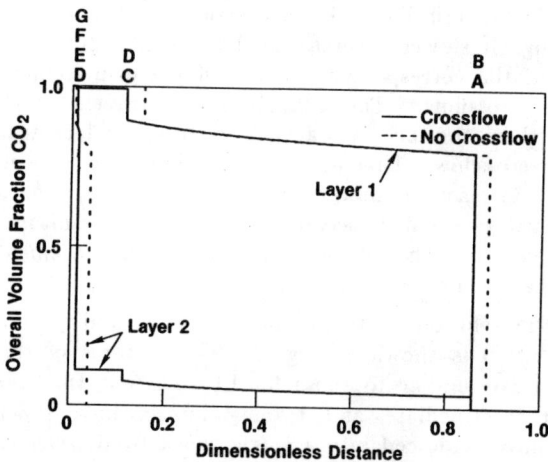

Figure 8. Composition Profiles for k_1/k_2 = 10.0

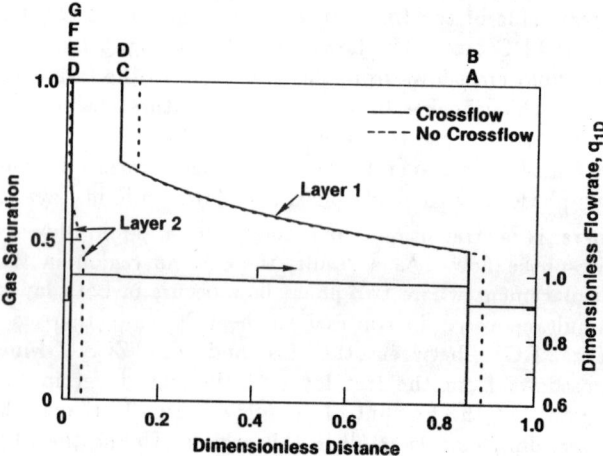

Fig. 9. Saturation and Flowrate Profiles for k_1/k_2 = 10.0

Comparison of the shock front positions with and without crossflow shows that the LS and TS are located near the corresponding shocks in the displacement without crossflow. Also, both intermediate shocks in the displacement with crossflow are located behind the corresponding shocks without crossflow. As in the LPC and MPC cases, a reduction in flowrate in Layer 2 due to crossflow of fluid into Layer 1 causes the TIS to be located well behind the leading shock in Layer 2 without crossflow.

5. DISCUSSION

MOC solutions for displacements with crossflow were developed using the idea that a composition in Layer 1 (C_1^1) and a composition in Layer 2 (C_1^2) travel together throughout the displacement (coherence). As a result, the solution route is independent of time and can be represented on a phase diagram. However, the solution route is a strong function of the degree of nonuniform flow. As the permeability contrast increases, the difference in composition between the layers also increases, and the composition route deviates from the 45 degree line (uniform flow) in phase diagrams like those shown in Figs. 1, 4, and 7. The phase regions traversed by the solution route were shown to be a function of permeability contrast. For each type of solution shown, there is some range of permeability contrasts for which the solution route traverses the same phase regions.

The no crossflow solutions have a solution route for each layer that is identical to that for uniform flow. However, if the solution route for the two layer displacement without crossflow were represented on a phase diagram, it would vary with time. That occurs because pairs of overall composition in Layers 1 and 2 do not travel with the same velocity throughout the displacement when there is no crossflow and the mobility ratio differs from unity. Thus, coherence applies to flow within a single layer but not to the coupled two-layer system in the displacement without crossflow.

A comparison of the leading and trailing shock heights for different permeability contrasts shows that the leading shock height in Layer 1 approaches that of a uniform displacement and the shock height in Layer 2 approaches zero, as the permeability contrast increases. The trailing shock height in Layer 2 approaches that of a uniform displacement and the shock height in Layer 1 approaches unity, as the permeability contrast increases. The leading shock height in Layer 1 and the trailing shock height in both layers are relatively insensitive to permeability contrast; however, the leading shock height in Layer 2 changes considerably with permeability con-

trast. These observations suggest that there is less interaction between the layers at the leading and trailing edges as the permeability contrast increases.

The effect of nonuniform flow on overall performance is summarized by comparing oil recovery curves for the various displacements, as shown in Fig. 10. As the permeability contrast increases, breakthrough occurs earlier, complete recovery takes longer, and crossflow contributes more to the recovery of oil from the low permeability layer during the early stages of the displacement. The recovery curves for displacements with and without crossflow cross several times before sweepout of the system, however, for medium and high permeability contrast crossflow improves recovery over that without crossflow over a practical time range (e.g., $t_D < 2$ PVI).

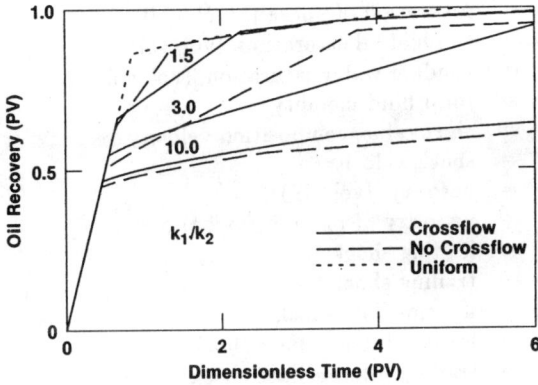

Figure 10. Oil Recovery Curves for k_1/k_2 = 1.5, 3.0, 10.0

6. CONCLUSIONS

(1) The method of characteristics can be applied to develop analytical solutions for two-phase, two component flow with vertical equilibrium viscous crossflow in a porous medium consisting of two layers.

(2) Fluid crossflow between layers generally speeds up oil recovery over that observed when crossflow is absent because oil is transferred from the slow layer to the fast layer where it can be recovered relatively quickly.

NOMENCLATURE

c_{ij} = concentration of component i in phase j
C_i = overall concentration of component i, (vol/vol)
f_j = fractional flow of phase j, (vol/vol)
F_i = overall fractional flow of component i, (vol/vol)
h = thickness, (L)
k = permeability, (L^2)
k_{rj} = relative permeability of phase j
k_{rj}^o = end point relative permeability of phase j
L = length, (L)
n_c = number of components
n_p = number of phases
p = pressure, (m/Lt^2)
q = volumetric flowrate, (L^3/t)
S_j = saturation of phase j, (vol/vol)
S_{or} = residual oil saturation, (vol/vol)
S_{wr} = residual water saturation, (vol/vol)
λ_{rt} = total fluid mobility
λ = eigenvalue, composition velocity
Λ = shock velocity
ϕ = porosity, (vol/vol)
μ_j = viscosity of phase j, (m/Lt)
LS = leading shock
TS = trailing shock
IS = intermediate shock
LIS = leading intermediate shock
TIS = trailing intermediate shock
RCV = region of continuous variation
LRCV = leading region of continuous variation
TRCV = trailing region of continuous variation
ZCS = zone of constant state
LZCS = leading zone of constant state
TZCS = trailing zone of constant state

ACKNOWLEDGEMENT

The research described here was supported by the U. S. Department of Energy under Contract No. DE-AC21-85MC22042 and by Stanford University Miscible Flooding Industrial Affiliates. That support is gratefully acknowledged.

REFERENCES

Buckley, S. E. and Leverett, M. C., 1942, Mechanism of Fluid Displacement in Sands, *Trans.*, AIME 146, 107-116.

Coats, K. H., Nielson, R. L., Terhune, M. H., and Weber, A. G., 1967, Simulation of Three Dimensional, Two-Phase Flow in Oil and Gas Reservoirs, *Soc. Pet. Eng. J.*, 377-388; *Trans.*, AIME, **240**.

Coats, K. H., Dempsey, J. R., and Henderson, J. H., 1971, The Use of Vertical Equilibrium in Two-Dimensional Simulation of Three-dimensional Reservoir Performance, *Trans.*, AIME **251**, 63-71.

Dumore, J. M., Hagoort, J., and Risseeuw, A. S., 1984, An Analytical Model for One-Dimensional Three Component Condensing and Vaporizing Gas Drives, *Soc. Pet. Eng. J.*, **24**, 169-179.

Helfferich, F. G., 1981, General Theory of Multicomponent, Multiphase Displacement In Porous Media, *Soc. Pet. Eng. J.*, 51-62.

Helfferich, F. G., 1982, Generalized Welge Construction for Two-Phase Flow in Porous Media in System With Limited Miscibility, paper SPE 9730 presented at the 57th Annual Technical Conference of SPE of AIME, Dallas, TX, Sept. 26-29.

Jeffrey, A., 1976, *Quasilinear Hyperbolic Systems and Waves*, Pitman Publishing, London, England.

Lohrenz, J., Bray, G., and Clark, C. R., 1964, Calculating Viscosities of Reservoir Fluids from Their Compositions, *J. Pet. Tech.* **6**, 1171-76.

Martin, J. C., 1968, Partial Integration of Equations of Multiphase Flow, *Trans.*, AIME **243**, 63-71.

Monroe, W. W., Silva, M. K., Larsen, L. L., and Orr, F. M., Jr., 1987, Composition paths in Four-Component Systems: The Effect of Dissolved Methane on CO₂ Flood Performance in One-Dimension, paper SPE 16172 presented at the 62nd Annual Technical Conference of SPE of AIME, Dallas, TX, Sept. 27-30.

Pande, K. K., 1989, *Interaction of Phase Behavior with Nonuniform Flow*, PhD Dissertation, Stanford University, Stanford, CA.

Pande, K. K. and Orr, F. M., Jr., 1989, Interaction of Phase Behavior, Reservoir Heterogeneity and Crossflow in CO₂ Floods, paper SPE19668 presented at the 64th Annual Technical Conference of SPE of AIME, San Antonio, TX, Oct. 8-11.

Pande, K. K. and Orr, F. M., Jr., 1990, Analytical Computation of Breakthrough Recovery for CO₂ Floods in Layered Reservoirs, paper SPE/DOE 20177 presented at the Seventh Symposium on Enhanced Oil Recovery, Tulsa, OK, April 22-25.

Peng, D. Y. and Robinson, D. B., 1976, A New Two-Constant Equation of State, *Ind. Eng. Chem. Fund.* **15**, 59-64.

Scheidegger, A. E., 1957, *Physics of Flow Through Porous Media*, MacMillian, London.

Zapata, V. J., 1981, *A Theoretical Analysis of Viscou Crossflow*, PhD Dissertation, University of Texas, Austin, TX.

Zapata, V. J. and Lake, L. W., 1981, A Theoretic: Analysis of Viscous Crossflow", paper SPE 101! presented at the 56th Annual Conference of SPE AIME, San Antonio, Texas, Oct. 5-7.

2nd European Conference on the Mathematics of Oil Recovery
© D. Guérillot, O. Guillon (Editors) and Éditions Technip, Paris 1990, pp. 219-234
27 rue Ginoux, 75015 Paris

Mathematical and Numerical Analysis of a Hyperbolic System Modeling Solvent Flooding

T. Johansen[1] and R. Winther[2]

Abstract

The fluid system under considerations in this paper consists of three chemical components. The phase properties of this system depends on its composition and we assume that a maximum of two phases can be formed. The mathematical model governs the purely convective transport of this fluid system through a one dimensional homogeneous porous medium. The model is discussed through mathematical analysis and numerical experiments.

1 Introduction

The recovery of hydrocarbons from reservoirs can be strongly improved by injecting a fluid with the ability to interact with residing hydrocarbons through the mechanism of partioning of individual components between the fluid phases. By an appropriate choice of injected fluid, a favourable phase behavior of the composite fluid system can be achieved in-situ, primarily represented by first contact miscibility and developed miscibility. The success of such an injection scheme first of all relies on an understanding of the phase behavior of the composite fluid system. From this, predictions of the flood performance may be determined on different levels of accuracy, ranging from more or less qualitative description in one space dimension and with fixed phase diagrams, through comprehensive full field simulations based on phase equilibrium calculations for many component systems. In the planning and accomplishment of a field project, both extremes play an important role.

The approach taken in this paper belongs to the former extreme. The objective here is to emphasize some new analytical results that may be useful in improving the accuracy in predicting one dimensional homogeneous flow for simplified systems, and also to investigate the performance

This research was supported by VISTA, a research cooperation between the Norwegian Academy of Science and Letters and Den norske stats oljeselskap a.s. (Statoil), and The Royal Council for Scientific and Industrial Research, Norway.

(1) Institute for Energy Technology, PO Box 40, 2007 Kjeller, Norway.
(2) Department of Informatics, University of Oslo, PO Box 1080, 0316 Oslo 3, Norway.

of standard numerical methods for the equations governing such flow. Though these experiments are performed in one dimensional homogeneous media subject to rather restrictive assumptions, we believe that such results are important as to disclose the limitations of standard numerical methods when applied to more complicated flow problems.

When dissipative effects are neglected the model is hyperbolic, but not strictly hyperbolic. Furthermore, due to the phase transition, the model is linear in one region of the state space and nonlinear in another. The numerical experiments will indicate that the solution of such models does not depend L_1-continuously on the initial data. Hence, in this respect the hyperbolic model is not a well-posed system. The calculations also show that the convergence of the finite difference solution to the solution of the continuous model is rather slow. In particular, this seems to be the case in regimes where the solution of the continuous model is unstable with respect to the initial data. In these regimes, the results obtained from a finite difference calculation may reflect the properties of the difference scheme more than the properties of the continuous model.

2   The mathematical model

A common approach when studying the coupling between flow and phase behavior for compositional systems is to impose a sufficient set of assumptions to guarantee hyperbolicity of the governing model, which will allow analytical construction in one dimensional homogeneous media. For three component systems this has for example been done by Welge [19], Walsh and Lake [18], Dumore et. al. [3], Larson [13] and Hirasaki [5]. Such constructions are also utilized by Pande and Orr [16], [17] to analyze nonuniform flow of three component systems involving cross-flow in layered reservoirs. In Monroe et. al. [14] analytical solutions for four-components systems are constructed. In Hellferich [4] extensions to

systems involving an arbitrary number of components are presented. However, constructions given there are performed only for three component systems.

The approach taken in this paper is similar to what is described above, and the set of assumptions imposed is:

i) Fluids and rock are incompressible.

ii) Processes are isothermal and isochoric.

iii) Variations in phase behavior caused by pressure gradients are negligible.

iv) Dissipative effects are negligible.

v) Chemical reactions and retention do not occur.

In particular, phase diagrams for the composite fluid system will be fixed. In addition, we assume that the medium is one dimensional and homogeneous. We will consider a three component system where for example two of the components (or pseudo-components) are hydrocarbons. The third component can be the injected solvent, for example carbon-dioxide, nitrogen or methane, to mention some of the most successfully applied bases of injected fluid. A typical phase diagram for this system is depicted in Fig. 1. Here u, v, w denote the overall volume fractions of the components. Thus we have $u + v + w = 1$. For compositions inside the region Φ_1 a single fluid phase is formed, while in the region Φ_2 the fluid system separates into two immiscible fluid phases in local equilibrium. The compositions and relative amounts of the phases are uniquely determined by the overall composition (u, v) and thermodynamical principles, which reduce to "tie-line rules" and "lever-rules". We also mention here that a three component system subject to the stated assumptions in general allow the formation of three distinguished phases (Gibbs phase rule). However, in this paper we consider systems which only allow one or two phases.

The two regions Φ_1, Φ_2 are separated by the binodal curve $\mathcal{B}$. For simplicity, we shall in this paper assume that $\mathcal{B}$ has the form

$$\mathcal{B} = \{(u,v)|v = \phi(u)\},$$

where $\phi : [u_0, u_1] \to [0,1]$ is a smooth concave function. Thus,

$$\Phi_2 = \{(u,v)|0 \leq v \leq \phi(u)\}.$$

We also assume that there exists a unique point $(u_c, v_c) = (u_c, \phi(u_c)) \in \mathcal{B}$ representing the composition for which the given pressure and temperature is the critical point (cf. assumptions *ii*) and *iii*) above).

The mass conservation of the three components subject to the assumptions stated above then yields the following model:

$$\begin{aligned} u_t + F(u,v)_x &= 0, \\ v_t + G(u,v)_x &= 0. \end{aligned} \tag{2.1}$$

Here $t \geq 0$ is dimensionless time and $x \in\, <-\infty, \infty>$ is dimensionless distance. The continuous functions F and G are the overall volumetric flux functions for the components u and v, respectively. If $(u,v) \in \Phi_1$, then $F(u,v) = u$ and $G(u,v) = v$. This reflects the single phase behavior in the one phase region Φ_1. In the two phase region Φ_2, F and G are nonlinear.

The tie-lines are phase equilibrium loci in Φ_2. For every $(u,v) \in \Phi_2$ there is exactly one tie-line passing through (u,v). Moreover, each tie-line intersects the binodal curve in exactly two points, one on each side of the critical point. Thus $v \equiv 0$ is an extended tie-line and the tie-lines approach the tangent to the binodal curve at the critical point. Let $(z, \phi(z)); z \in [u_0, u_c]$ and $(b(z), \phi(b(z))); b(z) \in [u_c, u_1]$, see Fig. 1, be the two points of intesection with the binodal curve $\mathcal{B}$ for a tie-line. This tie-line will be referred to as $L(z)$. The function b is a smooth decreasing function mapping $[u_0, u_c]$ onto $[u_c, u_1]$, such that $b(u_0) = u_1$ and $b(u_c) = u_c$.

Let $(u,v) \in \Phi_2$ and let $L(z)$ be the unique tie-line through (u,v). Then $z = z(u,v)$. Physically,

z and $b(z)$ are the concentrations of the component u in the phases into which the fluid system of overall composition (u,v) separates. We refer to these two phases as phase 1 and phase 2, respectively. Accordingly, the concentrations of component v in phase 1 is $\phi(z)$ and in phase 2, $\phi(b(z))$. Let $s \in [0,1]$ be the saturation of phase 1. For $(u,v) \in \Phi_2; (u,v) \neq (u_c, v_c)$, the "lever-rule" reads

$$\begin{aligned} u &= sz + (1-s)b(z), \\ v &= s\phi(z) + (1-s)\phi(b(z)). \end{aligned} \tag{2.2}$$

In particular, $s = s(u,v)$, $s(z, \phi(z)) = 1$ and $s(b(z), \phi(b(z))) = 0$. If f denotes the fractional flow function of phase 1, we also have the relations

$$\begin{aligned} F &= fz + (1-f)b(z), \\ G &= f\phi(z) + (1-f)\phi(b(z)). \end{aligned} \tag{2.3}$$

In Φ_2 it is natural to represent f as a function of s and z, since along a tie-line, phase composition is constant. In particular, we define $f(s, u_c) = s; s \in [0,1]$, which is consistent with the smoothness of F, G in Φ_2 and with the physical fact that the two phases are indistinguishable as the critical point is approached. In Fig. 2, typical shapes of F and f are plotted along a fixed tie-line.

In summary, we have in this section given a precise definition of the model (2.1). In the one phase region Φ_1, $F = u$ and $G = v$. In the two phase region Φ_2, F and G are given implicitly by (2.2) and (2.3) through the functions f, ϕ and b.

3 Riemann problems and elementary waves

The purpose of this section is to give a summary of the properties of the Riemann problem for the model (2.1); i.e., we consider initial value problems of the form

$$(u(x,0), v(x,0)) = \begin{cases} (u^L, v^L) & \text{if } x < 0, \\ (u^R, v^R) & \text{if } x > 0, \end{cases}$$

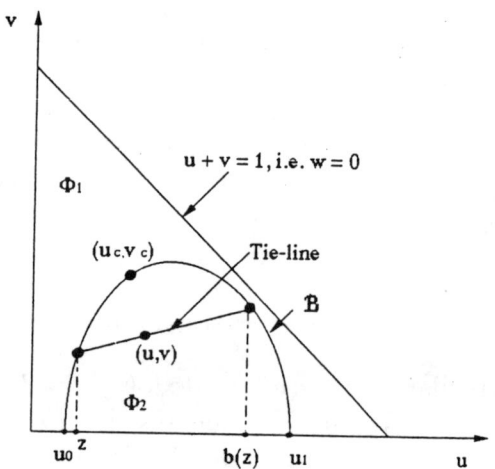

Figure 1: Composition space

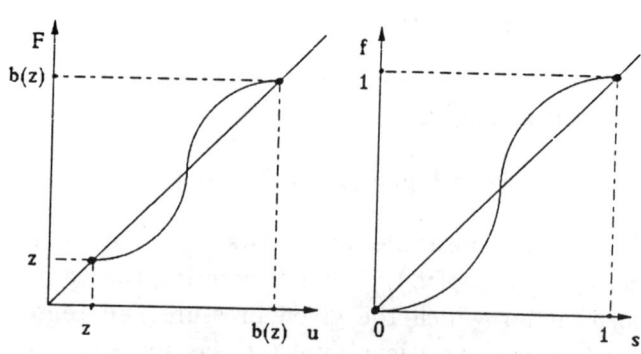

Figure 2: Flux functions on fixed tie-lines

where the left (injected) state (u^L, v^L) and right (initial) state (u^R, v^R) are given constant compositions. In particular, we shall discover that there are some unsolved uniqueness questions with this problem. These uniqueness problems will be studied by numerical experiments in the next section.

A detailed discussion of the Riemann problem for (2.1) is given in [7] under suitable assumptions on the model. Here we shall indicate how the results of [7] can be generalized by using the more general theory developed in [2].

We first consider the model (2.1) restricted to the two phase region Φ_2. Let $\eta = \eta(z)$ be the slope of the tie-line $L(z)$. Then,

$$v = \eta(z)u + a(z), \qquad (3.1)$$

where

$$a(z) = \phi(z) - \eta(z)z.$$

A simple manipulation of (2.3) gives

$$G(u, z) = \eta(z)F(u, z) + a(z). \qquad (3.2)$$

By substituting (3.1) and (3.2) into (2.1) we find, after expanding derivatives and canceling terms, that the second equation of (2.1) becomes

$$(\eta'(z)u + a'(z))z_t + (\eta'(z)F + a'(z))z_x = 0. \qquad (3.3)$$

Since the tie-lines do not intersect in Φ_2, v given by (3.1) for any fixed $u \in [u_0, u_1]$ is a strict monotonically increasing function of z. Hence,

$$\eta'(z)u + a'(z) > 0. \qquad (3.4)$$

If we also assume that $\eta'(z) \neq 0$, then (3.3) can be written

$$z_t + \frac{F + h(z)}{u + h(z)} z_x = 0,$$

where $h(z) = a'(z)/\eta'(z)$. Introducing a change of coordinates by

$$y = x - t,$$

the model (2.1) inside Φ_2, can be formulated in terms of u and z as independent variables:

$$\begin{aligned} u_t + \tilde{F}(u, z)_y &= 0, \\ z_t + \frac{\bar{F}}{u + h(z)} z_y &= 0. \end{aligned} \qquad (3.5)$$

Here, $\tilde{F}(u, z) = F(u, z) - u$. If we let U denote the vector $(u, v)^T$, the system (3.5) can also be written in the hyperbolic form

$$U_t + A(U)U_y = 0, \qquad (3.6)$$

where the 2×2-matrix $A(U)$ is given by

$$A(U) = \begin{pmatrix} \frac{\partial \tilde{F}}{\partial u} & \frac{\partial \tilde{F}}{\partial z} \\ 0 & \frac{\tilde{F}}{u+h(z)} \end{pmatrix}. \qquad (3.7)$$

In particular, the eigenvalues $\lambda_u = \frac{\partial \tilde{F}}{\partial u}$ and $\lambda_z = \frac{\tilde{F}}{u+h(z)}$ are real, positive and bounded. The model is therefore hyperbolic. However it is not strictly hyperbolic, since it is possible that $\lambda_u = \lambda_z$. This can be seen from a simple geometric inspection of the eigenvalues, cf. Fig. 3. Let $\Phi_2^- = \{(u,v) | \tilde{F} < 0\}$ and $\Phi_2^+ = \{(u,v) | \tilde{F} > 0\}$. The relation $\lambda_u = \lambda_z$ then defines two transition curves $T^- \subset \Phi_2^-$ and $T^+ \subset \Phi_2^+$ as illustrated in Fig. 4. The curve between T^- and T^+, defined by the relation $\lambda_u \equiv 1$, will be referred to as the unit velocity curve, and will be denoted $\mathcal{K}$.

In [9] the global Riemann problem for a model of the form (3.5) is solved. The model studied there is derived as a model for polymer flooding. However, the assumptions which are imposed in [9] are more restrictive than what can be assumed here. In particular, the analysis given there allows only one transition curve, and the function $h(z)$, which corresponds to the derivative of an adsorption isotherm in the context of [9], is assumed to be a strictly decreasing function.

The study of Riemann problems which was initiated in [9] has been extended in a series of papers [10], [8] and [2]. In particular [2] contains an analysis of Riemann problems of a rather general class of two phase, multicomponent displacement models. The analysis given there allows both two transition curves and a general function $h(z)$. Furthermore, it applies to the Riemann problem for (3.5), with minor modifications due to the particular shape of the two phase region Φ_2.

As long as the condition $\eta'(z) \neq 0$ is satisfied, the Riemann problem for (2.1), with data in the two phase region Φ_2, can therfore be uniquely solved with a solution completely contained in Φ_2. This remains true without making any other simplifying assumptions about the geometry of

the tie-lines or the binodal curve. We also remark that a Riemann solution with data in Φ_2 never can enter the one phase region Φ_1. This follows from entropy considerations.

A brief review of the properties of the Riemann solution for (2.1) in Φ_2 will now be given. A Riemann solution is in general composed of elementary waves and constant states. The elementary waves for the model (2.1) in the two phase region can be derived similarly as in [9]. The elementary waves corresponding to the eigenvalue λ_u correspond to elementary waves for a scalar equation of the form

$$u_t + F(u, z_0)_x = 0,$$

where $z_0 \in [u_0, u_c]$ is a constant. The associated wave curve in Φ_2 coincides with a tie-line, and therfore such waves are called tie-line waves. The nature of the elementary waves corresponding to the eigenvalue λ_z (non tie-line waves) depends on the sign of $h'(z)$. Consider first the simplest case when $h'(z) \equiv 0$. This will be satisfied if all the tie-line extensions pass through a common point $(u^*, 0)$. Then, $a(z) = -\eta(z)u^*$ and hence $h(z) = a'(z)/\eta'(z) = -u^*$. In many applications this is a reasonable approximation which will strongly simplify the construction of the Riemann solution. This is because in this case the λ_z-waves are contact discontinuities, and the associated wave curves constitute a single family of curves. The Riemann problem for (2.1) in Φ_2 in this case is analogous to the problems studied by Isaacson [6], Keyfitz,Kranzer [11] and Barkve [1].

If $h'(z) < 0$ or $h'(z) > 0$ the non tie-line waves are genuine rarefactions and shocks. Hence, in this case the rarefaction curves and the shock curves do not coincide. However, as mentioned above, the Riemann solution for (2.1) in Φ_2 can in this case be constructed using the procedure developed in [2].

From an application point of view, penetration into the two phase region is usually consider undesirable. However, it is shown in [16] that for layered reservoirs, the effect of crossflow between different layers outweighs the disadvantage of two phase flow. Even if Riemann solutions do not apply directly in non uniform flow, we believe that they can be utilized for improving accuracy in the analysis of such flow, provided the theory can be expanded to allow initial and injected states for the Riemann problem outside the two phase region Φ_2. The rest of this paper is devoted to the study of such extensions.

If $(u^L, v^L), (u^R, v^R) \in \Phi_1 \cup \mathcal{B} \cup \mathcal{K}$, we observe that the material balance relations for discontinuous waves for (2.1)

$$
\begin{aligned}
\sigma(u^L - u^R) &= F(u^L, v^L) - F(u^R, v^R), \\
\sigma(v^L - v^R) &= G(u^L, v^L) - G(u^R, v^R),
\end{aligned}
$$
(3.8)

are satisfied with a contact discontinuity of speed $\sigma = 1$; i.e., the waves have unit velocity.

In order to pick the physically relevant waves among these waves in the construction of Riemann solutions, an additional criterion (an entropy condition) is required. Otherwise, the system will in general not have a unique solution. However, for the present model involving phase transition, to our knowledge such a criterion has not been rigorously formulated. Therefore, we shall in the next section investigate the uniqueness question numerically. The approach is to perform numerical experiments by using a first order diffusive finite difference scheme to solve Riemann problems for (2.1), assuming that numerical diffusion models physical diffusion correctly. By this, vanishing diffusion is modelled through grid refinement. We remark here that comparison between such numerical solutions and results from slim tube experiments show good agreement, cf. [15].

Before we close this section, we investigate the possible shock waves joining states Φ_1 and Φ_2. (Rarefaction waves can of course never connect two such states.) Suppose for example that $(u^L, v^L) \in \Phi_1$ and $(u^R, v^R) \in \Phi_2$. Then, $F(u^L, v^L) = u^L$ and $G(u^L, v^L) = v^L$. Using the representations (3.1) and (3.2) for v^R and $G(u^R, v^R)$, respectively, substituting this into

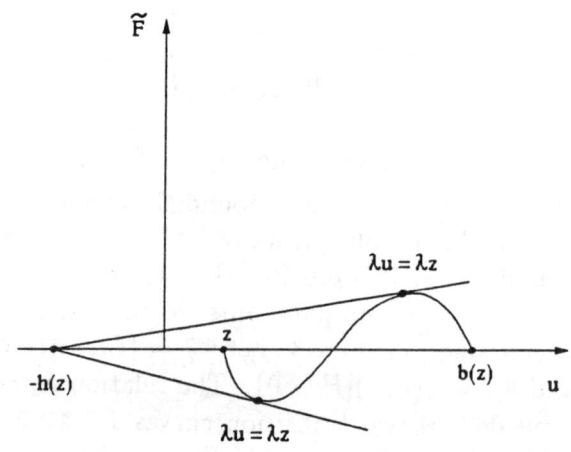

Figure 3: Coinciding eigenvalues

(3.8) and eliminating $F(u^L, v^L)$ yields

$$(\sigma - 1)v^L = (\sigma - 1)(\eta(z^R)u^L + a(z^R)). \quad (3.9)$$

One obvious solution of (3.9) is $\sigma = 1$; i.e., $(u^R, v^R) \in \mathcal{B} \cup \mathcal{K}$. For nonunit velocity shocks, we must have

$$v^L = \eta(z^R)u^L + a(z^R). \quad (3.10)$$

This means that (u^L, v^L) must lie on the extended tie-line passing through (u^R, v^R).

In conclusion, possible shocks joining Φ_1 and Φ_2 are either unit velocity waves or shocks joining states on the same extended tie-line. An immediate consequence of the latter is that when tie-line extensions are intersecting in the one phase region Φ_1, uniqueness problems in Riemann solutions may arise. Such problems are further discussed in the next section.

4   Experiments

The purpose of this section is to investigate the properties of the model (2.1) by numerical experiments. In particular we will consider examples when states from both the one phase region Φ_1

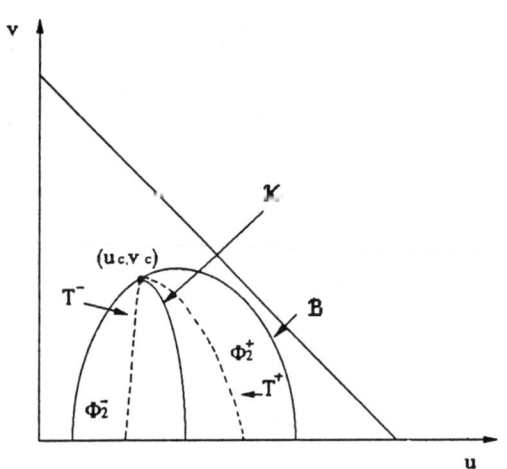

Figure 4: Subregions of state space

and the two phase region Φ_2 will be involved in the solution.

In all the experiments the two phase region Φ_2 is given by

$$\Phi_2 = \{(u,v)|u_0 \leq u \leq u_1, 0 \leq v \leq \phi(u)\},$$

where $u_0 = 1/4, u_1 = 3/4$ and the binodal curve $\mathcal{B}$ is defined by the concave function ϕ given by $\phi(u) = \frac{1}{4} - 4(u - \frac{1}{2})^2$. Furthermore, if $(u_c, \phi(u_c))$ is the critical point then the fractional flow function $f = f(\cdot, z)$ will be taken to be the quintic function

$$f(s,z) = \frac{1}{2} + (1 + \frac{\delta}{2})(s - \frac{1}{2}) - 2\delta(s - \frac{1}{2})^3 + 6\delta s^2(1-s)^2(s - \frac{1}{2}),$$

where $\delta = \delta(z) = 4(u_c - z)$. Observe in particular that $f(s, u_c) = s$.

In the Examples 1-3 below the critical value u_c is taken to be $9/20$. With this value of u_c the binodal curve $\mathcal{B}$ and the unit velocity curve $\mathcal{K}$ are plotted in Fig. 5 and the fractional flow function $f(\cdot, z)$, for two different values of z, is plotted in Fig. 6.

Two tie-line models will be used in the experiments. In the first model $u_c = 9/20$ and the function $b = b_1$ is simply given as a linear function from $[u_0, u_c]$ into $[u_c, u_1]$ such that $b_1(u_c) = u_c$ and $b_1(u_0) = u_1$. If these tie-lines are represented

on the form

$$v = \eta(z)u + a(z),$$

the function a have been plotted on Fig. 7. We observe that a is not a monotone function. This means that some of the tie-lines will indeed intersect in the one phase region Φ_1. The model $b = b_1$ for the tie-lines will be used in the Examples 1-3 below.

The second tie-line model is simply the function $b = b_2$ which satisfies the relation

$$\frac{\phi(z)}{z} = \frac{\phi(b_2(z))}{b_2(z)}.$$

Hence, in this case all the tie-lines will intersect at the origin. In Fig. 8 we have plotted different tie-lines in this case. This model for the tie-lines is used in Example 4 below. Approximations of solutions of the model (2.1) will be found from a finite difference approximation. Since all the characteristic speeds are assumed to be positive, the classical upwind scheme is useful. Hence, if Δx denotes the spatial grid size and Δt the time step, the approximations u_j^n, v_j^n of $u(j\Delta x, n\Delta t), v(j\Delta x, n\Delta t)$ are generated by the explicit difference scheme

$$\begin{aligned} u_j^{n+1} &= u_j^n - \mu(F_j^n - F_{j-1}^n), \\ v_j^{n+1} &= v_j^n - \mu(G_j^n - G_{j-1}^n), \end{aligned} \quad (4.1)$$

where $\mu = \Delta t/\Delta x$. We emphasize that it is not the only purpose of this paper to investigate the properties of the finite difference scheme (4.1). We primarily use the scheme as a tool in order to derive properties for the continuous model (2.1). However, we do indeed assume that the difference solutions converge to the desired entropy solution of the hyperbolic system (2.1). This is consistent we known theory and experiments for the upwind scheme, as long as the constant μ is chosen sufficiently small in order to satisfy the CFL-condition. We shall observe below however, that the convergence of the finite difference solutions may be rather slow.

In all the examples below we consider initial value problems for (2.1), or more precisely (4.1).

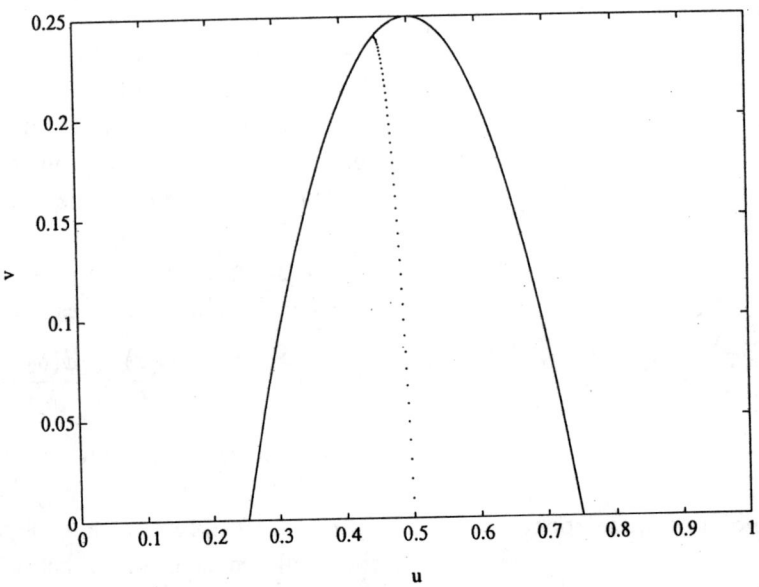

Figure 5: Solid line $\mathcal{B}$, dotted line $\mathcal{K}$

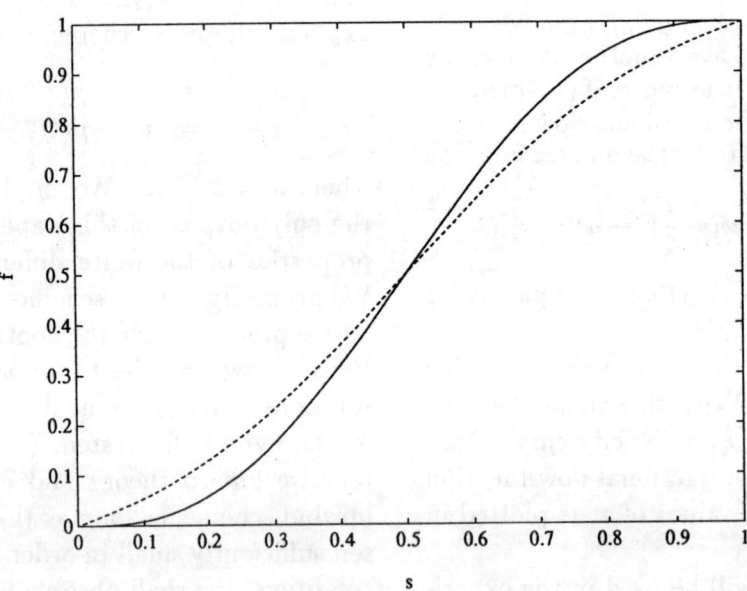

Figure 6: Solid line $z = 0.25$, dotted line $z = 0.35$

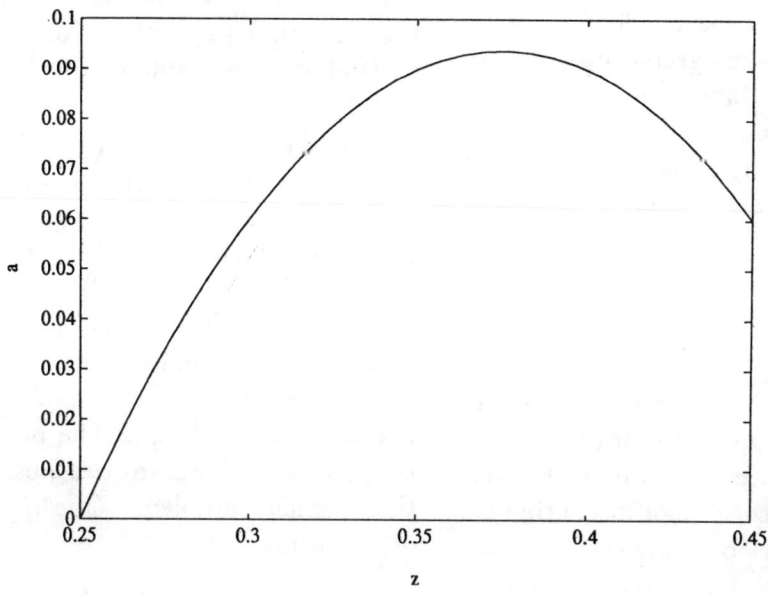

Figure 7:

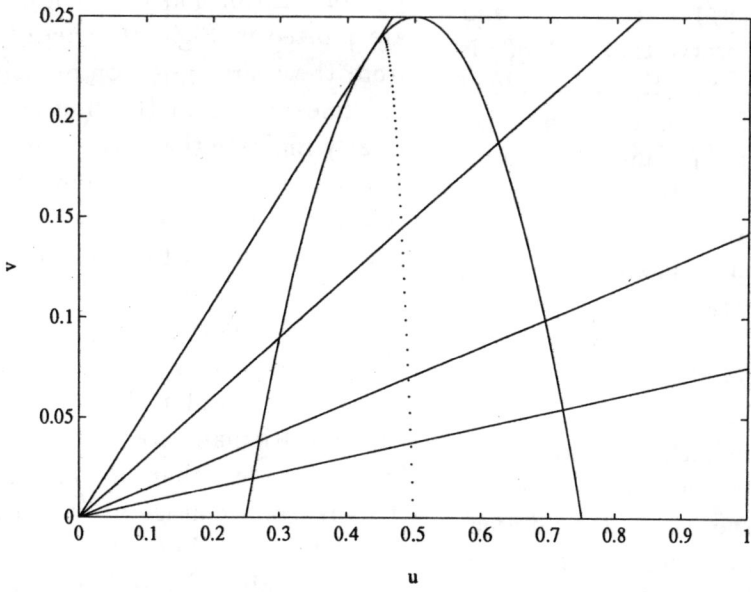

Figure 8:

In particular, if the initial functions consists of two constant states the initial value problem corresponds to a Riemann problem. The calculations will be done on the spatial domain $[0, 1]$ and the equations will be integrated up to time $t = T = 0.2$. The initial functions will be chosen constant close to the endpoints of the spatial domain such that essentially no boundary effects will occur in the calculations. The discretization parameters will be specified by the number of grid intervals $n = 1/\Delta x$ and the number of time steps $m = T/\Delta t$.

Example 1

The purpose of this example is to investigate the solutions of Riemann problems for (2.1) when the left and the right state both are located on the binodal curve $\mathcal{B}$. Hence, a single discontinuity with speed one will in fact be a weak solution of (2.1). The left state will be taken to be $u^L = 0.35$ and $v^L = 0.16$ throughout this example, while we consider three possible right states, (u_i^R, v_i^R) for $i = 1, 2, 3$. For $i = 1$ and $i = 3$ we choose the states where the binodal curve intersects the u-axis, while the third state is selected such that it lies on the same tie-line as (u^L, v^L). This gives $(u_1^R, v_1^R) = (0.25, 0)$, $(u_2^R, v_2^R) = (0.6, 0.21)$ and $(u_3^R, v_3^R) = (0.75, 0)$. We observe that (u_1^R, v_1^R) is located on the binodal curve on the same side of the critical point as the left state, while the critical point separates the left state and (u_i^R, v_i^R) for $i = 2, 3$. Hence, from the analysis given in [7], we expect the Riemann solution to consist of a single discontinuity, with path along the binodal curve, only for the right state (u_1^R, v_1^R). In the other cases we expect the solution path to be located strictly inside the two phase region. In this example the number of grid intervals n and the number of time steps m are both taken to be 500. The results of these experiments are plotted on Fig. 9. The upper left corner shows the different paths in state space. We observe, in particular, that the path from (u^L, v^L) to (u_1^R, v_1^R) is located on the binodal curve. The solutions u (solid line) and v (dotted line) as functions of x are presented on the three other plots. We observe that the ex-

periments confirm our expectations.□

Example 2

In this case we consider Riemann data where the left state $(u^L, v^L) = (0, 0.06)$ will be fixed throughout the example. The left state is chosen such that two tie-lines will intersect at this point. One of these lines are $L(u_c)$ which connects the left state with the critical point, while the other one corresponds to $L(z)$ for $z = 0.3$ (cf. Fig. 7). Four different right states are selected, $(u_1^R, v_1^R) = (0.5, 0)$, $(u_2^R, v_2^R) = (0.4875, 0.1087)$, $(u_3^R, v_3^R) = (0.4700, 0.2064)$ and $(u_4^R, v_4^R) = (u_c, v_c)$. All these right states are points on the unit velocity curve $\mathcal{K}$. Furthermore, the tie-line $L(0.3)$ connects the left state and (u_2^R, v_2^R). The point of this experiment is that there are two possible solutions of the Riemann problem. One possible solution is of the form

$$(u^L, v^L) \to (u_c, v_c) \to (u^R, v^R), \qquad (4.2)$$

while the other possible solution starts with a tie-line wave along $L(0.3)$. We have run the experiments for two grid levels, $n = m = 500$ and $n = m = 1000$. The different paths in state space are plotted in Fig. 10, where the solid lines are from the coarse grid computations and the dotted lines are from the fine grid. These results clearly indicate that, when the right state is different from the critical point, the proper entropy solution is the one which avoids the critical point and starts with a tie-line wave along $L(0.3)$. In Fig. 11 we have plotted the solutions u (solid line) and v (dotted line) as functions of x from the fine grid computation when the right state is (u_2^R, v_2^R). This plot clearly confirms that in this case the Riemann solution consist of waves with speed both less than one and greather than one. This is again consistent with the claim that the solution is a tie-line solution along $L(0.3)$.□

The results of the previous example already shows that the solution of (2.1) is discontinuous in L_1 (or in any other reasonable norm) with respect to the initial data. In order to see this we consider an initial value problem with initial data

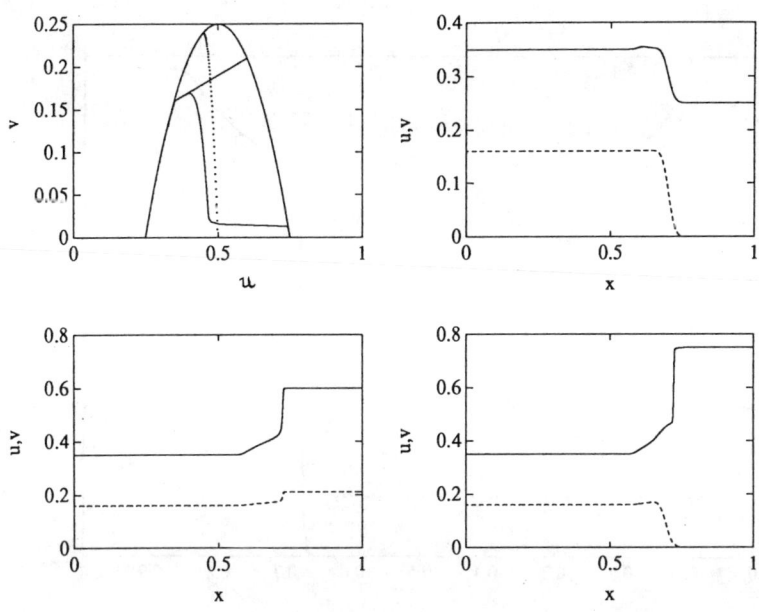

Figure 9: Example 1

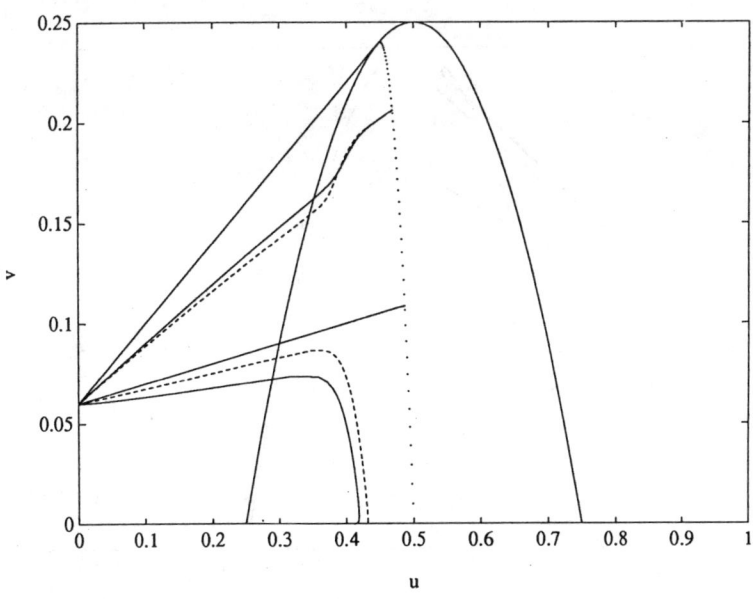

Figure 10: Example 2, paths in state space

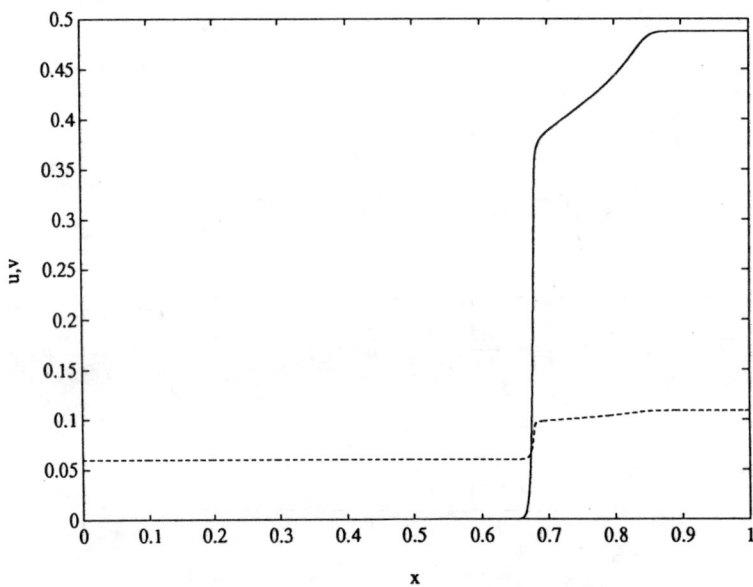

Figure 11: Example 2,u and v as functions of x

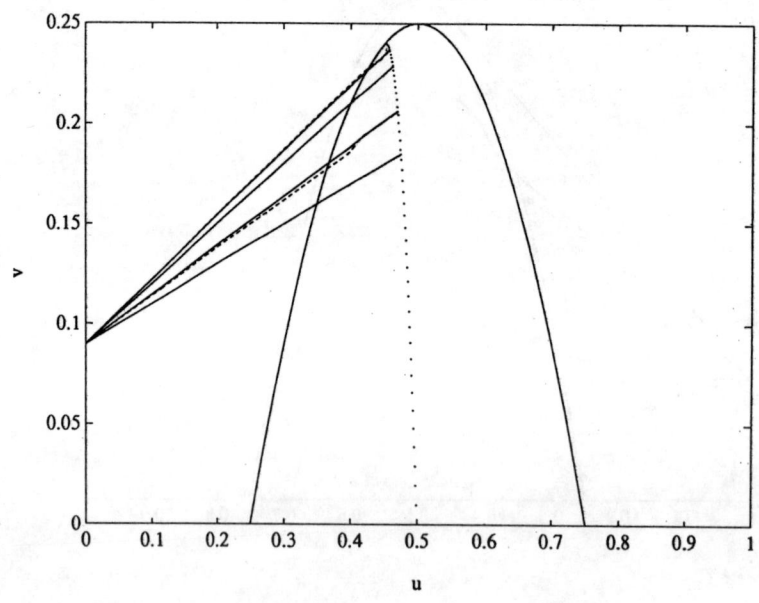

Figure 12: Example 3, paths in state space

of the form

$$(u(x,0),v(x,0)) = \begin{cases} (u^L,v^L) & x \leq A-\epsilon, \\ (u_c,v_c) & A-\epsilon < x < A+\epsilon, \\ (u_2^R,v_2^R) & x \geq A+\epsilon, \end{cases}$$

where A is an arbitrary real number, $\epsilon \geq 0$ and the states (u^L,v^L) and (u_2^R,v_2^R) are as in Example 2 above. The initial discontinuities both correspond to contact discontinuities with speed one. This is demonstrated in Example 2 above for the first discontinuity. The second discontinuity, along the unit velocity curve, can be checked similarly. The solution for $t > 0$ will therfore be given by

$$(u(x,t),v(x,t)) = (u(x-t,0),v(x-t,0)).$$

In particular, as ϵ tends to zero the solution converges in L_1 to a single discontinuity connecting (u^L,v^L) and (u_2^R,v_2^R), with speed one. However, this is not the solution of the initial value problem when $\epsilon = 0$. As observed in Example 2 this Riemann problem has a solution which consists of tie-line waves along the tie-line $L(0.3)$. Since this solution consists of waves with speed both less and greater than zero, this demonstrates that in general the solution of (2.1) is not continuous with respect to the initial data.

In the next example we will show that discontinuity phenomena of the form discussed above even can occur for Riemann problems.

Example 3

In this example we consider Riemann problems with a fixed left state $(u^L,v^L) = (0,0.09)$. This state corresponds to a point in state space where two tie-lines, $L(0.35)$ and $L(0.4)$, intersect. We choose four different right states on the, $(u_i^R,v_i^R), i = 1,2,3,4$, on the unit velocity curve $\mathcal{K}$. These states are given by $(u_1^R,v_1^R) = (0.4750,0.1850), (u_2^R,v_2^R) = (0.4700,0.2064), (u_3^R,v_3^R) = (0.4625,0.2288)$ and $(u_4^R,v_4^R) = (0.4575,0.2371)$. Here (u_1^R,v_1^R) and (u_3^R,v_3^R) are located on the tie-lines $L(0.35)$ and $L(0.4)$, respectively. Experiments are done with two grid levels, $n = m = 500$ and $n = m = 5000$. The different paths in state space are given in

Fig. 12, where the solid lines are from the coarse grid calculations and the dotted lines are from the fine grid. We observe that for the right states (u_1^R,v_1^R) and (u_3^R,v_3^R) the dotted lines are invisible, since they are located on top of the solid lines. This clearly indicates that the Riemann solution in these cases only consists of tie-line waves along one single tie line. Furthermore, the solutions for (u_2^R,v_2^R) seem to converge to a solution path which starts along $L(0.35)$. In order to get a better representation of the two paths corresponding to the right state (u_4^R,v_4^R) an enlarged version of the plot around this point is given Fig. 13. This plot clearly indicates that when the right state is given by (u_4^R,v_4^R) the proper Riemann solution is of the form (4.2). However, these calculations also indicate that the convergence of the finite difference to this limit solution is rather slow. The observed solution reflects strongly the properties of the finite difference scheme; i.e., it is strongly influenced by numerical diffusion.

From the discussion given in [7] it is reasonable to believe that the proper Riemann solution is of the form (4.2) for all right states on the unit velocity curve above the state (u_3^R,v_3^R). This means that in all these cases the desired Riemann solutions correspond to a single discontinuity with speed one. However, the limit problem with right state (u_3^R,v_3^R), which has a solution consisting of tie-line waves along $L(0.4)$, has waves with speed both less and greater than one. We have therfore again demostrated a lack of continuous dependence of the solution in L_1 with respect to the initial data.$\square$

In the Examples 1-3 above we have used the tie-line model, $b = b_1$. In Example 4 we use the second tie-line model, $b = b_2$; i.e., all the the tie-lines intersect in the origin.

Example 4

In this example we solve the Riemann problem with $(u^L,v^L) = (0,v_c)$ and $(u^R,v^R) = (1,0)$. We observe that both states are located in the one phase region Φ_1. Furthermore, the left state is located above all the tie-lines. Observe however, that the straight line from (u^L,v^L) to (u^R,v^R)

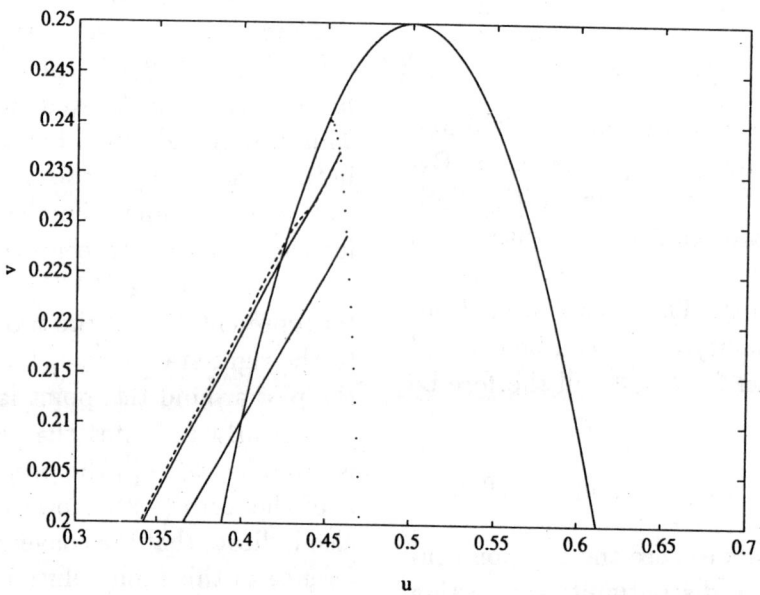

Figure 13: Example 3, enlarged plot

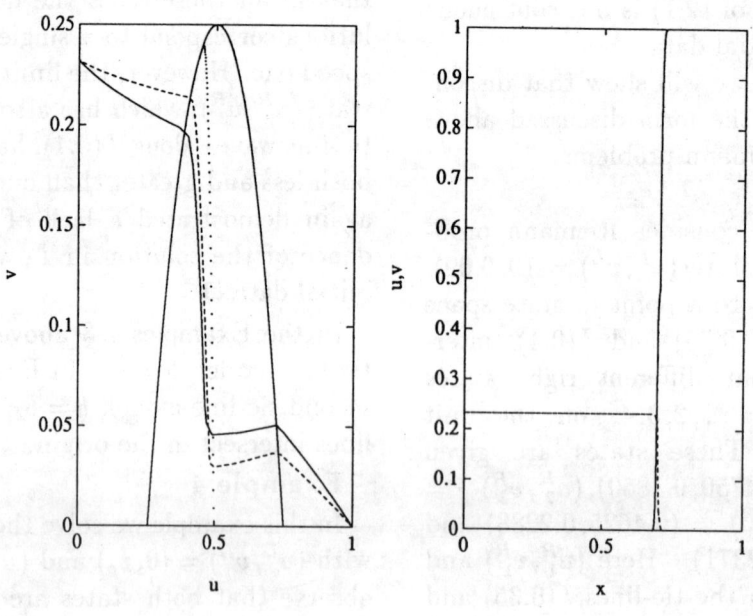

Figure 14: Example 4

passes through the two phase region Φ_2. The purpose of this experiment is to decide if the solution is a pure one phase solution which consists of a single discontinuity with speed one, or if the solution enters the two phase region and more complicated waves are created. This case corresponds to a multiple contact miscible displacement, where components condense from the injected fluid into the displaced oil (condensing gas drive). The experiment is done on a coarse grid, $n = m = 1000$, and on a fine grid, $n = m = 5000$. The two paths in state space are plotted on the left side of Fig. 14, where the solid line and the dotted line represent the paths from the coarse grid and the fine grid, respectively. The solutions u (solid line) and v (dotted line) as functions of x are plotted on the right side of Fig. 14.

The result seems to indicate that the proper limit solution is indeed a single jump from (u^L, v^L) to (u^R, v^R) with speed one. However, still the solution path enters the two phase region. The paths seem to converge to the Riemann solution

$$(u^L, v^L) \to (u_c, v_c) \to (1/2, 0) \to (u^R, v^R), \quad (4.3)$$

which consist of two contact discontinuities of speed one and one proper tie-line shock. We remark that the solution of the form (4.3) is predicted in the literature as the correct physical solution in this case (cf. for example Lake [12]).$\square$

5 Conclusion

Our experiments clearly indicate that the system (2.1) is not a well-posed model. Furthermore, the computations show that the convergence of a standard difference scheme is rather slow in certain regimes of initial data, and that the results obtained from calculations with such diffusive schemes may be more dependent of the scheme than of the properties of the continuous model. It is therefore reasonable to believe that, in these regimes, the solution will also be very sensitive to terms modelling physical diffusion. It

therefore seems to be necessary to include such terms in the model in order to get more proper results. Furthermore, in order to get accurate results from calculations in two or three space dimensions, it seems to be necessary to use numerical schemes where the influence of numerical diffusion is negligible.

References

[1] T. Barkve, The Riemann problem for a non-strictly hyperbolic system modelling non-iso-thermal, two-phase flow in a porous media, SIAM J. Appl. Math. 49 (1989), 784-798.

[2] O. Dahl, T. Johansen, A. Tveito, R. Winther, Multicomponent chromatography in a two phase environment, Preprint, Institute of Informatics, The University of Oslo, 1990.

[3] J.M. Dumore, J. Hagoort, A.S. Risseeuw, An analytical model for one dimensional three component and vaporizing gas drives, Soc. Pet. Eng. J. 24 (1984), 169-179.

[4] F.G. Helfferich, General theory of multicomponent, multiphase displacement in porous media, Soc. Pet. Eng. J. 21 (1981), 51-62.

[5] G.J. Hirasaki, Application of the theory of multicomponent, multiphase displacement to three-component, two-phase surfactant flooding, Soc. Pet. Eng. J. 21 (1981), 191-204.

[6] E. Isaacson, Global solution of a Riemann problem for a non-strictly hyperbolic system of conservation laws arising in enhanced oil recovery, Preprint, Rockefeller University.

[7] T. Johansen, The Riemann problem for a three component fluid system involving phase transition, Preprint, IFE 1990.

[8] T. Johansen, A. Tveito, R. Winther, A Riemann solver for a two-phase multicomponent process, SIAM J. Sci. Stat. Comp. 10 (1989), 846-879.

[9] T. Johansen, R. Winther, The solution of the Riemann problem for a hyperbolic system of conservation laws modelling polymer-flooding, SIAM J. Math. Anal. 19 (1988), 541-566.

[10] T. Johansen, R. Winther, The Riemann problem for multicomponent polymer flooding, SIAM J. Math. Anal. 20 (1989), 908-929.

[11] B. Keyfitz, H. Kranzer, A system of non-strictly hyperbolic conservation laws arising in elasticity theory, Arch. Rational Mech. Anal. 72 (1980), 219-241.

[12] L.W. Lake, Enhanced oil recovery, Prentice Hall (1989).

[13] R.G. Larson, The influence of phase behavior on surfactant flooding, Soc. Pet. Eng. J. 19 (1979), 411-422.

[14] W.W. Monroe, M.K. Silva, L.L. Larson, F.M. Orr, Composition paths in four-component systems: The effect of dissolved methane on CO_2 flood performance in one dimension, SPE 16712.

[15] F.M. Orr, Miscible flooding, SPE Lecture notes 1989.

[16] K.K. Pande, F.M.Orr, Interaction of phase behavior, reservoir heterogeneity and cross-flow in CO_2 floods, SPE 19668.

[17] K.K. Pande, F.M. Orr, Analytical computations of breakthrough recovery for CO_2 floods in layered reservoirs, SPE/DOE 20117.

[18] M.P. Walsh, L. W. Lake, Applying fractional flow theory to solvent flooding and chase fluids, Preprint.

[19] H.J. Welge et. al., The linear displacement of oil from porous media by enriched gas, J. Pet. Tech. 3 (1961), 787-796.

Gridding

2nd European Conference on the Mathematics of Oil Recovery
© D. Guérillot, O. Guillon (Editors) and Éditions Technip, Paris 1990, pp. 237-245
27 rue Ginoux, 75015 Paris

Mixed Methods, Operator Splitting, and Local Refinement Techniques for Simulation on Irregular Grids

M. S. Espedal[1], R. E. Ewing[2] and T. F. Russell[3]

ABSTRACT

The partial differential equations used to model multiphase and multicomponent fluid flows are convection-dominated, with important local properties. Operator-splitting techniques have been defined to address these different phenomena. Convection is treated by time stepping along the characteristics of the associated pure convection problem and diffusion is modeled via a Galerkin method for miscible displacement and a Petrov-Galerkin method for immiscible displacement. These ideas have been generalized to Eulerian-Lagrangian Localized Adjoint Method (ELLAM) formulations which conserve mass and allow more accurate treatment of boundary conditions. Accurate approximations of the fluid velocities needed in the characteristic time stepping are obtained by mixed finite-element methods.

Irregular grids can be extremely useful in treating faulting, dipping strata, pinchouts, and other complex lithological geometries. Corner-point geometry, which has been proposed for some of these problems, can yield inconsistencies, mass-balance errors, and other inaccuracies in simulators. Instead, local grid refinement can be introduced in an accurate manner to resolve the localized phenomena. Local grid refinement techniques, which closely resemble block-centered finite difference schemes, are discussed to approximate the pressure variable or the Darcy velocity directly via mixed finite-element methods. Although the composite-grid matrices then no longer have a banded structure, domain decomposition techniques can be used to obtain very efficient preconditioners for the composite-grid problem. Local time stepping can also be achieved using similar block preconditioning methods.

1. INTRODUCTION

The simulation of multiphase or multicomponent fluid flow often requires the solution of large, coupled systems of nonlinear partial differential equations of convection-diffusion type. The equations are usually convection-dominated, with small but important local diffusive effects. The partial differential equations that describe the component and phase balances are coupled to a system of equations for the total fluid pressure and velocity. Examples of such coupled systems appear in Section 2. Mixed finite element procedures, discussed in Section 3, are used to approximate the fluid velocities accurately even in the presence of heterogeneous porous media. These velocities are critical in the accurate treatment of the convection-dominated transport process.

In complex flow processes, two or more fluids can flow in an immiscible, or non-mixing, fashion at certain times and in a miscible mode, where mixing of the fluids takes place, at other times. For this reason, useful numerical techniques for multiphase or multicomponent reservoir simulators should be capable of treating both miscible and immiscible displacement phenomena.

Standard techniques for treating convective flow in the petroleum industry involve upstream weighting methods that produce artificial numerical dispersion of the order of the grid size and in directions parallel to the grid orientation. The effects of this dispersion when coarse grids are used for field-scale simulations have been quite serious. Techniques to control the numerical dispersion in traditional finite difference simulators have met with limited success. On the other hand, the combination of mixed finite element techniques and characteristic

(1) Department of Applied Mathematics, University of Bergen, Allegaten 55, 5007 Bergen, Norway.
(2) Enhanced Oil Recovery Institute, University of Wyoming, PO Box 3036, University Station, Laramie, Wyoming 82071, USA.
(3) Computational Mathematics Group, University of Colorado, Box 170, Denver, Colorado 80204, USA.

time stepping methods to follow the fluid flow has been extremely successful in essentially eliminating numerical dispersion and grid-orientation effects [30,39].

In order to retain the local diffusive/dispersive terms, which are critical to the physical processes [38], while accurately treating the dominant convective terms, operator-splitting techniques based on modified method of characteristics ideas have been developed [13,14,30,31,34,35,37]. These methods have been very effective in both miscible [15,30–32,34,35,37,39] and immiscible [7–10,14,15] displacement problems, and can span these regimes well. Recently, new Eulerian-Lagrangian Localized Adjoint Methods (ELLAM) have been developed [5,36] that conserve mass and which treat inflow and outflow boundary conditions in a consistent and accurate fashion. Extensions of these ideas have been made to multiple dimensions [37] and to multiphase flow [18]. These operator-splitting concepts are detailed in Section 4.

Although these characteristic methods greatly reduce the artificial numerical dispersion inherent in many upstream weighting techniques, they must be combined with adaptive local grid refinement methods if sharp, moving fluid interfaces that contain the essential physical interfacial interaction and mass transfer are to be resolved. We thus discuss accurate and efficient local grid refinement techniques for both static and dynamic applications in this paper.

Most petroleum engineering problems involve both macroscopic flow processes and highly localized phenomena that are often critical to the overall chemical and physical behavior of the flows. For field-scale applications, it is currently impossible to use a uniform grid that is sufficiently fine to resolve the important local phenomena. This resolution must be achieved either directly, by techniques of local grid refinement such as those described below, or indirectly, by the use of effective macroscopic parameters that take account of localized processes in an averaged manner. A study of the indirect approach, using the methods of Sections 3 and 4 in the context of miscible flow, has appeared in the petroleum literature [32]. Local grid refinement will be a vital tool in systematically deriving such effective parameters via simulations on various scales, so it is an integral part of either approach.

Although flow properties around features such as wells, faults, or pinchouts are fixed in space and can be treated via static refinements, the need to resolve the phase interactions along moving interfacial boundaries requires adaptive grid refinement methods. The local grid structure introduces more complicated communication links between the grid cells, thus destroying the banded structure of the solution matrices and their associated efficiency of solution on vector computers.

Domain decomposition techniques have been developed for local grid refinement applications [3,14,16,22] that retain the matrix solution for the coarse, quasi-regular grid as part of a preconditioner for the solution process on the composite grid obtained through local refinements. The block refinement strategy described in Section 5 allows efficient solution via block preconditioners for both static and dynamic grid refinement. Since many local phenomena are also localized in time, adaptive local time-stepping procedures that utilize domain decomposition techniques [21] are also described in Section 5.

2. EQUATIONS OF FLOW

The displacement of one incompressible fluid by another, completely miscible with the first, in a horizontal porous reservoir $\Omega \subset \mathbb{R}^2$ over a period $J = [T_0, T_1]$, is given by [11,12]

$$-\nabla \cdot \left(\frac{k}{\mu} \nabla p \right) \equiv \nabla \cdot \mathbf{u} = q, \qquad \mathbf{x} \in \Omega, \ t \in J, \quad (2.1)$$

$$\phi \frac{\partial c}{\partial t} - \nabla \cdot (\mathbf{D} \nabla c - \mathbf{u} c) = q\tilde{c}, \qquad \mathbf{x} \in \Omega, \ t \in J, \quad (2.2)$$

where p and $\mathbf{u}$ are the pressure and Darcy velocity of the fluid mixture, ϕ and k are the porosity and the permeability of the medium, μ is the concentration-dependent viscosity of the mixture, c is the concentration of the invading fluid, q is the external rate of flow, and $\tilde{c}$ is the inlet or outlet concentration. $\mathbf{D}$ is, in general, a diffusion-dispersion tensor which has two parts, molecular diffusion and a velocity-dependent dispersion term. Also, the viscosity μ in (2.1) is assumed to be determined by some mixing rule. In addition to (2.1) and (2.2), initial and no-flow boundary conditions are specified. The flow at injection and production wells is modeled in (2.1) and (2.2) via point sources and sinks.

The equations describing two-phase, immiscible, incompressible displacement in a horizontal porous medium are given by [6,14]

$$\phi \frac{\partial S_w}{\partial t} - \nabla \cdot \left(k \frac{k_{rw}}{\mu_w} \nabla p_w \right) = q_w, \qquad \mathbf{x} \in \Omega, \ t \in J, \quad (2.3)$$

$$\phi \frac{\partial S_o}{\partial t} - \nabla \cdot \left(k \frac{k_{ro}}{\mu_o} \nabla p_o \right) = q_o, \qquad \mathbf{x} \in \Omega, \ t \in J, \quad (2.4)$$

where the subscripts w and o refer to water and oil, respectively. S_i is the saturation, p_i is the pressure, k_{ri} is the relative permeability, μ_i is the viscosity, and q_i is the external flow rate, each with respect to phase i, $i = o$ or w.

The saturation constraint is given by summing the saturations to unity. Let $S = S_w = 1 - S_o$. The difference between the two phase pressures is described by a capillary pressure relationship.

Although formally the equations presented for miscible flow seem quite different from those in (2.3) and (2.4), the latter system may be rearranged in a form which very closely resembles the former system. In order to use the same basic simulator in sample computations to treat both miscible and immiscible displacement, we will briefly discuss a miscible/immiscible flow analogy.

Adding Equations (2.3) and (2.4) and performing some simple calculations, we obtain [6,14]

$$-\nabla \cdot (k\lambda(S)\nabla p) = q_w + q_o = q_t, \qquad (2.5)$$

$$\mathbf{v}_t = -k\lambda(S)\nabla p, \qquad (2.6)$$

where $k\lambda$ is the transmissibility, p is a global pressure of the fluid [6], and $\mathbf{v}_t$ is the corresponding fluid velocity. Taking the difference of Equations (2.3) and (2.4), the following equation is obtained:

$$\phi\frac{\partial S}{\partial t} - \nabla \cdot (D\nabla S) + \nabla \cdot (\overline{\lambda}_w \mathbf{v}_t) = q_w. \qquad (2.7)$$

Here the capillary diffusion term D and the fractional flow $\overline{\lambda}_w \mathbf{v}_t$ are defined in terms of relative permeabilities and capillary pressure terms as in [7–10,14].

With a special form of relative permeabilities and capillary pressure, the viscosity dependence in pressure equation (2.5) is the same as that specified for concentration dependence in (2.1). Therefore, $\mathbf{v}_o + \mathbf{v}_w$ (i.e., $\mathbf{v}_t$), solved from (2.5) and (2.6), will be identified to $\mathbf{u}$ in (2.1). Moreover, (2.7) has the same form as the concentration equation (2.2) if S is interpreted as a component concentration and q_w is equal to $\tilde{S}q_t$, where $\tilde{S}$ is the inlet or outlet saturation. This establishes the analogy between the two systems.

3. MIXED METHODS FOR ACCURATE VELOCITY APPROXIMATIONS

There are two major sources of error in the methods currently being utilized for the finite difference discretization of (2.5)–(2.7). The first occurs in the approximation of the fluid pressure and velocity. The second comes from the techniques for upstream weighting to stabilize (2.7). We first describe mixed finite element methods for the accurate approximation of the total velocity $\mathbf{u}$. We then discuss some alternatives to upstream weighting techniques in a finite element context in the next section for use in Equation (2.7).

Since the diffusion-dispersion and convection terms in (2.7) are governed by the fluid velocity, accurate simulation requires an accurate approximation of the velocity $\mathbf{u}$. Because the lithology in the reservoir can change abruptly, causing rapid changes in the flow capabilities of the rock, the coefficient k in (2.5) and (2.7) can be discontinuous. In this case, in order for the flow to remain relatively smooth, the pressure gradient changes extremely rapidly. Thus, standard procedures of solving (2.5) as an elliptic partial differential equation for pressure, differentiating or differencing the result to approximate the pressure gradient, and then multiplying by the discontinuous $k\lambda$ can produce very poor approximations to the velocity $\mathbf{u}$. In this section, a mixed finite element method for approximating $\mathbf{u}$ and p simultaneously, via a coupled system of first-order partial differential equations, will be discussed. This formulation allows the removal of singular terms in the equations [33] and accurately treats the problem of rapidly changing flow properties in the reservoir.

The coupled system of first-order equations used to define our methods arises from Darcy's Law and conservation of mass,

$$\mathbf{u} = -\frac{k}{\mu}\nabla p, \qquad \mathbf{x} \in \Omega, \qquad (3.1)$$

$$\nabla \cdot \mathbf{u} = q, \qquad \mathbf{x} \in \Omega, \qquad (3.2)$$

subject to the boundary condition

$$\mathbf{u} \cdot \mathbf{n} = 0, \qquad \mathbf{x} \in \partial\Omega. \qquad (3.3)$$

Clearly, in the incompressible case presented here, (3.1)–(3.3) will determine p only to within an additive constant. Thus, a normalizing constraint such as $\int_\Omega p(\mathbf{x})dx = 0$ or $p(\mathbf{x}_s) = 0$ for some $\mathbf{x}_s \in \Omega$ is required in the computation to prevent a singular system.

We next define certain function spaces and notation. Let $W = L^2(\Omega)$ be the set of all functions on Ω whose square is integrable. Let $H(\text{div}; \Omega)$ be the set of vector functions $\mathbf{v} \in \left[L^2(\Omega)\right]^2$ such that such $\nabla \cdot \mathbf{v} \in L^2(\Omega)$ and let

$$V = H(\text{div}; \Omega) \cap \{\mathbf{v} \cdot \mathbf{n} = 0 \text{ on } \partial\Omega\}. \qquad (3.4)$$

Let $(v, w) = \int_\Omega vw \, dx$ be the standard L^2 inner product on Ω. We obtain the weak solution form of (3.1)–(3.3) by dividing each side of (3.1) by k/μ, multiplying by a test function $\mathbf{v} \in V$, and integrating the result to obtain

$$\left(\frac{\mu}{k}\mathbf{u}, \mathbf{v}\right) = (p, \nabla \cdot \mathbf{v}), \qquad \mathbf{v} \in V. \qquad (3.5)$$

The right-hand side of (3.5) was obtained by further integration by parts and use of (3.4). Next, multiplying (3.2) by $w \in W$ and integrating the result, we complete our weak formulation, obtaining

$$(\nabla \cdot \mathbf{u}, w) = (q, w), \qquad w \in W. \qquad (3.6)$$

239

For a sequence of mesh parameters $h > 0$, we choose finite-dimensional subspaces V_h and W_h with $V_h \subset V$ and $W_h \subset W$ and seek a solution pair $(\mathbf{U}_h; P_h) \in V_h \times W_h$ satisfying

$$\left(\frac{\mu}{k} \mathbf{U}_h, \mathbf{v}_h \right) - (P_h, \mathrm{div}\, \mathbf{v}_h) = 0, \quad \mathbf{v}_h \in V_h, \quad (3.7)$$

$$(\mathrm{div}\, \mathbf{U}_h, w_h) = (q, w_h), \qquad w_h \in W_h. \quad (3.8)$$

We can now complete the description of our mixed finite element methods with a discussion of particular choices of V_h and W_h. Examples of these spaces are presented in [33].

For problems with smooth coefficients and smooth forcing functions, standard approximation theory results show that, by using basis functions of increasing polynomial degrees, correspondingly higher-order convergence rates can be obtained [12]. Superconvergence results for these methods have been obtained on certain loci in 2-D problems in [29]. For the fluid flow in porous media applications, the source and sink terms q are not smoothly distributed, but are sums of Dirac delta functions. As shown by Ewing and Wheeler [33], the resulting smoothness of $\mathbf{u}$ is reduced; $\mathbf{u}$ is not contained in the space L^2 and thus using the mixed methods, the velocity approximations *would not converge at the wells*. By removing the leading term of the singularities (the logarithm terms) from the linear equation (3.8) for the velocity, the remaining parts of the velocities are now in the Sobolev space $H^{2-\epsilon}$ for any $\epsilon > 0$. Thus the approximations to these parts will now converge at the wells since we have regained sufficient regularity for convergence.

We use mixed method techniques for accurately approximating the total fluid velocity $\mathbf{u}$. The transport equation (2.7) requires a phase velocity in regimes where two-phase flow occurs. In standard finite difference codes, the total velocity is partitioned among the phases by fractional flows that depend on upstream-weighted relative permeabilities. Ewing and Heinemann [20] discussed finite element weightings of the pressure obtained from standard finite difference codes for compositional problems; these resulted in significantly better phase velocities than via standard upstream weighting methods.

4. OPERATOR-SPLITTING TECHNIQUES

In finite difference simulators, the convection is stabilized via upstream weighting techniques. In a finite element setting, we use a possible combination of a modified method of characteristics and Petrov-Galerkin techniques to treat the transport separately in an operator-splitting mode. Analogous concepts can be formulated for finite differences as well [13].

In miscible or multicomponent flow models, the convective, hyperbolic part is a linear function of the velocity. An operator-splitting technique has been developed to solve the purely hyperbolic part by time stepping along the associated characteristics [13,30,34]. We first obtain the non-divergence form of (2.2) by using the product rule for differentiation on the $\nabla \cdot \mathbf{u} c$ term and applying (2.1) to obtain

$$\phi \frac{\partial c}{\partial t} + \mathbf{u} \cdot \nabla c - \nabla \cdot \mathbf{D} \nabla c = q(\tilde{c} - c). \quad (4.1)$$

Next, the first and second terms in (4.1) are combined to form a directional derivative along what would be the characteristics for the equation if the tensor $\mathbf{D}$ were zero. The resulting equation is

$$\nabla \cdot (\mathbf{D} \nabla c) + q(\tilde{c} - c) = \phi \frac{\partial c}{\partial t} + \mathbf{u} \cdot \nabla c \equiv \phi \frac{\partial c}{\partial \tau}. \quad (4.2)$$

The system obtained by modifying (2.1)–(2.2) in this way is solved sequentially. An approximation for $\mathbf{u}$ is first obtained at time level $t = t^n$ from a solution of (2.1) with the fluid viscosity μ evaluated via some mixing rule at time level t^{n-1}. Equation (2.1) can be solved as an elliptic equation for the pressure p, or via a mixed finite element method for a more accurate fluid velocity. Let $C^n(x)$ and $\mathbf{U}^n(x)$ denote the approximations of $c(x,t)$ and $\mathbf{u}(x,t)$, respectively, at time level $t = t^n$. The directional derivative is then discretized along the "characteristic" mentioned above as

$$\phi \frac{\partial c}{\partial \tau}(x, t^n) \approx \phi \frac{C^n(c) - C^{n-1}(\overline{x}^{n-1})}{\Delta t} \quad (4.3)$$

where $\overline{x}^{n-1}$ is defined for an x as

$$\overline{x}^{n-1} = x - \frac{\mathbf{U}^n(x) \Delta t}{\phi}. \quad (4.4)$$

This technique, first described for petroleum applications by Russell [35], is a discretization back along the "characteristic" generated by the first order derivatives from (4.2). Although the advection-dominance in the original Equation (4.2) makes it non-self-adjoint, the form with directional derivatives is self-adjoint and discretization techniques for self-adjoint equations can be utilized. This modified method of characteristics (MMOC) can be combined with either finite difference or finite element Galerkin spatial discretizations. If the "characteristic" is significantly curved, as it is in regions where the velocity changes rapidly, such as vicinities of wells, it can be tracked more carefully than Equation (4.4) suggests [30].

In immiscible or multiphase flow, the convective part is nonlinear. A similar operator-splitting technique to solve this equation needs reduced time steps because the pure hyperbolic part may develop shocks. Operator-splitting techniques have been developed for immiscible flows [7–10,14] which retain the long time steps in the characteristic solution without introducing serious discretization errors.

The operator splitting gives the following set of equations:

$$\phi\frac{\partial \overline{S}}{\partial t} + \frac{d}{dS}\mathbf{f}^m(\overline{S}) \cdot \nabla \overline{S} \equiv \phi\frac{d}{d\tau}\overline{S} = 0, \qquad (4.5)$$

$$\phi\frac{\partial S}{\partial \tau} + \nabla \cdot (\mathbf{b}^m(S)S) - \epsilon\nabla \cdot (D(S)\nabla S) = \mathbf{q}(\mathbf{x},t), \quad (4.6)$$

$t_m \leq t \leq t_{m+1}$, together with proper initial and boundary conditions. As noted earlier, the saturation S is coupled to the pressure/velocity equations, which will be solved by mixed finite element methods [12,30,31].

The splitting of the fractional flow function into two parts: $\mathbf{f}^m(S) + \mathbf{b}(S)S$, is constructed [14] such that $\mathbf{f}^m(S)$ is linear in the shock region, $0 \leq S \leq S_1 < 1$, and $\mathbf{b}(S) \equiv 0$ for $S_1 \leq S \leq 1$. Further, Equation (4.5) produces the same unique physical solution as

$$\frac{\partial S}{\partial t} + \nabla \cdot (\mathbf{f}^m(S) + \mathbf{b}(S)S) = 0 \qquad (4.7)$$

with an entropy condition imposed. This means that, for a fully developed shock, the characteristic solution of (4.5) always will produce a unique solution and, as in the miscible case, we may use long time steps Δt without loss of accuracy.

The solution of (4.6) via variational methods leads to Petrov-Galerkin equations as described in [7–10,14]. Localized test functions are defined which approximately symmetrize the bilinear form generated by (4.6). Explicit forms of these test functions are presented in [7–10]. As is discussed in [5], these methods are closely associated with localized adjoint methods or optimal test function schemes.

The modified method of characteristics techniques described above generally do not conserve mass. Also, the proper method for treating boundary conditions in a conservative and accurate manner using these techniques is not obvious. Recently, Celia, Russell, Herrera, and Ewing have devised Eulerian-Lagrangian localized adjoint methods (ELLAM) [5], a set of schemes that extend the characteristic methods and are defined expressly to address these issues. In particular, ELLAM schemes apply naturally to the conservative form (2.2) of the transport equation and do not need to pass to the nondivergence form (4.1) [37]. See [18,36,37] for extensions of these methods. The treatment of the multiphase flow case in [18] closely parallels the operator splitting from [7–10,14].

Although the ELLAM or MMOC schemes approximate the flow well without introducing the damaging truncation errors of upstream-weighting methods, they still cannot predict accurate transport if the important local physics in the moving fluid interface regions is not resolved. Specifically, if h represents the grid size, upstream weighting can only represent accurately fronts of width at least $O(h^{1/2})$, while characteristic methods can handle the desired $O(h)$, which in practice is about $3h$ [39]. Since the physics is often highly localized, these methods must be combined with dynamic adaptive local grid refinement techniques.

5. LOCAL GRID REFINEMENT STRATEGY

Many types of localized phenomena arise in petroleum simulation applications. Some are fixed in space and require only fixed, static refinement. Others, such as moving fluid interfaces, require moving, dynamic grid refinement techniques. Often the local properties arise from temporal transients which may require local time-stepping schemes. Each of these applications is discussed in this section.

A typical example of a fixed localized phenomenon which requires special treatment in simulation is fluid flow in the neighborhood of wells. Where fluid flow rates are specified at injection or production wells, the use of Dirac delta functions as point sources and sinks in the mathematical equations has been shown to be a good model for well-flow behavior beyond some minimal distance away from the wells. In this case, the pressure (which determines the flow) grows like $\ln r$ where r is the distance to that well. A different well model, involving specification of a bottom-hole pressure as a boundary condition, also gives rise to a logarithmic growth in pressure up to a finite specified pressure. Because of the rapidly changing behavior of the pressure in the vicinity of wells, accurate pressure approximations require some type of local treatment.

We will consider a simple example problem to illustrate our local refinement techniques. We want to approximate the pressure p of the fluid described by (2.1).

First, we consider the matrix A^c, generated by a finite element or finite difference approximation of (2.1) using a coarse quasi-uniform mesh. Let the solution P of the original coarse-grid problem be decomposed in the form $P = (P_1, P_2)^T$, where P_1 and P_2 are the parts of the coarse-grid solution in two separate domains Ω_1 and Ω_2, respectively. The corresponding decomposition of the matrix A^c can be described in

$$A^c \begin{pmatrix} P_1 \\ P_2 \end{pmatrix} = \begin{pmatrix} A_{11}^c & A_{12}^c \\ A_{21}^c & A_{22}^c \end{pmatrix} \begin{pmatrix} P_1 \\ P_2 \end{pmatrix}. \qquad (5.1)$$

We assume that a code exists or can be easily written to solve (5.1) for a quasi-uniform grid. Such a code can be highly vectorized to take advantage of the banded structure of the matrix $\overline{A}^c$, which is equivalent to A^c

except that it utilizes a standard lexicographical ordering of the unknowns.

Next, assume that due to some identified localized process, grid refinement is desired in Ω_2. Let P_r be the new approximation on the refined grid in Ω_2 and A_{rr} be the local matrix on Ω_2. Let A_{1r} and A_{r1} be the new connection matrices between Ω_1 and the refined grid on Ω_2. We can write the composite matrix problem in the form

$$\tilde{A}\tilde{P} = \begin{pmatrix} A_{11}^c & A_{1r} \\ A_{r1} & A_{rr} \end{pmatrix} \begin{pmatrix} P_1 \\ P_r \end{pmatrix} = \begin{pmatrix} f_1 \\ f_2 \end{pmatrix}. \quad (5.2)$$

Bramble *et al.* [3] defined a preconditioned conjugate gradient iterative procedure for the efficient solution of the composite problem (5.2). The preconditioner involves two local solutions on the refined grid in Ω_2 and one on the original coarse grid in Ω and yields a symmetric preconditioner. Below, we will describe a simpler iterative process which utilizes similar inversions on the refined grid Ω_2 and the coarse grid Ω.

Given a previous iterate for P_1, denoted P_1^n, we solve the local problem on Ω_2 with Dirichlet conditions on the interface between Ω_1 and Ω_2 (given by $A_{r1}P_1$):

$$P_r^n = A_{rr}^{-1}(f_2 - A_{r1}P_1^n). \quad (5.3)$$

This problem can be solved exactly or approximately by some iterative technique. This step could be considered as the first part of a block Gauss-Seidel iterative procedure for the solution of (5.2). The next step would be to use the approximation for P_r^n and then invert A_{11}^c to obtain an approximation for P_1^n. Since this block involves a complex region and may not be well-conditioned, we use an alternative solution method which involves a preconditioner, denoted by B, for the composite matrix $\tilde{A}$.

Using B, we define, for each iterate n and an iteration parameter τ,

$$\tilde{P}^{n+1} = \tilde{P}^n + \tau B^{-1}(\tilde{f} - \tilde{A}\tilde{P}^n). \quad (5.4)$$

Let Q be the residual vector given by

$$\tilde{f} - \tilde{A}\tilde{P}^n = \begin{pmatrix} f_1 - A_{11}^c P_1^n - A_{1r}P_r^n \\ f_2 - A_{r1}P_1^n - A_{rr}P_r^n \end{pmatrix} \equiv \begin{pmatrix} Q_1^n \\ Q_2^n \end{pmatrix} \quad (5.5)$$

Next, we solve the original coarse-grid problem with a perturbed right-hand side:

$$A^c \begin{pmatrix} W_1^{n+1} \\ W_2^{n+1} \end{pmatrix} = \begin{pmatrix} Q_1^n - A_{1r}A_{rr}^{-1}Q_2^n \\ 0 \end{pmatrix} \quad (5.6)$$

(or its rearranged equivalent problem using $\overline{A}^c$ to take advantage of banding of $\overline{A}^c$) for W_1^{n+1}. We have simply inverted A_{11}^c in an efficient and vectorizable manner. Then, using W_1^{n+1}, we complete the block Gauss-Seidel analogy on (5.2) and obtain W_r^{n+1} by solving

$$A_{rr}W_r^{n+1} = Q_2 - A_{r1}W_1^{n+1}. \quad (5.7)$$

Finally, from (5.4), we set

$$\tilde{P}^{n+1} = \begin{pmatrix} P_1^n \\ P_r^n \end{pmatrix} + \tau \begin{pmatrix} W_1^n \\ W_r^n \end{pmatrix}.$$

Since this algorithm only requires two separate solutions of mixed problems on the subregions (each subregion problem possibly being solved via a different parallel processor) and one solution on the original, uniform coarse grid, it is relatively easy to perform. Similarly, no complex data structure is required, and the algorithm can be implemented in existing large-scale codes without severely disrupting the solution process. Promising numerical results for the algorithm have appeared [3,21,26]. These results have also been extended to three-dimensional, three-phase reservoir simulation problems in a paper by Ewing, Boyett, Babu, and Heinemann [19].

As stated, the algorithm in its most general form involves two separate solutions on the subregions at each step. This iterative procedure is uniformly well-conditioned for finite element procedures such as those used in this paper or point-centered finite difference methods, but not for cell-centered finite differences [25–27]. For discretizations arising from cell-centered finite difference methods, a scaling of the iteration via the parameter τ in (5.4) may be necessary. The use of the algorithm as a preconditioner [17] for another iterative procedure such as conjugate gradient also involves two distinct solutions on the subregions at each step. This comes from the desire to have a symmetric preconditioner, which is important for conjugate gradient methods.

By considering the domain decomposition techniques presented by Bramble, Pasciak, and Schatz [4] that led to this algorithm, we can see that if the subregion problems ((5.3) and its sequels with updated guesses for P_1^n) are solved exactly, then Q_2^n in (5.5) and (5.6) is identically zero and the iterative method presented here requires only one subregion solution per iteration (from (5.7)). Preliminary computations indicate that if the subregion problem is solved iteratively with its own preconditioner, the full algorithm with two subregion solves will converge faster than the version with one subregion solve for some problems. Iterative solution of the unrefined region causes no difficulty with either version of the algorithm. This is an important consideration for reservoir simulation applications when iterative solution of the unrefined problems is essential due to their size, since direct solution of the refined region problems is usually not possible.

These methods were originally defined for elliptic or parabolic problems with fixed refinements for pressures. They were used in systems of equations for pressure approximation in [9,10]. If the velocity is approximated directly, local refinements are needed for mixed finite element methods; analysis and computations for locally refined mixed methods appear in [24]. Efficient implementation concepts are discussed in [23].

The techniques for following moving fluid interfaces are more complex than the static methods described above. The concepts described in [14] have been implemented efficiently in [9,10]. When two or more locally refined areas are adjacent to each other, multiplicative domain decomposition methods such as those presented in [3,4,8–10,14] require the solution on the full coarse grid, on the fine grids, and on the boundaries between fine grids. For elliptic problems, effective preconditioners for the edge problems are not too difficult [4]. For the transport-dominated petroleum applications, effective preconditioners are more difficult to obtain. See [10] for advances in this area. The use of additive Schwarz methods with overlapping grids removes this difficulty. See [2] for efficient implementation of these methods which has excellent potential for parallel architecture computers.

Another problem that has plagued large-scale reservoir simulators is the difficulty in treating local transients around wells in fully implicit codes. When one well is opened or shut to flow rapidly, the local fluid properties around the well change sufficiently quickly that the global Newton-Raphson method used to linearize the flow equations does not converge. The present industrial solution is to cut the time step over the whole reservoir to obtain convergence, even though the difficulty is highly localized. This is extremely wasteful and computationally intensive. The local domain decomposition techniques presented here have been incorporated in an industrial code to allow for accurate and efficient local time stepping. Results of local time stepping in an industrial multiphase simulator are presented in [1]. Analyses of these results appear in [28].

ACKNOWLEDGMENTS

This research was supported in part by Office of Naval Research Contract No. N00014-88-K-0370, by the Institute for Scientific Computation through NSF Grant No. RII-8610680, by the National Science Foundation under Grant No. DMS-8821330, and by the Pittsburgh and Minnesota Supercomputing Centers.

REFERENCES

1. B. A. Boyett, M. S. El-Mandouh, and R. E. Ewing, Local grid refinement for reservoir simulation, *Proceedings of the SIAM Conference on Mathematical and Computational Issues in Geophysical Fluid and Solid Mechanics*, SIAM, Philadelphia, Pennsylvania, (to appear).

2. J. H. Bramble, R. E. Ewing, R. Parashkevov, and J. Pasciak, Domain decomposition methods for problems with partial refinement, *Proceedings of the Copper Mountain Multigrid Conference*, Copper Mountain, Colorado, (to appear) and *SIAM J. Scientific and Statistical Computing*, (submitted).

3. J. H. Bramble, R. E. Ewing, J. E. Pasciak, and A. H. Schatz, A preconditioning technique for the efficient solution of problems with local grid refinement, *Comp. Meth. Appl. Mech. Eng.*, *67* (1988), 149–159.

4. J. H. Bramble, J. E. Pasciak, A. H. Schatz, An iterative method for elliptic problems on regions partitioned into substructures, *Math. Comp.*, *46* (1986), 361–373.

5. M. A. Celia, T. F. Russell, I. Herrera, and R. E. Ewing, An Eulerian-Lagrangian localized adjoint method for the advection-diffusion equation, *Advances in Water Resources*, (to appear).

6. G. Chavent, G. Cohen, and J. Jaffré, Discontinuous upwinding and mixed finite elements for two-phase flows in reservoir simulation, *Comp. Meth. in Appl. Mech. and Eng.*, *47* (1984), 93–118.

7. H. K. Dahle, Adaptive characteristic operator-splitting techniques for convection-dominated diffusion problems in one and two space dimensions, *Rep. No. 85*, Department of Applied Mathematics, University of Bergen, 1988.

8. H. K. Dahle, M. S. Espedal, and R. E. Ewing, Characteristic Petrov-Galerkin subdomain methods for convection diffusion problems, *IMA Volume 11, Numerical Simulation in Oil Recovery* (M. F. Wheeler, ed.), Springer-Verlag, Berlin, 1988, 77–88.

9. H. K. Dahle, M. S. Espedal, R. E. Ewing, and O. Sævareid, Characteristic adaptive sub-domain methods for reservoir flow problems, *Numerical Methods for Partial Differential Equations*, (to appear).

10. H. K. Dahle, M. S. Espedal, and O. Sævareid, Domain decomposition for reservoir flow problems, *Proceedings 1989 Conference on Domain Decomposition Methods*, SIAM Publications, (to appear).

11. B. L. Darlow, R. E. Ewing, and M. F. Wheeler, Mixed finite element methods for miscible displacement problems in porous media, *Proceedings Sixth SPE Symposium on Reservoir Simulation*, New Orleans, 1982, 137–146; and *Soc. Pet. Eng. J.*, *4* (1984), 391–398.

12. J. Douglas, Jr., R. E. Ewing, and M. F. Wheeler, A time-discretization procedure for a mixed finite element approximation of miscible displacement in porous media, *R.A.I.R.O. Analyse Numérique, 17* (1983), 249–265.

13. J. Douglas, Jr. and T. F. Russell, Numerical methods for convection-dominated diffusion problems based on combining the method of characteristics with finite element or finite difference procedures, *SIAM J. Numer. Anal., 19* (1982), 871–885.

14. M. S. Espedal and R. E. Ewing, Characteristic Petrov-Galerkin subdomain methods for two-phase immiscible flow, *Comp. Meth. Appl. Mech. and Eng., 64* (1987), 113–135.

15. M. S. Espedal, R. E. Ewing, T. F. Russell, and O. Sævareid, Reservoir simulation using mixed methods, a modified method of characteristics, and local grid refinement, *Proceedings of Joint IMA/SPE European Conference on the Mathematics of Oil Recovery*, Robinson College, Cambridge University, July 25–27, 1989.

16. R. E. Ewing, Domain decomposition techniques for efficient adaptive local grid refinement, *Domain Decomposition Methods* (T. F. Chan, R. Glowinski, J. Periaux, O. B. Widlund, eds.), SIAM, Philadelphia, Pennsylvania, 1989, 192–206.

17. R. E. Ewing, Adaptive grid refinements for transient flow problems, *Adaptive Methods for Partial Differential Equations* (J. E. Flaherty, P. J. Paslow, M. S. Shephard, and J. D. Vasilakis, eds.), SIAM, Philadelphia, Pennsylvania, Chapter 14, 1989, 194–205.

18. R. E. Ewing, Operator splitting and Eulerian-Lagrangian localized adjoint methods for multiphase flow, *Mathematical Analysis and Finite Elements in Applications (MAFELAP)*, Brunel University, London, England, (to appear).

19. R. E. Ewing, B. A. Boyett, D. K. Babu, and R. F. Heinemann, Efficient use of locally refined grids for multiphase reservoir simulation, SPE 18413, *Proceedings Tenth SPE Symposium on Reservoir Simulation*, Houston, Texas, February 6–8, 1989, 55–70.

20. R. E. Ewing and R. F. Heinemann, Incorporation of mixed finite element methods in compositional simulation for reduction of numerical dispersion, SPE 12267, *Proceedings Seventh SPE Symposium on Reservoir Simulation*, San Francisco, California, November 15–18, 1983, 341–347.

21. R. E. Ewing, P. Jacobs, R. Parashkevov, and J. Shen, Applications of adaptive grid refinement methods, *Proceedings of Fifth IIMAS Workshop on Numerical Analysis*, SIAM, Philadelphia, Pennsylvania, (to appear).

22. R. E. Ewing and R. D. Lazarov, Adaptive local grid refinement, SPE 17806, *Proceedings of 1988 Rocky Mountain Regional Meeting of SPE*, Casper, Wyoming, May 1988, 643–652; and *Proceedings Fourth Wyoming Enhanced Oil Recovery Symposium*, Casper, Wyoming, May 1988, 87–102.

23. R. E. Ewing, R. D. Lazarov, P. Lu, and P. S. Vassilevski, Preconditioning indefinite systems arising from mixed finite element discretization of second-order elliptic systems, *Proceedings Conference on Preconditioned Conjugate Gradient Methods*, Nijmegen, The Netherlands, June 15–17, (to appear).

24. R. E. Ewing, R. D. Lazarov, T. F. Russell, and P. S. Vassilevski, Local refinement via domain decomposition techniques for mixed finite element methods in rectangular Raviart-Thomas elements, *Proceedings of 1989 Conference on Domain Decomposition Methods*, SIAM Publications, (to appear).

25. R. E. Ewing, R. D. Lazarov, and P. S. Vassilevski, Local refinement techniques for elliptic problems on cell-centered grids, I: Error analysis, *Math. Comp.*, (submitted).

26. R. E. Ewing, R. D. Lazarov, and P. S. Vassilevski, Local refinement techniques for elliptic problems on cell-centered grids, II: Two-grid iterative methods, *Math. Comp.*, (submitted).

27. R. E. Ewing, R. D. Lazarov, and P. S. Vassilevski, Local refinement techniques for elliptic problems on cell-centered grids, III: Algebraic multilevel BEPS preconditioners, *Numerische Mathematik*, (submitted).

28. R. E. Ewing, R. D. Lazarov, and P. S. Vassilevski, Finite difference schemes for parabolic problems on grids with local refinement in time and in space, *Computing*, (submitted).

29. R. E. Ewing, R. D. Lazarov, and J. Wang, Superconvergence of the velocities along the Gaussian lines in the mixed finite element methods, *SIAM J. Numer. Anal.*, (to appear).

30. R. E. Ewing, T. F. Russell, and M. F. Wheeler, Simulation of miscible displacement using mixed methods and a modified method of characteristics, SPE 12241, *Proceedings Seventh SPE Symposium on Reservoir Simulation*, San Francisco, California,

November 15–18, 1983, 71–82, in SPE Reprint Series No. 20, Numerical Simulation II, Society of Petroleum Engineers, Dallas, Texas, and *Soc. Pet. Eng. J.*, (to appear).

31. R. E. Ewing, T. F. Russell, and M. F. Wheeler, Convergence analysis of an approximation of miscible displacement in porous media by mixed finite elements and a modified method of characteristics, *Comp. Meth. Appl. Mech. Eng., 47,* 1984, 73–92.

32. R. E. Ewing, T. F. Russell, and L. C. Young, An anisotropic course-grid dispersion model of heterogeneity and viscous fingering in five-spot miscible displacement that matches experiments and fine-grid simulations, SPE 18441, *Proceedings Tenth SPE Symposium on Reservoir Simulation*, Houston, Texas, February 6–8, 1989, 447-466; and *SPE Res. Eng.*, (to appear).

33. R. E. Ewing and M. F. Wheeler, Computational aspects of mixed finite element methods, *Numerical Methods for Scientific Computing* (R. S. Stepleman, ed.), North Holland Publishing Co., 1983, 163–172.

34. T. F. Russell, Finite elements with characteristics for two-component incompressible miscible displacement, SPE 10500, *Proceedings Sixth SPE Symposium on Reservoir Simulation*, New Orleans, Louisiana, January 31–February 3, 1982, 123–135.

35. T. F. Russell, Time stepping along characteristics with incomplete iteration for a Galerkin approximation of miscible displacement in porous media, *SIAM J. Numer. Anal, 22* (1985), 970–1013.

36. T. F. Russell, Eulerian-Lagrangian localized adjoint methods for advection-dominated problems, *Proceedings 13th Biennial Conference on Numerical Analysis*, Pitman Pubishing Company, Dundee, Scotland (1989).

37. T. F. Russell and R. V. Trujillo, Eulerian-Lagrangian localized adjoint methods with variable coefficients in multiple dimensions, *Proceedings 7th International Conference Computer Methods in Water Resources*, Venice, Italy, (to appear).

38. T. F. Russell and M. F. Wheeler, Finite element and finite difference methods for continuous flows in porous media, in *The Mathematics of Reservoir Simulation* (R. E. Ewing, ed.), Frontiers in Applied Mathematics, Vol. 1, Society for Industrial and Applied Mathematics, Philadelphia, 1983, 35–106.

39. T. F. Russell, M. F. Wheeler, and C. Chiang, Large-scale simulation of miscible displacement by mixed and characteristic finite element methods, in *Mathematical and Computational Methods in Seismic Exploration and Reservoir Modeling* (W. E. Fitzgibbon, ed.), Society of Industrial and Applied Mathematics, Philadephia, Pennsylvania, 1986, 85–107.

245

2nd European Conference on the Mathematics of Oil Recovery
© D. Guérillot, O. Guillon (Editors) and Éditions Technip, Paris 1990, pp. 247-253
27 rue Ginoux, 75015 Paris

Domain Decomposition Methods in Reservoir Simulation Coupling Well and Full Field Models

O. Gosselin[1] and J. M. Thomas[2]

ABSTRACT

In industrial reservoir simulators wells are usually with just a few large discrete cells and simplified source terms. The complex flow mechanisms that arise around wells are thus not accurately represented. This can have serious consequences on results.

A natural idea to obviate these defects would be to use a finer grid mesh around the wells. But such local grid refinements intoduces mesh irregularities with an excessive contrast in the mesh sizes between the grids. Conventional numerical schemes and conventional solvers to handle such irregularities are often inadequate and considerbly degrade the computational performance of codes.

This papers considers an other approach by decomposing the above problem over two overlapping or non overlapping subdomains: reservoir and wells. For each time step, we solve the differential equations in separate mesh resolutions and iterate between subdomains until convergence is reached at the internal boundary. The boundary conditions are provided by results of the adjacent domain (pressures, saturations and fluxes).

We present some techniques of decomposed modelling applied to model equations of diphasic immiscible flows. We use overlapping subdomains or alternately non overlapping domains with relaxation of interface conditions to achieve convergence.

These algorithms have been developped for a three-dimensional Dead-Oil model with slight compressibility, under "fully implicit" formulation. We compare different strategies related to imposed boundary conditions on the interface of subdomains (Dirichlet/Neumann) and their influence on number of global iterations.

The numerical experiments prove to be efficient, for an appropriate treatment of wells in a reservoir simulator, with regards to accuracy and computing time.

1. MOTIVATIONS

1.1 Improving Well Modelling in the Full Field Model

In reservoir simulation we are faced with a difficulty due to a large constrast between far field size and wells size that leads to use discrete cells 500 times greater than the wells. We are faced also with the complexity and specificity of flow mechanisms in the vicinity of wells. Usually, in order to improve the computing accuracy, two solutions are used: specific well models in a local simulation or local grid refinement in a global simulation.

1.2 Dynamical Coupling

The main and big defect of using local models is the non-coupling with a full field model and non integration in a global simulation. How to obviate these difficulties?

1.3 Efficient Computing

If we want to keep a regular mesh, we are obliged to decrease the space-step discretization near the wells. But a lot of others small cells are thus generated in areas where they are not necessary.

In the opposite, the disadvantage of using a local grid refinement is the irregularity of systems that are produced and the difficulty to handle efficiently such structures with conventionnal numerical schemes and solvers.

---

(1) Elf Aquitaine, 26, avenue des Lilas, 64018 Pau, France.
(2) Laboratoire de Mathématiques Appliquées (URA CNRS 1204), Université de Pau et des Pays de l'Adour, avenue de l'Université, 64000 Pau, France.

In fact, in industrial reservoir engineering, studies are treated on vector supercomputers and local grid refinement options inhibits the vectorization. The same concern leads to examine the bests techniques improving reservoir simulator efficiency on multiprocessor computers.

2. DOMAIN DECOMPOSITION APPROACH

This mathematical method is useful in solving large and complex problems, possibly on parallel computers.

2.1 The Idea of Application

Simply, it consists first in a partition of field into two subdomains: (1) one is the wells area, (2) the other one for the other part of the field. Then during simulation, for each time-step, we solve successively reservoir and wells problems on respective domain. The solution of one subdomain problem provides boundary conditions in order to solve the other subdomain problem. We alternate these resolutions until convergence is reached at the internal boundary between subdomains.

2.1 A Priori Advantages

This technique, as in local grid refinement, does not increase the global number of cells very much, but in contrast it saves in each subdomain the regularity of meshes and of sytems to be solved.

The well subdomain (including their vicinity) is obviously refined with respect to the size of the well and gives a greater accuracy. The reservoir subdomain keeps a coarse grid with dead cells.

In order to improve more the representation of the well phenomena, we are quite free to use a different model for the well, because the resolution in well domain is independant.

Finally, regardless of individual well treatment, the same technique applied to reservoir domain would allow a new partition of this area in several independent subdomains in order to optimize the performances in terms of elapsed time on multiprocessor computers by parallelising separate resolutions.

2.3 Sensitive Points

In spite of its simplicity, we must be careful about some difficulties:

a/ use of different and incompatible meshes is not standard

b/ definition of geometrical links between subdomains and choice of convergence criterium

c/ choice of boundary conditions on the artificial internal interface

d/ computing of passed informations between different size grids

e/ guiding of external iterations related to internal iterations due to linearization, with respect to the global convergence.

3. METHOD PRINCIPLE

3.1 Symetric formulation

The original method belongs to Schwarz (1869) and it consists in decomposing a complex problem on a large domain Ω into easier problems on smaller domains, in order to spent less computing time or memory place.

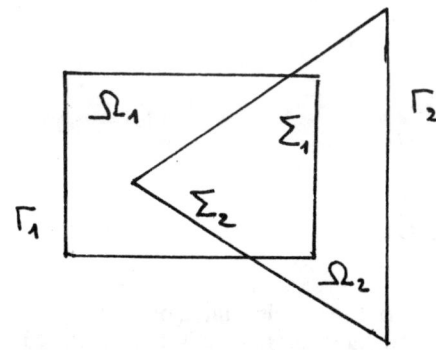

We give a general formulation of an extension of this method:

Consider the following problem: find 'u 'such

$$\text{(P)} \quad Au = f \text{ on } \Omega, \quad Bu = g \text{ on } \partial\Omega$$

where $\Omega = \Omega_1 \cup \Omega_2$, $\Sigma_i = \partial\Omega_i \setminus \partial\Omega$, and $\Gamma_i = \partial\Omega_i \setminus \Sigma_i$.

Possibly $\Omega_1 \cap \Omega_2 = \Sigma_1 = \Sigma_2 = \Sigma$.

The initial problem (P) is replaced by the following two problems :

$$\text{(Pi)} \quad A\, u_i = f \text{ on } \Omega_i, \quad B\, u_i = g \text{ on } \Gamma_i, \quad B_i u_i = B_i u_j,$$
$$\text{on } \Sigma_i$$
$$\text{with } i \neq j$$

such as $u_i = u \vert \Omega_i$.

The third condition on Σ_i expresses the continuity of $B_i u$ on the internal boundary Σ_i in Ω.

For instance, $B\, u$ and $B_i\, u$ may be a linear combination of u and $\partial u / \partial n$ (Robin-Fourier condition), incuding the case of Dirichlet and Neumann conditions.

That leads to the following iterative procedure:

$$\begin{cases} Au_i^{n+1}=f \text{ on } \Omega_i, \; Bu_i^{n+1}=g \text{ on } \Gamma_j \\[2mm] \dfrac{\partial u_i^{n+1}}{\partial n_i}+\lambda_i u_i^{n+1}=\dfrac{\partial u_j^n}{\partial n_i}+\lambda_i u_j^n \text{ on } \Sigma_i \\[2mm] 0 \le \lambda_i \le +\infty, \; i=1,2 \end{cases}$$

With sufficient overlapping, the convergence is proved in the following cases, for an elliptic operator (A=-Δ):

$\lambda_1=\lambda_2=+\infty$ (classical Schwarz procedure)

$\lambda_1=\lambda_2=0$

Without overlapping, the convergence is also guarantied in the case:

$0< \lambda_1=\lambda_2<+\infty$

(cf PL. Lions in [1])

3.2 Dissymetric formulation

In the case with non-overlapping subdomains ($\Sigma_1=\Sigma_2=\Sigma$), we can choose another distribution of the internal boundary condition in each subdomain: $B_1 u= u|\Sigma$ and $B_2 u= \partial u/\partial n_2|\Sigma$ respectively.

So we obtain this iterative procedure:

$$\begin{cases} \begin{cases} Au_1^{n+1}=f \text{ on } \Omega_1, \; Bu_1^{n+1}=g \text{ on } \Gamma_1 \\[2mm] u_1^{n+1}=\lambda^n \text{ on } \Sigma \end{cases} \\[6mm] \begin{cases} Au_2^{n+1}=f \text{ on } \Omega_2, \; Bu_2^{n+1}=g \text{ on } \Gamma_2 \\[2mm] \dfrac{\partial u_2^{n+1}}{\partial n}=\dfrac{\partial u_1^{n+1}}{\partial n} \text{ on } \Sigma \end{cases} \\[6mm] \lambda^{n+1}=F(u_2^{n+1},\lambda^n) \text{ on } \Sigma \end{cases}$$

We can prove the convergence:

$$u_i^n \to u|_\Omega$$

if the suit (λ^n) is convergent.

But the simplest choice $\lambda^{n+1}=u_2^n|_\Sigma$ does not ensure it in any case of geometry. A.Quarteroni & al [2,3] has shown that we can apply a relaxation method on λ :

$$\lambda^{n+1}= \theta \, u_2^n+ (1-\theta) \lambda^n$$

for some values $\theta < \theta^*$ (for the model equation - $\Delta u+\mu u=f$).

The same authors have given an algorithm which avoids estimating the parameter θ, but just using this variant scheme:

$$\lambda^{n+1}= \theta^n \, u_2^n + (1-\theta^n) \lambda^n$$

such as $\theta^n \to \theta^*$ if it is computed as follows:

$$\theta^n =\frac{(e_1^n, e_1^n - e_2^n)}{\left\| e_1^n - e_2^n \right\|^2}$$

and where Ω_1 is the Dirichlet domain.

3.3 Others methods

The above Schwarz method, Dirichlet-Dirichlet, is sequential between domain 1 and 2, but another variant, called additive, can be parallelised if we take as boundary condition:

$$u_i^{n+1}= u_j^{n-1}$$

this technique allows a parallel resolution on both subdomains but the convergence speed up is less.

An other method, well adapted to linear problems, consists of solving simultanately both subproblems with the same Dirichlet condition:

$$u_1^n= u_2^n= \lambda^n$$

and after minimizing the difference on Neumann

conditions: $\dfrac{\partial u_i^n}{\partial n} - \dfrac{\partial u_j^n}{\partial n}$

In the discret field, that corresponds to the so called Schur matrix where the interface problem is solved by conjugate gradient method (if symmetric).

There exits also more and more variants but theoretical difficulties exist for non elliptic, non linear problems and incompatible meshes.

4. APPLICATION TO RESERVOIR EQUATIONS

The rigourous application of these methods to equations governing flows in hydrocarbon reservoir conflicts with their mathematical complexity and with the missing of theoretical bases for approximation schemes used. Let us pose the main aspects of the problem. More details or questions on this subject are touched on in [8]

249

4.1 Diphasic Immiscible Flows

For studying the validity in application of Domain Decomposition Methods and for developping software prototypes a first reasonable model seems to be the Dead-Oil model slightly compressible for two immiscible phases, one wetting phase (w) and another non wetting (n):

$$(1) \quad \frac{\partial}{\partial \tau}\Big[\rho_\alpha(P_\alpha)S_\alpha\Phi(P)\Big]+div\Big[\rho_\alpha(P_\alpha)Q_\alpha\Big]=f_\alpha$$

$$(2) \quad Q_\alpha=\frac{kr_\alpha(S_\alpha)}{\mu_\alpha(P_\alpha)}K\Big[\nabla P_\alpha-\rho_\alpha(P_\alpha)g\Big]$$

$$(3\text{-}4) \quad S_n+S_w=1, \quad P_n-P_w=Pc(S_w)$$

$$\alpha = n, w$$

with initial values and boundary conditions on $\partial\Omega$ (Q.n= 0 on $\partial\Omega$), where P and S are pressures and saturations of phases, Φ the rock porosity, ρ and μ volumic masse and viscosity, kr the relative permeability and K the absolute permeability tensor (diagonal). The dependance of prosity, volumic mass and viscosity are affine with respect to pressure. Usual choice of unknows is:

$$P= Pn \text{ and } S= Sw$$

4.2 Time Semi Discretized Equations

For one time-step (tm= m.Δt) these continuous equations are becoming:

$$(5) \quad \frac{1}{\Delta t}\overset{\bullet}{\rho_\alpha^m}\overset{\bullet}{\Phi^m}S_\alpha^{m+1}+div\Big[\overset{\bullet}{\rho_\alpha^m}Q_\alpha^{(m)}\Big]=F_\alpha^m$$

$$(6) \quad Q_\alpha^{(m)}=\frac{kr_\alpha(\overline{S_\alpha^m})}{\overset{\bullet}{\mu_\alpha^m}}K\Big[P_\alpha^{m+1}-\overset{\bullet}{\rho_\alpha^m}g\Big]$$

$$(7) \quad P_w^{m+1}=P_n^{m+1}-Pc(\overline{S_n^m})$$

a/ if m*= m, or if the model is incompressible, and if $\overline{m}$ = m+1, the sytem (5-6-3-7) can be written:

$$(8) \quad a_n(1\text{-}S) - div\Big\{b_n(S) K\Big[\nabla P\text{-}g_n\Big]\Big\}= C_n$$

$$(9) \quad a_w S - div\Big\{b_w(S) K\Big[\nabla(P\text{-}Pc(S))\text{-}g_w\Big]\Big\}= C_w$$

where $S=S_w^{m+1}$ and $P=P_n^{m+1}$

b/ if $\overline{m}$ =m the problem is becoming linear and can be decoupled into one elliptic equation in pressure and an explicit calculation for the saturation.

The system (8-9) sets up our initial non linear problem that we decompose into subdomains with overlapping and Dirichlet-Dirichlet method (continuity of pressures, saturations and fluxes) or without overlapping using a θ-method (Quarteroni & al) and so a Dirichlet domain and the other Neumann domain (continuity of fluxes).

4.3 Geometrical Decomposition

The reservoir is divided into (1) the "reservoir domain" which corresponds to the full field cells minus the cells of the wells and (2) the "well domain" which corresponds to the set of cells from the wells in a coarse grid (plus, with overlapping, some adjacents cells). This choice is related with a given usual discretization. Two possibilities of coupling the domains are:

4.3.1 without overlapping

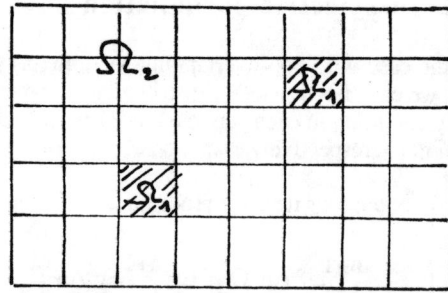

We believe that only one cell of the coarse reservoir grid is sufficiently large to define the well and its vicinity, and so we are minimizing the computing cost. Coarse grid is a cartesian mesh and fine grids are either cartesian or radial.

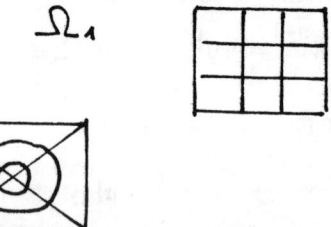

In fact, the discretization in space is made by a finite volume method and so we can consider that the overlapping is existent but minimal. This remark can allows us to use a Dirichlet-Dirichlet method in that case.

4.3.2 with overlapping subdomains

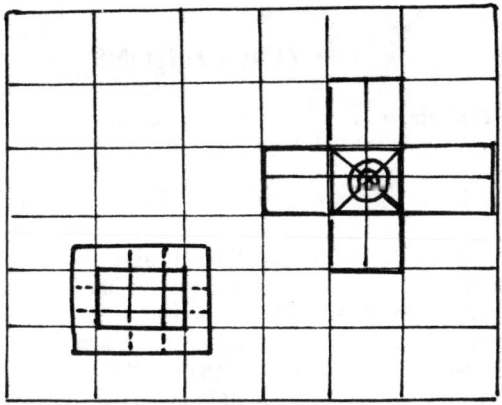

In order to ensure more strongly the convergence for a Dirichlet-Dirichlet method, in this case, the well domain is increased with 4 adjacent cells in radial or with 8 neighbouring cells in a cartesian mesh. This choice allows to keep a regular structure for the matrices.

4.4 Domain Decomposition Strategies

The first work is to prove by numerical experiment the ability of these techniques. Afterwards we can compare possibilities of using or not using overlapping. In the non overlapping case, an important topic is the choice of "Dirchlet domain": a) well resolution with Dirichlet condition and reservoir resolution with Neumann condition, b) the opposite of above reservoir resolution with Dirichlet condition and well resolution with Neumann condition. In both cases we can use a fixed θ parameter (sufficiently small) or an adaptative parameter. Another important question to be examined is the balance in terms of iteration numbers between internal iterations of Newton method and subdomains external iterations (choice of stop criterium: precision and maximal number for each kind of iterations). Particuly the convergence criterium at the interface can consist of jump values (extrapoled) or relative stagnation on each sideof the interface. Also different intergrid operators can be compared (constant by cells or linear or, for fluxes, weighted by permeabilities). At last we can improve the convergence with a preconditioner at the beginning of each time-step with use of a very few Newton iterations for a global resolution on the reservoir coarse grid.

4.5 Other numerical approximation choices

In developped codes the hypothesis of compressibility are kept, and so the linearized Newton iterations are used to update the both unknown P and S. As we said above the space discretization scheme used is a finite volume approximation as made in industrial simulators. We have only been developping fully implicit formulation, but the decomposition technique would allow to replace it in well resolution by IMPES formulation. Which solver one.implement in the code , in the first step, is not an important choice to prove the ability of such a method, because one preserves the regularity of mesh but, naturally, if one aims for a maximal efficiency, in the further step, on vector and parallel computers, one must make up the best choice.

5. NUMERICAL RESULTS

The numerical experiments we are presenting here are obtained with three codes: a) one is two-dimensional without capillary pressure effect and with a global or decomposed grid, b) another three-dimensional with overlapping, c) with non-overlapping domain but relaxation θ-method on the interface. The data correspond with a homogenous medium, for 7x7x1 or 7x7x3 grids ($\Delta x=\Delta y=900ft$), the simulations are for 10 years., and the two injectors and producors wells are refined in 3x3 cartesian mesh and 3x4 radial.

The obtained results were compared with reference industrial simulator and with results from global fine grid with success.

5.1 Computing Time Gain

The table (1) shows results from a 2D code over three kinds of non decomposed grids (fine grid 15x15, "scotish" grid 9x9 for refining the two wells and coarse grid 5x5) and over a decomposed local refined grid 5x5 + 2x(3x3).

GRIDS	# UNKNOWN	# ITERATIONS	CPU/ratio
5×5	50	977	0.38
9×9	162	1065	3.8
15×15	450	1063	22.6
5×5 3×3 3×3	84	10244	1.

The average number of external iterations for one time-step was approximately 3. That seems to prove the increase of iteration numbers is largely compensated by the small number of additional cells.

5.2 Overlapping Or Not

The comparison from table (2) concerns the overlapping of subdomains (D= without and R= with), in a two and three-dimensional cases for a cartesian and radial refinement in overlapping case over a 10 year simulation.

	# TIME STEP	# NEWTON ITER	# EXT. ITER	S CPU time
R2	267	3811	716	176
	177	2786	428	62
D2	231	3367	333	27
R3	158	2033	334	169
	153	2257	311	151
D3	159	2736	208	76

The convergence is ensured with an equivalent number of iterations but the non overlapping technique decreases the cost of computing and provides always a regular mesh in well subdomains.

5.3 Dirichlet/Neumann Conditions Domains Choice

The other main topic concerns the choice of Dirichlet domain, in the non overlapping case. The table (3) makes comparison between the two possibilities: (1) Reservoir as the Dirichlet domain and Wells as Neumann domain and (2) the opposite case.

	(1)	(2)
# TIME-STEP	7866	177
# NEWTON ITER.	98 869	3372
# EXTERN. ITER.	11718	318
% BALANCE ERROR	.99 E-9	.96 E-10
% OIL RECOVERY	22.73	22.75
S. CPU TIME	1186	27

The second choice is far the best, but actually, the choice of parameter θ is not the same: for the second case adaptative algorithm is better, in contrary it is inefficient in the first and only a fixed choice of a small value for θ (0.3 eg) ensures the (very slow!) convergence.

5.4 Others Remarks

In all of these examples a preliminar phase of global coarse grid resolution have preconditionned the problem with just one Newton iteration. We must note also that with these kind of prototype codes the average time-step and the absolute CPU time are not very significant.

Others tests have shown that, over 4 external iterations, we do not improve the convergence.

The parameters and options we can adjust are many and quite a lot of others choices would be tested.

6. CONCLUSIONS

The above results are very encouraging and promising and we can bring out the following trends:

6.1 Accuracy and Efficiency Benefits

Our study confirms the ability, for Domain Decomposition techniques, to improve the computing accuracy locally around the wells, without greatly increasing the computing operations (a few additional cells and iterations) and without degrading the possibility of vectorization which is important in industrial applications.

If one want to compare the efficiency of this method more theoreticaly with a global approach, one can notice the consuming time

$$A = IT . M^o$$

for a global simulation, where IT is the global number of iterations, M the number of unknown with whatever grid structure and a the order of a solver well adapted with the kind of mesh;

and the similar time

$$B = itc . N^\beta + itext (itr . N^\beta + itw . n^\beta)$$

for a decomposed approach, where 'itc' is the number of preconditionner coarse grid iterations, 'itr' and 'itw' this number on reservoir and wells domains respectively, 'itext' the average external iterations number, 'N' and 'n' the numbers of unknown on coarse and fine grids respectively and 'b' the order of the used solver adapted, eg for 5 or 7 diagonal matrices.

In the main cases one can observe that A > B, if itext is not too large related with number of subdomains (with consequences on 'n'). In particulary we have observed that ' IT > itr or itw ' and this consolidates our argument.

6.2 Recommanded Strategies

An minimal overlapping seems sufficient to ensure the convergence. Better, the Dirichlet/Neumann Quarteroni method preserves an equivalent speed up of convergence with fewer numbers of cells. The relaxation of boundary conditions on the interface is necessary. But the wells domain seems to be chosen as Dirichlet domain. The number of external iterations can be limited at 4 or 5 and the exact Newton convergence is not necessary: 10 iterations or less are sufficient.

6.3 New future developments

We are intending to test a time step subdivision in the refined domains as suggered by R.Ewing [4] in order not to penalize the resolution on coarse grid.

The next step in our work is trying to couple a specific well model to further improve the representation of well phenomena and so the accuracy of results over the whole field.

All this efforts would not be useful if a good representation of front would not be avalaible: it is possible to reach this aim by dynamical local refinement using also domain decomposition methods.

At last, we must not forget the iminent rise of multiprocessor computers in reservoir simulation. These techniques are well suited to profit from the evolution of multiprocessor hardware technology. We can introduce easily prallelizatiion of certain tasks as resolution on each wells simultanately and by decomposing the reservoir domain also.
Another possibility, in the same direction, can be found in the application of Schur complement techniques, partialy for the linearised problems on reservoir domain (cf [5]).

ACKNOWLEDGMENTS

All this work takes place in a collaboration with *IFP,GDF* and *CFP* in *ARTEP* association,
We also acknowledge the participation of C.Lepretre who wrote the main parts of code when he was working at Stanford as VSNS for *Elf Aquitaine* .

REFERENCES

[1] The proceedings of the first three International Domain Decomposition Conferences (87,88,89) organized by SIAM.

[2] D.Funaro, A Quarteroni, P.Zanolli: An Iterative Procedure with Interface Relaxation for Domain Decomposition Methods. SIAM J. NUMER. ANAL. (88).

[3]L.D.Marini, A.Quarteroni: An Iterative Procedure for Decomposition Methods: A Finite Element Approach (87) in [1]

[4] R.E. Ewing: Domain Decomposition Techniques for Efficient Adaptative Local Grid Refinement in [1]

[5] Y.H De Roeck: A Local Preconditioner in a Domain-Decomposed Method. CERFACS Report, (89)

[6] J.M. Thomas: Methodes de decomposition de domaines. Second congres franco-chilien et latinoamericain de Mathematiques Appliquees, Santiago du Chili (Dec 89)

[7] J.M.Thomas: Domain Decomposition and Primal-Dual Finite Element Method. Fourth International Symposium on Domain Decomposition Methods for Partial Differential Equations, Moscow May 90)

[6] O.Gosselin, J.M.Thomas: Domain Decomposition Techniques Applied in Reservoir Simulation (in preparation).

2nd European Conference on the Mathematics of Oil Recovery
© D. Guérillot, O. Guillon (Editors) and Éditions Technip, Paris 1990, pp. 255-262
27 rue Ginoux, 75015 Paris

A Characteristic Finite Element Method for Solving Non-Linear Convection-Diffusion Equations on Locally Refined Grids

R. H. J. Gmelig Meyling[1]

ABSTRACT

A method for solving nonlinear convection-diffusion equations in two or three space dimensions is described. These equations play an important role in the numerical simulation of immiscible, two-phase flow through porous media. All computations are performed on locally refined and dynamically adapted grids. This increases efficiency and ensures an optimal representation of shock fronts. Operator-splitting is used to decouple convection and diffusion, which reduces the problem to an alternating sequence of hyperbolic and elliptic equations. An accurate characteristic method deals with the hyperbolic equations. Nonlinearities in the convection term are treated by solving Riemann problems along streamlines. Elliptic equations are discretised by mixed finite elements and solved by multi-grid. Gravity effects are included by a spatial splitting of the convection term. The method induces almost no numerical diffusion. It also permits to use large time steps and it conserves mass exactly. Numerical results are presented which demonstrate the performance of the method for some multi-dimensional test problems.

1. INTRODUCTION

The displacement of oil by water in a porous medium is described by a system of nonlinear partial differential equations. If the fluids are immiscible and if capillary pressure effects are present, then the conservation equation for water is a nonlinear convection-diffusion equation. This parabolic equation generally has a strong convection term. Hence, its properties are similar to those of first-order, hyperbolic equations. Solutions for the water saturation may thus exhibit sharp fronts.

Reservoir simulators based on standard finite differences and regular spatial grids often introduce a considerable amount of numerical diffusion. This seriously limits the accuracy of the predicted flow patterns as well as the hydrocarbon recovery.

In this paper, a combination of local grid refinement and a special front-tracking technique is used to overcome the problem of numerical diffusion. Locally refined and dynamically adapted grids permit an accurate representation of sharp, moving fluid fronts. The use of these grids also reduces computational

---

(1) Koninklijke/Shell Exploratie en Produktie Laboratorium, PO Box 60, 2288 GD, Rijswijk, (ZH), the Netherlands.

costs associated with large-scale reservoir simulations. The effects of convection and diffusion are treated separately by using operator-splitting.

Convection equations are very accurately solved by using the method of characteristics. This approach almost completely eliminates the artificial smearing of sharp fluid fronts. The method avoids severe restrictions of the time step by taking nonlinear wave interactions into account.

An implicit, first-order time discretisation applied to diffusion equations leads to elliptic problems. These are conveniently discretised by mixed finite elements and solved by a multi-grid method specially designed for locally refined grids.

Some computational results will be presented for two- and three-dimensional waterflooding problems. These results demonstrate the efficiency and accuracy, that can be obtained by solution methods based on locally refined grids.

2. TWO-PHASE IMMISCIBLE DISPLACEMENT

Consider two-phase flow of immiscible fluids (oil(o) and water(w)) in a heterogeneous, porous reservoir. The displacement process is described by a coupled system of nonlinear, time-dependent partial differential equations.

The flow $\vec{u}$ and pressure p of the total incompressible fluid are governed by Darcy's law and the principle of fluid conservation

$$\vec{u} = - k(\vec{x})\lambda(s)\nabla p , \quad \nabla.\vec{u} = q(\vec{x}) , \quad (1)$$

where k is the (spatially varying) absolute permeability tensor. The total mobility $\lambda = \lambda_w + \lambda_o$ of the fluid is a function of the water saturation s. In this formulation, wells are modelled as point sources/sinks with a given flow rate q.

The water saturation is defined by a nonlinear convection-diffusion equation

$$\phi(\vec{x})s_t + \nabla.(f(s)\vec{u}) - \nabla.(D(s)\nabla s) = c(s)q(\vec{x}) , \quad (2)$$

with ϕ the rock porosity; f(s) the water fractional flow function; and c(s) the water fraction of the fluid at the wells. The diffusion tensor D(s) reflects the capillary pressure effects, i.e.

$$D(s) = k(\vec{x}) (\lambda_w\lambda_o/\lambda) dp_c/ds.$$

These equations are decoupled for each time step. This gives the elliptic equation (1) for $\vec{u}$, p, and the parabolic equation (2) for s, each subject to the appropriate initial/boundary conditions. The function f(s) is nonlinear and nonconvex. Equation (2) may therefore give rise to solutions having steep fronts travelling through the reservoir.

3. DYNAMIC LOCAL GRID REFINEMENT

In the presence of certain local phenomena (e.g. wells, fluid fronts), fine grid blocks are required to achieve a proper resolution of the local physical behaviour. In order to solve large reservoir simulation problems in an efficient manner, local grid refinement is advantageous. Here, small grid blocks are used only in special areas (for example, around active wells and near saturation fronts). In other areas of the reservoir, coarse-grid blocks are sufficient. The local grid size is chosen as a trade-off between accuracy and computational efficiency.

Since the front travels through the reservoir and wells may be shut down or switched on, the grid is dynamically adapted during simulation of the waterflooding process.

In n space dimensions, the discretisation grid is built from a regular base grid, which is refined by repeatedly dividing selected blocks into 2^n identical subblocks (see Figs 3, 4 and 5).

The fluid front moves along the characteristic directions. This fact is exploited to identify the regions where fine blocks will be required at the next time step. Grid blocks are thus refined near active wells and at the new location of the front. On the other hand, fine blocks are removed near inactive wells and at the old position of the front. In this manner, local grid refinement allows to follow the moving front. The mass balance is preserved when grid blocks are refined or removed.

Grid smoothness can be controlled by bounding differences in size between adjacent grid blocks. The data structure supports local grid refinement in up to three space dimensions.

Even for complex reservoir geometries the grid generation can be done automatically. The local grid refinement is controlled by the flow pattern, the water saturation, permeability contrasts, rock types, etc.

In this way, numerical solutions of high accuracy can be obtained with considerable savings in computer time and storage.

4. MIXED FINITE ELEMENTS AND MULTI-GRID

An efficient multi-grid algorithm has been developed to solve elliptic equations on locally refined grids (Schmidt and Jacobs, 1988). The construction procedure for a locally refined grid defines in a natural way a sequence of nested grids of increasing fineness as required for the multi-grid process. Multi-grid is used to solve a scalar field ψ and a vector field $\vec{v}$ from a mixed finite element discretisation of the elliptic system

$$c\,\psi + \nabla \cdot \vec{v} = c\,\psi_R$$

$$\nabla \psi + W \cdot \vec{v} = W \cdot \vec{v}_R \tag{3}$$

subject to a general boundary condition $a\psi - \beta\vec{v}\cdot\vec{n} = \gamma$, with $\vec{n}$ the unit outward

normal to the domain. All coefficients may be arbitrary functions of the space coordinate $\vec{x}$, as long as a, β, $c \geq 0$ and the tensor W is positive definite. The lowest-order elements of Raviart-Thomas are used, i.e. piecewise constant approximation of ψ and piecewise linear approximation of the components of $\vec{v}$.

Every time step, the multi-grid method solves $\vec{u}$ and p from equation (1), which is of the basic elliptic type (3). The mixed finite element approach gives a particularly accurate estimate for the fluid velocity $\vec{u}$, which is then substituted into the convection-diffusion equation (2) for the water saturation. The computed fluid velocity largely determines the local speed and direction of the fluid front.

5. OPERATOR-SPLITTING

The saturation equation is generally convection-dominated, but the diffusive effects due to capillary pressures are still important. Operator-splitting is used to treat convection and diffusion separately, each by the most effective numerical technique. Every time step, two equations are solved, namely

$$\phi\, s_t + \nabla \cdot (f(s)\vec{u}) = c(s)q \ ,$$

$$\phi\, s_t - \nabla \cdot (D(s)\nabla s) = 0 \ , \tag{4}$$

where the solution of the first equation serves as initial condition for the second. The first equation is solved by the method of characteristics. The solution procedure will be described in the next section. Let Δt be the length of a time step from t^n to t^{n+1} and let $\hat{s}^{n+1}$ denote the solution of the convection equation. A first-order time discretisation turns the second (diffusion) equation into an elliptic problem

$$\phi/\Delta t \ s^{n+1} \ + \ \nabla \cdot \vec{w}^{n+1} = \phi/\Delta t \ \hat{s}^{n+1}$$

$$\nabla s^{n+1} \ + \ \tilde{D}^{-1} \vec{w}^{n+1} = \vec{0} \qquad (5)$$

The final solution s^{n+1} for the saturation is determined from this system using mixed finite elements and multi-grid.

Iteration is used to handle the nonlinearity in the diffusion coefficient $\tilde{D}(\vec{x}) = D((\hat{s}^{n+1} + s^{n+1})/2)$. Hence, for a realistic time step Δt, operator-splitting will not introduce undue large errors.

6. NONLINEAR CONVECTION

Consider the first-order, nonlinear, nyperbolic conservation law

$$\phi \ s_t \ + \ \nabla \cdot (f(s)\vec{u}) = c(s)q \ , \qquad (6)$$

subject to an initial condition $s(\vec{x},t^n) = s^n(\vec{x})$. The function s^n is constant on grid blocks. The fluid velocity $\vec{u}(\vec{x})$ is assumed to be constant during the entire time step. The problem is now to compute $s(\vec{x},t^{n+1})$ at any given point $\vec{x}$ (e.g. the center of a grid block).

The multi-dimensional problem (6) can be reduced to a one-dimensional problem along a streamline. Let $\vec{x}(v)$ be the streamline passing through the point $\vec{x}$. It is defined by the system of ordinary differential equations

$$d\vec{x}/dv = \vec{u}(\vec{x})/\phi(\vec{x}) \ , \text{ for } v < 0$$
$$\vec{x}(v=0) = \vec{x} \qquad (7)$$

Here, v is a parameter which runs backwards along the streamline. The system (7) can be solved analytically if we replace $\vec{u}$ by its mixed finite element representation. Suppose that $f'(s) \geq 0$

and that the streamline does not pass through a well ($q=0$). Backward integration along the streamline turns (6) into a one-dimensional Cauchy problem

$$s_t + (f(s))_v = 0 \ , \text{ for } t > t^n$$
$$s(v,t^n) = s^n(v) \ , \qquad (8)$$

from which we will now solve $s(v=0,t^{n+1})$ by the method of characteristics. The initial data in (8) equals the piecewise constant restriction of $s^n(\vec{x})$ along the streamline through $\vec{x}$, i.e. $s^n(v) = s^n_j$, for $v_j < v < v_{j+1}$ ($j = 0,1,\ldots,m$). The two end points of the v-range are given by $v_0 = - \Delta t \max f'(s)$, $v_{m+1} = 0$.

A weak solution $s(v,t)$ of the hyperbolic equation (8) contains a finite number of discontinuities joined by smooth segments. The smooth segments (or rarefaction waves) are self-similar solutions $s(\xi)$, with $\xi = (v-v^*)/(t-t^n)$, of the equation $f'(s(\xi)) = \xi$. Hence, $s(\xi) = s^n(v^*)$ is constant along every characteristic ray $\xi = $ constant. Let s_L and s_R be two constant states on the left and right of a discontinuity. The speed of propagation S of the discontinuity must satisfy the Rankine-Hugoniot jump condition

$$S = (f(s_L)-f(s_R)) \ / \ (s_L - s_R) \qquad (9)$$

In addition, an entropy condition (Lax, 1973) is needed to determine physically relevant discontinuities. This condition ensures that entropy of fluid passing through a discontinuity increases. It is sufficient if one requires that the characteristics do not diverge from the line of discontinuity, i.e.

$$f'(s_L) \geq S \geq f'(s_R) \qquad (10)$$

A discontinuity satisfying both (9) and

(10) is called a shock. If the shock speed S equals the characteristic speed at s_L or at s_R (or at both), the shock is termed a contact discontinuity.

The piecewise constant initial data $s^n(v)$ defines a sequence of Riemann problems

$$s_t + (f(s))_v = 0 \ , \ \text{for} \ t > t^n \tag{11}$$

$$s(v,t^n) = \begin{cases} s^n_{j-1} & , \ \text{for} \ v < v_j \\ s^n_j & , \ \text{for} \ v > v_j \end{cases}$$

$$(j = 1,\ldots,m)$$

The solution of each Riemann problem depends strongly on the shape of $f(s)$ and the values of s^n_{j-1} , s^n_j . In reservoir simulation, the flux function $f(s) = \lambda_w/\lambda$ is typically nonconvex. It has an S-shape and a single inflection point (as shown in Fig. 1). The solution of the Riemann problem consists of the two constant states separated by a combination of rarefaction waves and shocks (or contact discontinuities). Methods that apply some type of linearisation to the flux function $f(s)$ may fail to correctly predict such a solution.

There are two cases to consider, namely $s^n_{j-1} > s^n_j$ and $s^n_{j-1} < s^n_j$. In the first case, the solution of (11) can be found analytically by constructing the convex envelope $f_c(s)$ of $f(s)$ on the interval $[s^n_j, s^n_{j-1}]$ (Concus and Proskurowski, 1979). A region where $f_c > f$ leads to a shock, while a region where $f_c = f$ gives rise to a rarefaction wave. Figure 1 shows the well-known Buckley-Leverett profile that will result from a Riemann problem with $s^n_j = 0$ and $s^n_{j-1} = 1$. In the case where $s^n_{j-1} < s^n_j$, the solution of (11) is obtained by constructing

the concave envelope of $f(s)$. This solution procedure is used for each of the m discontinuities in the initial data.

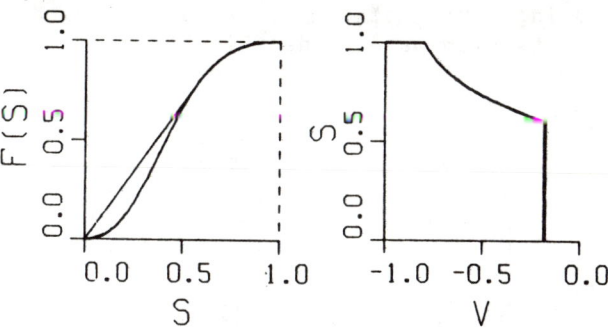

FIG. 1. Convex envelope of flux function and Buckley-Leverett profile.

In our method, interactions between the solutions of neighbouring Riemann problems are fully taken into account. Hence, large time steps are now feasible (Chang and Hsiao, 1988), (Holden et al., 1988), (Leveque, 1982).

The numerical method represents a shock as a single discontinuity between two constant states s_L, s_R moving with speed $S = (f(s_L)-f(s_R))/(s_L-s_R)$. A rarefaction wave is represented by a sequence of small discontinuities between consecutive constant states $\{s_i\}_{i=1}^{N+1}$. Here, every discontinuity between s_i and s_{i+1} travels at its own characteristic speed $f'((s_i+s_{i+1})/2)$. The s_i-values making up a single wave are either monotone increasing or monotone decreasing. The number of discontinuities used to represent a wave is determined by accuracy considerations.

In order to properly handle wave interactions, all discontinuities (either shocks or parts of waves) each moving at their own speed are followed by the method. If two or more discontinuities coalesce, they are merged into a new discontinuity for which we solve the Riemann problem. The discontinuity is then decomposed into shocks/waves and the procedure continues until the time step of length Δt is over.

Figure 2 shows an example of shock-

shock and shock-rarefaction interaction in the (v,t)-plane for a non-monotone initial saturation $s^n(v)$. The shock propagation speed S changes gradually during interaction of a shock with a rarefaction wave. The line of discontinuity in the (v,t)-plane is thus curved.

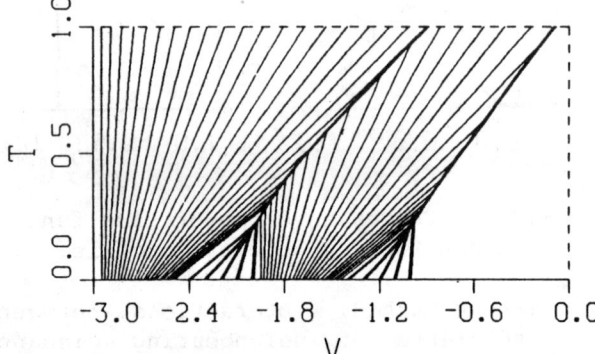

FIG. 2. Interaction of shocks and waves in the (v,t)-plane for a non-monotone initial saturation.

7. MASS CONSERVATION

The solution method for the nonlinear convection equation described in the previous section does not conserve mass exactly. The mass balance is restored by projecting the convection solution $\tilde{s}^{n+1}$ onto a nearby conservative solution $\hat{s}^{n+1}$. Let $\tilde{v}^{n+1}$ be the predicted total water flux passing through faces of grid blocks during the time step. A solution $\hat{s}^{n+1}$ conserving the total water mass can then be found from the elliptic system

$$\phi(\hat{s}^{n+1} - \tilde{s}^{n+1}) + \nabla.(\hat{v}^{n+1} - \tilde{v}^{n+1}) =$$
$$q^{n+1} + \phi(s^n - \tilde{s}^{n+1}) - \nabla.\tilde{v}^{n+1}$$
$$\tag{12}$$
$$\nabla(\hat{s}^{n+1} - \tilde{s}^{n+1}) + W.(\hat{v}^{n+1} - \tilde{v}^{n+1}) = \vec{0}$$

Here, W is a tensor that weighs the deviations in saturations and fluxes. The source term $q^{n+1} = q(\vec{x}) \int c(s)dt$ and the flux $\tilde{v}^{n+1} = \vec{u}(\vec{x}) \int f(s)dt$ are computed by using a 1D characteristic procedure along streamlines as described in section 6. The flow $\vec{u}$ and the well flow rate q are assumed to be independent of t during the time step.

8. GRAVITY

The method can be extended to two-phase, immiscible flow problems affected by gravitational forces. In this case, the flow $\vec{u}$ and pressure p are obtained from the system

$$\vec{u} = - k(\vec{x})\lambda(s)[\nabla p - (\lambda_w \rho_w + \lambda_o \rho_o)/\lambda \nabla z]$$

$$\nabla.\vec{u} = q(\vec{x}) , \tag{13}$$

with ρ_w, ρ_o the densities of water/oil and $z(\vec{x})$ the reservoir depth. These equations are of the basic elliptic type (3) and can thus be solved by mixed finite elements and multi-grid.

In the presence of gravity, the water saturation is defined by

$$\phi(\vec{x})s_t + \nabla . (f(s)\vec{u} + h(s)\vec{v}) \tag{14}$$
$$- \nabla . (D(s)\nabla s) = c(s)q(\vec{x}) ,$$

where $h(s) = \lambda_w \lambda_o/\lambda$, $\vec{v} = (\rho_w - \rho_o)gk\nabla z$, with g the gravity acceleration.

The convection term in (14) contains two (different) directions $\vec{u}$ and $\vec{v}$. It is treated by a splitting technique, where one solves an alternating sequence of convection problems either in the $\vec{u}$ - or $\vec{v}$ - direction. Let the operator $C_{\vec{u}}(\tau)$ denote the solution of equation $\phi s_t + \nabla.(f(s)\vec{u}) = cq$ over a time step of length τ. Similarly, let $C_{\vec{v}}(\tau)$ be the solution operator for $\phi s_t + \nabla.(h(s)\vec{v}) = 0$.

A simple alternating direction scheme for the solution of the convection part of (14) would then be $s^{n+1} = C_{\vec{u}}(\Delta t)C_{\vec{v}}(\Delta t)s^n$. One could also use a second-order splitting, such as $s^{n+1} = C_{\vec{u}}(\Delta t/2)C_{\vec{v}}(\Delta t)$ $C_{\vec{u}}(\Delta t/2)s^n$ (Strang, 1968). This asymmetric scheme is more accurate in the $\vec{u}$-direction than in the direction of $\vec{v}$. It would thus be particularly useful in situations where convection in the direction of $\vec{u}$ is dominant. Splitting of convection is efficient and generally gives good results.

The function h(s) is nonconvex and it has in general two inflection points (Proskurowski, 1981). The solution of a Riemann problem (with $s^n_{j-1} > s^n_j$) consists of two shocks propagating in opposite directions and separated by a rarefaction wave. The sign changes in h'(s) force us to integrate both backwards and forwards along streamlines. In other words, we take $v_0 = -\Delta t \max |h'(s)|$, $v_{m+1} = - v_0$.

9. NUMERICAL RESULTS

In this section, we present some numerical results obtained by our solution method for nonlinear convection-diffusion problems both in two and three space dimensions. We consider problems consisting of a symmetry element with one injector and one producer located at opposite corners of the domain. On the boundary of the reservoir, we impose the no-flow conditions

$$\vec{u} \cdot \vec{n} = 0 \ , \ (D(s)\nabla s) \cdot \vec{n} = 0 \qquad (15)$$

The mobilities of water and oil are defined by the simple identities

$$\lambda_w(s) = s^2 \ , \ \lambda_o(s) = (1 - s)^2 \qquad (16)$$

All computations were performed on locally refined grids. Grid blocks have four different levels, i.e. the finest grid blocks can be obtained from coarsest blocks by four consecutive refinements.

All the figures showing the water distribution in a reservoir were produced by POSTSCRIPT. A gray-scale has been used to indicate the water saturation, with black representing water (s=1) and white representing oil (s=0).

9.1. Two-dimensional results

In two space dimensions, simulations were performed to investigate the effect of permeability variations on the flow.

Figure 3 shows the solution of the quarter five-spot problem at 0.3 PVI with a high-permeability channel running from injector to producer. The permeability contrast between the channel and the rest of the reservoir is a factor 10. A total of 879 discretisation blocks was used for the solution of this problem.

In figure 4 the solution at 0.5 PVI is displayed of a problem having two low-permeability barriers. The permeability contrast is 200. The problem was solved using 885 discretisation blocks.

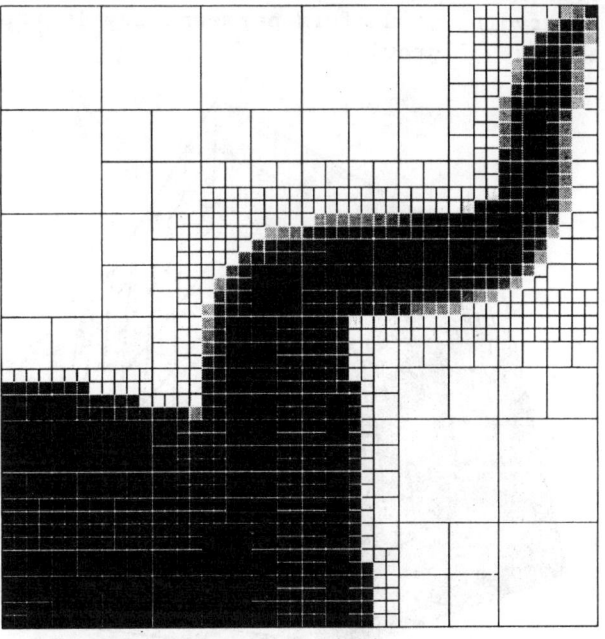

FIG. 3. Solution of the quarter five-spot problem with a high-permeability channel.

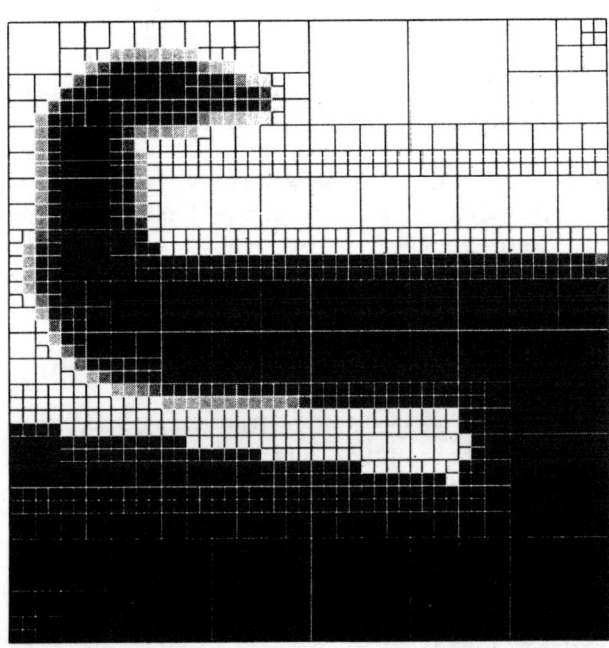

FIG. 4. Solution of the quarter five-spot problem with two low-permeability barriers.

9.2 Three-dimensional results

Figure 5 illustrates a solution at 0.15 PVI with 7188 grid blocks for a 3D-reservoir of uniform permeability in the absence of gravity.

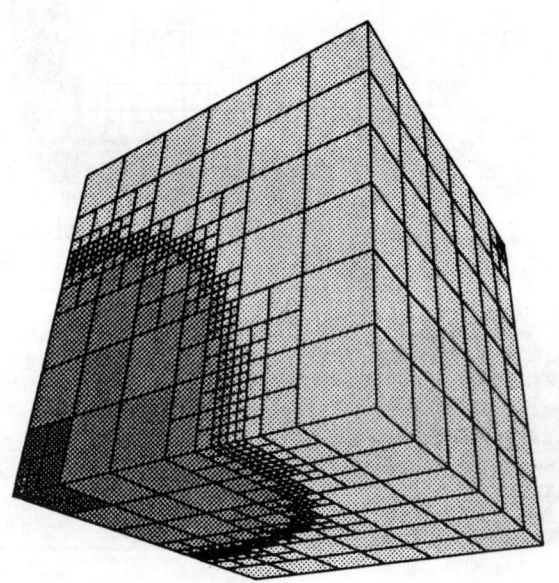

FIG. 5. Solution for a 3D-reservoir with uniform permeability.

10. CONCLUSIONS

The computational technique, based on adaptive local grid refinement, mixed finite elements, multi-grid, and the method of characteristics, is well-suited to solve convection-diffusion equations with complicated nonlinearities. Integration along streamlines allows an accurate treatment of nonlinear convection.

The method introduces almost no numerical diffusion. Compared to standard methods, large time steps can be used.

When applied to reservoir simulation problems, this method leads to considerable savings in computer time and storage. It also enhances the resolution of sharp fluid fronts.

REFERENCES

Chang, T. and Hsíao, L., 1988, The Riemann problem and interaction of waves in gas dynamics, John Wiley, New York.

Concus, P. and Proskurowski, W., 1979, Numerical solution of a nonlinear hyperbolic equation by the random choice method, J. Comput. Phys. , 30, p. 153-166.

Holden, H., Holden, L. and Hoegh-Krohn, R., 1988, A numerical method for first order nonlinear scalar conservation laws in one dimension, Comput. Math. Applic., 15, p. 595-602.

Lax, P.D., 1973, Hyperbolic systems of conservation laws and the mathematical theory of shock waves, SIAM, Philadelphia.

Leveque, R.J., 1982, Large time step shock-capturing techniques for scalar conservation laws, SIAM J. Numerical Anal., 19, p. 1091-1109.

Proskurowski, W., 1981, A note on solving the Buckley-Leverett equation in the presence of gravity, J. Comput. Phys., 41, p. 136-141.

Schmidt, G.H. and Jacobs, F.J., 1988, Adaptive local grid refinement and multi-grid in numerical reservoir simulation , J. Comput. Phys. , 77, p. 140-165.

Strang, G., 1968, On the construction and comparison of difference schemes, SIAM J. Numerical Anal., 5, p. 506-517.

2nd European Conference on the Mathematics of Oil Recovery
© D. Guérillot, O. Guillon (Editors) and Éditions Technip, Paris 1990, pp. 263-270
27 rue Ginoux, 75015 Paris

A Coordinate System for Local Grid Refinement Close to Wells

S. Ekrann[1]

ABSTRACT

The construction of an orthogonal curvilinear grid is described, suitable for local refinement close to wells. The grid is obtained by conformal mapping. It is approximately polar close to the well, and provides for a smooth transition to a surrounding cartesian grid. It is shown that this grid has several advantages over competing grids. Examples illustrate that strongly improved accuracy, over coarse grid simulations, is obtainable with relatively few extra grid blocks.

1. INTRODUCTION

Reservoir simulation must often, by practical necessity, proceed with rather coarse numerical grids. In areas with large saturation gradients or rapidly varying pressure gradients, large numerical errors may result. An attractive remedy is local grid refinement. The present paper considers local grid refinement close to wells.

Cartesian refinement (see Fig.1) has been studied by several authors (Ewing et al.,1988; Forsyth et al.,1985; Han et al.,1987; Heineman et al.,1983; Quandalle et al.,1983; Quandalle et al.,1985; Rosenberg, 1982; Wasserman, 1987). Significant loss of accuracy tends to result at the transition between fine and coarse grids. Two main effects seem to be at work. For block centred grids, the local truncation error becomes of order zero due to the different grid block sizes in fine and coarse grids (Aziz et al., 1979), even when block centres are aligned. Normally, block centres will not be aligned (compare Fig. 1). One coarse block will couple to several fine blocks, implying that the simplest finite difference approximation will generally be of order -1 (Ewing et al.,1988; Forsyth et al.,1985; Han et al., 1987; Heineman et al.,1983; Quandalle et al.,1985). Interpolation between coarse grid blocks can produce a higher order approximation in this situation, at the expense of involving more grid blocks (Ewing et al.,1988; Heineman et al.,1983; Quandalle et al., 1985; Rosenberg, 1982; Wasserman, 1987). This has the significant disadvantage of further complicating the structure of the resulting coefficient matrices, and even sometimes destroying useful properties like symmetry (Ewing et al.,1988).

---

(1) Rogaland Research Institute, PO Box 2503, Ullandhang, N 4001 Stavanger, Norway.

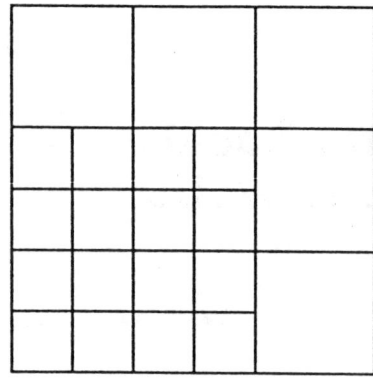

FIG. 1. Example cartesian local grid refinement.

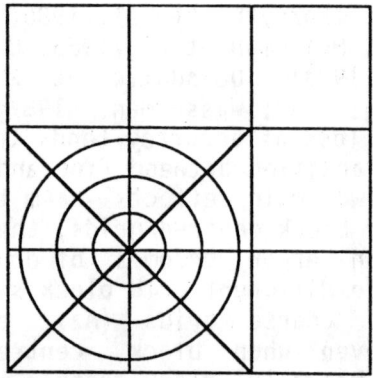

FIG. 2. Example hybrid grid as suggested by Pedrosa & Aziz (1985).

Pedrosa & Aziz (1985) suggested to use a polar grid in the regions to be refined (well regions). A hybrid grid results (see Fig.2), with several potential advantages. Firstly, one has the possibility to roughly match the radial grid block lengths of the outermost blocks with those of the surrounding cartesian grid, thereby largely avoiding one of the problems mentioned above. Secondly, fine and coarse blocks are matched angularly, reducing the need for interpolation between coarse blocks. Finally, and perhaps most importantly, the polar grid lines up with the dominant flow direction close to the well. One would expect, therefore, that fewer refined blocks are necessary to obtain a given accuracy, than what is the case with cartesian refinement.

There are some drawbacks inherent in the Pedrosa & Aziz approach, however. In computing transmissibilities for the irregularly shaped blocks at the transition between grids, they resort to fictious blocks of regular shape. This is heuristic and somewhat unsatisfactory. No analysis of accuracy was given. More importantly, perhaps, it is difficult to see how one could practically implement their grid with non-square well regions or off-centre wells.

The present paper builds on the hybrid grid idea of Pedrosa & Aziz, while trying to alleviate the shortcomings outlined. The construction of an orthogonal curvilinear grid is described, which is approximately polar close to the well and provides for a smooth transition to the surrounding cartesian grid. The grid is constructed by conformal mapping.

2. A COORDINATE SYSTEM BASED ON CONFORMAL MAPPING

The desired mapping is conveniently formulated via complex variables. Mappings defined by analytic functions are automatically conformal, i.e. angle preserving (Kreyzig, 1967). The construction proceeds in two steps: The Schwarz-Christoffel transformation (Spiegel, 1964) maps the upper half space onto the interior of a polygon. For ease of implementation, we have limited attention to rectangular well regions. For mapping onto a rectangular region, the Schwarz-Christoffel transformation takes the simple form

$$w(z) = B \int_0^z \frac{dt}{\sqrt{(1-t^2)(1-k^2 t^2)}} \qquad (1)$$

The mapping is illustrated in Fig.3. The real axis maps onto the boundary of the rectangle. z_0 maps onto w_0. B is a parameter determining the size, and k a parameter determining the length/height ratio of the rectangle. Note that the mapping is not easily invertible. For a given rectangle length/height ratio, therefore, k must be determined numerically. Similarly, for a given well position w_0, z_0 must be determined numerically.

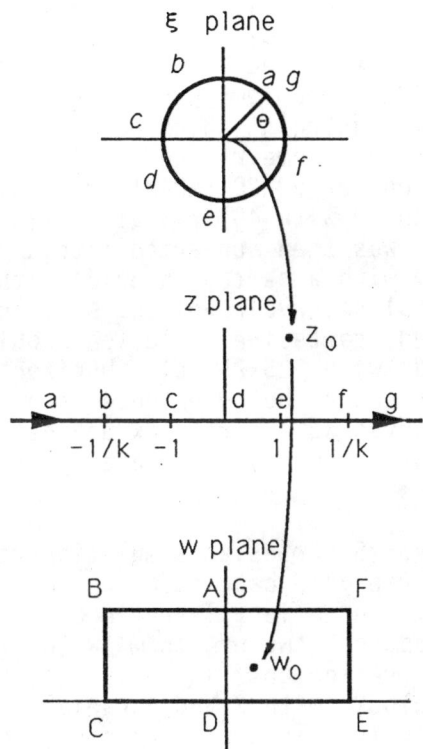

FIG. 3. Illustration of the conformal mapping.

To complete the mapping,

$$\xi(z) = e^{i\theta}\,\frac{z - z_0}{z - \bar{z}_0} \qquad (2)$$

maps the upper half space onto the interior of the unit circle. The real axis maps onto the circumference, and z_0 maps onto the origin, as illustrated in Fig.3. θ is an additional mapping parameter, determining the circumferential positions of the images of points on the real axis. This mapping is readily invertible:

$$z(\xi) = \frac{e^{i\theta} z_0 - \xi\,\bar{z}_0}{e^{i\theta} - \xi} \qquad (3)$$

Combining eq.s (1) and (3), we obtain a conformal mapping $w(z(\xi))$ from the interior of the unit circle onto the interior of a given rectangle, such that the origin $(0,0)$ maps onto a given point w_0 (well position), and such that the circumference maps onto the rectangle boundary. By the angle preservation property, if two curves intersect at a right angle in the ξ-plane, so will their images in the w-plane. A curvilinear orthogonal grid in the rectanglular well region can be constructed, therefore, as the image of a polar grid in the unit circle. The polar coordinates (ρ,α) in the ξ-plane can serve as (curvilinear) coordinates in the well region. Close to the well position w_0, the grid will be approximately polar. Grid lines will still be orthogonal at the rectangle boundary, i.e. constant α curves will meet the rectangle boundary at right angles (except in corners, which are socalled critical points of the mapping). This provides for a smooth transition to a possible cartesian surrounding grid.

Example grids are shown in Figs. **4** and **5**. A surrounding cartesian grid is also indicated. As can be seen, transition between grids is smooth, and blocks in the two grids match up exactly along the intersection between grids. The well region can have any (rectangular) shape, and the well position is arbitrary. In practice, one would probably not wish to operate with strongly off-centre wells in strongly non-square regions, because of very narrow blocks close to the well, and very elogated blocks at the far rectangle edge (see Fig. 5).

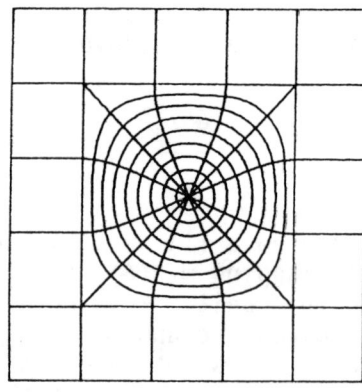

FIG. 4. Example curvilinear grid.

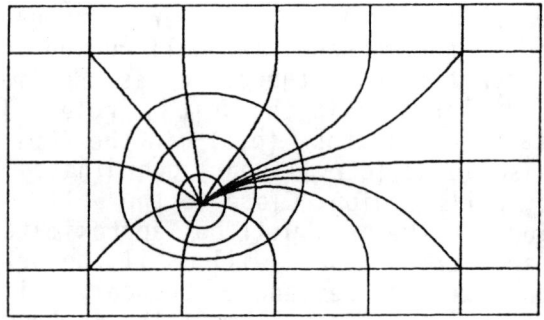

FIG. 5. Example curvilinear grid.

3. EXAMPLE SIMULATIONS

The local grid refinement outlined above has been implemented into a reservoir simulator. In this implementation the innermost blocks are coalesced into one single numerical block, such that there is only one well block in each vertical layer. The implementation is for vertical wells only. No refinement is performed vertically. The horizontal refinement runs from the top to the bottom of the reservoir. Computation of transmissibilities is described in Appendix A. A brief truncation error analysis is given in Appendix B.

Some example simulations will be reported. Unfortunately, neither Pedrosa & Aziz' method nor cartesian refinement has been conveniently available. Thus, direct comparison between refinement methods was not performed. It is demonstrated, however, that dramatically increased accuracy, above unrefined simulations, is obtainable with the present technique, with relatively few additional grid blocks.

Our examples are based on the second SPE comparison project (Chappelear et al., 1982). This is a one-well gas and water coning problem with 15 vertical layers. Prescribed rate variations are strong. The problem has a bottom-hole pressure constraint which becomes active. The origical geometry had radial symmetry. A converged 2D radial grid solution was produced with 40 radial cells. The problem was then converted into a square geometry with a cartesian grid, with the same total reservoir volume. A reasonably converged cartesian solution could be produced with 25x25 cells horizontally. The discrepancy between these two solutions serves as a yardstick against which to measure accuracy of locally refined solutions.

A 5x5x15 cartesian simulation was run as a coarse grid base case on which to improve. Three locally refined solutions were produced, two of them with (horizontal) refinement in only the central coarse block, with 2 and 5 radial blocks respectively. In the third case, the 3x3 central region was refined with 9 radial blocks, as illustrated in Fig. 4. In all cases, refinement was equidistant in ρ.

Displayed in Fig. 6 are the gas-oil ratios for all six cases. The abrupt fall in GOR at 720 days is caused by a tenfold decrease in the prescribed oil production rate, reducing GOR to solution levels. The base case coarse grid simulation is unable to cone gas at all, the GOR curve staying essentially flat throughout the entire period. Local refinement in the central block with only 2 radial blocks gives a dramatic improvement, although accuracy is not quite satisfactory. 5 radial blocks produce an answer well within the expected, as do the 9 radial blocks in the 3x3 region.

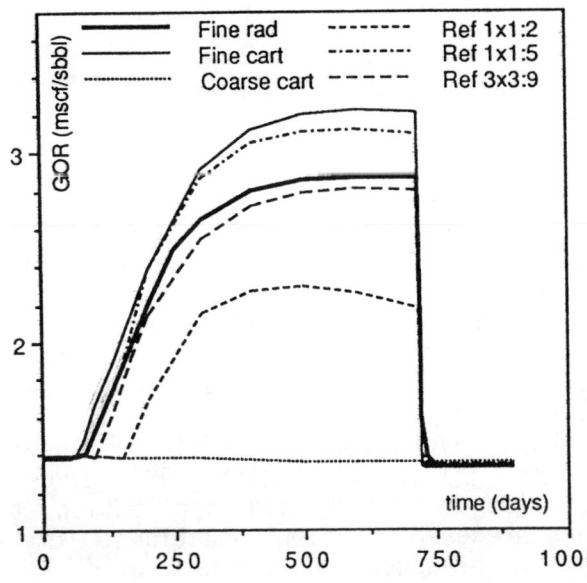

FIG. 6. Simulated GOR vs. time with different grid configurations. Data from SPE second comparison project.

4. SUMMARY AND CONCLUSIONS

The construction of a curvilinear orthogonal grid, suitable for local grid refinement in rectangular well regions, has been reported. The grid is a distorted polar grid, obtained by conformal mapping. The grid is approximately polar close to the well. It provides for a smooth transition to a surrounding cartesian grid, in the sense that grid line tangents are continous across the intersection between grids. Along the intersection, a given grid block has only one geometrical neighbour in the opposite grid.

Such a grid can be constructed for any well region height/length ratio, and for any well position inside the well region.

It was demonstrated that local refinement based on this technique can produce strongly improved accuracy as compared to unrefined simulations, with relatively few additional grid blocks.

Arguments were given why one should expect the present technique to be more accurate, for a given number of blocks in the refined region, than are competing local refinement methods, such as cartesian refinement or the undistorted polar refinement of Pedrosa & Aziz (1985).

5. ACKNOWLEDGEMENT

The work on which this paper is based, was performed under contract with BP Petroleum Development (Norway) Limited, whose permission to publish is gratefully acknowledged. I am also very grateful to Hans Kleppe for performing the simulations reported.

6. BIBLIOGRAPHY

Aziz, K. and Settari, A., 1979, Petroleum Reservoir Simulation, Applied Science Publishers, London, sec. 3.5.1.

Chappelear, J.E. and Nolen, J.S., 1982, Second Comparative Solution Project: A Three-Phase Coning Study, paper SPE 10489 presented at the Sixth SPE Symposium on Reservoir Simulation, Feb. 1982.

Ewing, R.E. and Lazarov, R.D., 1988, Adaptive Local Grid Refinement, paper SPE 17806 presented at the SPE Rocky Mountain Regional Meeting, May 1988, p. 643-651.

Forsyth, P.A. and Sammon, P.H., 1985, Local Mesh Refinement and Modelling of Faults and Pinchouts, paper SPE 13524 presented at the 1985 SPE Reservoir Simulation Symposium, Feb. 1985, p. 267-278.

Han, D.K., Han, D.L., Yan, C.Z., and Peng, L.T., 1987, A More Flexible Approach of Dynamic Local Grid Refinement for Reservoir Modeling, paper SPE 16014 presented at the Ninth SPE Symposium on Reservoir Simulation, Feb. 1987, p. 243-252.

Heineman, Z.E., Gerken, G., and von Hantelmann, G., 1983, Using Local Grid Refinement in a Multiple-Application Reservoir Simulator, paper SPE 12255 presented at the SPE Reservoir Simulation Symposium Nov. 1983, p. 205-218.

Kreyzig, E., Advanced Engineering Mathemetics, 1967, John Wiley and Sons, New York, p. 565.

Nghiem, L.X., 1988, An Integral Approach To Discretizing the Reservoir Flow Equations, SPERE, May 1988, p. 685-690.

Pedrosa Jr., O.A. and Aziz, K., 1985, Use of Hybrid Grid in Reservoir Simulation, paper SPE 13507 presented at the SPE 1985 Middle East Oil Technical Conference and Exhibition, March 1985, p. 99-111.

Quandalle, P. and Besset, P., 1983, The Use of Flexible Gridding for Improved Reservoir Modeling, paper SPE 12239 presented at the SPE Reservoir Simulation Symposium, Nov. 1983, p. 51-60.

Quandalle, P. and Besset, P., 1985, Reduction of Grid Effects Due to Local Sub-Gridding in Simulations Using a Composite Grid, paper SPE 13527 presented at the SPE 1985 Reservoir Simulation Symposium, Feb. 1985, p.295-305.

von Rosenberg, D.A., 1982, Local Mesh Refinement for Finite Difference Methods, paper SPE 10974 presented at the 57th Annual Fall Technical Conference and Exhibition of the SPE, Sept. 1982.

Spiegel, M.R., 1964, Complex Variables, Schaum's Outline Series, McGraw-Hill, New York.

Wasserman, M.L., 1987, Local Grid Refinement for Three-Dimensional Simulators, paper SPE 16013 presented at the Ninth SPE Symposium on Reservoir Simulation, Feb. 1987, p. 231-241.

Appendix A. DISCRETIZATION

In discretizing the governing equations, an integral (control volume) approach was adopted. Integration is performed over each grid block. Application of the divergence theorem implies introduction of interblock fluxes as variables. The technique is well known, and perhaps particularly useful in curvilinear grids. L.X. Nghiem (1988) gives a detailed account of its application to the reservoir flow equations, in a general orthogonal curvilinear grid.

The way the interblock fluxes are approximated largely determines the numerical method. The approximations used in the present implementation will be briefly described. A very simple scheme was chosen. There is nothing to prevent implementation of other schemes with the present coordinate system, of course, like the ones advocated by Nghiem (1988) and Pedrosa & Aziz (1985).

With one phase, and unit viscosity, an interblock flux F can be expressed as

$$F = - \int \int k \, \partial\Phi/\partial n \, dA \qquad (A.1)$$

The integration is performed over the common boundary of the two blocks in question, with dA being the surface element. Φ is the pressure potential and k the appropriate permeability component (it is assumed for simplicity that the principal axes of the permeability tensor are always aligned with the coordinate system). $\partial\Phi/\partial n$ is the normal component of the pressure potential gradient. In an orthogonal grid, this is equivalent to the directional derivative along the appropriate constant coordinate curve. In approximating (A.1), it is therefore natural to set

$$- F \sim k A \frac{\Phi_{s'} - \Phi_s}{l_{ss'}} \qquad (A.2)$$

Subscripts s and s' denote grid blocks. $kA/l_{ss'}$ becomes the transmissibility. $l_{ss'}$ is the distance, along the proper constant coordinate curve, between the computational points in the neighbouring blocks. Computational points are assumed to be positioned on the same constant coordinate curve. A is the area of the common boundary. Numerical integration is necessary to produce A and $l_{ss'}$, due to the curvilinearity of the grid. If k is different in the two blocks, harmonic averaging is used. In a cartesian grid, (A.2) would reduce to the standard block centered grid approximation.

The approximation (A.2) is unmodified at the transition between grids, i.e. when s is in the curvilinear and s' is in the surrounding cartesian grid. In that case, the two component length increments making up $l_{ss'}$ would belong to different grids. Computational points are chosen such that corresponding constant coordinate curves in the curvilinear and cartesian grids meet exactly at the grid interface (see Fig. 7)

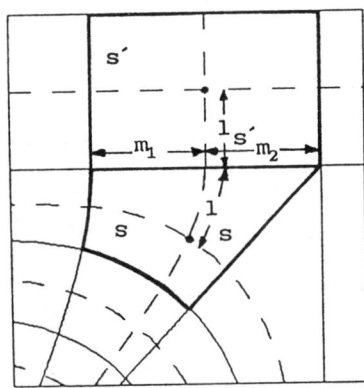

FIG. 7. Detail of composite grid, at intersection between grids.

Appendix B. TRUNCATION ERROR ANALYSIS

The local accuracy of the above approximations can be studied using the standard Taylor expansion techniques. Of particular interest is accuracy at the transition between grids, which will be covered in some detail. For convenience, reasoning will be carried out in 2D. Let now Φ be a sufficiently differentiable test function. Refer to Fig. 7. The curve joining the two computational points has a curvilinear and a rectilinear part. On that curve, Φ can be regarded as a function of l, arc length along the curve. On the interblock boundary over which integration is to be performed, Φ (and $\partial\Phi/n$) are sufficiently differentiable functions of m, arc length along the boundary. Expanding $\partial\Phi/\partial n$ about the point of intersection (l=0, m=0), indicated by subscript $_0$, and letting a dot indicate differentiation with respect to m, one obtains

$$\frac{\partial\Phi}{\partial n}(l=0,m) = (\frac{\partial\Phi}{\partial n})_0 + (\frac{\dot{\partial\Phi}}{\partial n})_0 \, m$$

$$+ (\frac{\ddot{\partial\Phi}}{\partial n})_0 \, \frac{1}{2} \, m^2 + .. \qquad (B.1)$$

A similar expansion can be carried out along l, to express $\Phi_{s'}$ and Φ_s in intersection point variables. Some care must be exercised in this case, however, since continuity of higher order derivatives is generally not guaranteed across the boundary. Collecting terms, one obtains

$$(\Phi_{s'} - \Phi_s) =$$

$$\Phi_{0+}{}' \, l_{s'} + \Phi_{0+}{}'' \, \frac{1}{2} \, l_{s'}^2 + \cdots$$

$$+ \Phi_{0-}{}' \, l_s - \Phi_{0-}{}'' \, \frac{1}{2} \, l_s^2 + \cdots \qquad (B.2)$$

269

Here, ' signifies differentiation with respect to arc length 1. Derivatives are limiting values, as the intersection point is approached. Subscripts $_{+-}$ indicate whether the approach is from the cartesian or from the curvilinear side. By assumption, Φ is sufficiently smooth. Any discontinuities must therefore be caused by discontinuity of curve tangents, curvatures etc. By construction, the curve has continous tangents. Thus

$$\Phi_{0+}' = \Phi_{0-}' = \left(\frac{\partial\Phi}{\partial n}\right)_0 \qquad (B.3)$$

The curvature is generally not continous, except in some cases with extensive symmetry. Thus, the r.h.s. second order terms in Eq. (B.2) will generally not cancel, and one may write

$$\left(\frac{\partial\Phi}{\partial n}\right)_0 = \frac{\Phi_{s'} - \Phi_s}{1_{ss'}} + O(1_{s'}, 1_s) \qquad (B.4)$$

with $1_{ss'} = 1_{s'} + 1_s$. Using (B.4) in (B.1) and inserting into (A.1), one obtains after performing the integration

$$-f \equiv -\frac{F}{m_1 + m_2} = k\frac{\Phi_{s'} - \Phi_s}{1_{ss'}} +$$

$$O(1_{s'}, 1_s) + O(m_1{}^2, m_2{}^2) \qquad (B.5)$$

since $m_1 = m_2$ by construction. f is the flux density. The discussion on truncation errors given in the Introduction pertained to the full second order differential operator. The results mentioned are therefore not directly comparable with those above. Generally, i.e. with no special symmetry to cause cancellation, one expects the order of an approximation to be reduced by one when the order of the differential operator is increased by one. The indications are, therefore, that the above approximations are of order zero, in the language of the Introduction.

Except when helped by symmetry, the Pedrosa & Aziz (1985) grid will have discontinuous tangents at the intersection between grids, in which case (B.3) is no longer true, and the approximation (B.5) becomes of order zero in $(1_{s'}, 1_s)$. Similarily, one can reason that the simplest (i.e. no interpolation between coarse blocks) schemes with cartesian refinement are equivalent to non-continous tangents, again producing a zero order approximation.

2nd European Conference on the Mathematics of Oil Recovery
© D. Guérillot, O. Guillon (Editors) and Éditions Technip, Paris 1990, pp. 271-280
27 rue Ginoux, 75015 Paris

Data Structure and Algorithms for Adaptive Mesh Refinement

T. Hermitte[1,2] and D. Guérillot[1]

ABSTRACT

Fluid flow in an oil and gas reservoir is governed by a system of nonlinear partial-differential equations with different types of boundary conditions. To compute an approximate solution to this evolving problem, the integration domain (geometry of the reservoir) is discretized. Given the large size of this three-dimensional geometry (several kilometers in areal extent and sometimes up to 100 meters thick), and the cost of fluid flow simulation depending on the number of unknowns for each grid block, the time spans evolved (several years of production) and the features of current computers, the mesh used for solving these partial-differential equations is not refined enough.

So as not to have too many grid blocks, adaptative mesh techniques are used to obtain a discretization that evolves during simulation. The local mesh refinement is done solely in some parts of the reservoir, e.g. in the vicinity of fronts, near the wells, etc., to improve the accuracy of the results. The main difficulties in this method stem from the dynamic aspect of the problem. Changing the mesh by the creation and disappearance of some grid blocks at given times causes the structures of current static data to be poorly suited for correctly accounting for this constant evolution. This paper describes an original dynamic data structure and new algorithms for mesh management (refining, enlarging, searching for neighboring cells, propagating). The way that the mesh structure is stored by dynamic memory allocation programming in ADA language is described for a three-dimensional case together with the different components (new numbering, geometric and physical data, etc.) required for the management
of each grid block. To test the evolution of the mesh, an experiment was performed on an analytical solution giving the exact position of the front in time.

1 INTRODUCTION

In a reservoir model, fluid flow in a porous medium is governed by a system of nonlinear partial-differential equations. To compute an approximate solution to this problem, the domain is discretized into N parts (called cells or grid blocks) in which all the unknowns of the problem (saturations, pressures, temperature for thermal flow, etc.) are associated. This collection of cells is called a mesh. From the mathematical standpoint, each unknown is represented by an N dimensional vector (where N is the number of cells). Given the large size of the three-dimensional geometry of a reservoir and the number of unknowns taken account in fluid flow simulators, several authors have considered the problem of adaptive mesh refinement (Ewing[9], Quandalle[8], Berger[6], Bank[7], Wasserman[4], etc.). With this method, we can increase the number of cells (or unknowns) in the areas where some difficulties are expected (often arround the wells) and decrease this number in others. Generally, this mesh does not change in time. Therefore, it may be worthwhile to have more unknowns in some regions where temporary local phenomena appear such as a large variation of fluid saturation, for example. For this, a new adaptive data structure must be elaborated. The purpose of this study is to develop a new dynamical data structure for a

(1) Institut Français du Pétrole, 1 et 4, avenue de Bois-Préau, 92506 Rueil Malmaison, France.
(2) Université de Provence Aix-Marseille I, 3, place Victor Hugo, 13331 Marseille Cedex, France.

three-dimensional geometry. With a programming language like Fortran usually used today in simulators, the mesh modification over time becomes very laborious and also not efficient. Here, the programming language chosen is Ada. The reasons and the criteria for this choice are described in Section 2.2.

The different aspects and techniques used to develop the new data structure (representation, storage, numbering, etc.) are given in Sections 2.3 and 2.4. Section 3 describes the several algorithms used for mesh management such as refinement and clustering because mesh modification, a new fast and efficient search of surrounding cells, is studied.

The good behavior of the new dynamic data stucture and the efficiency of mesh management algorithms are illustrated by a typical example in the last section.

The data structure and all the mesh management algorithms have been programed in Ada and run for the three-dimensional case.

2 DATA STRUCTURE

For the adaptive mesh refinement technique, the data structure plays an important role in simulation (Berger[4]). Modifications of the mesh by creation and deletion of some grid blocks over time require the data structure used to be dynamic (its memory space varies from one moment to another) and to ensure rapid and effective management of the data.

Present static data structures, in which the mesh is fixed permanently when the simulation is initialized, lend themselves poorly to such techniques. A modification of these methods to make this structure in some way "dynamic" would be unfavorable and too awkward for convenient management.

Since the purpose of adaptive mesh refinement is to solve evolution problems more accurately, but at reasonable cost, a data structure must be worked out that is appropriate to the problem posed, and one which does not jeopardize the rest of the code for existing simulators.

In this study, we shall first provide some standard definitions made to pinpoint the problem. We shall then give the reasons for and the choice of the data processing language and introduce some specific concept used for developing the data structure. We shall then describe the data structure concerning the mesh as well as the cells making it up.

2.1 Definitions

• A cell is said to be refined if it is subdivided into r^d cells, where d is the dimension space.

• A cell is said coarse if it belongs to the base mesh (mesh at time t=0).

• The level of a cell is the level of refinement to which it belongs. Coarse grid blocks are allocated to level 1. If a cell with level (m) is refined, level $(m + 1)$ is allocated to the r^d subcells.

• Two cells are said to be neighbors if they have a face, an edge or a point in common. The cell itself is considered to be a neighbor. Thus, for a coarse grid block that is not on the side of the domain, the number of its neighbors is 3^d.

2.2 Programming language

The programming language must respond to the **dynamic aspect of the problem.** The change in the number of cells over time implies that the memory space used to store the mesh also varies, because some cells are destroyed whereas others are created. With the tools available in a programming language like Fortran, a rather large memory space would have to be reserved for the initialization of the mesh, since the maximum number of cells over time is totally unknown in advance. Thus it is perfectly conceivable that, for a given simulation, this static space allocated for storing the mesh could be too large (in the best possible case) or insufficient, which would stop the simulation and require recreating a version runnable with a much larger space. The idea would be to be able to **allocate the memory space dynamically** so as to occupy only the space required for the mesh at the present moment.

Acces to the cell-related data must be rapid and efficient. At each modification of the mesh, a number of processing operations must be repeated (search for neighboring cells, for example). Acces must therefore be possible to the data that have to be processed as rapididly as possible, implying a simple, effective link between the cells. From the mesh management standpoint, the algorithms can be defined recursively, i.e. some procedures or functions inherently contain calls on themselves. To simplify their writing, **the language must be recursive.**

We also need a language that can be compatible with Fortran in order to avoid rewriting resolution codes that already exist.

Ada language (Barnes[14], for example) appears to be the ideal programming language. It is compatible with Fortran and is beginning to be used in the petroleum industry (Ramamurthy[10]). In addition to being recursive, the concept of a dynamic object introduced by "pointers" also makes it possible to work out the data structure in relation to the mesh. Among modern languages such as C and Pascal for example, the enables us Ada language has been chosen because the software permit to program more clear and concise codes.

2.2.1 Pointers

In most programming languages such as Fortran and Basic, the variables are referred to by a name (simple variable) or by a name and a shift (arrays, records). At compilation, all the occurrences of the identifier of the variable in the program are replaced by an address. Since when running the program all the addresses are fixed, this is termed **static memory allocation**. However, programming languages exist in which variables of any type can be allocated during execution, when the need arises, under the control of the program, by means of special instructions. This is referred to as **dynamic memory allocation**. In Ada, as in Pascal, these variables are designated by their address, deposited in a variable of a special type, called a "pointer".

2.2.2 Chained lists

A chained list is a collection of elements, called links, each consisting of:

- the information to be processed (the cells in this case),

- a chaining link with other elements of the list performed by "pointers".

The use of a chained list also requires an isolated pointer containing the address of the first link of the list. Each link then "points" to its succesor.

2.2.3 Recursivity

Like in all modern programming languages (such as C and Pascal), Ada is a **recursive language**. We can write a subroutine, which inherently has a call on itself. Such a subroutine is said to be recursive. This mechanism is made possible, because each procedure or function contains an environment and variables specific to it. Thus, at each call of the subroutine, a new environment and new variables are attributed to them.

2.3 Structure of a cell

A cell is represented by a "pointer" of which the records serving to identify it are the following (Fig. 1):

- **Num**: cell number.

- **Status**: cell status (refined or not).

- **Geometry**: contains the geometric parameters associated with the cell: size in three main directions, depth, etc.

- **Physical**: contains the physical parameters associated with the cell: porosity, permeability, etc.

- **Unknowns**: saturations, pressures, etc.

- **Ant**: "pointer" performing chaining with the previous cell.

- **Suc**: "pointer" performing chaining with the next cell.

N	U	M	Status	Unknowns
Geometry			Physical	
• Ant			Suc •	

Fig. 1 - Structure of a cell

This form of a structure for cells remains very flexible. A component can be added or deleted without having to change the entire structure.

2.3.1 Cell numbering

The choice of numbering is very important because it influences the structure of the linear system and hence, directly the performance of the solver used. The literature offers many cell numbering methods, each having its advantages and drawbacks. Brand[11] uses the RRB method (Repeated Red Black) in which the cell numbering is based on the parity of the sum of the column number with that of the line to which the cell belongs, while Heinemann[3] uses a more sophisticated method.

One of the simplest methods consists in numbering the cells by increasing level of refinement (Forsyth[5]), but this method must be repeated at each important modification of the mesh, and, in addition, the matrix loses its band structure. The idea therefore would simply be to number the basic mesh and to shift this numbering for local cell refinement. Thus the matrix preserves a good structure, but the shift is expensive in computer time.

The cell numbering adopted in this article is a compromise between the above two methods (Hermitte[12]).

Let L_{max} be the maximum level authorized for refinement. A cell number is composed of a series of integers a_k where the subscript k corresponds to a refinement level ($Lev(a_k) = k$). We can write:

$$\mathbf{Num} = a_1 a_2 ... a_{L_{max}}$$

where $1 \leq a_1 \leq (Nx * Ny * Nz)$
and $0 \leq a_k \leq r^d$ $\quad k\epsilon[2, L_{max}]$

$$\mathbf{Num} = \sum_{k=1}^{L_{max}} a_k * 10^{(d-1)(L_{max}-k)}$$

The numbering of the coarse grid block, i.e. of the integers a_1, is achieved by first scanning the X-axis, then the Y direction, and finally the vertical axis (Fig. 4). The distribution of the integers a_k such that $k > 1$ is given in Fig. 2.

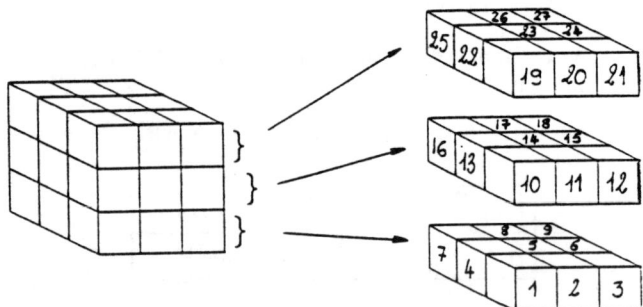

Fig. 2 - Distribution of integers for a refined cell

The number allocated to a cell also gives the cell's refinement level. It corresponds to the last subscript k such that $a_k \neq 0$.

Based on the number of a cell, all the numbers of its ancestors can be determined.

Thanks to this cell numbering, for an integer corresponding to a refinement level higher than 1, its location in the cell where it originates is known. For example (Fig. 3), the cell whose number is 320 is located to the south of cell 300.

2.4 Structure of the mesh

The mesh is represented by a two-directional chained list, i.e. a shift can be made in this list from a current node along two direction: down or up. The mesh can be represented by a tree-like structure in which the roots are given by the coarse grid blocks, the branches by the two-directional links, and the leaves by the cells created during refinement (Fig. 3).

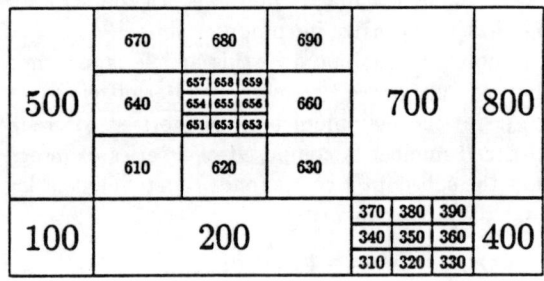

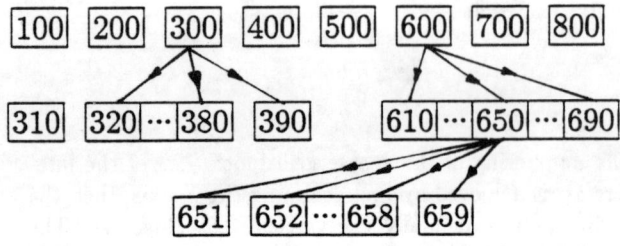

Fig. 3 - Tree-like structure for the mesh

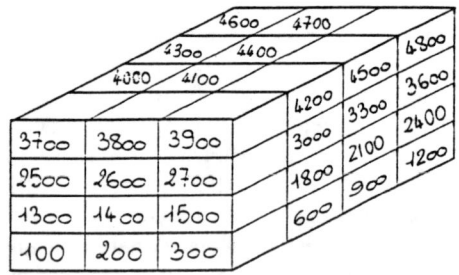

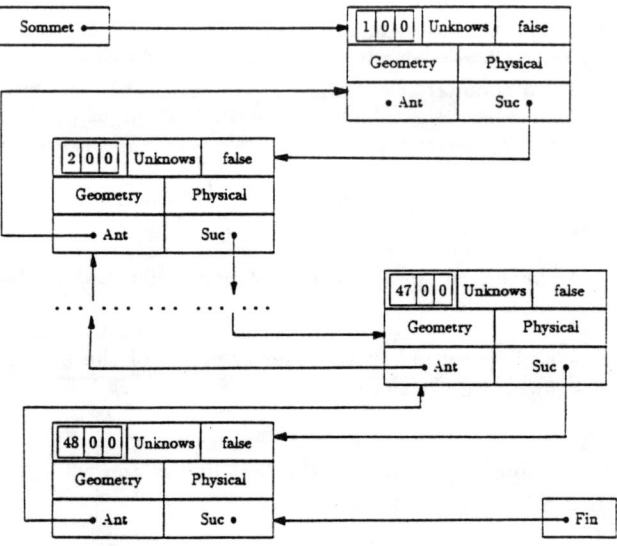

Fig. 4 - Structure of the mesh

3 MESH MANAGEMENT

For the adaptive mesh refinement technique, the cells are managed by means of algorithms such as refinement and clustering. These procedures, which manipulate the mesh on the basis of clearly-defined criteria (not developed here), directly influence the number of grid blocks and hence the neighborhood of certain cells. From the numerical standpoint, the schemes normally used, like the finite volume method, take into account the data and discrete unknows of the neighboring cells. Hence it appears clear that, if the number of neighbors for a cell is too large, the writing of the scheme will be very difficult. To avoid the occurrence of such cases, the procedure used propagates the refinement according to certain criteria, so as to increase the maximum number of neighbors for a cell. This section describes the algorithms for mesh management procedures, such as refinement and clustering. The criteria involved in propagation are then described, followed by the search for the neighboring cells, which plays an important role in the adaptive mesh refinement

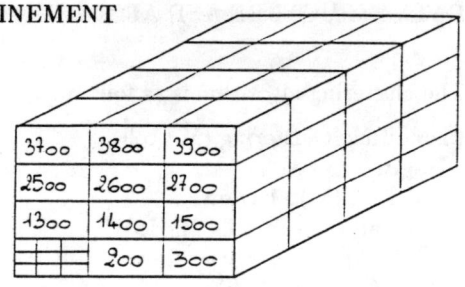

technique. The modification of the mesh at given times disturbs the neighborhood of some cells. The neighbors of these grid blocks must therefore be found. The new algorithm is based on the cell numbers. No intersection computation is used. **This speeds up the search considerably.**

3.1 Refinement

Refinement is performed by dividing a grid block into r^d cells having the same volume, in other words the size of the cell in each direction is divided by r (Fig. 5). In this study we take $r = 3$. Thus, to refine a cell containing a well, the well is always found to be at the middle of a cell, and the productivity index does not have to be calculated again. **The subdivision into an uneven number of cells is also well suited for multigrid methods** (the prolongation and restriction operators).

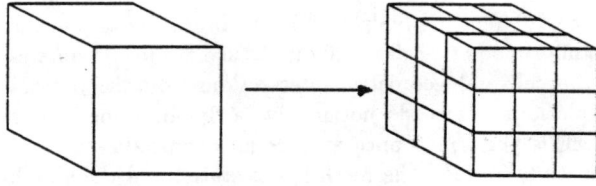

Fig. 5 - Refinement

Note that cell refinement is possible only if the level of this cell is not equal to L_{max}.
The mesh refinement algorithm is as follows:

```
procedure Refine (M : cell);
  begin
    if M.status = true then
      for all the sub-cells Mi do
        Refine(Mi)
      end for
    else
      M.status = true
      Lev = Level(M)
      for i= 1 to 3d do
        creation of the sub-cell Mi
        Mi.num = M.num + i*10^((d-1)(Lmax-Lev-1))
        Mi.status = false
        links
      end for
      Insert (M1...M3d)
    end if
  end Refine
```

The last two chaining operations must be performed in the order indicated because a cell is accesssed only by means of the chaining. If the order were reversed, the last subcell created would be chained with its elder, because it is the next one on the list. The refined cell is not deleted (Fig. 6). This helps to avoid its recreation during clustering.

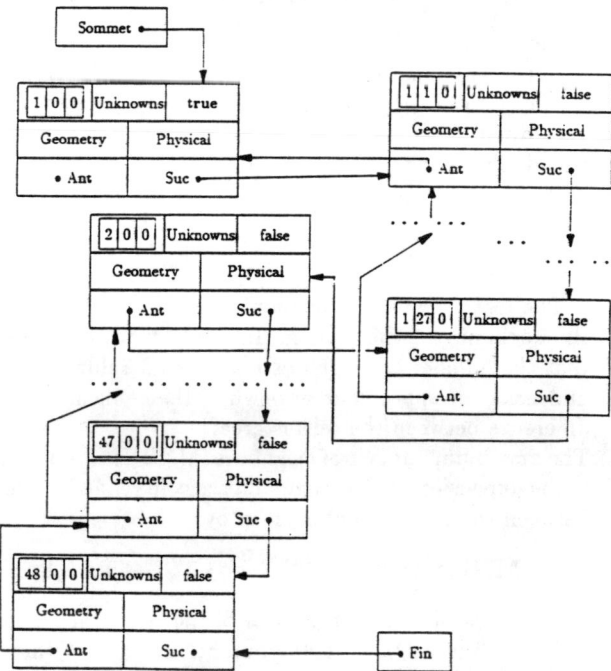

Fig. 6 - Example of mesh management

3.2 Clustering

The principle of this algorithm is to return to the previous state of the cell, i.e. this procedure can be considered as the reverse of refinement (Fig. 7).

Fig. 7 - Clustering

Note that only a cell that has been refined can be enlarged. In practice, this restriction doesn't seem to be too troublesome.
Always keeping in memory a cell that has been refined helps to refer to it for clustering (Fig. 6).
The subcells are suppressed automatically by the program. In the Ada language, if a pointer can no longer be reached, it is systematically destroyed. This happens as soon as the chaining with the other cells is broken.

275

The clustering algorithm is as follows.

procedure Clustering (M : cell)
  begin
    if Level(M)>1 **then**
    for all the sub-cells M_i **do**
      if M_i.status = true **then**
        Clustering(M_i)
      end if
    end for
    M.status = false
    deletion of the sub-cells
    end if
  end Clustering

3.3 Propagation

The reason for a propagation procedure were stated at the beginning of the section. In brief, it has been found that the number of neighbors rises considerably with the dimension of space, but also when the refinement level difference between the cells is great.

The maximum number of neighbors (MNBN) as a function of the dimension of space and difference in refinement level between the cells ($Dlev$) is given by:

$$\text{MNBN} = nf * 3^{2Dlev} + ne * 3^{Dlev} + nc + 1$$

where nf is the number of faces of the cell (0 in 1-D and 2-D, and 6 in 3-D), ne is the number of edges of the cell (0 in 1-D, 4 in 2-D, and 12 in 3-D), and nc the number of corners of the cell (2 in 1-D, 4 in 2-D, and 8 in 3-D).

Dim	Dniv	NBVM
	1	17
2-D	2	41
	3	113
	1	99
3-D	2	603
	3	4707

According to this table, it seems reasonable to set a $Dlev$ between the neighboring cells, even if these extreme cases only occur very rarely. A $Dlev$ limit of 1 is applied to fix the number of tolerable neighbors.

Criteria: If a cell has a neighbor such that the difference in refinement level between these two cells is greater than 1, the one with the lower level is refined (Fig. 8).

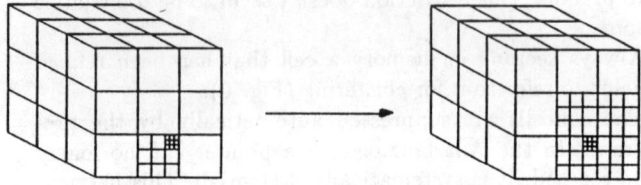

Fig. 8 - Criteria for propagation

For propagation, only cells whose level of refinement is strictly higher than 2 can be performed.
The propagation algorithm is as follows:

procedure Propagation (M : cell)
  begin
    if (M.status = false) and ($Level(M) > 2$) **then**
      Neighborhood(M)
      for all the neighbors V_i of M **do**
        if $Dlev(M, V_i) > 1$ **then**
          Refine(V_i)
          Propagation(V_i)
        end if
      end for
    end if
  end Propagation

3.4 Search of surrounding cells

The different methods used in the literature to find the neighborhood of a cell mainly make use of the coordinates of the cell and geometric intersections. In the present simulations, which do not involve a dynamic mesh, this search is performed once and for all at initialization. In this study, in fact, the mesh is not static, and the search for neighbors must be repeated at each mesh modification. The methods used thus far seem to be rather costly in time and computation. Therefore an attempt must be made to use neither the coordinates nor the geometric intersections. **The new algorithm given here takes into account only the data contained in the cell number.** The principle of the method is a two-step procedure: first, the search for the neighbors with the same refinement level as the cell. These cells are then qualified as "potential neighbors". The proper neighborhood is then determined from these as follows:

- if the "potential neighbor" exists and is not refined, it is a "good" neighbor;

- if the "potential neighbor" is refined, the "good" neighbors corresponding to its place in the neighborhood must be found among these direct sub-cells and not beyond (thanks to propagation);

- if the "potential neighbor" does not exist, the "good" neighbor is the one that has the number of the cell in which it originates; this cell still exists thanks to propagation.

To find the "potential neighbors" the processing of a cell is slightly different according to whether or not it belongs to the coarse level.

For a coarse cell, the neighborhood is determined from integer a_1 as a function of the number of cells along the main directions (Fig. 9).

a_1+Nx-1	a_1+Nx	a_1+Nx+1
a_1-1	a_1	a_1+1
a_1-Nx-1	a_1-Nx	a_1-Nx+1

Fig. 9 - 2-D example of neighborhood for a coarse cell located in the center of the mesh

If the coarse cell is located on the edge or a corner of the mesh, its "potential neighbors" are a sublist of the one given in the table. For a cell with a level higher than 1, its position in the cell from which it derives is observed. Based on this information, its neighboring "brothers" are first placed, then the other "potential neighbors" are determined by means of two very simple permutations ("vertical" and "horizontal") of the integers a_k making up the numbers of the cells already placed. Consider the cell with number $ab10$ (with $L_{max}=4$ in 2-D).

• first step: placement of the "brothers" (Fig. 10).

......	ab40	ab50
......	ab10	ab20
......	......	......

Fig. 10 - Placement of brothers for cell $ab10$

• second step: placement of the missing "potential neighbors" (Fig. 11).

hp(ab40)	ab40	ab50
hp(ab10)	ab10	ab20
vp(hp(ab10))	vp(ab10)	vp(ab20)

Fig. 11 - Potential neighborhood for cell $ab10$

$vp(i)$ and $hp(i)$ are respectively the "Vertical Permutation" and the "Horizontal Permutation" of the number i.

Let k be a refinement level. The processing of a cell by permutations is given in Fig. 12.

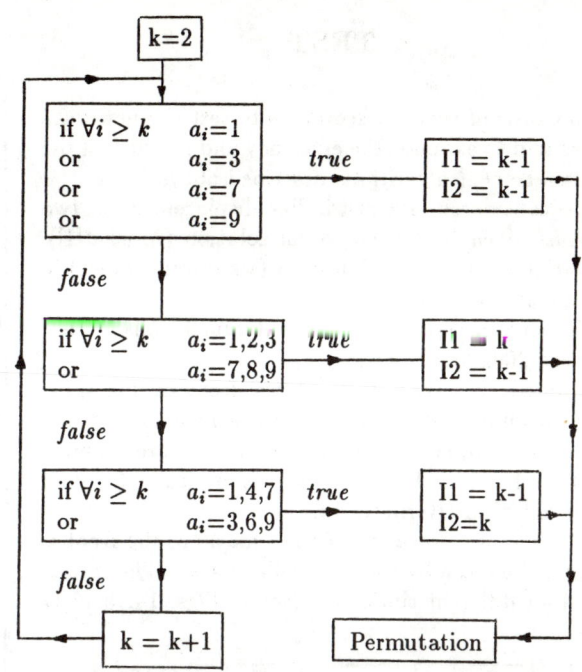

Fig. 12 - 2-D permutation processing

where $I1$ is the subscript for "horizontal permutation", and $I2$ is the subscript for "vertical permutation".

The algorithms for both permutations are similar.

The number $b_1..b_{L_{max}}$ of the "potential neighbor" is determined as follows.

procedure Permutation (I: permutation subscript
                    direction : "v" or "h"
                    $a_1..a_{L_{max}}$: cell number)
    begin
      for k=1 **to** (I-1) **do**
        $b_k = a_k$
      end for
      for k=I **to** L_{max} **do**
        $b_k = Change(a_k, direction)$
      end for
    end Permutation

The *Change* function is based on the following remark: a mesh that is totally refined indicates a periodicity of the integers corresponding to the most refined level, according to the lines and the columns (Fig. 13).

..7				..8				..9
..4	.5	.6	..4	.5	.6	..4	.5	..6
..1				..2				..3
..7	.8	.9	..7	.8	.9	..7	.8	..9
..4				..5				..6
..1	.2	.3	..1	..2	..3	..1	.2	..3

Fig. 13 - 2-D example of periodicity for the last integers

277

4 TEST

The purpose of this last section is to test the new data structure and to examine the efficiency and rapidity of the mesh management algorithms described above.

The case considered is a piston-like displacement in two dimensions given by an analytical solution (Muskat[1]). The position of the saturation front (water-oil contact) in space and time is known.

The initial mesh consists of 13x20x1 coarse cells. Their volume are 200x200x50 m^3.

The intended exercise is to track the front during time. The algorithm for analyzing the front is the following: if a mesh has an occupied intersection with the representative curve of the front at the present time step, it is refined. If not, it is enlarged (Hermitte[13]).

Thus, at each new iteration of the time step, the front is located on the mesh by the local refined area. The results obtained for different times are given in Figs. 14,15,16,17 and 18.

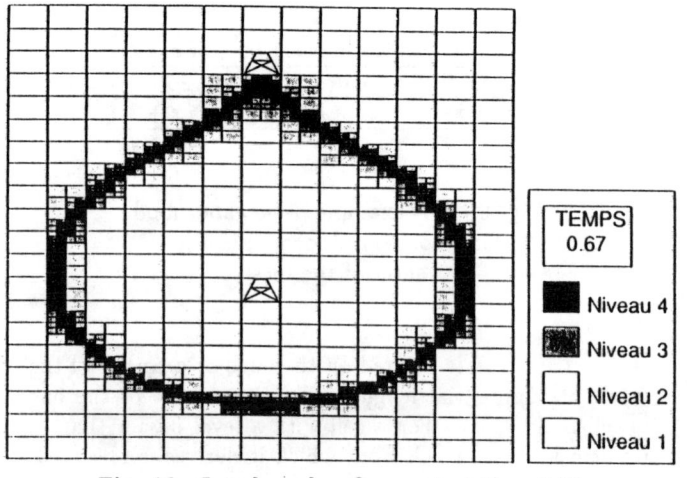

Fig. 16 - Local mesh refinement at time 0.67

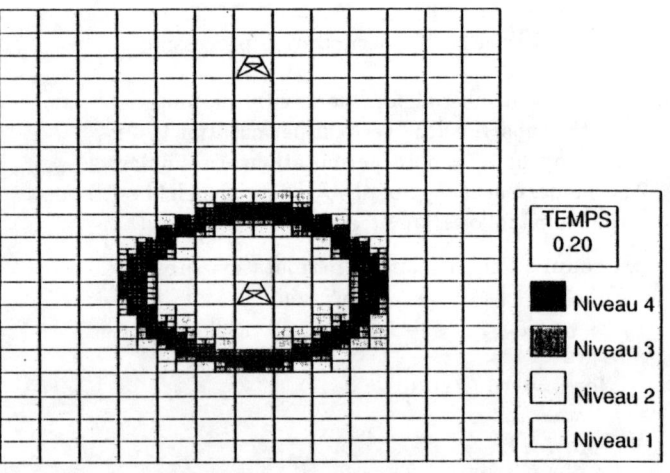

Fig. 14 - Local mesh refinement at time 0.20

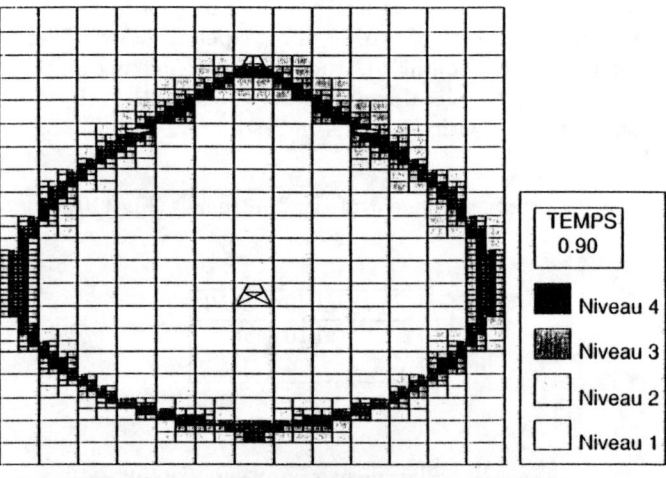

Fig. 17 - Local mesh refinement at time 0.90

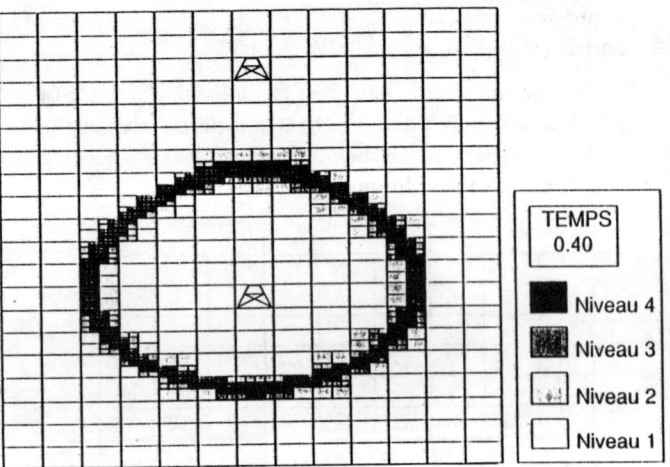

Fig. 15 - Local mesh refinement at time 0.40

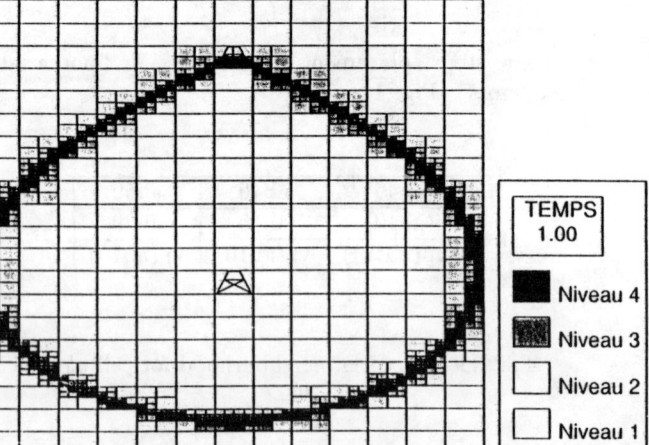

Fig. 18 - Local mesh refinement at time 1.00

4.1 Conclusion

Tests have shown that the new data structure described here is very well suited to the dynamic nature of the adaptive mesh method:

1. The memory requirements used during iteraions are reduced to what the present mesh occupies. Through this dynamic allocation of memory, no place is either lost or wasted.

2. Concerning the structure adopted for the grids using pointers and records, it takes heterogeneous media into consideration. Thus by adding on a new record, a temporarily refined grid can be differentiated because of the presence of the front, and another initially refined grid can be differentiated because of a heterogeneity.

3. The redundance of the "Ant" and "Suc" pointers linking the grid to the chaining list representing the mesh has proven to be very useful for covering this list as quickly as possible.

4. As for the new numbering used in this study, it makes it possible both to preserve a good structure for the linear system (also depending on the numerical scheme used) and to recognize multiple data on the grid such as its position in the mesh, its level of refinement, its genealogy and to find its neighborhood quickly.

5. For mesh management algorithms, recursivity makes the procedures more legible. We find that refinement and clustering algorithms can be avoiding in parallel. Propagation, which is very useful for avoiding to much contrast in size between grids and too many neighbors, greatly slows down iterations when L_{max} increases.

6. From the point of view of implementation and the Ada programming language, its use has provided to be effective, i.e. rigid, clear and legible programming with a wide choice of possibilities. In particular, error processing is very advantageous. The graphic outputs written in Fortran show that the interfacing of two languages is actually possible.

Acknowledgments

This research was done within the "Association de Recherche sur les Techniques d'Exploitation du Pétrole" (ARTEP) with the companies Elf - GdF - TOTAL CFP and IFP and financially supported by the "Fond de Soutien des Hydrocarbures" (FSH).

References

[1] Muskat, Physical Principles of Oil Production, McGraw-Hill Book Company, INC. 1949 (chap. 12.10).

[2] R.E. Bank and A.H. Sherman, "A Refinement Algorithm and Dynamic Data Structure for Finite Element Meshes", Center for Numerical Analysis, University of Texas, Austin (October 1980).

[3] Z.E. Heinemann, "Using Local Grid Refinement in a Multiple Application Reservoir Simulation", SPE 12255, SPE Reservoir Simulation symposium, San Francisco, CA, Nov. 15-18, 1983.

[4] M.J. Berger, "Data Structures for adaptive mesh refinement", in Adaptive Computational Methods for Partial Differential Equations, I. Babuska, J. Chandra, and J.E. Flaherty, eds., SIAM, Philadelphia (1983) pp. 237-251.

[5] P.A. Forsyth and P.H. Sammon, "Local Mesh Refinement and Modelling of Faults and Pinchouts", SPE 13524, SPE Reservoir Simulation Symposium, Dallas, TX, Feb. 10-13 1985.

[6] P. Quandalle and P. Besset, "Reduction of Grid Effects Due to Local Sub-Gridding in Simulations Using a Composite Grid", SPE 13527 , SPE Reservoir Simulation Symposium, Dallas, TX, Feb. 10-13, 1985.

[7] W.L. Wasserman, "Local Grid Refinement for Three-Dimensional Simulators", SPE 16013, Ninth SPE Symposium on Reservoir Simulation, San Antonio, Tx, Feb 1-4, 1987.

[8] G.F. Carey, M. Sharma and K.C. Wang, "A Class of Data Structure for 2-D and 3-D Adaptive Mesh Refinement", International Journal for Numerical Methods in Engineering, vol 26, 2607-2622 (1988).

[9] R.E. Ewing and R.D. Lazarov, "Adaptive Local Grid Refinement", SPE 17806, SPE Rocky Mountain Regional Meeting, Casper, WY, May 11-13 1988

[10] G. Ramamurthy and V.L. Ward "An Ada-Based Software Development Environment as Applied to Completion and Production Operations", SPE 17772, SPE Symposium on Petroleum Industry Applications of Microcomputers, San Jose, California, June 27-29, 1988.

[11] C.Brand and Z.E. Heinemann, "A new Iterative Solution Technique for Reservoir Simulation Equations on Locally Refined Grid", SPE 18410, SPE Symposium on Reservoir Simulation, Houston, Tx, Feb. 6-8 1989.

[12] T.Hermitte, "Affinage et Grossissement de Maillage Tridimensionnel : Structure de données et Algorithmes", (in french) I.F.P. Report (ref. 37 165) June 1989.

[13] T. Hermitte, "Test de la Structure de Données pour l'Affinage et le Grossissement d'un Maillage Tridimensionnel", (in french) I.F.P. Report (ref. 37 482) October 1989.

[14] J.G.P. Barnes, Programming in Ada, third edition, International Computer Science Series, 1989.

Poster-Conferences

2nd European Conference on the Mathematics of Oil Recovery
© D. Guérillot, O. Guillon (Editors) and Éditions Technip, Paris 1990, pp. 283-286
27 rue Ginoux, 75015 Paris

Two Dimensional Stochastic Modelling of Flow in Non-Uniform Confined Aquifers. Correction of the Systematic Bias Introduced by Numerical Models when They Are Used Stochastically

P. Lachassagne[1], E. Ledoux[1] and G. de Marsily[2]

1 AVERAGING PERMEABILITIES

One may first recall that the averaging of permeabilities is not necessarily a straightforward operation: it is well known, for instance, that when different parallel layers of a porous medium with permeability K_i and thickness e_i are assembled, the average permeability K of the medium is the harmonic mean ($\Sigma e_i / K = \Sigma (e_i / K_i)$) if the flow is orthogonal to the layers, or the arithmetic mean ($(\Sigma e_i) K = \Sigma (e_i K_i)$) if the flow is parallel to the layers. In a more general way, if N measurements of permeability K_i are available in an aquifer, we will call the harmonic mean $K_h = N / \Sigma (1/K_i)$, the arithmetic mean $K_a = (1/N) \Sigma K_i$, and the geometric mean $K_g = \exp (1/N \Sigma \ln K_i)$, i.e. the arithmetic mean taken in the log space. One always has the following inequality between the three means: $K_h < K_g < K_a$. It is also possible to define a unique averaging formula, as, for example, in Journel et al. (1986), the power averaging formula:

$$K^m = (1/N) \Sigma M_i^m$$

with $m = -1$ for the harmonic average, $m = +1$ for the arithmetic average, and $m \to 0$ for the geometric average (at the limit), but all other possible values of m will also produce an average.

The issue to be discussed is the type of average which is needed to predict flow. It is, however, necessary to first discuss the type of distribution that is generally observed in the field. Several authors, e.g. Neuman (1982), have observed that permeabilities are generally lognormally distributed in a given formation; this means that if the logarithm of the permeability is taken, then this magnitude has, in general, a Gaussian frequency if a large number of measurements are available. If we now consider that the permeability in space is a random function, i.e. that in two different locations, the permeabilities will in general be different and that they are independent of each other, then some theoretical results can help to determine the type of average needed.

Matheron, for instance, has shown that:

(1) Centre d'Informatique Géologique, École des Mines de Paris, 35, rue Saint-Honoré, 77305 Fontainebleau, France.
(2) Laboratoire de Géologie Appliquée, Université Pierre et Marie Curie, Paris VI, 4, Place Jussieu, 75005 Paris, France.

(a) if the flow is "macroscopically uniform" (parallel flow lines on the average), whatever the number of dimensions of the space, the distribution of the permeability and its spatial correlation, the average permeability always ranges between the harmonic mean and the arithmetic mean of the local permeabilities;

(b) if in addition the probability density function of the permeability is lognormal and unvarying by rotation, in two dimensions, the average permeability is exactly equal to the geometric mean;

(c) it is not possible to define an average permeability in the steady state for radial flow.

Similarly, Bakr et al. (1978) have given linearized approximations of the average permeability in uniform flow for a lognormal distribution function of the permeability:

- in one dimension: $K = K_g(1 - \sigma^2/2)$,
- in two dimensions: $K = K_g$,
- in three dimensions: $K = K_g(1 + \sigma^2/6)$,

where K_g is the geometric mean and σ^2 the variance of $\ln K$.

A last result concerns the dependence or independence of the different permeability measurements in space. If a large number of permeability measurements is available from different locations in space in the same formation, it is easy to observe that the permeability measurements are not independent: close to a location where a high permeability has been measured, the probability is greater that a high value will be measured again, and *vice versa*. However, this dependence is a function of the distance between the measurement points; after a certain distance, the measurements appear in general to become independent. The theory of geostatistics is a means of estimating regionalized variables from local measurements and of quantifying this spatial dependence.

It can be shown that the above results on the averaging of permeabilities still apply even if they are spatially correlated: the geometric mean is the correct average for two dimensional uniform flow, and has to be weighted as a function of σ^2 for other dimensions, to the first order.

2 USE OF A TWO DIMENSIONAL NUMERICAL MODEL WITH PROBABILISTIC THEORY

An attempt has been made to represent heterogeneous media with a mathematical groundwater simulation code.

The study was to verify if the different simulation codes commonly used at the Centre d'Informatique Géologique do represent in an accurate way flow in a two dimensional porous medium, i.e. that for "macroscopically uniform" flow, the effective transmissivity of a medium is equal to the geometric mean of its local transmissivities (see above).

Both a finite elements and a finite differences simulation codes were therefore tested (Marsily et al., 1978). In order to calculate the effective permeability (or transmissivity) of the domain represented by the grid, a reproduction of the Darcy experiment was chosen. Thus different prescribed head boundaries were assumed at the two opposite limits of the square grid studied (Fig. 1) with zero flow boundaries assumed at the other two. The square grid consisted of square elements of ten metres side.

A transmissivity value taken from a randomly generated log normal distribution was applied to each element. This distribution is defined by its geometrical mean and its standard deviation. The effective transmissivity T_{ef} of the discrete block medium is then calculated by measuring the steady state flow rate Q through the model:

$$T_{ef} = Q/iA$$

where i is the hydraulic gradient between the two extremities of the grid and A is the area of the section of the model perpendicular to the flow.

Thus both the equivalent transmissivity and an estimation of the round off and computing errors can be inferred from the measurements of the inflow and outflow along the two opposite prescribed head boundaries of the grid.

Making the assumption that the flow is "macroscopically uniform", the calculated effective transmissivity can then be compared with the geomet-

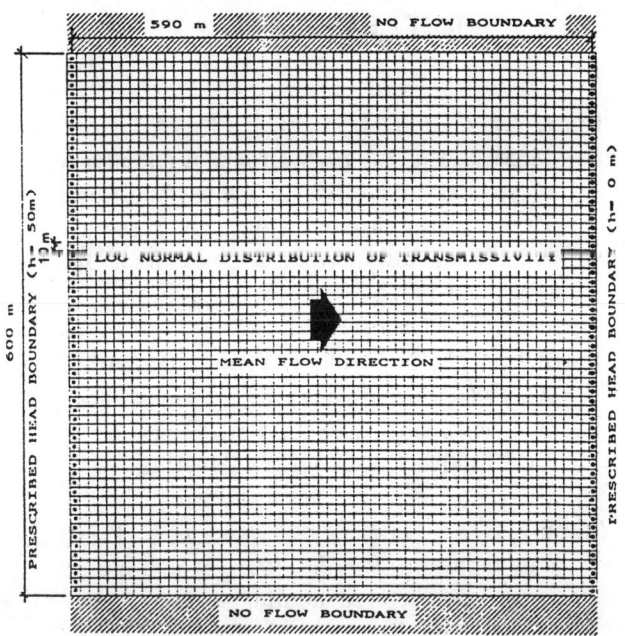

FIG. 1. Grid for the calibration
of the model

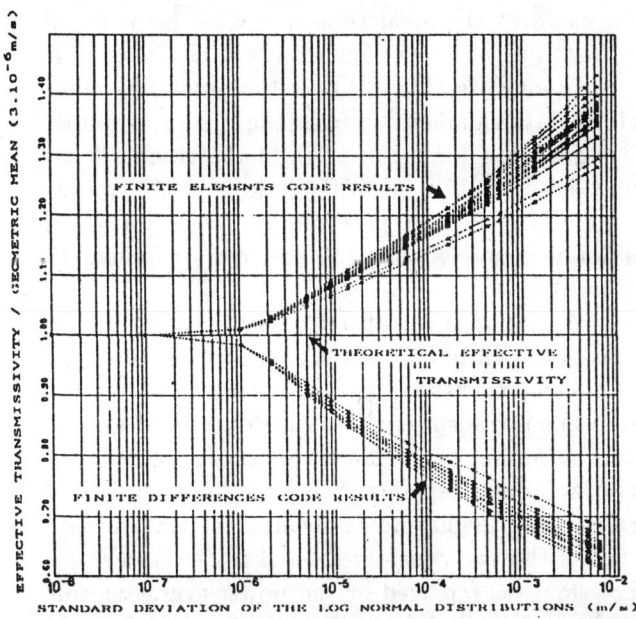

FIG. 2. Results of the macroscopically uniform
flow test for the finite difference and the finite
element models (one set of 170 realizations)

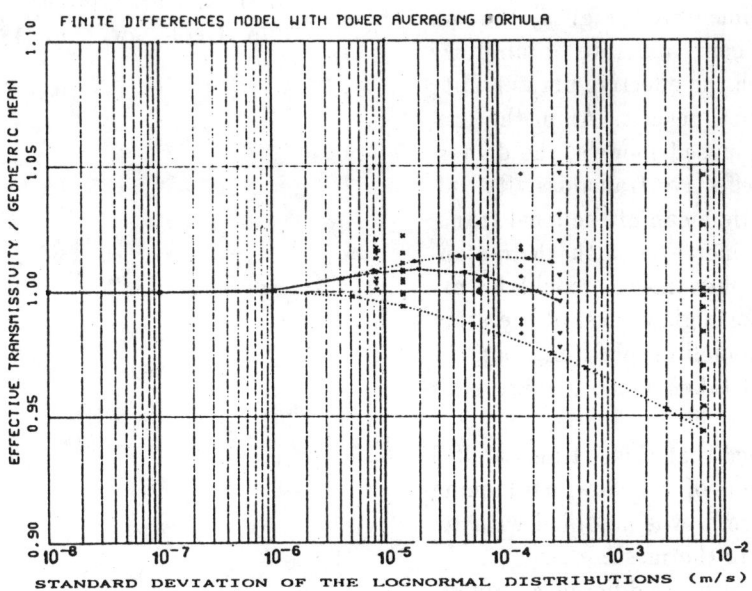

FIG. 3. Results of the "macroscopically uniform"
flow test with the correct finite difference model
(85 realizations)

ric mean of the local transmissivites for a set of realisations using various standard deviations. It clearly appears (Fig. 2) that the numerical models deviate significantly from the theory proposed by Matheron for high values of the standard deviation: the finite elements code tends to overestimate the effective permeability while the finite differences model shows a tendency to underestimate it.

An attempt was therefore made to correct the bias of the finite differences code. In fact this type of model presents an explicit way of computing the interblock transmissivities which are used for the determination of the flow rate exchanged between two adjacent elements. This method consists of calculating the harmonic mean of the transmissivities of the two considered blocks. The former expression was replaced by the power averaging formula described earlier. A value of m which could minimize the bias observed was then sought. In fact, a calculation with an m value greater than -1 (harmonic mean) will increase the local flow rate exchanged between each element of the grid, the global flow rate through the model, and thus its effective transmissivity.

It appeared experimentally (Fig. 3) that an m value of -0.23 is the more accurate to minimize the discrepancy between the effective permeability curves and the theoretical results. As in the preceeding cases with the biased models, the dispersion of the calculated effective transmissivity values around the geometric mean of the local transmissivity distributions increases with their standard deviation. This phenomenon could be explained by the effect, for high values, of the standard deviation of a channelling phenomenon: the majority of flow would taken place in particular pathways.

Further tests showed that this fitting was still valid when the direction of the flow, still being "macroscopically uniform", was modified with respect to the directions of the mesh.

The two-dimensional finite difference numerical model has therefore been corrected and now globally respects the probabilistic theory established by Matheron.

REFERENCES

Bakr, A., Gelhar, L.W., Gutjahr, A.L. and McMillan, J.R., 1978, Stochastic analysis of spatial variability in subsurface flow. Part 1: Comparison of one and three dimensional flows. Wat. Resour. Res. 14, (2), p. 263-271.

Journel, A.G., Deutsch, C.V. and Desbarat, A., 1986, Power averaging for block effective permeability. Soc. Pet. Engrs. SPE 15128, 56th Cal. Regional Meeting.

Marsily, G. de, Ledoux, E., Levassor, A., Poitrinal, D. and Salem A., 1978, Modelling of large multilayered aquifer systems: theory and applications. J. Hydrol., 36, p. 1-34.

Neuman, S.P., 1982, Statistical characterization of aquifer heterogeneities; an overview. In: Recent Trends in Hydrogeology (ed. by T.N. Narashiman), p. 81-102. Spec. Pap. Geol. Soc. Am. 189.

2nd European Conference on the Mathematics of Oil Recovery
© D. Guérillot, O. Guillon (Editors) and Éditions Technip, Paris 1990, pp. 287-290
27 rue Ginoux, 75015 Paris

An Estimator for the Effective Permeability

L. Holden[1], J. Høiberg[2] and O. Lia[1]

ABSTRACT

An estimator for the effective permeability, based on one-phase incompressible flow, is presented. The method gives accurate estimates for all types of heterogeneous blocks. It is considerably faster than a full simulation and also provides a measure of the error involved.

1 INTRODUCTION

In an oil reservoir the reservoir performance is influenced by heterogeneities on all scales. The size of fluctuation in the absolute permeability can be severe, ranging over many orders of magnitude. This makes it difficult to perform homogenization of the absolute permeabilities in a fine-scale reservoir description, i.e. to assign to each large grid-block in a reservoir performance simulator one single effective value which gives the same mean flow as in the fine-scale model. Many attempts have been made to address this problem, but so far all the proposed fast methods have been limited to certain classes of heterogeneity distributions.

In layered reservoirs the effective permability can be computed using a simple analytic formulae, in most other reservoirs we have to be content with numerical approximations. The geometric mean is a good estimator if there is no spatial correlation in the fine scale permeability variation (Warren and Price, 1961). Even better estimates for such reservoirs can be computed using a renormalization technique (King, 1988). Fast and accurate methods have also been proposed for reservoirs consisting of homogenous sand with shale barriers (Haldorsen and Lake, 1984).

There will always be some spatial correlation in the fine scale variation of the permeability, and there may be several other dominant sources of heterogeneity than shale barriers. Thus none of the estimators listed above are general.

2 A GENERAL ESTIMATOR

The permeability is a tensor and consequently the effective permeability is also a tensor. The effective permeability may be a non-diagonal

---

(1) Norwegian Computing Center, PO Box 114, Blindern, 0314 Oslo 3, Norway.
(2) Østfold Regional College, ØDH-EDB, Os Allé 9, 1750 Halden, Norway.

tensor, even if the fine scale permeability is diagonal. We assume here that the permeability is a diagonal tensor, but the estimator presented may be used to estimate off-diagonal elements just by rotating the block.

The flow in the reservoir is modeled by Darcy's law,

$$v = -k\nabla p,$$

where v is the volumetric flow velocity, k is the permeability and p is the pressure. The viscosity is set equal to 1. Because the mass of the fluid is conserved, v satisfies the differential equation

$$\nabla \cdot v = 0.$$

We define the effective permeability in a block for a fixed direction as the permeability which gives the same flow through a homogeneous block as the flow through the heterogeneous block when the boundary condition is constant pressure on the inflow and outflow sides and no flow through the other sides of the block. Thus the equation system is

$$\nabla \cdot (k\nabla p) = 0$$

with boundary conditions $p = 1$ at the inflow boundary, $p = 0$ at the outflow boundary and $\partial p / \partial n = 0$ at the boundary sides parallell to the flow.

The effective permeability is defined from one-phase incompressible flow. When studying only one block at a time, it is not possible to know the boundary condition the block has during a full reservoir simulation. The boundary condition above is in some sense a neutral choice, occuring when the neighbouring block is a mirror image of the considered block.

Assume that the absolute permeability $k = \{k_{i,j,l}\}_{i,j,l}$ is known on a $n \times n \times n$ fine-scale grid. The pressure p can then be computed using finite differences or finite element methods, which involves solving a linear equation system with n^3 (or n^2 in two dimensions) unknowns.

If the grid is uniform and the dimensions of the block is 1 in each direction, the flow through

layer number l is given as

$$Q_l = \frac{-1}{n^2} \sum_{i=1}^{n} \sum_{j=1}^{n} k_{i,j,l} \nabla p_{i,j,l} \cdot e_3,$$

where $\nabla p_{i,j,l}$ is the pressure gradient in block i, j, l and e_3 is the unit vector in the flow direction. Since there is no flow through the sides parallell to the flow, the flow through each layer must be the same, i.e. $Q_l = Q_1$ for all l. In a homogeneous block the flow is equal to the permeability of the block if the pressure is constant equal to 1 (0) at the inflow (outflow) side. When the pressure is known, an estimate of the effective vertical permeability is

$$k_{eff} = Q_1 = \frac{-1}{n^2} \sum_{i=1}^{n} \sum_{j=1}^{n} k_{i,j,1} \nabla p_{i,j,1} \cdot e_3.$$

This estimate is accurate, but for large n the numerical solution of the equation system is very time-consuming.

The uncertainities in a reservoir simulation are so large that it often suffices to know the effective permeability up to one significant digit. If we solve the equation system completely, we do an unnecessary good job. This fact is used in the estimator proposed below.

A large number of methods for solving linear equation systems are available and some of the fastest of these methods are iterative. These iterative methods start with an approximation to the solution, and in each iteration a new and better approximation is computed. The basic idea behind our estimator is to *stop the iteration process before it has converged*, and then *use the approximation to the pressure distribution to estimate the effective permeability.*

This approach is valid for all types of heterogeneities because the estimator converges to the definition of the effective permeability. Although the number of iterations needed to get sufficient precision may vary with the types of heterogeneities involved, the estimate for the effective permeability seems to converge much faster than the pressure distribution. This may be explained by the well-known fact that some iterative methods reduce the high frequency variation fastest (Stuben and Trottenberg, 1982).

Error bounds can easily be computed by using the change in the estimate from one iteration to the next. The iterative process can be stopped either when the error is sufficiently small or when the flow through each layer is approximately the same. For a large class of heterogenous reservoirs only a small, fixed number of iterations (e.g. 2) is needed. In some cases, e.g. if the block consists of several channels touching each other, it may be necessary to increase the number of iterations, but the reduction in computing time is still significant when compared to complete numerical solution for the pressure.

After stopping the iterations, the effective permeability can be estimated from the approximate pressure distribution as the harmonic mean of the flow in each layer normal to the flow,

$$k_{eff} = \frac{n}{\sum_{j=1}^{n} \frac{1}{Q_j}}.$$

In our current implementation of this method we use a conjugated gradient method preconditioned by relaxed, incomplete Cholesky factorization (Axelsson and Lindskog, 1986) as our equation solver. The starting approximation for p is chosen as a piecewise linear function having the value 1 at the inflow side and 0 at the outflow side, representing the exact pressure if the reservoir is layered with layers normal to or parallell to the flow.

3 TEST EXAMPLE

In our testing of the method we have generated heterogeneities stochastically, using the model described in (Aasen et al., 1989). A typical cross section generated by this model is shown in Fig. 1.

FIG. 1. Permeability in a cross-section.

The example uses a realization with a $15 \times 15 \times 20$ grid. In order to compute the pressure distribution exactly (up to machine precision), 20 iterations were needed. Table 1 shows the estimated vertical permeability obtained by stopping the process after a different number of iterations. For each estimate the corresponding computing time on a MicroVAX 3600 is shown. Note that the relative error is halved in each iteration. After two iterations the error is 6%. This is probably an acceptable error in most applications.

TABLE 1

Estimates for succesive iterations

Iterations	Estimate	CPU sec.
0	197.12	.01
1	43.63	2.7
2	39.50	3.2
3	38.13	3.8
4	37.61	4.3
5	37.46	4.7
10	37.30	8.0
20	37.30	13.5

In Figs. 2 and 3 contour plots of the pressure after 1 and 20 iterations are shown. Note that the qualitative description of the pressure close to the barriers is remarkably good after one iteration.

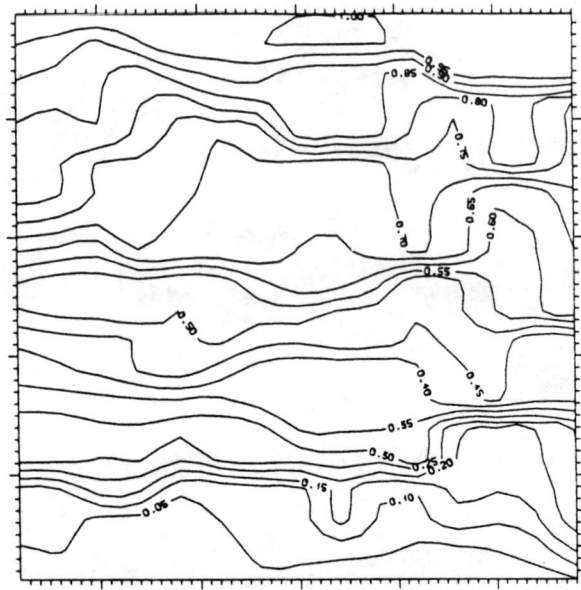

FIG. 2. Pressure after 1 iteration.

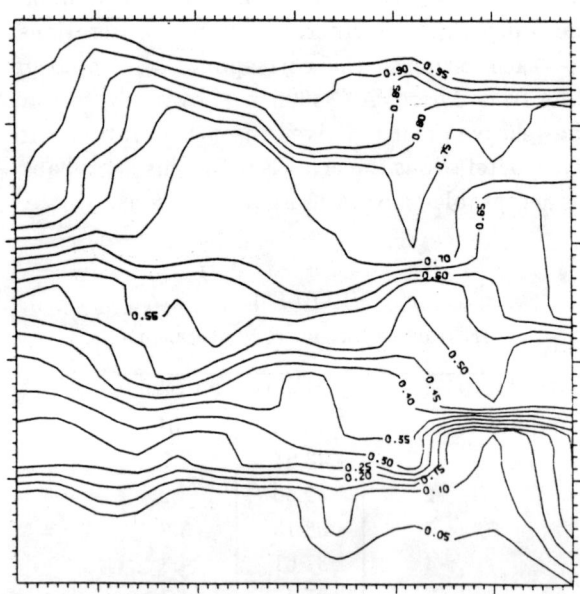

FIG. 3. Pressure after 20 iterations.

4 CONCLUDING REMARKS

The method described above provides valid estimates for the effective permeability for all kinds of heterogenous blocks and also gives an indication of the error in the estimate. It is easy to implement and is based on well-known numerical methods.

The number of iterations needed to get sufficient precision in the estimate depends on the type of heterogeneities, but the method will always run significantly faster than a full numerical solution.

Further development of this estimator will be done, both theoretically and by testing and improving the implementation.

REFERENCES

Aasen, J.O. et al., 1989, A Stochastic Reservoir Model and Its Use in Evaluations of Uncertainties in the Results of Recovery Processes, North Sea Oil and Gas Conference, Trondheim.

Axelsson O. and Lindskog G., 1986, On the Eigenvalue Distribution of a Class of Preconditioning Methods, Numerische Mathematik 48, p. 479-498.

Haldorsen, H.H. and Lake, L.W., 1984, A New Approach to Shale Management in Field-Scale Models, SPE Journal, August, p. 447-457.

King, P.R., 1988, The Use of Renormalization for Calculating Effective Permeability, Transport in Porous Media.

Stuben K. and Trottenberg U., 1982, Multigrid Methods: Fundamental Algorithms, Model Problem Analysis and Applications, Lecture Notes in Mathematics, 960, Springer Verlag, Berlin.

Warren J.E. and Price, H.S., 1961, Flow in Heterogeneous Porous Media, SPE Journal 1, p.153-169.

2nd European Conference on the Mathematics of Oil Recovery
© D. Guérillot, O. Guillon (Editors) and Éditions Technip, Paris 1990, pp. 291-296
27 rue Ginoux, 75015 Paris

Numerical Simulation of Hydraulic Fracturing in a Discrete Element System

S. Thallak[1], L. Rothenburg[1], M. Dusseault[1] and R. Bathurst[2]

Abstract

Hydraulically induced fracture in an assembly of co-hesionless discs is numerically simulated using a discrete element model, consisting of discrete particles coupled with an inter-void fluid flow model. Grains are represented by circular discs; to simulate flow, a geometrically coupled channel network is created by assigning nodes to pores, and flow channels to pore throats. Flow rates in channels are assumed to be proportional to the pressure gradient according to the Hagen-Poiseuille equation. The paper describes the main features of the model and explains the fracture initiation due to fluid injection and the propagation process at a grain level.

1. INTRODUCTION

The majority of heavy oil and oil sands in-situ projects utilize hydraulic fracturing as a necessary first step in the operation (Hsu et al., 1984). Oil sands are cohesionless, low porosity sands composed of discrete particles interacting through contacts (Dusseault and Morgenstern, 1978). There has been much effort directed towards modelling hydraulic fracture in oil sands (Hsu et al., 1984; Kular et al., 1988). A conventional approach to fracture propagation is usually based on continuum mechanics principles using poro-elasticity with a tensile rupture propagation criterion. These models are phenomenological and are primarily concerned with mathematical modelling of observed phenomenon without detailed attention to their fundamental physical significance. In granular materials, it is advantageous to treat the medium as an assemblage of particles, rather than as a continuum, as this permits exploration of fundamental mechanisms.

Numerical simulation with many discrete elements is one means of accomplishing this. The Discrete Element Method (DEM), developed by Cundall (1971) for rock mechanics problems, is a numerical technique capable of handling particles of any shape. Until recently, DEM approaches considered only dry particulate materials. For hydraulic fracture modelling in discrete systems, the pore fluid response must be incorporated.

We describe the mode of coupling the DEM to a conforming network of nodes and channels that defines the pattern of flow in the porous medium. The coupled model, HYDROFRAC, is used to ex-

(1) University of Waterloo, Waterloo, Ontario, N2L 3G1, Canada.
(2) Royal Military College of Canada, Kingston, Ontario, K7K 5LO, Canada.

amine hydraulic fracturing of a 1000-disc assembly. Fracturing is initiated by increasing injection pressure at the center of the assembly, maintaining constant total stress and pore pressure at the assembly boundary. The objective of this paper is to show that a flow-coupled DEM realistically emulates the behaviour of injection induced deformations. The program HYDROFRAC is proposed as a new tool for fundamental research in the petroleum recovery processes in porous media.

2. DISCRETE ELEMENT MODELLING

The DEM employs an explicit time-finite difference scheme in which each calculation cycle includes the application of the laws of motion to each disc, followed by application of two simple force-displacement laws at all disc contacts. A detailed description is given in Cundall and Strack (1979), only the principles are summarized here:

1. DEM can be imagined as a network of lumped-mass dashpot elements in which linear springs connect disc-shaped masses.

2. Each element satisfies the equation of motion.

3. A physical contact exists and carries a contact force only when there is an overlap between the discs.

4. Disc accelerations and velocities calculated from Newton's laws are assumed to be constant over each time step.

5. Coulomb's friction law applies at contacts, limiting the tangential component of contact force.

The DEM actually models a highly-damped dynamic transient mechanical system. Although the system is dynamic, static equilibrium is obtained by keeping loading rates low, thus inertial forces are always a small fraction of the average contact forces, and kinetic energy is dissipated through damping.

We limit our analysis to 2-D assemblies of frictional discs in contact which can separate if normal contact force is reduced to zero. The finite

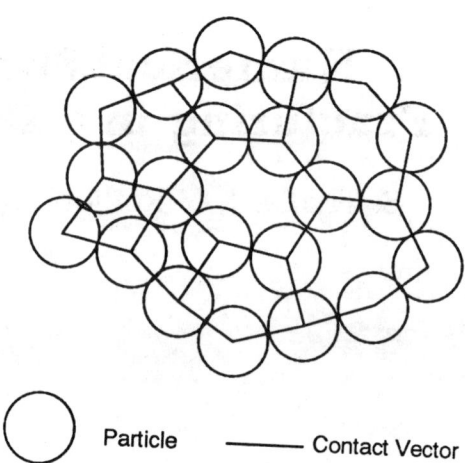

○ Particle —— Contact Vector

Figure 1: Polygon Network Formed by Particles in Contact

difference expressions used include body forces, but for simplicity, in the case we present, gravitational body forces are ignored. We introduce a new body force, the seepage force, associated with pressure drops in porous granular media. This is the critical element required to couple force-displacement behaviour to fluid flow in DEM.

2.1 Simulation of Fluid Flow

A physically precise flow simulation in a 2-D disc model is not possible. However, the approach used has been shown to fully emulate Darcy flow in a porous medium, assuming pore connectivity in the 2-D representation (Thallak et al.,1990).

A flow network geometrically coupled to the disc assemblage is created by assigning nodes to pores, and flow channels to throats. First, each contact which carries a normal force is identified as a link of a closed loop or polygon (Fig 1). All polygons are identified in the assembly, and a polygon network formed. Second, the centroid of each polygon is determined, and the flow network defined by joining centroids of neighbouring polygons which have a common face (Fig 2).

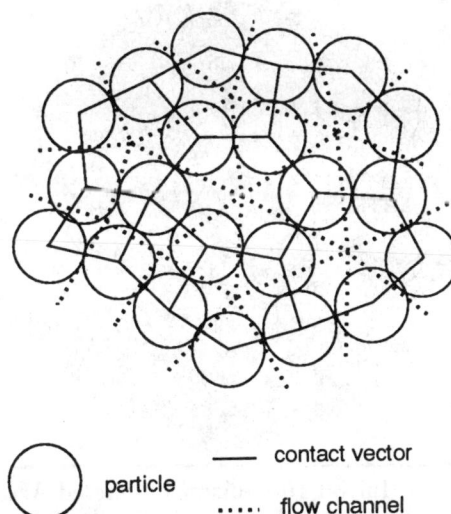

Figure 2: Identification of Associated Flow Network

To solve the flow network, we use a model of pipe elements with a single degree of freedom (pressure) at each node. Nodal pressures are obtained by bringing net inflows to zero in all non-boundary nodes. The flow rate q is assumed to be proportional to the pressure gradient $g = \Delta p/l$ according to the Hagen-Poiseuille equation, which is as follows:

$$q = -Kg$$

where,

$$K = \frac{\pi D^4}{128\mu} \tag{1}$$

Here, D is channel diameter, μ is fluid viscosity. The linear system of equations obtained is solved for nodal pressures, which are interpreted as average fluid pressures in pores.

The injection pressure at the assembly center is assumed to dissipate to zero at the boundary, under static conditions of steady-state flow. As the increasing pressure reduces contact normal forces, loss of contacts results in a simulation of fracturing in the porous medium. Once a flow network is prepared, it is not necessary to reformulate it at each time step during calculations; this is done only if a contact between particles is lost, or when particle locations change markedly.

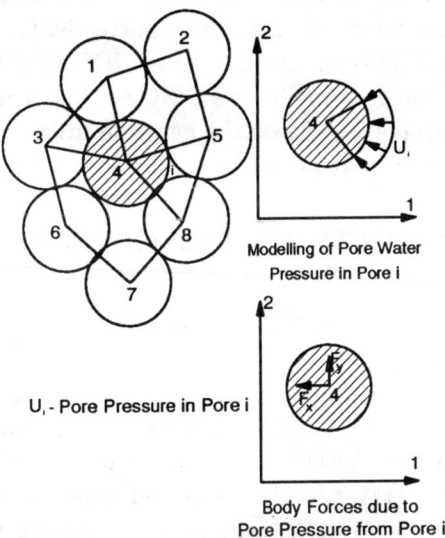

Figure 3: Calculation of Body Forces Due to Pore Pressure

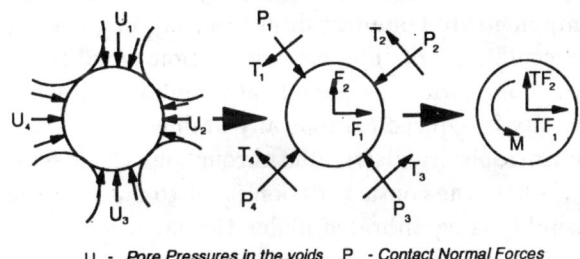

U_i - Pore Pressures in the voids P_i - Contact Normal Forces
T_i - Contact Tangential Forces TF_1 & TF_2 - Net Force Components
M - Net Moment acting at Disc Centroid

Figure 4: Outline of the Mode of Coupling in DEM

2.2 Mode of Coupling

For DEM fracture simulations, particle movements must be coupled with flow.

Consider a disc sorrounded by several pores, each with a different uni-valued isotropic pressure U_k. This pressure is an applied boundary stress over the particles that define that pore and it is resolved as forces acting on each particle (Fig 3). The resolved forces are included as body forces in the force-displacement solutions for the disc assembly, as illustrated in Fig 4.

The current investigation uses a modified version

of the program DISC developed by Bathurst (1985) which implements the DEM as reported by Cundall and Strack (1979). The program HYDROFRAC was developed by modifying DISC so that it incorporates proper flow laws; the general outline of the coupling is described in Fig 4.

3. RESULTS AND DISCUSSIONS

3.1 Numerical Simulation

The numerical experiments in this investigation used an assembly of 1000 particles having 20 different disc radii. Assembly generation uses a random number generator to place non-overlapping discs in a circular region until there are no remaining voids sufficiently large to accept a particle. The result is a loose assembly of non-contacting discs whose density diminishes with distance from the sample center, and we compact it by applying velocity or force components to boundary discs. During compaction, the coefficient of interparticle friction is taken as zero, boundaries are periodically updated and the porosity is reduced isotropically until contact forces are isotropically distributed throughout the assembly. Then, the contact friction is set to 0.3, and the assembly is equilibrated under the same boundary conditions.

Figure 5 shows the compacted assembly used for hydraulic fracture simulation. The line thickness corresponds to the contact force magnitude. Hydraulic fracturing is simulated by monotonically increasing injection pressure at the center of the assembly, maintaining constant total stress and pore pressure at the boundary, until the contact force between two particles becomes zero. When this occurs, the two adjacent pores are coalesced, and this corresponds to fracture initiation. Repeated iterations with slowly increasing pressures result in a progressive coalescence of pores and a propagating fracture is created.

3.2 Observations

Results of injection-induced deformations in the

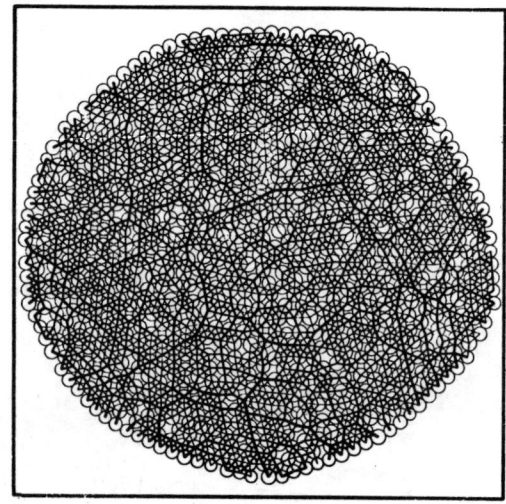

Figure 5: Initial 1000-disc Compacted Assembly

disc assemblage are shown in Figs 6(a), (b)and, (c). Contact vectors joining neighbouring disc centroids are shown as lines with thickness proportional to contact force magnitude at different stages of propagating fracture. Disc trajectories are shown in Figs 7(a), (b)and, (c), corresponding to stages of a propagating fracture in Figs 6. The vector length is scaled to magnitude of particle movements. Deformations are all directed outwards from the injection point, and on average correspond precisely to the seepage forces associated with pressure drops. Large reduction in the effective normal forces around the fracture surface is observed. The particles away from the injection area are subjected to modest changes in normal contact forces and remain approximately in their original positions, with a small outward displacement component.

Contact force redistribution takes place due to reduction in contact forces around the injection point which causes concentrated deformations, dominated by plastic slip and parting. This occurs only around the injection point and leads to fracture initiation and propagation. Contacts close to injection point have low normal contact forces. This is physically correct as the increase in pore pressure due to injection causes the reduction in the effective normal force, which reduces the local effective normal stress under constant total stress boundary conditions.

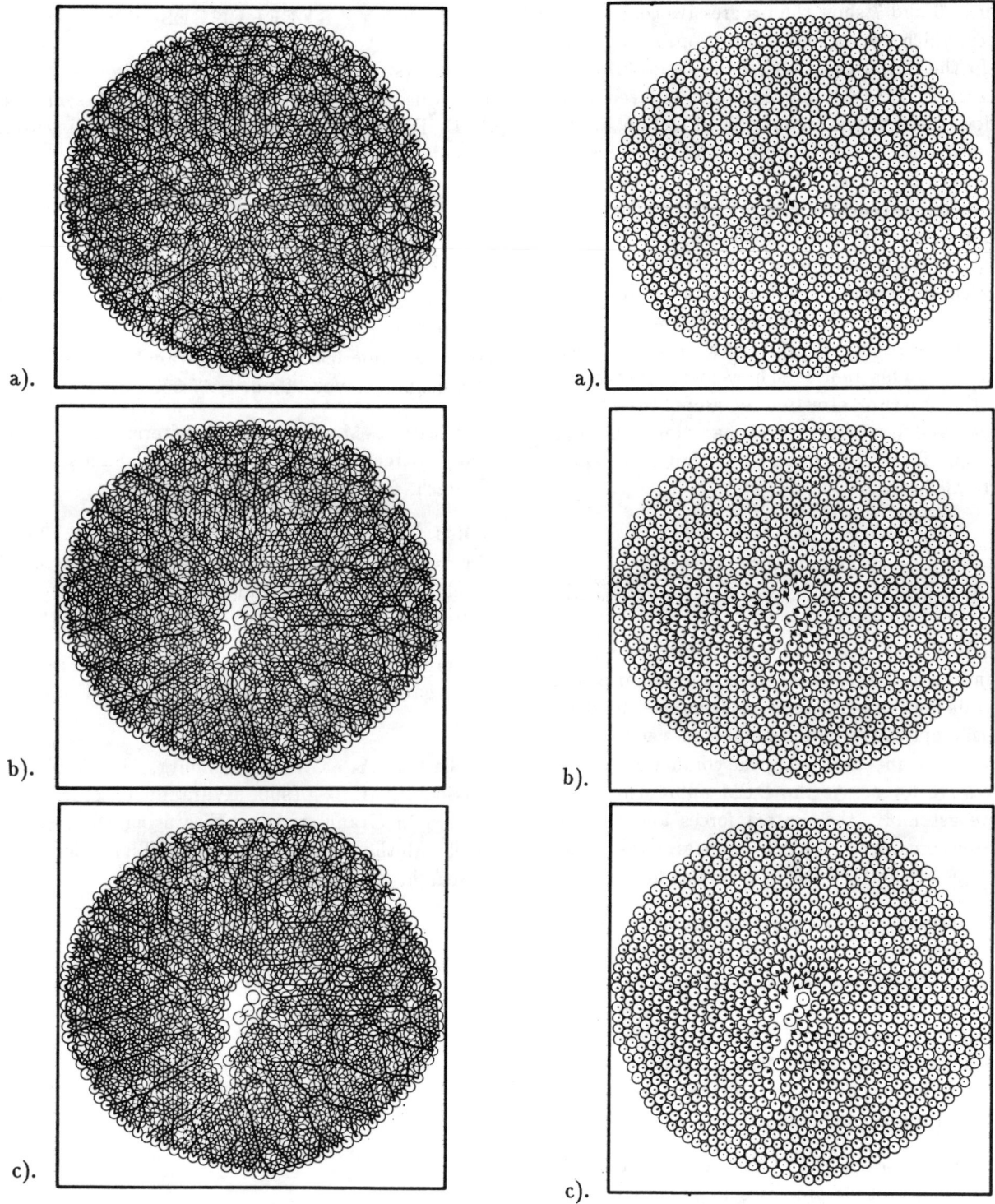

Figure 6: Contact force in the Assembly- a). Fracture initiated, b). Propagated fracture at an intermediate stage, and c). Large fracture.

Figure 7: Disc trajectories in the Assembly- a). Fracture initiated, b). Propagated fracture at an intermediate stage, and c). Large fracture.

Figures 6 and 7 show the progressive coalescence of pores which corresponds to a propagating fracture. In the initial assembly, the force distribution was isotropic (Fig 5); and thus the induced fracture does not have a preferential direction for its propagation at a macroscopic level. If one carefully studies Figs 6(a) and (b), it can be noted that the induced fracture in Fig 6(a) has changed its direction of propagation. As the fracture tip propagates, it intersects a contact carrying a larger interparticle force and can not propagate through it. The fracture changes its direction, following a path of local least resistance, but with an overall planar aspect. This indicates how local defects can affect the fracture direction in globally isotropic medium, and how fracture propagation generates anisotropic force and contact distributions in particulate medium.

4. CONCLUSIONS

HYDROFRAC is a tool for developing better understanding of fracture behaviour in granular media at a grain mechanisms level. An induced fracture in an isotropic medium propagates in a path of local least resistance, following the contact with least compressive force. As the DEM approach uses a discrete assembly, the contact forces and the applied pressures are finite, which ensures the forces or stresses at the tip of the fracture as finite.

This approach can clearly tackle anisotropy and heterogenity issues from a more fundamental approach, as the fabric evolution can be observed and mapped in the assembly. The model is a true departure from the continuum mechanics approach, emphasizes the physics of the fracture processes, and helps in understanding the limitations of conventional fracture propagation models.

It may be noted that DEM can be used to simulate many other processes in particulate and jointed reservoirs, including sand production, injection/production strains, subsidence, and so on. Further developments will be in models of 3-D particles, 2-D ellipses, transient, and multiphase flow.

REFERENCES

Bathurst, R. J., 1985, A Study of Stress and Anisotropy in Idealized Granular Assemblies, Ph.D. Dissertation, Queen's University, Kingston, Canada.

Cundall, P. A., 1971, A Computer Model for Simulating Progressive Large Scale Movements in Blocky Rock Systems, Symp. ISRM, Nancy, France, Proc. 2, 129-136.

Cundall, P. A. and Strack, O. D. L., 1979, A Discrete Numerical Model for Granular Assemblies, Geotechnique, Vol. 29,no. 1, 47-65.

Dusseault, M. B. and Morgenstern, N. R., 1978, Shear Strength of Athabasca Oil Sands, Can. Geotech. Journal, 15, 216-238.

Hsu, T. R., Pizey, G., and Ashour, H. A., 1984, On An Analytical Method for Insitu Fracture of Oil Sands Formation, 35th Annual Technical Meeting of Petroleum Society of CIM, no.84-35-44, 727-738.

Kular et.al., 1988, Multiple Hydraulic Fracture Propagation in Oil Sands, SPE, no.17534, pp.477-491.

Thallak, S. G., Rothenburg, L. R., and Dusseault, M. B., 1990, Hydraulic Fracture Simulation In Granular Assemblies using Discrete Element Method, submitted to AOSTRA Journal of Research, Edmonton, Canada.

2nd European Conference on the Mathematics of Oil Recovery
© D. Guérillot, O. Guillon (Editors) and Éditions Technip, Paris 1990, pp. 297-300
27 rue Ginoux, 75015 Paris

A 3-D Network Simulating Two Phase Immiscible Displacements in Porous Media

D. Zhou and E. H. Stenby[1]

1 INTRODUCTION

As a continuation of the work on simulation of multiphase flow in porous media, associated with Enhanced Oil Recovery, the previously developed two dimensional network model (Zhou and Stenby, 1989) was extended to a three dimensional version. The model can simulate immiscible two phase displacement in porous media using invasion percolation theory. In this paper, the structure of the 3D model, the stability, and the boundary conditions at the outlet of the medium will be discussed briefly. The simulated results are compared with experimental data with respect to the capillray end effects, viscosity ratio and the wettability effects on the displacement process.

2 THE NETWORK MODEL

The porous medium is represented by a 3D array of pore bodies and pore throats, which follow given distribution functions. All pores have six neighbours and are connected by the pore throats. To each pore body and each pore throat a random number is assigned, which follows the pore body and pore throat size distribution functions respectively. The sizes (L x X x Y) of the samples in this work are 100 x 10 x 10 and 30 x 10 x 10.

3 THE DISPLACEMENT PROCESS

3.1 The Governing Equations

In the previous paper, we have derived the governing equation, combining Darcys law, Hagen-Poiseuille equation and the Laplace equation under the following assumptions:

1) By-passing is the only procedure for entrapment of the displaced phase.
2) The clusters of the displaced phase, isolated by the displacing phase, can not be moved and forms the residual oil saturation.
3) The flow in the pores is viscous Newtonian flow.
4) The fluids are incompressible.
5) Darcy's law applies.

For a fully developed interface, the maximum value of the pressure difference over the interface through pore i can be calculated using the following equation:

$$P_{max} = \frac{4\,\sigma}{D_{ab}}\left\{ -\frac{D_{ab}}{D_{ti}}g(i,x,di)_{max} + Bn.\left[32\,c\frac{D_{af}^4}{D_{if}^4}\left[x + \frac{\mu_o}{\mu_w}(1\text{-}x) \right]_{max} + \left(\frac{n1}{K_{rw}} + \frac{n2}{K_{ro}}\frac{\mu_o}{\mu_w} \right) \right] \right\} \quad (1)$$

(1) Institut for Kemiteknik, Technical University of Denmark, Building 229 DTH, DK2800 Lyngby, Denmark.

4.2 The Relative Permeabilities

In the governing equation (1), the relative permeabilities of the oil and water phases are requried. The same procedure as in the previous paper (Zhou and Stenby, 1989) was used.

$$K_{ri} = \frac{\sum D_i^4 f}{\sum D_j^4 f} \qquad (6)$$

where the relative permeability K_{ri}, for phase i is calculated as the ratio of the summed flowing diameter of the pores filled with phase i to the fourth power and the total sum of all diameters also to the fourth power, indicated with j.

4.3 Programming

The model requires the accessibility of a pore to the displacing phase. This involves a global search and uses most of the CPU time of the program. In order to be accessible to the displacing phase a pore should be: untrapped and directly connected to the displacing phase. The details of the programing is given by Zhou (1990). The calculation time is found to be proportional to the cube of the number of pores in the medium. The computer program is written in Pascal and the simulations were carried out on a HP9000/370 microcomputer.

In the following discussion, the medium is considered as either completely water-wet or oil-wet, i.e. $\theta = 0$ for imbibition processes, or $\theta = 180$ for drainage processes.

5 RESULTS AND DISCUSSION

5.1 Capillary End Effects

Capillary end effects result from the discontinuity of the capillary forces at the outlet of the medium, especialy for the imbibition process. When the displacing phase arrives at the outlet of the medium, it has a tendency to stay at the outlet and to force the oil out of the system. Fig.1 shows the water saturation distributions at water arrival and water breakthrough with capillary numbers of 10^{-6} and 10^{-3}. The dimensions of the sample are 30x10x10, and the viscosity ratio is 1.6. It is clearly shown that at low capillary number, a large fraction of oil is produced after water arrival. While the water breakthrough occurs as soon as water

arrives at the outlet for high capillary numbers. Simulations show that the recovery at water arrival is proportional to the logarithmic value of the capillary number , while the recovery at water breakthrough is a constant value within the low capillary number range. There is only slight differences between the recoveries at water arrival and water breakthrough within the high capillary number range. This is in very good agreement with the results reported by Kyte and Rapoport (1958). The simulations with this model also show that, with low capillary numbers, the capillary forces help the imbibition process to produce more oil from a long medium than that from a short. The opposite is true for drainage processes.

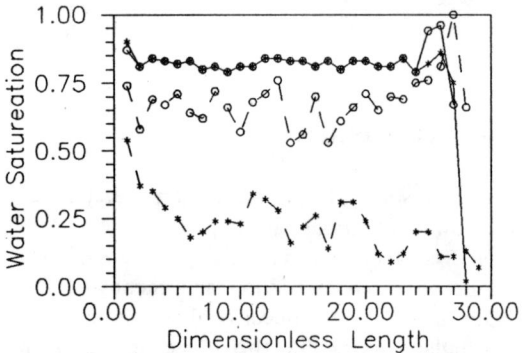

Fig. 1. Simulated Water Saturation Distributions at Water Arrival(*) and Water Breakthrough (o). Nc = 10^{-6} (Dashed) and 10^{-3} (Solid).

5.2 Viscosity Ratio Effects

The stability of immiscible displacement depends on the mobility ratio between the fluids, as shown in section 4. The variation of the viscosity ratio between the fluids leads to a change of the mobility ratio. The dependence of the capillary desaturation curve on the viscosity ratio has been discussed by Abrams(1975) based on experimental data from several samples. The residual oil saturation was empirically correlated with the 0.4th power of the viscosity ratio for both stable and unstable displacement. Fig. 2 shows the capillary desaturation curves with various viscosity ratios. The experimental results were produced from a sinthered glass core, 10 cm long and 2.54 cm in diameter (Zhou,1990). The desaturation curve with a viscosity ratio of 3.0 is very much different from that with unity viscosity ratio. The simulations were carried out on a sample of 100x10x10. The viscosity ratios of the fluids are 1.0,3.0, and 100 resectively. The simulations can satistactorily

$$Bn = \frac{D_{ab}^2 \; B}{D_{af}^2 \; 4.c} \; Nc \qquad (2)$$

$$Nc = \frac{v \; \mu_w}{\sigma \; \phi} \qquad (3)$$

where:

B = the ratio of the pore length to the pore body diameter of the medium,

c = a constant for a given medium,

D_{ab} = the average pore body diameter of the medium,

D_{af} = the average flow diameter of the medium,

D_{if} = the flow diameter of pore i,

D_{ti} = the pore throat diameter of pore i,

g = a function representing the wettability of the system,

K_r's = the relative permeabilities of the fluids,

Nc = the capillary number,

v = the Darcy flow rate,

σ = the interfacial tension between the two phases,

μ's = the viscosities of the fluids,

φ = the porosity of the medium,

As shown earlier (Zhou and Stenby, 1989), the maximum value of the g function can be described in the following way:

When the wetting phase displaces the nonwetting phase.

$$g_{max} = \frac{cos(\theta_1)}{A_i} \qquad (5)$$

when the nonwetting phase displaces the wetting phase,

$$g_{max} = -cos(\theta_2) \qquad (6)$$

where

A_i = the ratio of the pore body diameter to the pore throat diameter of pore i,

θ_1 = the contact angle for imbibition,

θ_2 = the contact angle for drainage.

3.2 Stability Criterion

Since the pores are ramdomly distributed in the medium, the factor determining whether a displacement is stable or not is the relative location of the favorable pores. The pressure drop P_{max} should decrease with increasing n2. The criterion for stable displacement is:

$$\frac{\mu_o}{\mu_w \; k_{ro}} - \frac{1}{k_{rw}} \leqslant 0 \qquad (7)$$

or

$$\frac{\mu_o \; k_{rw}}{\mu_w \; k_{ro}} \leqslant 1.0 \qquad (8)$$

The same criterion was obtained from the linear analysis of the immiscible displacements in a sample of infinite sizes.

3.3 The Boundary Condition

The boundary condition at the outlet of a core was discussed in an earlier work (Stenby and Zhou, 1989). The conclusion is that the capillary force is zero at the end of the medium, and then we can write g_{max} for the case of a completely water-wet medium as follows:

$$g_{max} = \begin{cases} 1/A_i & i < t_i \\ 0 & i = t_i \end{cases} \qquad (9)$$

where t_i is the number of pore rows in the medium. When i is equal to t_i, it means the displacing phase has reached the outlet of the medium.

The boundary condition for the drainage process can be derived as:

$$g_{max} = \begin{cases} -cos(\theta_2) & i < t_i \\ 0 & i = t_i \end{cases} \qquad (10)$$

4 SIMULATION

4.1 The Distribution Functions

The model allows us to use any function as pore size distribution function. In this work the simulated results are based on using $\alpha(R_t) = 2R_t \times exp(-R_t^2)$ as pore throat size distribution function and $\alpha(R) = 2R/3 \times exp(-R^2/9)$ as pore body size distribution function. So that the mean pore body is 3 times larger than the mean pore throat. These have been sucessfully used in the simulation of relative permeabilities by some investigators.

predict the experimental observation with viscosity ratios of unity and 3.0. There is no data available yet for the viscosity ratio as high as 100. The reason to include the simulated results here is to show the tendency of the variation of the capillary desaturation curve with viscosity ratio.

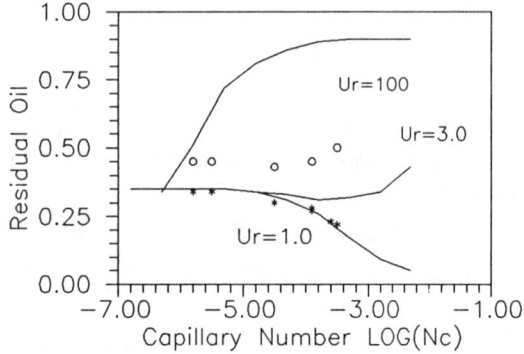

Fig. 2. Simulated Residual Oil Saturation with Various Viscosity Ratios.

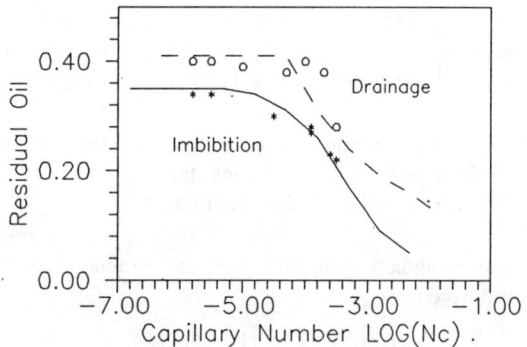

Fig. 3. Comparison of the Experimental (dots) and Simulated (curves) Residual Saturations for Imbibition and Drainage Processes (100x10x10, Ur=1.0)

5.2 Wettability Effects

It has been experimentally proven that the wettability of the medium affects the behaviour of the capillary desaturation curves (Gupta,1979, Zhou, 1990). The capillary forces in the imbibition process help the displacing phase to enter the pores, while it gives a negative effect to the drainage process. Fig.3 shows the capillary desaturation curves for both imbibition and drainage processes. The critical capillary number above which the residual oil saturation begins to decrease with

increasing capillary number, is larger for the drainage process than that for the imbibition process. Simulations show that the structure of the pores in the medium is responsible for this difference. The experimental results are from the same core as mentioned above. The core is considered as completely water-wet. As the imbibition processes are used water displacing n-decane, while the drainage process is n-decane displacing water. The viscosity ratio for both cases can be consideried as unity. The simulated results are for samples of 100x10x10, and the viscosity ratio is 1.0. The simulated results excellently agree with the experimental data.

6. Conclusion

From the above discussion, it can be concluded, that the three dimentional model developed is capable of simulating two phase displacement taking the viscosity ratio , wettability and capillary end effects into consideration. The computing time still limits the size of porous media to be studied, however the size is large enough to evaluate the end effects.

7 References

Abrams, A., 1975 " The Influence of Fluid Viscosity, Interfacial tension, and Flow Velocity on residual Oil saturation Left by Waterflood ", *Society of Petroleum Engineers Journal*, Oct., pp 437-447

Gupta, S.P. and Trushenski,S.P., 1979 " Micellar Flooding --compositional Effects on Oil Displacement", *Society of Petroleum Engineers Journal*, April, pp 116-128

Kyte, J.R. and Rapoport,L.A., 1958 "Linear waterflood Behavior and End Effects In Water-wet Porous Media," *J. Pet. Tech.* Oct., pp 47-50.

Stenby, E.H. and Zhou,D.,1989 " Capillary End Effects in Short Core Immiscible Displacement," IEA/EOR Workshop at Stanford University, 4-6 Oct..

Zhou,D.,1990, Ph.D thesis, Department of Chemeical Engineering, DTH.

Zhou,D. and Stenby, E. H., 1989 " Immiscible Displacement in a Porous Medium Simulated by a Statistical M odel," Proceedings from 2nd International Conference on North Sea Oil and Gas Reservoirs, May 8-11, Trondheim, (in print).

2nd European Conference on the Mathematics of Oil Recovery
© D. Guérillot, O. Guillon (Editors) and Éditions Technip, Paris 1990, pp. 301-304
27 rue Ginoux, 75015 Paris

An Analytical and Numerical Study of the Three-Phase Surfactant Displacement Problem

J. W. Barker[1]

1 SUMMARY

Numerical simulation has been used to investigate the effects of middle phase mobility, dispersion, and a salinity gradient, on the performance of a tertiary surfactant flood involving Type III (three phase) phase behaviour. The BPOPE simulation model[1] was used.

The results indicate that the relative mobility of the middle phase is an important consideration in the design of a surfactant flood. This mobility should be low compared to the mobility of the oil phase.

Unfortunately, the effects of dispersion on the low mobility flood are severe, and "self-sharpening" behaviour must be introduced, for example by means of a salinity gradient. In the salinity gradient flood, the middle phase mobility again has a strong influence on the solution, and the salinity is not necessarily optimal at any point within the surfactant bank.

In a future paper, we shall demonstrate that inclusion of an alcohol in the surfactant slug can also produce self-sharpening behaviour, provided the properties of the alcohol are chosen correctly. No additional benefit is gained by varying the alcohol concentration within the chemical slug. Varying the composition of the surfactant within the surfactant slug does not produce self-sharpening behaviour.

2 ASSUMPTIONS

Incompressible, isothermal, nonreacting, one dimensional flow in a homogeneous porous medium is simulated, neglecting capillary pressure and ion exchange. A classical progression from Type II(-) through Type III to Type II(+) phase behaviour, with increasing salinity, is assumed. The residual oil saturation is assumed to decrease to zero, and the relative permeabilities to become straight lines, as the capillary number increases.

Low, intermediate and high middle phase mobility cases are considered, differing only in the middle phase viscosity (values of 2.4, 0.6 and 0.15 cP respectively). Initially, adsorption is neglected. Other parameters are representative of North Sea reservoir conditions. Although a small (0.15 PV) slug size is used, the conclusions drawn do apply to larger slugs when realistic levels of adsorption are allowed for.

3 RESULTS

3.1 Effect of Middle Phase Mobility

Figure 1 illustrates the effect of the middle phase mobility on oil recovery for continuous surfactant injection. Constant (optimal) salinity is assumed. Sufficient grid blocks are used that the effects of numerical dispersion are negligible.

When the middle phase mobility is low, piston-like displacement of all the residual oil is achieved. Recovery is complete by just over 1.0 PV injected, implying that, in the absence of dispersion and adsorption, a slug of less than 0.1 PV would be required to recover all the residual oil.

As the middle phase mobility increases, the time to complete oil recovery, and hence the slug size required, increases (up to 0.7 PV for the high mobility case).

---

(1) BP Research Centre, Chertsey Road, Sunbury-on-Thames, Middlesex, TW16 7LN, UK.

Figure 2 shows the oil and surfactant concentrations after 0.45 PV for the intermediate middle phase mobility case. The solution is of the form described by Aanonsen[2]. An oil bank, a chemical front, a chemical bank (where there is three phase flow), and a solubilisation front, are all visible. The chemical front is dispersed, rather than being sharp, since adsorption has been neglected.

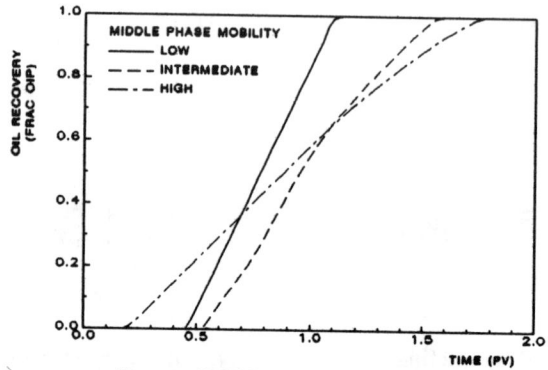

Figure 1. EFFECT OF MIDDLE PHASE MOBILITY
CONTINUOUS SURFACTANT INJECTION
CONSTANT (OPTIMAL) SALINITY

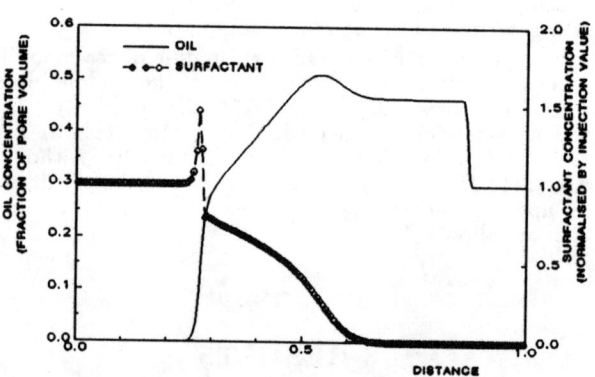

Figure 2. OIL AND SURFACTANT CONCENTRATIONS
AFTER 0.45 PV
INTERMEDIATE MIDDLE PHASE MOBILITY

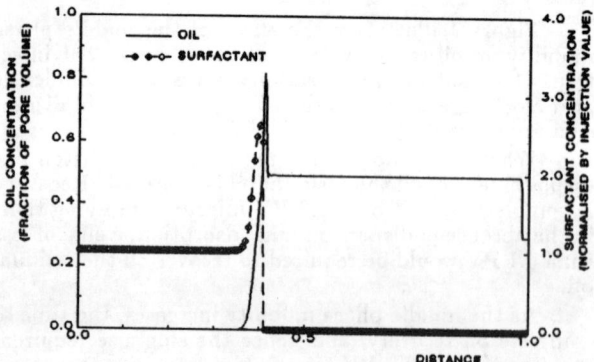

Figure 3. OIL AND SURFACTANT CONCENTRATIONS
AFTER 0.45 PV
LOW MIDDLE PHASE MOBILITY

Figure 3 shows the corresponding plots for the low middle phase mobility case. All the residual oil is displaced by the surfactant in a piston-like manner. The solubilisation front and the chemical front have merged, and the three-phase flow region has disappeared (a small region of three-phase flow remains, owing to the presence of a small amount of numerical dispersion). The combined front advances at unit dimensionless velocity (the maximum allowed by material balance considerations).

This change in the solution may be understood by considering material balance of the oil component at the solubilisation front. Since there is no oil behind the solubilisation front, the speed of this front is given by:

$$V_s = F_o^{(+)}/C_o^{(+)} \qquad (1)$$

where $F_o^{(+)}$ and $C_o^{(+)}$ are the flux and concentration of the oil component immediately ahead of the front, i.e. in the three-phase flow region. If the middle phase mobility is high, the fractional flow of the excess oil phase will be low, resulting in a low value of $F_o^{(+)}$ and hence also of V_s. As the middle phase mobility decreases, the fractional flow of the excess oil phase increases, and hence V_s increases (i.e. the time to complete oil recovery decreases). Eventually, when the middle phase mobility becomes low enough, the speed of the solubilisation front approaches that of the chemical front and the two fronts merge. A sufficient condition for piston-like displacement to occur is that V_s as given by equation (1) should be greater than or equal to unity for all points in the three-phase triangle, since this precludes any jump from the injection composition to a composition in the three-phase region.

3.2 Effect of Dispersion

Figure 4 shows the severe effect of dispersion on oil recovery for a 0.15 PV slug application of the low middle phase mobility surfactant system (dispersion is represented in the simulations by numerical dispersion). The dispersion causes the overall composition to enter the three-phase region, in which the surfactant lies in the middle phase, which propagates slowly on account of its high viscosity.

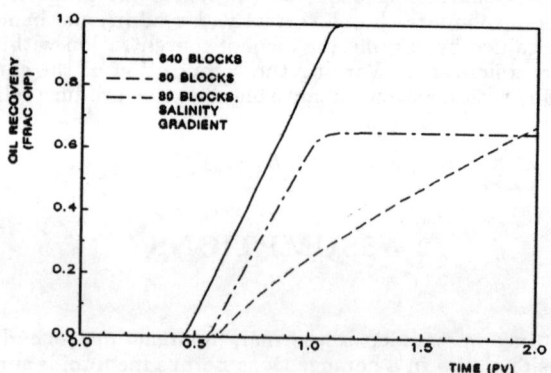

Figure 4. EFFECT OF DISPERSION AND SALINITY GRADIENT
0.15 PV SLUG, LOW MIDDLE PHASE MOBILITY

302

The effect of dispersion on the intermediate mobility case is less severe (Figure 5), while in the high mobility case (Figure 6), it actually increases oil recovery. This unusual behaviour relies on the independence of interfacial tension on surfactant concentration, and thus may not be realistic.

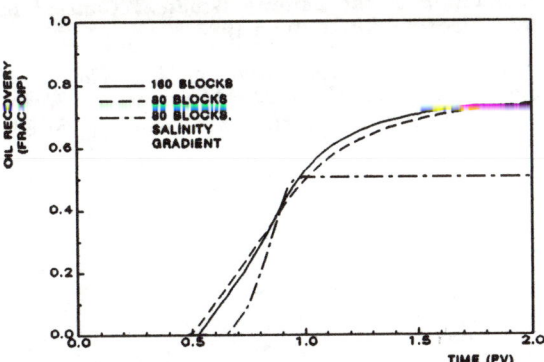

Figure 5. EFFECT OF DISPERSION AND SALINITY GRADIENT
0.15 PV SLUG, INTERMEDIATE MIDDLE PHASE MOBILITY

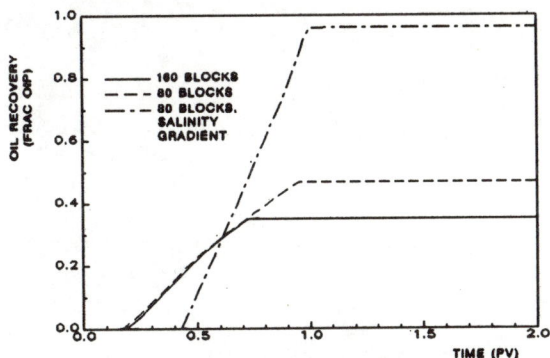

Figure 6. EFFECT OF DISPERSION AND SALINITY GRADIENT
0.15 PV SLUG, HIGH MIDDLE PHASE MOBILITY.

3.3 Effect of a Salinity Gradient

In the salinity gradient floods, a slug salinity equal to the optimal salinity of the surfactant is used, together with a drive salinity just below the salinity at which the phase behaviour becomes Type II(-), and an initial reservoir salinity above the salinity at which the phase behaviour becomes Type II(+).

For the low mobility case (Figure 4), the salinity gradient restores the piston-like nature of the displacement, though the ultimate oil recovery is lower than the 100% value obtained in the absence of dispersion. Figure 7 shows the profiles of salinity and surfactant concentration after 0.45 PV (0.15 PV chemical slug plus 0.3 PV drive). Dispersion has reduced the salinity profile to a smooth continuous transition from the initial formation

value to the drive water value, defining a region in which the phase environment is Type III. On account of the low middle phase mobility, the surfactant piles up at the rear of this region. Note that the peak surfactant concentration is almost twice the injected concentration.

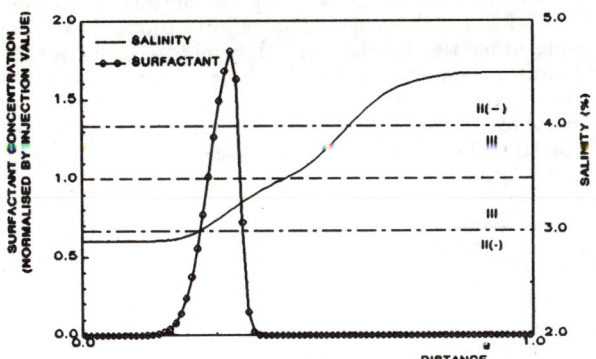

Figure 7. SALINITY AND SURFACTANT CONCENTRATION
AFTER 0.45 PV
SALINITY GRADIENT FLOOD, LOW MIDDLE PHASE MOBILITY

Note also that almost all the surfactant is located at points where the salinity is well below optimum. This is contrary to the assertion of Hirasaki et al[3] that the salinity gradient ensures that the salinity is optimal at some point within the surfactant slug.

In the intermediate mobility case, the salinity gradient actually reduces oil recovery, from 73.2% to 50.6% (Figure 5). The oil recovery curve is less dispersed, however, and recovery is complete by 0.9 PV, at which time the recovery in the constant salinity case is also about 50%. Figure 8 shows that the salinity profile at 0.45 PV is similar to the low mobility case (Figure 7). However, the surfactant is spread over most of the Type III region, with a peak concentration near the middle of this region, where the salinity is near to optimal. The peak concentration is somewhat below the injection concentration.

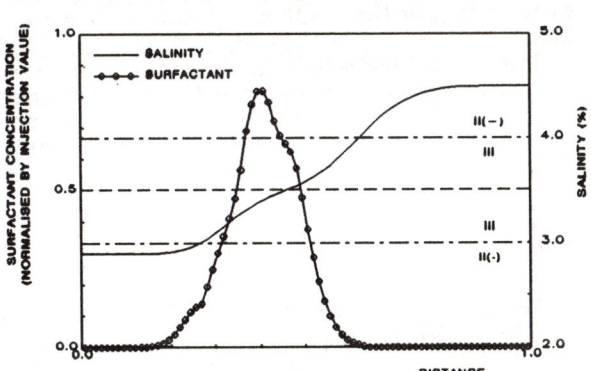

Figure 8. SALINITY AND SURFACTANT CONCENTRATION
AFTER 0.45 PV
SALINITY GRADIENT FLOOD, INTERMEDIATE MIDDLE PHASE MOBILITY

The greater recovery in the constant salinity case again relies on the independence of interfacial tension on surfactant concentration, which is not realistic. The salinity gradient design would also be preferable on account of its greater robustness to variations in reservoir conditions[3].

Figure 6 shows that the salinity gradient has a remarkable effect on the high mobility flood. Recovery is increased to 96%, and the displacement is nearly piston-like. Figure 9 shows the salinity and surfactant profiles at 0.45 PV for this case. The salinity profile is similar to the other two cases. The high middle phase mobility causes the surfactant to bank up at the front of the Type III region (although some has been delayed in an oleic phase formed as the overall composition enters the Type II(+) lobe of the Type III phase diagram at low oil concentration).

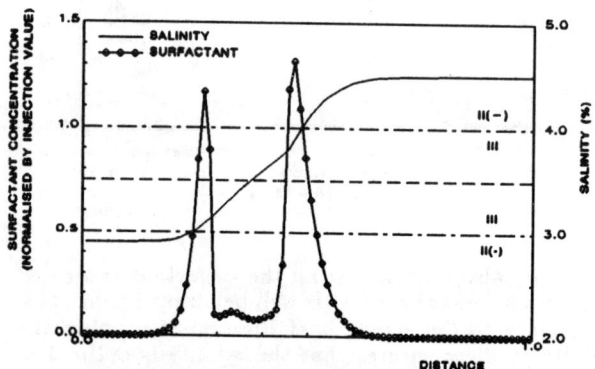

Figure 9. SALINITY AND SURFACTANT CONCENTRATION
AFTER 0.45 PV
SALINITY GRADIENT FLOOD. HIGH MIDDLE PHASE MOBILITY

This flood is so efficient because it has the character of a high concentration Type II(+) flood for which piston-like displacement occurs in the absence of dispersion[4]. The salinity gradient and very high middle phase relative mobility have combined to nullify the adverse effects of dispersion which would be expected for the Type II(+) flood. Further investigation is necessary to ascertain whether the high mobility, salinity gradient flood would be robust to viscous fingering, gravity override and heterogeneities in a three-dimensional field-scale setting.

Note that in these salinity gradient simulations, it is only in the intermediate mobility case that the surfactant is actually at or near optimal salinity. Despite this, the intermediate mobility case is the least effective in terms of oil recovery.

Acknowledgements

We thank the British Petroleum Co. plc for permission to publish this paper.

References

[1]Barker, J.W., and Fayers, F.J., *Factors Influencing Successful Numerical Simulation of Surfactant Displacement in North Sea Fields*, In Situ, **12**, no.4, pp.275-318, 1988-89.

[2]Aanonsen, S.I., *Application of Fractional Flow Theory to 3-Phase, 1-Dimensional Surfactant Flooding*, Proc. Euro. Conf. on Math. of Oil Recovery, Cambridge, UK, July 1989.

[3]Hirasaki, G.J., van Domselaar, H.R. and Nelson, R.C., *Evaluation of the Salinity Gradient Concept in Surfactant Flooding*, SPEJ, June 1983.

[4]Hirasaki, G.J., *Application of the Theory of Multicomponent, Multiphase Displacement to Three-Component, Two-Phase Surfactant Flooding*, SPEJ, April 1981.

2nd European Conference on the Mathematics of Oil Recovery
© D. Guérillot, O. Guillon (Editors) and Éditions Technip, Paris 1990, pp. 305-308
27 rue Ginoux, 75015 Paris

A Triangular Model for Three-Phase Flow

L. Holden[1,*] and A. Tveito[2,**]

1   Introduction

In this paper some numerical experiments with a recently proposed model for the flow of water, oil and gas in a porous medium are discussed. We consider the case of incompressible flow and neglect cappilary effects. In many oil reservoars the three phases are in contact with each other. The most important parameters describing the flow are the three-phase relative permeabilities. Unfortunately, these data are typically very difficult to measure and is often not available.

There has been a considerable interest in the mathematics of three-phase flow, see [9] and references therein. Bell, Trangenstein and Shubin [2] showed that there may be an elliptic region in the three-phase state space. Models with elliptic regions are mathematically very complicated. In the elliptic region the problem seems to be unstable and we can not expect to find a reasonable solution. Most of the interest in three-phase flow problems from mathematicans focus on the problem with elliptic regions.

The model we discuss is proposed in [6]. This model is based on a simple observation. Since the gas viscosity usually is much smaller than the vis-

*This research has been supported in part by the Royal Norwegian Council for Technical and Industrial reseach (NTNF).

**This research has been supported by VISTA, a reseach cooperation between the Norwegian Academy of Science and Letters and Den norske stats oljeselskap a.s. (Statoil).

cosities of oil and water c.f. [3], the fractional flow function of gas may be approximated by a function which only depend on the gas saturation. The error in this approximation seems to be smaller than the uncertainities in the data. If the fractional flow function of gas only depend on the gas saturation, then the model is triangular, see [6]. Gimse [5] has proved that the solution of triangular three-phase Riemann problems exists and satisfies some stability properties. Thus the model behaves more reasonable on Riemann problems than models with elliptic regions.

2   The mathematical model

The well-known three-phase flow model is

$$\varphi \frac{\partial s_g}{\partial t} + K v_t \frac{\partial f_g}{\partial x} = 0$$
$$\varphi \frac{\partial s_w}{\partial t} + K v_t \frac{\partial f_w}{\partial x} = 0 \qquad (1)$$

see e.g. Peaceman [7] and Aziz and Settari [1]. The notation is

- w, o, g in supscripts denotes water, oil and gas respectively,

- φ is porosity

- s_i is the saturation, $i = w, o, g$

- K is the permeability

- v_t is the total velocity

(1) Norwegian Computing Center, PO Box 114, Blindern, 0314 Oslo 3, Norway.
(2) Department of Informatics, University of Oslo, PO Box 1080, Blindern, 0316 Oslo 3, Norway.

- $f_i = \frac{\lambda_i}{\lambda_t}$ is the fractional flow, $i = w, o, g$

- $\lambda_i = \frac{k_{ri}}{\mu_i}$ is the mobility, $i = w, o, g$

- $\lambda_t = \lambda_w + \lambda_o + \lambda_g$ is the total mobility

- k_{ri} is the relative permeability, $i = w, o, g$

- μ_i is the viscosity, $i = w, o, g$

Notice that $s_w + s_g + s_o = 1$. The relative permeabilities depend on the saturations, i.e. $k_{ri} = k_{ri}(s_w, s_g)$ for $i = w, o, g$. Usually relative permeability functions are interpolations of two phase water/oil and oil/gas data.

Delshad and Pope [4] discuss several different models for the three-phase relative permeabilities. All their models satisfy the Stone's assumption, i.e. that the relative permeabilities of water and gas only depend on the water and gas saturation respectively, i.e. $k_{r,w} = k_{r,w}(s_w)$ and $k_{r,g} = k_{r,g}(s_g)$. A frequently used model is due to Stone [8] and normalized by Aziz and Settari [1]. In this model

$$k_{ro} = \frac{s_o k_{row} k_{rog}}{(1 - s_w)(1 - s_g)}$$

where the residual saturations are set equal to zero and k_{row} and k_{rog} denote the two-phase relative permeabilities of oil. In this and similar models, see [4], there may be elliptic regions, see [9].

Some models satisfy the additional assumption that also the relative permeability of oil only depend on the oil saturation, i.e. $k_{ro} = k_{ro}(s_o)$. With this assumption the model is always hyperbolic also when gravity is included, [9]. Unfortunately this model does not satisfy usual two-phase data. Trangenstein [9] has proved that all other models which satisfies the Stone's assumption have elliptic regions for some data when gravity is included.

The fractional flow of gas is

$$f_g(s_g, s_w) = \frac{k_{rg}(s_g)}{k_{rg}(s_g) + \frac{\mu_g}{\mu_w}(k_{rw}(s_w) + k_{ro}(s_o)\frac{\mu_w}{\mu_o})} \quad (2)$$

In many cases $\frac{\mu_g}{\mu_w} < .02$, c.f. [3]. This implies that the the partial derivative of f_g with respect to s_w is small . This is illustrated in Fig. 1. The corresponding fractional flow of water is illustrated in Fig. 2. The curves in the figures are the fractional flow functions for different fixed water and gas saturation respectively. Notice that curves of

the fractional flow of gas are insensitive to changes in s_w wheras the fractional flow of water are quite sensive to changes in s_g almost identical.

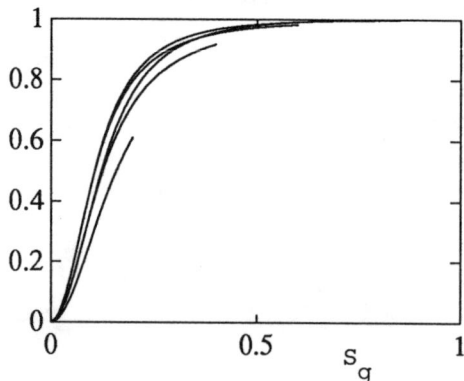

Fig. 1. The fractional flow of gas

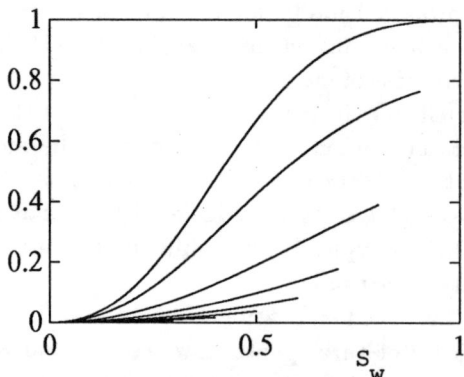

Fig. 2. The fractional flow of water

We have used Stone's model with the following data: $\mu_o = 1.$, $\mu_w = 0.5$, $\mu_g = 0.02$, $k_{rw} = s_w^2$, $k_{rg} = s_g^2$, $k_{row} = (1 - s_w)^2$ and $k_{rog} = (1 - s_g)^2$. Notice that all the Corey exponents are set equal to 2.0

Since the the partial derivative of f_g with respect to s_w is small, it is reasonable to let the fractional flow of gas be only a function of the gas saturation, i.e. $f_{T,g} = f_{T,g}(s_g)$. This results in a triangular model because the first equation in (1) then becomes independent of the water and oil saturation. Thus the gas flow is independent of whether the gas is in contact with oil or water. We propose to set

$$f_{T,g}(s_g) = \frac{k_{rg}(s_g)}{k_{rg}(s_g) + \frac{\mu_g}{\mu_w}(k_{rw}(0) + k_{ro}(1-s_g)\frac{\mu_w}{\mu_o})} \quad (3)$$

Then the model satisfies the same two-phase data as (2).

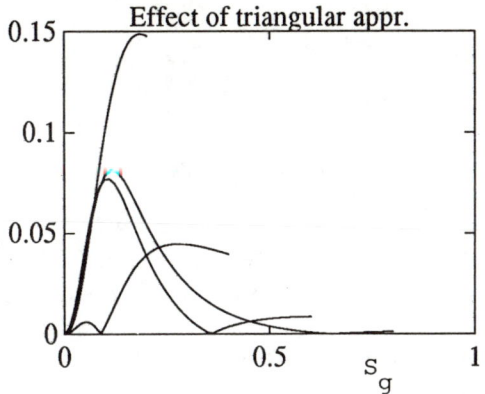

Fig. 3. Difference in f_g due to the triangular approximation

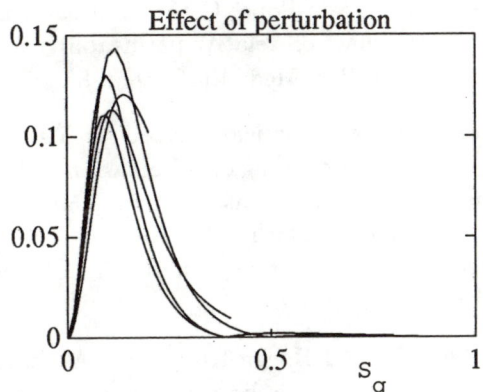

Fig. 4. Difference in f_g due to perturbation in the Corey exponents

Fig. 3 shows the absolute value of the difference between the fractional flow of gas defined in (2) and the fractional flow in the triangular model. The different curves are for different fixed water saturations. Fig. 4 shows the corresponding difference between two fractional flow curves of gas, one with all Corey exponents equal to 2.0 and one with all Corey exponents equal to 2.3. Notice that the triangular model is closer to the original curve than the pertubed model (Corey exponents equal to 2.3).

3 Experiments

In this section we will show that the difference between a numerical solution of the Stone's model

and a numerical solution of the triangular model is quite small. In the example the numerical solution with three different models are compared. The three models are the Stone's model with the data described in the previous section, the Stone's model with pertubed data and the triangular model. The only difference between the pertubed data and the original data is that all the Corey exponents are set equal to 2.3 instead of 2.0. The authors believe that this perturbation is smaller than typical uncertainities in real data. As proposed above we set the fractional flow function of gas equal to the fractional flow of gas in (1) with $s_w = 0$ in the triangular model, i.e. $f_{T,g}(s_g) = f_g(s_g, 0)$. Then the two models have the same two-phase data. The following initial data have been used

$$s_o = \begin{cases} \frac{1}{2} & x \leq 0 \\ \frac{1}{4} + \frac{1}{4}\cos^2(\frac{\pi x}{2}) & x \in (0,1) \\ \frac{1}{4} & x \geq 1 \end{cases}$$

and

$$s_g = \begin{cases} \frac{1}{4} & x \leq 0 \\ \frac{1}{4} + \frac{1}{4}\sin^2(\frac{\pi x}{2}) & x \in (0,1) \\ \frac{1}{2} & x \geq 1 \end{cases}$$

The numerical method choosen is an upwind scheme, see e.g. [2], with $\Delta t = .002$ and $\Delta x = .016$.

In Fig. 5 the saturations are plottet for $t = 1$. Notice that all three models behave quite similary. Defining the norm

$$\| S \| = \Delta x \sum | s_o | + \Delta x \sum | s_g |,$$

where the summation is over the grid blocks, we get

$$\frac{\| S - S_T \|}{\| S \|} = 0.0040$$

and

$$\frac{\| S - S_P \|}{\| S \|} = 0.0094$$

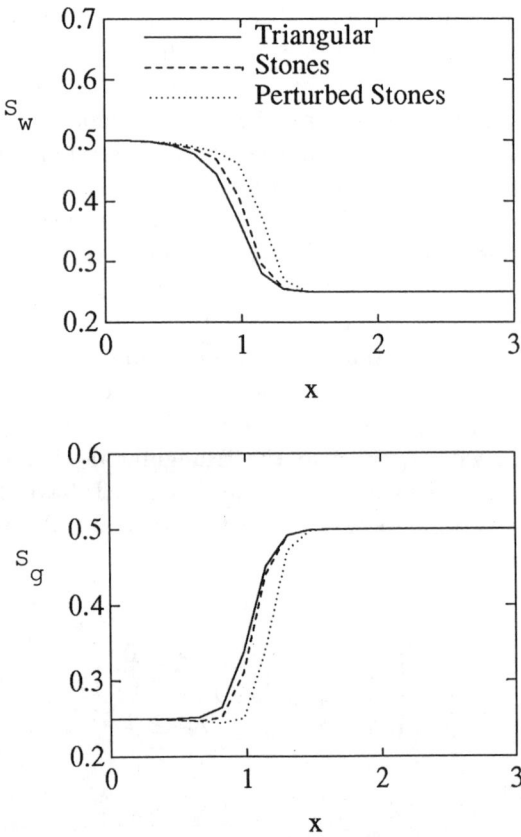

Fig. 5. The numerical solutions with three different models

where S, S_T and S_P are the Stone's model with original data, the triangular model and the Stone's model with pertubed data respectively. Observe that the distance from S_T to S is smaller than the distance from S_P to S.

4 Conclusion

Most traditional models for flow of water/oil/gas have elliptic regions which implies serious mathematical difficulties. If $\mu_g \ll \mu_o, \mu_w$ a small perturbation in the fractional flow function, results in a triangular model with reasonable mathematical properties. The model satisfies usual two-phase data. The numerical experiment shows that the solutions of the different model are quite close. In fact the solution of the triangular model seems to be within the uncertainity of the solution of the Stone's model. The uncertainity is due to the large error in the estimation of the relative permeabilities.

References

[1] Aziz, K. and Settari, A., Petroleum reservoir simulation, Applied Science (1979)

[2] Bell, J. B., Trangenstein, J. A. and Shubin, G. R., Conservation laws of mixed type descibing three-phase flow in porous media, Siam J. of Appl. Math., 46 (1986) 1000-1117

[3] Dake, L. P., Fundamentals of resrvoir engineering, Elsevier, (1978)

[4] Delshad, M. and Pope, G. A., Comparison of the three-phase oil relative permeability models, Transp. Por. Med. 4 (1989) 59-83

[5] Gimse, T., A numerical method for a system of equations modelling one-dimensional three-phase flow in a porous medium, 159-168 in Ballmann and Jeltsch (eds.) Nonlinear hyperbolic equations - theory, computational methods and applications.

[6] Holden, L. and Høegh-Krohn, R., A class of N nonlinear hyperbolic conservation laws, J. diff. eq. 84 (1990) 73-99

[7] Peaceman, D. W., Fundamentals if numerical reservoir simulation, Elsevier Scientific Publ. Comp. New York (1977)

[8] Stone, H. L., Probability model for estimating three-phase relative permeability, J. Pet. Tech. (1970) 214-218

[9] Trangenstein, J. A., Three-phase flow with gravity, Contemp. Math. 1990, 100 (1989) 147-160

2nd European Conference on the Mathematics of Oil Recovery
© D. Guérillot, O. Guillon (Editors) and Éditions Technip, Paris 1990, pp. 309-313
27 rue Ginoux, 75015 Paris

A Practical Front Tracking Technique for Control of Numerical Diffusion

M. Halilu and R. I. Issa[1]

ABSTRACT

A front tracking scheme capable of drastically reducing the numerical diffusion which inhibits resolution of sharp contact fronts is developed and implemented in a multi-dimensional simulation technique. The most attractive feature of the scheme is its ease of incorporation into standard reservoir simulation methods, whereby only the finite difference equations in the vicinity of the contact front need to be altered. Implementation of the scheme in conjunction with an IMPES method is outlined here, and the results of one and two dimensional flow applications are presented. These show marked improvements in the ability to capture sharp fronts.

1 INTRODUCTION

One of the main factors limiting the usefulness of most existing reservoir simultion techniques is numerical diffusion that results in the severe smearing of sharp fluid contact front over the computational grid. This arises when the single point upstream weighting scheme which is of first order accuracy is used. Unlike higher order differcing schemes, this scheme gives realistic solutions that are bounded by the physical limits of the problem(thus explaining the reason for its very wide use). As a consequence of numerical smearing, the interface between the injected and resident fluids cannot be accurately located. The practical consequences of this in reservoir engineering is that, the prediction of breakthrough time and the subsequent production profile will be far from accurate. Hence good production planning will be difficult. For this reason, several schemes have been developed in the past to reduce smearing and produce sharp contact fronts. They include several filtered higher order differencing schemes[14,8,9,15,16], moving mesh and moving point methods[2,4,5], front tracking methods[1,3,4,10,11,12] and various methods of modifying relative permeability curves[6,7,17]. These schemes have met with various degrees of success when implemented. The major limitations have been :

(i) some are only applicable to flows in one dimension
(ii) others are computationally too complex and expensive to apply to multidimensional flows and
(iii) most of the schemes cannot be incorporated into existing simulators.

In this work, the simple line interface calculation(SLIC)[3,10,11] method of front tracking is used as the basis of our approach. The contact front is tracked explicitly and all spatial differencing within the vicinity of the front is replaced by more accurate flux calculations. The rest of the domain can be solved for by any of the existing schemes; in this case, single point upstream weighting is used.

1.1 Equations

For a two phase flow of incompressible fluids in an isothermal porous medium the governing equations are:

(1) Department of Mineral Resources Engineering, Imperial College, London SW7 2 BP, UK.

Water saturation

$$\frac{\partial}{\partial x}\left\{\frac{kk_{rw}}{\mu_w}\cdot\frac{\partial p}{\partial x}\right\} = A\phi\frac{\partial s_w}{\partial t} \qquad (1)$$

Oil saturation

$$\frac{\partial}{\partial x}\left\{\frac{kk_{ro}}{\mu_o}\cdot\frac{\partial p}{\partial x}\right\} = A\phi\frac{\partial s_o}{\partial t} \qquad (2)$$

A pressure equation can be derived from the above equations as:

$$\frac{\partial}{\partial x}\left\{\left(\frac{k_{rw}}{\mu_w}+\frac{k_{ro}}{\mu_o}\right)\cdot\frac{\partial p}{\partial x}\right\} = 0 \qquad (3)$$

The phase equations in (1) and (2) can be written in descretised form[9] as:

$$-\frac{Q}{A\phi}f'\frac{\Delta s}{\Delta x}+\left(\frac{s^{n+1}-s^n}{\Delta t}\right) = 0 \qquad (4)$$

where Δs is the spatial difference operator on saturation, the form of which depends on the type of spatial and temporal weighting schemes used. Since a front tracking technique is to be employed here in order to capture the sharp contact front, it is perfectly justfiable to use first order spatial upwind scheme everywhere else.

As the front tracking technique is ideally suited to a temporarily explicit calculation, an explicit treatment is accorded to the saturation equations for the sake of simplicity of its implementation.

2 FRONT TRACKING

2.1 Outline

The method is based on a simple line interface calculation(SLIC) originally proposed by Noh et al[11]. A parameter Φ is assigned to each cell corresponding to the fraction of the cell lying behind the front. Thus a cell has an Φ value of 1 if it is entirely behind the front, 0 if it is entirely ahead, and between 0 and 1 if it straddles the front. These Φ values are updated during each time step in accordance with appropriate velocity field and the known speed of the front. The front is then constructed by geometrical calculations[4,6] based on the inspection of frontal cells and their neighbours.

2.2 The present implementation

So far, there has been only one attempt to apply the SLIC method in porous media flow[6]. Since a practicable scheme is sought, our method will differ from the previous one in two fundamental aspects. These are:

(i) instead of using characteristics based methods[13,10] to solve for the flow equations away from the front, the standard upstream weighting is used. The complicated computations involved in characteristics based schemes make them unattractive for practical reservoir simulation.

(ii) the geometrical calculation for locally constructing the front is completely eliminated.

For the sake of clarity, a one dimensional two phase flow of incompressible fluids in a horizontal reservoir will be used to outline the method. The two dimensional case follows the same reasoning with operator splitting.

2.2.1 Description of the front

Consider Fig 1.

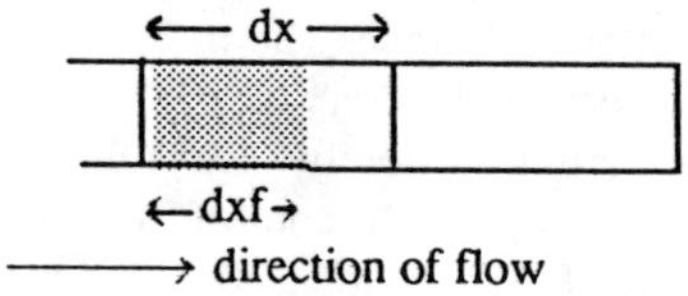

FIG 1

We define Φ as $\Phi=dxf/dx$, where dxf = location of front from the upstream boundary of the cell. In two dimensional flow, Φ is defined as $\Phi=dxf/dx + dyf/dy$.

2.2.2 Advancement of the front

The velocity with which the front is advanced is derived from the Darcy and continuity equations together with the Rankine-Hugoniot jump condition. Let s_l and s_r be the values of saturation to the left and the right of the front. The velocity of the front is given by:

$$V_f = q.\ df/ds = q.\ \{f(s_r)-f(s_l)\}/\{s_r-s_l\}$$

where q= total flowrate across the cell boundary
f=fractional flow function.

In one dimension, q can be directly determined from the injection rate. In multidimensions, q can be determined from the pressure gradients using known saturations from the previous time step for the calculation of the coefficients. The saturation behind the front, s_l can be replaced by the analytical value, swf, and that ahead of it by the initial saturation swi. Hence the Rankine-Hugoniot jump condition becomes:

$$\{f(s_r)-f(s_i)\}/\{s_r-s_i\} = df/ds \quad \text{at swf,}$$

which can be directly satisfied from the fractional flow curve. The front is then advanced by a distance $V_f dt$. The new position of the front is given by $X_{new} = X_{old} + V_f dt$ from which Φ values are updated.

2.2.3 Computation of fluxes

Fluid fluxes in the neighbourhood of the front are calculated in a manner consistent with the flood front advance such that exact material balance is maintained at all times. Consider Figs 2 and 3.

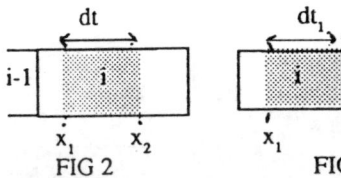
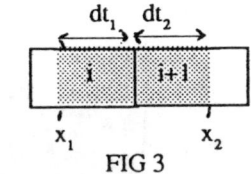

FIG 2   FIG 3

Suppose the front is at x_1 at a given time t_n. For a given velocity, the location of the front after a time step dt can be determined. Two cases are possible: either the front moves to a location within the same block as in Fig 2 position x_2, or it moves to the block downstream as in Fig 3 position x_2. The relevant fluxes are computed accordingly. For example, across the boundary $i-1/2$ in Fig 2, the fluxes are computed from the equations:

$$Fw_{i-1/2} = V_f dt.A(Swf-Swi)/dt$$
$$Fo_{i-1/2} = \lambda o_{i-1} \Delta Po_{i-1/2}$$

and at $i-1/2$ in Fig 3 from:

$$Fw_{i-1/2} = V_f dt_1.A(Swf-Swi)/dt + \lambda w_{i-1} \Delta Po_{i-1/2}.dt_2/dt$$
$$Fo_{i-1/2} = \lambda o_{i-1} \Delta Po_{i-1/2}$$

where :- dt_1 = time taken by the front to move
                from x_1 to $i+1/2$
            dt_2 = time taken by the front to move
                from $i+1/2$ to x_2

Under all circumstances these are the only two possible situations around the front in an IMPES simulation. We can see that flux computations are required in no more than three cells. This is because each frontal cell is coupled to its neighbours by the fact that influx from one cell is exactly equal to outflux from its neighbour. It is this which always guarantees material balance automatically.

2.2.4 The Flow Equations

The saturation and pressure equations are derived in exactly the same way as the IMPES method in the whole domain. The only modification is that the computed fluxes in the region of the front now replace the standard upstream weighting scheme at the appropriate cell faces.

The general form of the equation is :

$$(Ft_{1,i+1/2}-Ft_{1,i-1/2})dt_1/dt + (Ft_{2,i+1/2}-Ft_{2,i-1/2})dt_2/dt$$
$$= A.\phi.dx.ds/dt + \text{sources.}$$

where Ft_1 and Ft_2 are the computed fluxes during the times dt_1 and dt_2 respectively.

Hence the modified saturation equation for block i in Fig 2 are as follows:

$$Sw_i^{n+1} = [Fw_{i-1/2} + A/dx.\lambda w_{i+1}.(Pw_{i+1}-Pw_i) + Qw_i]dt/dv + Sw_i^n$$
$$So_i^{n+1} = \{[A/dx.\lambda o_{i+1}.(Po_{i+1}-Po_i)] - [A/dx.\lambda o_{i-1}.(Po_i-Po_{i-1})] + Qo_i\}dt/dv + So_i^n$$

Similar equations can be derived for the case in Fig 3. The pressure equation is derived in the usual IMPES manner by adding the two saturation equations.

3 RESULTS AND DISCUSSION

Validation tests were made in horizontal reservoirs in one and two dimensions with oil to water viscosity ratio of 10.

Figs 4 (a) to (d) show the results for a one-dimensional reservoir obtained with 50 grid blocks. Time steps used vary from 1 day to the maximum possible time step dictated by stability.

It can be seen that the new method gives a marked improvement on standard upwinding. Comparison with the analytical solution shows that the smearing of the front is limited to two cells at most and the frontal speed is seen to be predicted very well. The Figs also show that the method is very robust to the time step size. There is very little difference in the saturation profiles when time steps in the range 1 to 11.4 days (which is the maximum possible) are used.

We also note that the difference between the proposed method and standard upwinding reduces for bigger time steps. This is because the diffusion-like numerical error in the upwind scheme is directly proportional to $\{1-V_f dt/dx\}$ and this becomes zero when dt is exactly equal to dx/v_f. However, such large time steps are seldom possible in practical calculations as this value of dt corresponds to the stability limit of explicit methods. Perhaps the best illustration of the improvement achieved by the new method is shown in Fig 5. As the number of grid blocks increases from 10 to 50, smearing is still limited to no more than two blocks. As shown in Figs 5(b) and 5(d), the mesh density can be reduced to less than half, yet good results are still obtained.

The two-dimensional results are shown in Figs 6 and 7(the wrinkles in Fig 6(a) are not representative of actual data values but are due the plotting routine used). For better clarity, profiles along cross-sections A-A, B-B and C-C are shown in Figs 6 (b) to (d). These figures show similar features to the one-dimensional results. The value of our new method can best be seen in Fig 7, where results obtained with a 5x5 grid appear to better than those obtained with a 20x20 grid using standard upwinding alone. The implications of the insensitivity of the method to both time step size and number of grids is that the prediction of fluid flow and associated parameters can be much more reliable in practical simulation.

Two dimensional flow

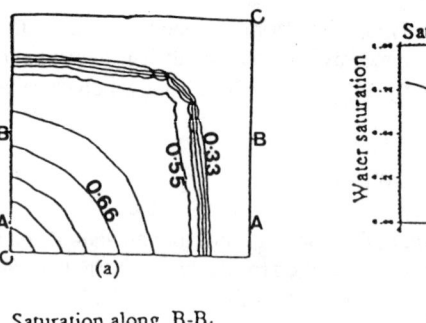

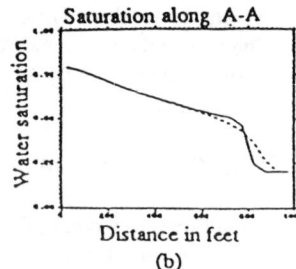
Saturation along A-A

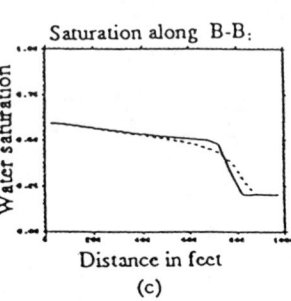
Saturation along B-B

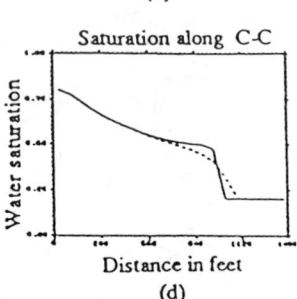
Saturation along C-C

FIG. 6. (a) Saturation contour plot using 20x20 grid cells and time step of 1 day. (b) Cross section along A-A in 6(a). Cross section along B-B in 6(a). (c) Cross section along C-C in 6(a).

One dimensional flow

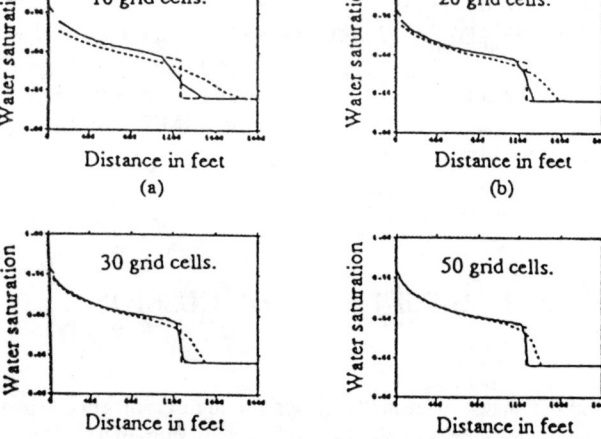

FIG. 5. Saturation profiles at 1 year using different number of grid cells and a time step of 1 day.

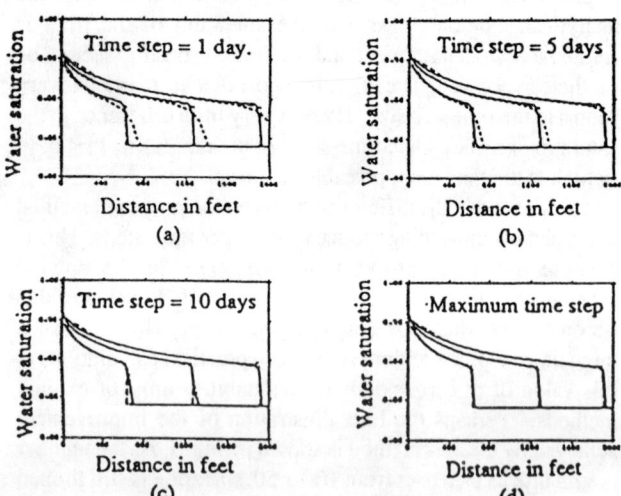

FIG. 4. Saturation profiles at 0.5, 1 and 1.5 years using 50 grid cells and different time steps.

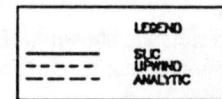

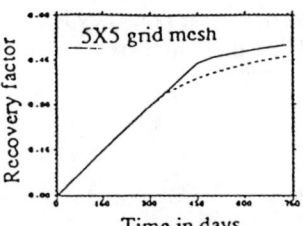
5X5 grid mesh

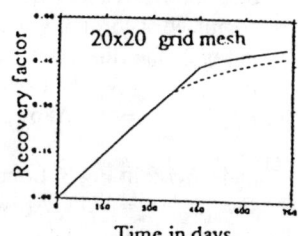
20x20 grid mesh

FIG. 7 Oil production profile (a) Using a 5x5 grid mesh. (b) using a 20x20 grid mesh.

LEGEND
——— SLIC
- - - - UPWIND

4 REFERENCES

(1)A.J.Chorin., "Flame advection and propagation algorithms."Journal of Computational Physics,35,1980,pp.1-11.

(2)C. L. Farmer and R. A. Norman. "The implementation of moving point methods for convection-diffusion equations.", Numerical methods for fluid dynamics, K.W. Morton and M.J. Baines, eds, Academic Press, Oxford, 1985.

(3) A. F. Ghoneim. "Effect of large scale structures on turbulent flame propagation.", Combustion and Flame, 64, 1986, pp321-336.

(4)C. W. Hirt and B. D. Nichols "Volume of fluid (VOF) method for the dynamics of free boundaries." Journal of Comp. Physics, 39,1981, pp201-225.

(5)J.M. Hyaman. "Numerical methods for tracking interfaces.", Los Alamos national laboratory, Los Alamos, NM87545, USA, pp 396-407.

(6)J. R. Kyte and D. W. Bery. "New pseudo functions to control numerical dispersion.", SPE AIME, August 1975, pp269-276.

(7)Laprea-Bigott and R. A. Morse."Improved pressure response representation and reduction of numerical dispersion effects in reservoir simulation.", SPE 8875, pp1-10.

8)R. C. Larson. "Controlling numerical dispersion by variably imed flux updating in two dimensions.", SPE 9374, pp1-12.

(9)R. C. Larson. "A novel method for controlling numerical dispersion in finite difference simulation of flow in porous media." SPE 8027, pp1-16.

(10) P. Lostedt. "A front tracking method applied to Burgers equation and two phase porous flow.", Journal of Comp. Physics, 47,1982, pp211-228.

(11)W. F. Noh and P. Woodward. "SLIC(simple line interface calculations).", Lecture notes in physics, 59 Springer-Verlag, Berlin,1976, pp330-340.

(12)I. S. Partrom. "Application of VOF method to sloshing of a luid in a partially filled cylindrical container.", International journal for numerical methods in fluids., vol7, 1987, pp535-550.

(13) P. L. Roe "Characteristics based schemes for Euler equations.",Annual review of fluid mechanics.,1986 ,vol 18, pp337-365.

(14)T. C. Tan."PhD dissertation, Min. Res. Eng. Imperial College, London., 1982.

(15)M. R. Todd, P. M O'dell and G. J. Hirasaki. "Methods for increased accuracy in numerical reservoir simulation.", SPEJ, Dec. 1972, pp515-530.

(16) P. K. Vinsome and A. K. Au. "One approach to grid orientation problem in reservoir simulation.", SPEJ, April 1985. pp160-161

(17)M.S.Young,P.C.Casinder and D.C.Wilson.,"The use of higher order finite difference methods in reservoir simulation.", paper Eur180, proc. European off shore petroleum conference, London, Oct. 1980, pp 337-348.

Nomenclature

A	=	cross sectional area.
dx	=	block length
f	=	fractional flow function
F	=	flux
K	=	permeability
Kr	=	relative permeability
n	=	time level
o	=	oil
p	=	pressure
Q	=	total flowrate
s	=	saturation
swi	=	initial water saturation
swf	=	frontal saturation
t	=	time
V_f	=	frontal velocity
w	=	water
λ	=	mobility
ϕ	=	porosity
Φ	=	fraction of cell behind front

2nd European Conference on the Mathematics of Oil Recovery
© D. Guérillot, O. Guillon (Editors) and Éditions Technip, Paris 1990, pp. 315-318
27 rue Ginoux, 75015 Paris

FAC Method for Reservoir Simulation

R. Boyer, B. Martinet and K. Saïkouk[1]

ABSTRACT

The simulation of a bidimensional two-phase immiscible flow is considered using a static composite grid (locally refined).

This composite grid consists of a global, rectangular coarse grid and a set of locally refined patches around wells. For the spatial discretisation, we use cell centered finite volume approximation and for time stepping, the IMPES scheme with upstream weighting.

In order to solve the pressure equation on composite grid efficiently, we use the FAC (Fast adaptive composite) method proposed by Mc. CORMICK [6]. This technique has analogies with BEPS method used by EWING and al. [2], [3].

A numerical application (oil-water flow) is developed to illustrate the FAC method.

1 INTRODUCTION

In reservoir simulation, many problems involve both large scale processes and localized phenomena. The use of a uniform fine grid would be very expensive, indeed even impossible. It is then useful to introduce local refined patches that will be static or dynamic. Many authors used local refinement techniques for petroleum problems. In particular, EWING and al [2], [3] suggest solving the pressure equation by a preconditioned conjugate gradient algorithm with an efficient preconditionning based on a decomposition in coarse global and fine local problems.

We use the FAC method as an autonomous iterative procedure. This technique is efficient, can be easily implemented and enables interesting extensions.

2 MODELISATION

Let us now briefly review the basic relations (cf. AZIZ-SETTARI [1]).

The equations for conservation of the oil and water phases are :

$$div(\lambda_i(\vec{grad}\, p_i - \rho_i g \cdot \vec{grad}\, z)) = \frac{\partial}{\partial t}\left(\phi \frac{S_i}{B_i}\right) + q_i \quad (i = 0, w)$$

The gas equation is :

$$div(\lambda_g(\vec{grad}\, p_g - \rho_g \cdot g \cdot \vec{grad}\, z)) + div(R_S \lambda_0(\vec{grad}\, p_0 - \rho_0 g \cdot \vec{grad}\, z))$$

$$= \frac{\partial}{\partial t}\left(\phi\left(\frac{S_g}{B_g} + R_S \frac{S_0}{B_0}\right)\right) + q_g + R_S q_0$$

With $\qquad \lambda_i = \frac{kr_i}{\mu_i B_i} k$ for i=0, w, g.

The saturations are related by :

$$\sum_i S_i = 1$$

---

(1) Université de Provence, Aix-Marseille I, 3, place Victor Hugo, 13331 Marseille, France.

and the pressures are related to the capillary pressures by :

$$p_{c0w} = p_0 - p_w$$

$$p_{cg0} = p_g - p_0$$

3 DISCRETISATION ON COMPOSITE GRID.

3.1 local refinement – composite grid.

In order to obtain an accurate approximation, a static refinement around wells is introduced by using a convenient criterion. The refinement is brought about by dividing the steps of the concerned cells by a factor of three. This enables us to preserve the wells in the centers of cells (see fig.1 in the case of a quarter of a five spot problem). The grid obtained after local refinement is called a composite grid.

Irregular grids are thus introduced at the interface between refined and unrefined areas.

3.2 Cell centered finite volume approximations on composite grid.

The case of regular cells is effected in a standard way (cf. AZIZ-SETTARI [1]). For irregular cells we use a conservative and symmetric scheme proposed by EWING and al. [3].

4 FAC METHOD.

4.1 Description :

We use the two-level FAC method, which is an iterative procedure. At each iteration, a coarse global problem (which can be done by using the existing code) and fine local problems on refined areas must be solved. These local problems can be often solved in parallel.

This technique can be interpreted as a particular multigrid method (cf. HACKBUSCH [4]) where the coarse global correction step is customary and where the "smoothing" step consists of local fine resolutions (often small problems). An important feature is the choice of interpolation and restriction operators and specially for strongly discontinuous coefficients (heterogeneities).

In the numerical applications presented here, we use bilinear piecewise interpolation and as restriction operator the adjoint of the interpolation relative to a particular scalar inner product (cf. WESSELING [7]).

The B.E.P.S. method used by EWING [3] is based on a different approach but leads to a similar implementation.

The FAC method can be extended to a multilevel refinement cases. One important aspect is the possibility of parallelization (AFAC method cf. Mc. CORMICK [6]).

4.2 Implementation in the WFLOOD code :

The WFLOOD code is an experimental simulator of water-oil flow in the two dimensional case which was developped by the "Institut Français du Pétrole (I.F.P) cf. JAIR [5]). We implemented the FAC technique in this code in accordance with its structure and its logic, and adding subroutines for the fine local resolution. The data structure used was chosen in order not to modify the initial one. It however enables the extension to the multi-level case and to the dynamic refinement.

5 NUMERICAL EXAMPLE – RESULTS

We present here the simulation of a quarter of a five spot reservoir with an injection well in the southwest corner and a production well in the north-east one. The results obtained show a significant improvement in comparison to the regular grid case. Fig.4 gives the WOR (water oil ratio) evolution according to time for several grids.

These results shows that the procedure requires 4 to 5 iterations per time step to actieve a residual L_2-norm of 10^{-4}, which leads to an average reduction factor for an iteration of about 0.11 (cf. Fig.2). This factor is almost independant of space and time steps (cf. Fig.3).

6 CONCLUSION

The proposed technique and data structure are easely implemented in the existing code.

The convergence is fast. The extensions currently developped are :
1) oil-gas flow. 2) local time stepping procedures.

We also intend to treat the multilevel case and to implement the method on a parallel computer.

Nomenclature :
- For every phase i=o, w, g :
ρ_i : density
p_i : pressure
μ_i : viscosity

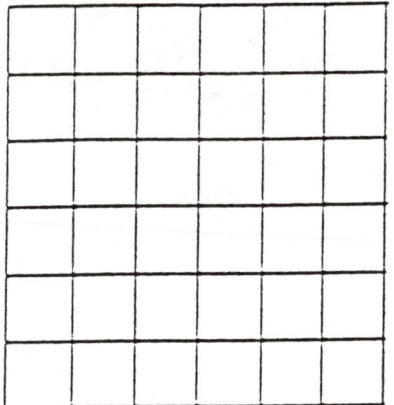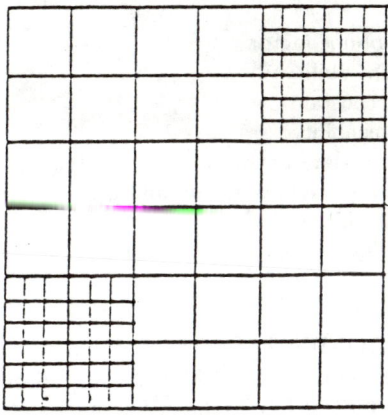

Fig.1 Coarse and Composite Grids

h (coarse)	h=1/9	h=1/18	h=1/30
average factor	k=0.11	k=0.13	k=0.13

Fig.2 The average reduction factor k
for several grids

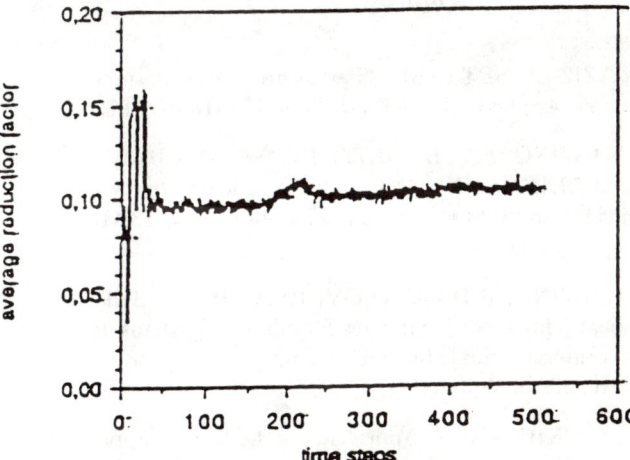

Fig.3

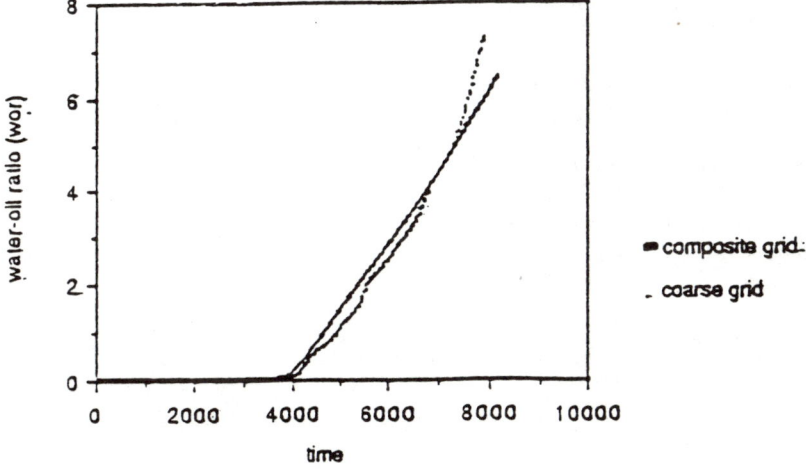

Fig.4 Water-Oil Ratio

λ_i : mobility

S_i : saturation

q_i : flow rate

B_i : formation volume factor

k_{r_i} : relative permeability

g : gravitational constant

k : absolute permeability

p_{cow} : capillary pressure between oil and water

p_{cog} : capillary pressure between oil and gas

R_s : solution gas - oil ratio

t : time

z : depth.

ACKNDWLEDGMENTS : The authors wish to thank MM. D.GUERILLOT, P. LEMONNIER and other colleagues of Département Recherche et Gisements of Institut Français du Pétrole.

References

[1] K. AZIZ, A. SETTARI : "Petroleum reservoir simulation". Applied science Publishers. LONDON. 1979.

[2] R.E.EWING, B.A.BOYETT, D.K.BABU, R.F.HEINEMANN : "Efficient use of locally refined grids for multiphase reservoir simulation" SPE 18.413 February 1989.

[3] R.E.EWING, R.D.LAZAROV, P.S.VASSILEVSKI : "Local refinement techniques for elliptic problems on cell centered grids". Inst. Sci. Comp. Rep. # 1988.16. University of Wyoming.

[4] W.HACKBUSCH : "Multigrid methods and applications". Springer Series in Computational mathematic. Vol.4 Springer Verlag 1985.

[5] JAIR : "Logiciel WFLOOD" Rapport interne IFP . Paris (1981).

[6] S.Mc.CORMICK : "Multilevel adaptive methods for partial differential equations". to appear.

[7] P.WESSELING : "Cell centered multigrid for interface problems". p.631-641 in Multigrid methods ed. by S.Mc.CORMICK. Lecture Notes in pure and applied mathematics vol. 110. 1988.

2nd European Conference on the Mathematics of Oil Recovery
© D. Guérillot, O. Guillon (Editors) and Éditions Technip, Paris 1990, pp. 319-322
27 rue Ginoux, 75015 Paris

Mixed Finite Elements for Multiphase Flow in Porous Media Consisting of Different Rock Types

Ø. Bøe[1] and G. E. Fladmark[2]

1 INTRODUCTION

Modelling of heterogeneous reservoirs has received increased attention during the last years. In high permeable zones, the flow could be dominated by advection, while processes such as imbibition or drainage might dominate in other parts. To illustrate the ideas presented in this paper, we consider two phase incompressible, immiscible flow in two spatial dimensions. The ideas could be generalized to more than two phases. The relative permeabilities $k_{r_l}, l = w, nw$ (w: wetting phase, nw: nonwetting phase) and capillary pressure data P_c are highly dependent upon the rock type. Indeed, we shall define a rock type as a porous medium for which such data are dependent upon the saturations only. Let the reservoir domain Ω concist of N_r rock types. We assume throughout the paper that the following characterization holds:

$$\Omega = \Omega_1 \cup \Omega_2 ... \Omega_r \cup \Omega_{r+1} ... \cup \Omega_{N_r} \qquad (1)$$

$$k_{r_l}(S_w, \vec{x}) = k_{r_l}^{(r)}(S_w), \ \vec{x} \in \Omega_r \qquad (2)$$

$$P_c(S_w, \vec{x}) = P_c^{(r)}(S_w), \ \vec{x} \in \Omega_r \qquad (3)$$

$$S_{wc}(\vec{x}) = S_{wc}^{(r)}, \ \vec{x} \in \Omega_r \qquad (4)$$

$$S_{nwr}(\vec{x}) = S_{nwr}^{(r)}, \ \vec{x} \in \Omega_r \qquad (5)$$

The boundary conditions at the interface separating two rock types i and ii are continuity of normal components of fluxes and pressures:

$$\vec{u}_l^{(i)} \cdot \vec{n}^{(i)} = -\vec{u}_l^{(ii)} \cdot \vec{n}^{(ii)}, \ l = w, nw \qquad (6)$$

$$p_l^{(i)} = p_l^{(ii)}, \ l = w, nw \qquad (7)$$

Continuity of the total Darcy flux $\vec{u}$ now follows by adding equations (6) for the two phases. By subtracting the wetting phase pressure p_w from the nonwetting phase pressure p_{nw} we get continuity of the capillary pressure:

$$P_c^{(i)}(S_w^{(i}) = P_c^{(ii)}(S_w^{(ii)}) \geq 0 \qquad (8)$$

(8) is a nonlinear condition for the saturation. It generally leads to a discontinuity in the saturations. Such a discontinuity could not be modelled directly by conventional field simulators which are based upon continuity of pressures and saturations (finite difference models).

2 MATHEMATICAL MODEL

In order to deal with discontinuous behavior, we turn to a weak formulation of the partial differential equations. (Chavent et al., 1986) derived a weak formulation for two phase incompressible flow. Our formulation will be related to this one, but it is not formulated in terms of the global pressure. Darcy's law and the mass conservation equations for each phase give the following first order system governing two phase immiscible, incompressible flow:

(1) IBM Bergen Scientific Centre, Thormøhlensgate 55, Bergen High Technology Centre, N-5006 Bergen, Norway.
Now with : Norsk Hydro, PO Box 4314, Nygårdstangen, N-5028 Bergen, Norway.
(2) Department of Applied Mathematics, University of Bergen, Allegaten 55, 5007 Bergen, Norway. *Also with :* Norsk Hydro, same address as above.

$$\nabla \cdot \vec{u} = 0 \qquad (9)$$

$$\frac{\vec{u}}{M(S,\vec{x})} = -\mathbf{K}(\vec{x})[\nabla p - H(S,\vec{x})\nabla P_c(S,\vec{x})$$
$$- G(S,\vec{x})\vec{g}] \qquad (10)$$

$$\phi(\vec{x})\frac{\partial S}{\partial t} + \nabla \cdot \vec{f}(S,\vec{x}) = 0 \qquad (11)$$

$$\vec{f}(S,\vec{x}) = \vec{f}_c(S,\vec{x}) + \vec{f}_d(S,\vec{x}) \qquad (12)$$

$$\vec{f}_c(S,\vec{x}) = A(S,\vec{x})\vec{u}$$
$$+ H(S,\vec{x})\mathbf{K}(\vec{x})(\rho_w - \rho_{nw})\vec{g} \qquad (13)$$

$$\vec{f}_d(S,\vec{x}) = \mathbf{K}(\vec{x})H(S,\vec{x})\nabla P_c(S,\vec{x}) \qquad (14)$$

$$\lambda_l = \frac{k_{r_l}}{\mu_l}, \ l = w, nw \qquad (15)$$

$$M(S,\vec{x}) = \frac{1}{\lambda_w(S,\vec{x}) + \lambda_{nw}(S,\vec{x})} \qquad (16)$$

$$A(S,\vec{x}) = \frac{\lambda_w(S,\vec{x})}{M(S,\vec{x})} \qquad (17)$$

$$G(S,\vec{x}) = \frac{\lambda_w(S,\vec{x})\rho_w + \lambda_{nw}(S,\vec{x})\rho_{nw}}{M(S,\vec{x})} \qquad (18)$$

$$H(S,\vec{x}) = \frac{\lambda_w(S,\vec{x})\lambda_{nw}(S,\vec{x})}{M(S,\vec{x})} \qquad (19)$$

$\vec{u} \cdot \vec{n}$	given on Γ_u	(20)
$\vec{f} \cdot \vec{n}$	given on Γ_w	(21)
p	given on Γ_p	(22)
S	given on Γ_s	(23)
$p(\vec{x})|_{t=0}$	given in $\bar{\Omega}$	(24)
$S(\vec{x})|_{t=0}$	given in $\bar{\Omega}$	(25)
p	continuous across Γ_{int}	(26)
$P_c^i(S^i) = P_c^{ii}(S^{ii})$	on Γ_{int}	(27)
$\vec{u}_l \cdot \vec{n}_{\text{int}} \ \ l = o, w$	continuous across Γ_{int}	(28)

If $\Gamma_p = \emptyset$ (empty), then the pressure is not unique, and one has to specify the pressure at some point in Ω.

Because of the boundedness of the relative permeabilities and the fact that their sum do not vanish simultaneously, the mobility functions (15)-(19) are bounded, i.e. members of $L^\infty(\Omega)$.

(9)-(19) together with the conditions (20)-(28) give a first order partial differential equation system for the nonwetting phase pressure p, the wetting phase saturation S and the fluxes $\vec{u}$ (total) and $\vec{f}$ (wetting). Substituting (10) into (9) and (12) into (11) leads to a system of two quasilinear second order partial differential equations. The pressure equation is of elliptic

type and the saturation equation is of parabolic type. The mobility fraction, $A(S,\vec{x})$, is an S-shaped function of saturation and increases monotonic from zero at $S = S_{wc}(\vec{x})$ to unity at $1 - S_{nwr}(\vec{x})$ as a function of S. The harmonic average of the mobilities $H(S,\vec{x})$ is a bell-shaped function of saturation and vanishes for both $S_{wc}(\vec{x})$ and $1 - S_{nwr}(\vec{x})$, causing singular behavior at $1 - S_{nwr}(\vec{x})$ as the saturation equation becomes hyperbolic.

In each rock type r we define the capillary potentials (Aavatsmark, 1989):

$$\Phi^{(r)}(S) = \int_{S_{wc}^{(r)}}^{S} \mathcal{A}^{(r)}(S) P_c^{\prime(r)}(S)dS \qquad (29)$$

$$\Psi^{(r)}(S) = \int_{S_{wc}^{(r)}}^{S} \mathcal{H}^{(r)}(S) P_c^{\prime(r)}(S)dS \qquad (30)$$

The capillary potentials are decreasing functions of S since the integrand is negative.
By the chain rule we have:

$$\nabla \Phi^{(r)}(S) = A^{(r)}(S) P_c^{\prime(r)}\nabla S \qquad (31)$$

$$\nabla \Psi^{(r)}(S) = H^{(r)}(S) P_c^{\prime(r)}\nabla S \qquad (32)$$

We will use the above concept of capillary potentials in order to derive a weak formulation that accounts for discontinuous behavior. Inside each rock type the capillary potentials (29)-(30) are functions of saturation only, according to the assumptions forced by (2)-(5). At the boundaries separating different rock types they are generally discontinuous. The functions $\Phi(S)$ and $\Psi(S)$ may be calculated and tabulated before simulation starts.

Consider the test functions $\vec{s} : \bar{\Omega} \to \mathbf{R}^2$, and $v : \bar{\Omega} \to \mathbf{R}$ representing regular mappings into $\mathbf{R}^2$ and $\mathbf{R}$ respectively.

$$\forall \text{ regular } \vec{s}: \Omega \to \mathbf{R}^2 \text{ set}:$$
$$\vec{s}_r = \text{RESTRICTION}(\vec{s}, \Omega_r) \qquad (33)$$
$$\forall \text{ regular } v: \Omega \to \mathbf{R} \text{ set}:$$
$$v_r = \text{RESTRICTION}(v, \Omega_r) \qquad (34)$$

To obtain a weak formulation of the system (9)-(28), the equations (9)-(10), (11)-(12) are multiplied by the test functions (33)-(34) and integrated by parts. We apply partial integration to the terms containing gradients of pressure and capillary potentials. This process takes place in each rock type. The equations for each rock type are then added together, and the continuity conditions (26)-(28) are taken into account. This process leads to the following weak formulation of the partial differential equations (9)-(12):

$$(\nabla \cdot \vec{u}, v)_\Omega = 0 \qquad (35)$$

$$\left(\frac{\mathbf{K}^{-1}\vec{u}}{M(S,\vec{x})}, \vec{s}\right)_\Omega = (p, \nabla \cdot \vec{s})_\Omega - (\Phi(S,\vec{x}), \nabla \cdot \vec{s})_\Omega$$
$$+ (G(S,\vec{x})\vec{g}, \vec{s})_\Omega - \langle p, \vec{s}\rangle_\Gamma \qquad (36)$$
$$+ \langle \Phi(S,\vec{x}), \vec{s}\rangle_\Gamma + \langle [\Phi(S,\vec{x})]_{ii}^i, \vec{s}\rangle_{\Gamma_{int}}$$

$$\left(\phi(\vec{x})\frac{\partial S}{\partial t}, v\right)_\Omega = -(\nabla \cdot \vec{f}, v)_\Omega \qquad (37)$$

$$(\mathbf{K}^{-1}(\vec{x})\vec{f}, \vec{s})_\Omega = (A(S,\vec{x})\mathbf{K}^{-1}(\vec{x})\vec{u}, \vec{s})_\Omega$$
$$+ ((\rho_w - \rho_{nw})H(S,\vec{x})\vec{g}, \vec{s})_\Omega$$
$$- (\Psi(S,\vec{x}), \nabla \cdot \vec{s})_\Omega + \langle \Psi(S,\vec{x}), \vec{s}\rangle_\Gamma$$
$$+ \langle [\Psi(S,\vec{x})]_{ii}^i, \vec{s}\rangle_{\Gamma_{int}} \qquad (38)$$

3 MIXED FINITE ELEMENTS FOR INTERIOR BOUNDARIES

Taking into account the conditions (20)-(28) and assuming that the permeability tensor is invertible, the inner products in (35)-(38) exist if $S, P \in L^2(\Omega)$ and $\vec{u}, \vec{f} \in H(\mathrm{div}, \Omega)$. The weak formulation of the problem is posed as the variational problem of finding the pairs $(p, \vec{u})$ and $(S, \vec{f}) \in L^2 \times H(\mathrm{div}, \Omega)$ such that (35)-(38) holds for all pairs $(v, \vec{s})$ in this product space. The Raviart-Thomas mixed finite element method (Raviart et al., 1977) is a natural choice for the discrete solution of the variational formulation because the solution is sought in discrete subspaces of $L^2(\Omega) \times H(\mathrm{div}, \Omega)$. The method was originally designed for second-order elliptic problems. We use the method to develop a numerical scheme for the coupled system of pressure and saturation equations with solutions in the Raviart-Thomas mixed finite-element spaces. There results a simultaneous solution of the nonwetting phase pressure p, the total darcy flux $\vec{u}$, the wetting phase saturation S and the flux $\vec{f}$ of the wetting phase. The scalar variables p and S need not to be continuous across the element boundaries, thus allowing for discontinuous approximations of the saturation S at the boundaries separating different rock types.

We design a discretization of problem (35)-(38) together with the conditions (20)-(28) that allows several degrees of freedom (saturations) at each node. (Chavent et al., 1986) introduced a discontinuous finite element scheme for the saturation equation based on linear interpolation. At each node there are as many

global degrees of freedom as there are neighbouring elements. We modify as follows: Instead of having as many degress of freedom as there are neighbouring elements, one could have the number of rock types that meet at the node. Consequently, we force a 'continuous' discrete solution at the interior of each rock type, and a 'discontinuous' solution at the boundaries separating different rock types.

Let N_i rock types meet at node i. We then have N_i global degrees of freedom associated with that node, one degree for each of the N_i rock types. One of the degrees of freedom is used to build the discrete element (saturation) equation associated with node i. This equation is formed by taking the sum:

$$F_i = \sum_{j=1}^{N_i} f_i^{e_j} \qquad (39)$$

$f_i^{e_j}$ is the local equation on element e_j associated with node i. In the summation (39) continuity of the normal component of the flux given by (28) is accounted for. The remaining $N_i - 1$ degrees of freedom are used to impose the $N_i - 1$ independent relations that exist between the capillary pressure curves:

$$P_c^{k_j}(S_i^{k_j}) = P_c^{l_j}(S_i^{l_j}), \quad j = 1, 2, ... N_i - 1 \qquad (40)$$

(The pairs (k_j, l_j) are chosen to form $N_i - 1$ independent relations.)

The reservoir domain Ω is triangulated into finite elements (triangles) such that the interior boundaries coincide with the edges of the finite element mesh. On each element T we use constant interpolation of the pressure p, linear interpolation of the saturation S and the components of the fluxes $\vec{u}$ and $\vec{f}$. The local degrees of freedom for the discrete pressure is the constant value on each triangle, and the degrees of freedom for the saturation is the values at each vertice of the triangle. The degrees of freedom for the fluxes will be the normal components integrated along each edge of element T, $\Gamma_{j,T}$, $j = 1, 2, 3$.

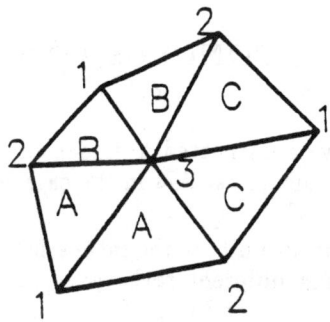

Fig. 2. The number of degrees of freedom (saturation) for three rock types A, B and C.

To illustrate the capabilities of the model we simulate imbibition from a fracture (as rock type 1, to the left) into a matrix block (as rock type 2). Data is given in (Bøe et al., 1989). Only the two ends of the fracture are open to flow. Saturation contours plotted on top of the computational grid are shown in Fig. 2. Except for numerical broadening, the solution is well-behaved. Notice the discontinuity in the saturation at the fracture-matrix boundary.

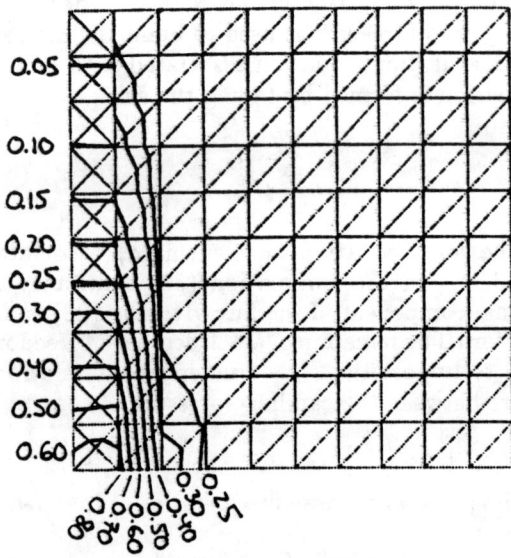

Fig. 2. Saturation contours

The numerical scheme resulting from the discretization uses streamline diffusion to stabilize convection dominated flows. Euler-backwards is used for discretization in time. The fully implicit nonlinear scheme is solved by Newton-Raphson's method. The linear system is solved by either a sparse gaussian solver or orthomin with incomplete LU-factorization as a preconditioner.

4 CONCLUSIONS

We have constructed a model for numerical simulation of multiphase flow in the case of several rock types.

The discontinuity in the saturation at the boundary separating different rock types satisfies the capillary continuity condition.

The introduction of multiple degrees of freedom in certain equations increases the bandwidth of the matrix. (This may be dealt with by domain decomposition techniques.)

NOTATION NOT DEFINED IN THE TEXT

$$(U, V)_\Omega = \int_\Omega U \cdot V \, d\Omega$$

$$\langle U, \vec{V} \rangle_\Gamma = \int_\Gamma U\vec{V} \cdot \vec{n} \, d\Gamma$$

$\vec{n}$: outward unit normal from Γ

$[U]_{ii}^i = U_{\Gamma_{int}}^i - U_{\Gamma_{int}}^{ii}$

(the trace differences taken over all Γ_{int})

$L^2(\Omega)$: space of square integrable functions

$L^\infty(\Omega)$: space of bounded functions

$H(\mathrm{div}, \Omega)$: space of vectors with components and divergence in $L^2(\Omega)$

$\vec{x}$: spatial vector

t: real time

Γ: outer boundary

Γ_{int}: interior boundaries

Γ_u: part of Γ with (20) specified

Γ_w: part of Γ with (21) specified

Γ_p: part of Γ with (22) specified

Γ_s: part of Γ with (23) specified

$\phi(\vec{x})$: porosity

$\mathbf{K}(\vec{x})$: permeability tensor

μ_l: viscosities

ρ_l: densities

S_{wc}: connate wetting phase saturation

S_{nwr}: residual nonwetting phase saturation

REFERENCES

Aavatsmark, I., 1989, Kapillarenergie als Entropiefunktion, Z. angew. Math. Mech 69, p. 319-127.

Bøe, Ø. and Fladmark, G.E., 1989, Mixed Finite Elements for Simulation of Multiphase Flow in Fractured Porous Media, IBM Bergen Scientific Centre Report series 89/12.

Chavent G. and Jaffre, J., 1986, Mathematical Models and Finite Elements for Reservoir Simulation, North Holland.

Raviart P.A and Thomas, J.M, 1977, A Mixed Finite Element Method for 2-nd Order Elliptic Problems, Lecture Notes in Mathematics 606, Springer-Verlag, Berlin.

2nd European Conference on the Mathematics of Oil Recovery
© D. Guérillot, O. Guillon (Editors) and Éditions Technip, Paris 1990, pp. 323-325
27 rue Ginoux, 75015 Paris

Radial Transport in Porous Media with Dispersion and Adsorption

L. Ci-Qun and Y. Jie[1]

MATHEMATICAL MODEL

A transport equation characterizing dispersion and absorption of a chemical solution for radial porous flow can be derived by a mass balance as follows:

$$\frac{\partial C}{\partial t} + \frac{1-\phi}{\phi}\,\rho_r\,\frac{\partial S}{\partial t} + v\,\frac{\partial C}{\partial r} = \frac{1}{r}\frac{\partial}{\partial r}\left(r.D.\frac{\partial C}{\partial r}\right) \quad (1)$$

The dynamical formula for chemical adsorption is:

$$\frac{\partial S}{\partial t} = k\,(mc - s) \quad (2)$$

The initial and boundary conditions are:

$$C\,(r,o) = S\,(r,o) = 0 \quad (3)$$

$$\left(v\,.\,C - D\frac{\partial c}{\partial r}\right)_{r\,=\,rw} = C_o V \quad (4)$$

$$\lim_{r\,\to\,\infty} C\,(r,t) = 0 \quad (5)$$

Eqs. (1)-(5) constitute the closed set of transport equations charactirizing dispersion and adsorption for radial porous flow of ground-fluids with the inner third type boundary condition.

Because of the simple radial geometry, the Darcy's velocity varies inversely with the radius:

$$v = \frac{Q}{2\pi h r} \quad (6)$$

Neglecting the molecular diffusion coefficient the hydro-dynamic dispersion coefficient is expressed as:

$$D = \lambda\,\frac{v}{\phi} = \lambda\,\frac{Q}{2\pi h r} \quad (7)$$

Substituting Eq. (7) into Eq. (1), the Eqs. (1) - (5) can be written in dimensionless form:

$$\frac{\partial C_D}{\partial t_D} + N_A\frac{\partial S_D}{\partial t_D} + \frac{1}{r_D}\frac{\partial C_D}{\partial r_D} = \frac{1}{r_D}\frac{\partial^2 C_D}{\partial r_D^2} \quad (8)$$

$$\frac{\partial C_D}{\partial t_D} = N_k\,(m C_D - S_D) \quad (9)$$

$$C_D(r_D, 0) = S_D(r_D, 0) = 0 \quad (10)$$

(1) Institute of Porous Flow Mechanics, Academia Sinica, PO Box 4U, Wan-Zhuang, Lanfang, Hebei, China.

$$\left(C_D - \frac{\partial C_D}{\partial r_D}\right)_{r_D = r_{WD}} = 1 \qquad (11)$$

$$\lim_{r_D \to \infty} C_D(r_D, t_D) = 0 \qquad (12)$$

where: $\quad C_D = \frac{C}{C_0}$; $\quad N_A = \frac{1-\phi}{\phi}\rho_r S_0$;

$$N_k = \frac{k \cdot Q}{2\pi h \phi \lambda^2} ; \quad r_D = \frac{r}{\lambda} ; \quad r_{WD} = \frac{r_W}{\lambda} ;$$

$$S_D = \frac{S}{S_0} ; \quad t_0 = \frac{Qt}{2\pi h \phi \lambda^2}$$

SOLUTION IN LAPLACE TRANSFORM DOMAIN

Eqs. (8) - (12) can be inverted as follow by using the method of Laplace transformation:

$$p\overline{C}_D + N_A p\overline{S}_D + \frac{1}{r_D}\frac{d\overline{C}_D}{dr_D} = \frac{1}{r_D}\frac{d^2\overline{C}_D}{d r_D^2} \qquad (13)$$

$$p\overline{S}_D = N_k\left(m\overline{C}_D - \overline{S}_D\right) \qquad (14)$$

$$\left(\overline{C}_D - \frac{d\overline{C}_D}{dr_D}\right)_{r_D = r_{WD}} = \frac{1}{p} \qquad (15)$$

$$\lim_{r_D \to \infty} \overline{C}_D = 0 \qquad (16)$$

where: $\quad \overline{C}_D(r_D, p) = \int_0^\infty C_D(r_D, t_D) \cdot e^{-p t_D} dt_D$

$$\overline{S}_D(r_D, p) = \int_0^\infty S_D(r_D, t_D) \cdot e^{-p t_D} dt_D$$

Solving for $\overline{S}_D$ from Eq. (14) and substitution into Eq. (13) will yield:

$$\frac{d^2\overline{C}_D}{dr_D^2} - \frac{d\overline{C}_D}{dr_D} = r_D \cdot q \cdot \overline{C}_D \qquad (17)$$

where: $\quad q = p\left(1 + N_A \frac{m N_k}{p + N_k}\right)$

The solution of ordinary differential Eq. (17) subjecting to boundary conditions (15) and (16):

$$\overline{C}_D = \frac{1}{p} \exp\left(\frac{r_D - r_{WD}}{2}\right) \frac{\left(\frac{Z}{Z_W}\right)^{\frac{1}{2}} K_{\frac{1}{3}}(\xi)}{\frac{1}{2}K_{\frac{1}{3}}(\xi_W) + (qZ_W)^{\frac{1}{2}} K_{\frac{1}{3}}(\xi_W)} \qquad (18)$$

where: $\quad Z = r_D + \frac{1}{49}$; $\quad Z_W = r_{WD} + \frac{1}{49}$;

$$\xi = \frac{2}{3} q^{1/2} \cdot Z^{3/2} ; \quad \xi_W = \frac{2}{3} q^{1/2} \cdot Z_W^{3/2}$$

On may compute the concentration behaviour for radial dispersion problem with adsorption by using the Eq. (18).

When $N_R \to \infty$, the dynamical adsorption reduces to the equilibrium adsorption:

where $\quad q = p\left(1 + m N_A\right)$

NUMERICAL INVERSION OF LAPLACE TRANSFORMATION

The values for the modified Bessel function of the second kind of order may be computed according to the following integral form:

$$K_\upsilon(Z) = \int_0^\infty ch\upsilon t \cdot \exp\left(-Z ch t\right) dt \qquad (18)$$

Eq. (19) shows that $K_u(Z)$ is monotically decreased funtion with u, i.e.

$$K_0(Z) < K_{1/3}(Z) < K_{2/3}(Z) < K_1(Z) \qquad (19)$$

The concentration results can be obtained with the data: $r_{wo} = 1$, $m = 0.2$, $N_k = 0$ or 100 on Micro-Vax II ; by substituting Eq. (18) into the formulae for numerical inversion of Laplace transformation.

Fig. 1. represents the concentration profiles for radial dispersion problem with the third inner boundary condition. The agreements between the results of numerical inversion with that obtained analytically are good.

Fig. 2. represents the concentration profiles at different time for convection-dispersion-adsorption problem. Comparing Fig. 1. with Fig. 2., we know that the effects of adsorption cause the concentration profile delayed and decreased.

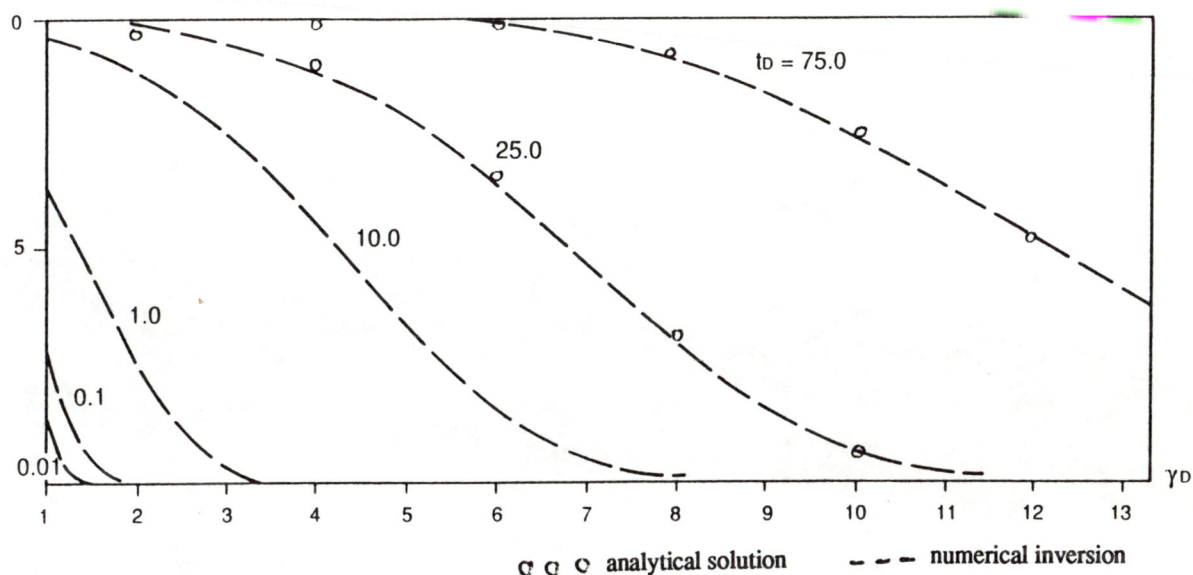

FIG. 1. Concentration distribution at different for radial convection-dispersion problem.

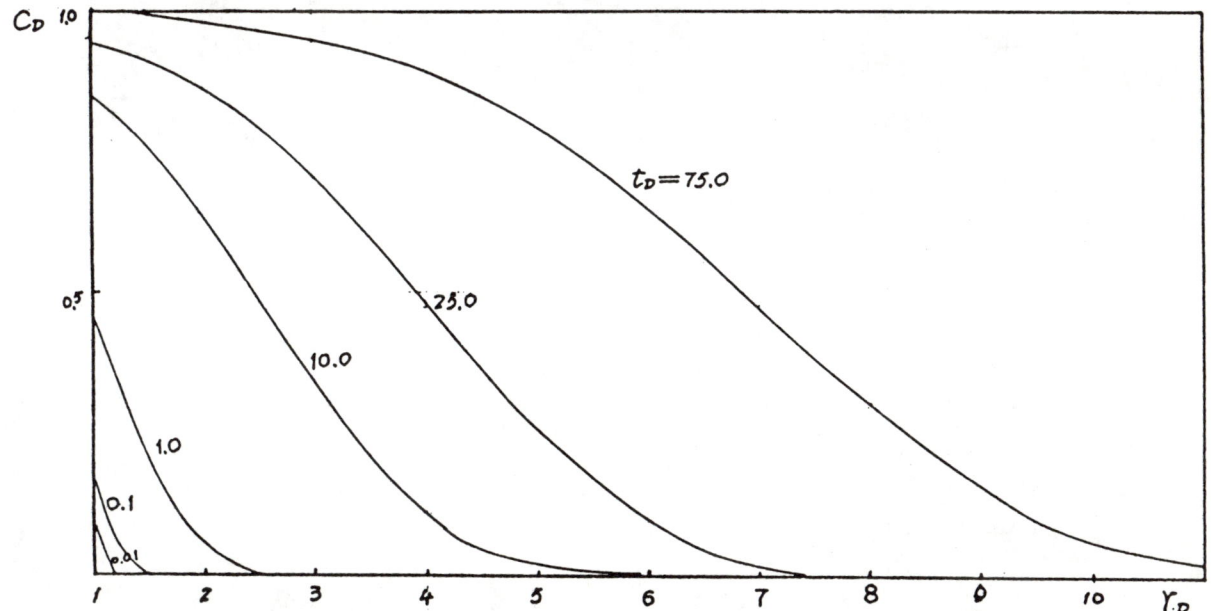

FIG. 2. Concentration distribution at different time for radial convection-dispersion-adsorption problem.

2nd European Conference on the Mathematics of Oil Recovery
© D. Guérillot, O. Guillon (Editors) and Éditions Technip, Paris 1990, pp. 327-330
27 rue Ginoux, 75015 Paris

Constant-Time Step Deconvolution Model for Variable-Rate Well Test Pressure Data

S. Buitrago, G. Gedler and R. Manzanilla[1]

1 FORMULATION OF THE PROBLEM

The transient pressure well test analysis is a technique which allows the petroleum engineer to determine reservoir properties, such as permeability, porosity, the drainage volume of the reservoir, static pressure and, in general, to characterize or describe the reservoir-well system in order to indicate damage or stimulation of well, fracturing or not of well, the existence of faults or flow barriers, the approximate shape of the drainage area of the reservoir or the change of the reservoir lithological properties.

The transient pressure well test analysis is based on the solution of the diffusivity equation for the case of constant flow rate. The diffusivity equation, which describes the flow of slightly compressible fluids through a porous media, is obtained from the conservation of mass law, Darcy's law and the equation of state (Kucuk et al., 1985).

The general form of the diffusivity equation is as follows:

$$\Delta p = \frac{\phi \mu c}{k} \frac{\partial p}{\partial t},$$ (1)

with the boundary conditions

$$\lim_{r \to \infty} p(r,t) = p_I \text{ and } 2\pi r_w \frac{\partial p(r,t)}{\partial r}\Big|_{r_w} = \frac{q\mu}{kh}.$$ (2)

One of the solution of equation (1), with the condition (2), is

$$p(r,t) = p_I - \frac{q\mu}{4\pi kh}\left[-E_i\left(-\frac{\phi \mu c r^2}{4kt}\right) + S\right]$$ (3)

Equation (3) is called the line-source solution of the diffusivity equation.

The typical analysis consists of comparing the test data with a set of type-curves, such as those proposed by Gringarten and Bourdet, each of which represents characteristic well parameters. This technique assumes a constant flow rate during the transient test, since the type-curves are based on the line–source solution of the diffusivity equation.

In general, the wellbore storage effect and the phase segregation effect, among others, cause the flow rate to vary during the transient test. Therefore, we need a method to generate an equivalent constant-rate set of data from the measured flow rate and pressure data, in order to use the type-curves to characterize or define the reservoir-well system. This is where the superposition principle play an important role.

Applying the superposition principle in time we can replace a variable-flow-rate well by several constant-flow-rate wells, all of them located at the same point.

Consider a well, with a skin factor S, producing at rate q_1 from t_0 to t_1, at rate q_2 from t_1 to t_2 ,......., at rate q_n from t_{n-1} to t_n. This scheme gives the following equations

$$p_I - p_{wf} = \frac{\mu}{2\pi kh}\left[\sum_{i=1}^{n}(q_i - q_{i-1})p_D(t - t_{i-1}) + q_n S\right],$$ (4)

where the time variations are dimensionless and $q_0 = 0$, which is called the discrete form of the superposition principle in time.

---

(1) Analisis Matematico y Estadistica, INTEVEP SA, Apartado 76343, Caracas 1070-A, Venezuela.

In general, taking into account that the flow rate q is a continuous function of time, equation (4) can be written as follows,

$$p_I - p_{wf} = \frac{\mu}{2\pi k h} \left[\int_0^t q'(\alpha) \, p_D(t-\alpha) \, d\alpha + q(t) \, S \right]. \quad (5)$$

Equation (5) is called the continuous form of the superposition principle in time. Hence, the general problem to solve can be written as follows:

$$g(t) = \int_0^t f'(t-\alpha) \, K(\alpha) \, d\alpha + c f(t), \quad (6)$$

where f and g are observed data and K is the kernel function to be determined.

2 DECONVOLUTION MODEL, VARIABLE FLOW–RATE PRESSURE WELL TEST

The physical problem described in early section can be written as follows:

$$g(t) = c f(t) + \int_0^t f(t-\alpha) \, K'(\alpha) \, d\alpha. \quad (7)$$

The methodology presented here is a modification of that suggested by Kucuk, 1985.

a. The interval $[0, T]$ is divided in N sub-intervals $[t_i, t_{i+1}]$, $i = 0, ..., N-1$ of length Δt, with

$$\Delta t = \frac{T}{N}; \qquad t_i = i\Delta t, \; i = 0, ..., N. \quad (8)$$

b. f and g at the points t_i, $i = 0, ..., N$ are known.
c. The kernel function K is represented by $at + b\sqrt{t} + c$ in $[t_0, t_1]$ and by a polynomial of degree two in each interval $[t_i, t_{i+1}]$, $i = 0, ..., N-1$. We also require K and its derivative to be continuous in $[0, T]$ and the second derivative to be continuous in $[t_0, t_2]$.
d. The function f is considered to be linear in each interval $[t_i, t_{i+1}]$, $i = 0, ..., N-1$.

Using (a) and (b) the solution K of the problem (7) is obtained through

$$c, K_1', K_i'', \; i = 1, ..., N-1 \quad (9)$$

with

$$K_1' = K'(t_1) \quad \text{y} \quad K_i'' = K_+''(t_i). \quad (10)$$

Replacing t_L with t, for $L = 1, ..., N$ in (7) we get an N×(N+1) linear system. Thus, if the system has a solution, it is not unique.

In order to reduce the set of solutions, we impose the following constraints on K:

$$K(t) \geq 0; \quad K'(t) \geq 0; \quad K''(t) \leq 0 \quad \text{if } t \in [0, T]$$
$$K''(t) \quad \text{increasing monotonically} \quad \text{if } t \in [0, T] \quad (11)$$

This constraints correspond to the fact that the pressure is an increasing continuous function, and that the changes in pressure decrease in time.

Now we use:

$$c, K_1', K_1'', K_i'' - K_{i-1}'', \; i = 1, ..., N-1 \quad (12)$$

instead of (9) as the most convenient variables for the problem (7) to make the representation of the constraints (11) easier.

By using (12), the constraints in (11) become:

$$c \geq 0$$
$$K_i'' - K_{i-1}'' \geq 0, \; i = 1, ..., N-1$$
$$K_1'' + \sum_{i=2}^{N-1} \left(K_i'' - K_{i-1}'' \right) \leq 0$$
$$K_1'' + \Delta t[(n-2) K_1'' + \sum_{i=2}^{N-2} (n-i-1) \left(K_i'' - K_{i-1}'' \right)] \geq 0. \quad (13)$$

The constraints given by Kucuk et al., 1985, do not yield the ones stated in (11), therefore we add additional constraints to generate the form of the kernel function K.

Replacing t_L with t, for $L = 1, ..., N-1$ in (7), taking into account the representation of f in (d) and the representation of K as a function of the variables in (12), yield the following linear system of equations

$$Mx = b, \quad (14)$$

with

M an N×(N+1) matrix
$$x^t = (c, \Delta t K_1', \Delta t^2 K_1'', \Delta t^2 \left(K_2'' - K_1'' \right), ..., \Delta t^2 \left(K_{N-1}'' - K_{N-2}'' \right))$$
$$b^t = (g(t_1), g(t_2), ..., g(t_N))$$

Hence, we formulate the following minimization problem

$$\min_{x \in \Omega} (Mx - b)^t (Mx - b), \qquad (15)$$

where Ω is the set of vectors x such that their components yield (13), to solve the problem (14) with the conditions in (13). The problem (15) can now be written as:

$$\min_{x \in \wedge(x)} \frac{1}{2} F(x), \qquad \wedge(x) = \{x : Ax \geq d\}, \qquad (16)$$

where:

$$F(x) = x^t M^t M x - 2 x^t M^t b + b^t b,$$

$$
M = \begin{pmatrix}
m_{11} & m_{12} & m_{13} & 0 & \cdots & 0 \\
m_{21} & m_{22} & m_{23} & 0 & \cdots & 0 \\
m_{31} & m_{32} & m_{33} & m_{34} & \cdots & 0 \\
m_{41} & m_{42} & m_{43} & m_{44} & \cdots & 0 \\
\vdots & \vdots & \vdots & \vdots & \ddots & \vdots \\
m_{N1} & m_{N2} & m_{N3} & m_{N4} & \cdots & m_{N(N+1)}
\end{pmatrix}
$$

with $m_{(i+1)(j+1)} = m_{ij}$ for $i = 3, ..., N-1$ and $j = 4, ..., N$,

$$
A^t = \begin{pmatrix}
1 & \cdots & 0 & -1 & \cdots & 0 & 0 & 0 \\
0 & \cdots & 0 & 0 & \cdots & 0 & 0 & 1 \\
0 & \cdots & 0 & 0 & \cdots & 0 & -1 & N-2 \\
\vdots & \ddots & \vdots & \vdots & \ddots & \vdots & \vdots & \vdots \\
0 & \cdots & 1 & 0 & \cdots & -1 & -1 & 0
\end{pmatrix}
$$

$$\underbrace{\qquad\qquad}_{N+1} \underbrace{\qquad\qquad}_{N+1} \underbrace{\qquad}_{2}$$

and

$$d^t = (0, 0, -10^9, 0, ..., 0,$$
$$-10^9, -10^9, 0, -10^9, ..., -10^9, 0, 0).$$

Finally, we solve the problem (16) using the Powell's algorithm (Powell, 1983), which is based on the active set methodology.

3 TEST EXAMPLES

The numerical model presented in this work has been implemented for an IBM3090 in the FORTRAN program NUMDEC.

The observed flow-rate and pressure values, were used to generate three data sets. These data sets were used to analyse the effect of the numerical deconvolution technique on the data. We can observe that numerical deconvolution minimizes noise existing in the data, as well as the wellbore storage effect produced at the beginning of the pressure well test.

Using this technique an equivalent constant-rate data set can be generated from the measured flow-rate and pressure

data, thus allowing the use of conventional interpretation techniques. As a result we have a better estimated well characteristic parameter and reservoir properties from the matching of the deconvoluted pressure with type–curves, such as Gringarten and Bourdet.

In figures 1 to 2 we present the pressure drop against time from observed and deconvoluted data in cartesian, semilog and log-log coordinates. Also, we present the change in observed flow-rate and pressure in log-log coordinates. Figures 1 to 2 demonstrate the effectiveness of the deconvolution technique presented for pressure well test analysis.

Example 1 is an injectivity test carried out in the water injection well of a reservoir in Lake Maracaibo. The objective of the test was to determine the length of a waterflood-induced fracture. Previous tests conducted in this well revealed that it was effectively fractured. The data is from a report by Prado et al., 1987.. The flow-rate and pressure in the wellbore were measured simultaneously, using high-precision tools. Water was injected into the well at a constant rate of 3000 B/D for a period 45 hours, then the well was closed for 42 hours and finally a variable-flow-rate injectivity test of short duration (1.5 hours) was performed. Fig.1 shows the results from the fall off test analyzed using NUMDEC.

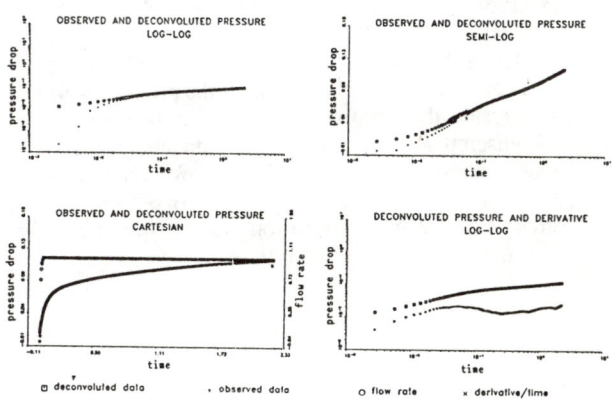

FIG.1. Injectivity tests in Lake Maracaibo.

Example 2 is a buildup test carried out in a low permeability gas well of the Atascosa reservoir, Texas. The data was taken at the surface and has been reported by Ahmed et al., 1987.. The numerical deconvolution technique results are presented in Fig.2.

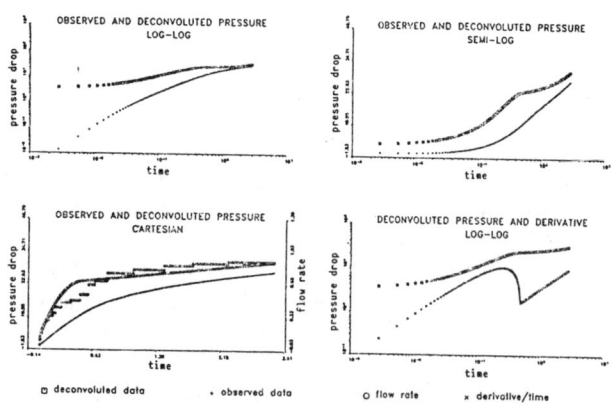

FIG.2. Buildup test in Atascosa.

4 CONCLUSIONS AND RECOMMENDATIONS

1. We have developed a program using the deconvolution technique for interpretation of pressure well test data where the flow-rate and pressure are measured simultaneously.
2. We have indicated some of the mistakes that can be made by the direct use of the type-curves methodology on crude data, especially in the case of a natural or artificially fractured well.
3. The technique presented minimized the wellbore storage effect produced in the beginning of the pressure well test. As a result, the duration of the test can be reduced.
4. The discretization of the convolution integral is based upon a constant time step scheme. We recommend the development of an algorithm for a variable time step in order to decrease the dimension of the matrices involved and to reduce the execution time.

NOMENCLATURE

h : reservoir thickness.

k : reservoir permeability.

K : kernel function.

K' : time derivative of K.

$K''_+(t)$: second right side time derivative of K.

$p(r,t)$: pressure at distance r of the well center at time t.

$p_D(r_D, t_D)$: dimensionless pressure.

p_I : reservoir pressure.

p_{wf} : wellbore flowing pressure.

q : flow-rate.

r_w : wellbore radius.

S : skin factor.

Δp : laplacian of p.

μ : viscosity.

REFERENCES

KUCUK, F.; CARTER, R. G.;AYESTERAN, L. Numerical Deconvolution of Wellbore Pressure and Flow Rate. SPE 13960. 1985 SPE Reservoir Simulation Symposium.

BUITRAGO, S.; GEDLER, G.; MANZANILLA, R. A Deconvolution Technique for Well Test Pressure Analysis. ASOVG and SEG First Symposium on Reservoir Characterization . Maracaibo, Venezuela. March, 1990.

HAMMING, R. W. Numerical Methods for Scientists and Engineers. Mc Graw-Hill, 375–377, 1973.

POWELL, M. J. D. ZQPCVX a Fortran Subroutine for convex Quadratic Programming. Department of Applied Mathematics and Theoretical Physics, University of Cambridge. England. November 1983.

PRADO, L. R.; VAN KRUYSDIJK, C. P. J. W.; VROOM, J. K. A. B. Interpretation of the VLA-440 Fall-Off Test in the M4.2 Sand- unit of the Horst Reservoir Block I, Lake Maracaibo. Status Report 49. Exploratie en Produktie Laboratorium. Rijswijk. The Netherlands. 1987.

AHMED, U.; KUCUK, F.; AYESTERAN, L. Short-Term Transient-Rate and Pressure–Build–Up Analysis of Low–Permeability Reservoirs. SPE Formation Evaluation. Diciembre 1987.

EARLOUGHER, R. C. Jr. Advances in Well Test Analysis. Soc. Petr. Eng. of AIME. 1977.

2nd European Conference on the Mathematics of Oil Recovery
© D. Guérillot, O. Guillon (Editors) and Éditions Technip, Paris 1990, pp. 331-334
27 rue Ginoux, 75015 Paris

Inverse Modeling for Compressible Flow. Application to Gas Reservoirs

B. Bréfort and V. Pelcé[2]

ABSTRACT

Classical methods to determine permeability distribution in a gas reservoir are sometimes inadequate: for a gas field, only core measurements are available and scale problems always appear when applying these values to the whole reservoir.

An inverse modeling procedure is proposed to determine automatically the permeability field for a dry-gas reservoir. This method consists in the minimization of the difference between a historical record of real pressures and pressures computed from discretized equations of monophasic gas flow in porous medium. A Lagrangian method combined with geostatistical information is used.

Two applications are presented: a history matching of a gas field giving a permeability distribution then used for prediction and a transient draw-down test on a gas well which gives an estimate of the local permeability in the surrounding matrix.

Acknowledgements
The authors gratefully acknowledge permission granted by *Gaz de France* and *Texas Gas Transmission Corporation* to publish this paper.

1. THE INVERSE MODELING PROCEDURE

Such methods had already been developed in theory for general equations (Chavent, 1973) or with industrial applications for slightly compressible flow (Marsilly, 1983; Fasanino, 1986). We try to apply such methods for highly compressible flow (gas) with the additional difficulty of a non-linear equation.

1.1 Description

The aim of the inverse method is to minimize an "objective function" defined as the sum of the square differences between observed and calculated pressures (Chavent, 1973). The several steps of the method are

1. initialize the parameters
2. calculate the transmissivity field using geostatistics
3. solve the flow equation
4. calculate the objective function and stop if satisfactory

---

(1) DETN, Gaz de France, 361, avenue du Président Wilson, 93211 La Plaine-Saint-Denis Cedex, France.

5. calculate the gradient of objective function with respect to the parameters
6. modify the parameters and go back to 2.

1.2 Use of Geostatistics

The unknowns of the problem are the transmissivity in each cell of the discretised reservoir space. It may be a rather high number. The idea is to reduce this quantity using geostatitical information (Matheron, 1971). The parameters are then the transmissivity at points chosen by the user (called "pilot points"). Then the complete transmissivity field is determined by kriging, using the parameters and local transmissivity measurements.

1.3 Optimization procedure

The aim of the procedure is to minimize the objective funtion defined as the sum of the square differences between observed and calculated pressures. For that purpose, we use a method of gradient. The gradient calculation utilizes an approach called method of "adjoint state" (Chavent, 1973).

The pressures are calculated from numerical solution of the flow equation by a finite difference method. The gas flow equation is expressed by

$$\text{div} \left[\text{ hk } \overrightarrow{\text{grad}} \, m(P) \right] + \frac{2 \, h \, RTe}{M} q = a \frac{\partial m(P)}{\partial t}$$

(gravity gradient is neglected) where

P pressure
Te temperature
T duration of observation
M molecular weight
μ gas viscosity
Z compressibility factor
R universal gas constant
q mass flow rate per unit of volume
β_g gas compressibility
S_g gas saturation
k permeability
Φ porosity
h thickness of the layer
a $= h \, \mu \, \Phi \, S_g \, \beta_g$
$m(P)$ is the pseudo pressure

The coefficient "a" depends on the pressure, thus the first equation is non- linear. The pseudo pressure is a function of the pressure

$$m(P) = 2 \int_{P_m}^{P} \frac{\lambda}{\mu(\lambda)Z(\lambda)} \, d\lambda$$

The adjoint state "y" is defined as the solution of a partial differential equation similar to the gas flow equation:

$$-a \frac{\partial y}{\partial t} - \text{div}(hk \, \overrightarrow{\text{grad}} \, y) = - \sum_{j=1}^{\text{Nobs}} 2 \, e_j$$

The derivative of the objective function with respect to the parameter "hk" is then given by

$$\frac{\partial J}{\partial hk} = \int_0^T \int_\Omega \overrightarrow{\text{grad}} \, p \, \overrightarrow{\text{grad}} \, y \, dx \, dt$$

where Ω is the reservoir space. This value is then used to modify the parameter hk iteratively using the steepest descent method.

2. APPLICATIONS

This inverse modeling technique has been implemented in a two dimensional computer program and has proven its efficiency by solving several real world cases. Two of them are presented in this paper.

2.1 History matching of a gas field

This utilization of the method concerns a whole underground gas storage: the aim is to obtain a good transmissivity field of the reservoir to allow prediction.

The reservoir considered here is a depleted gas field, located in the United States, converted into a gas storage. The small water-drive at the border of the gas bubble is propitious to monophasic modeling during a limited period.

Thickness and porosity distributions are known for each cell. Vertical refeeding contributes to the flow mechanism. The initial field of transmissivity is built from the few known values of some wells. Although data, i.e. observed pressures and history of injection and production, are available over a longer period, the period of optimization is reduced to 480 days, to keep the efficiency and the rapidity of the method.

The results are then introduced in a three dimensional diphasic reservoir simulator, in order to use the data over a longer period (1000 days) in the configuration of the future utilization for prediction. The good agreement between calculated and observed pressures (see fig.1) validates the predictions which can be made with the model.

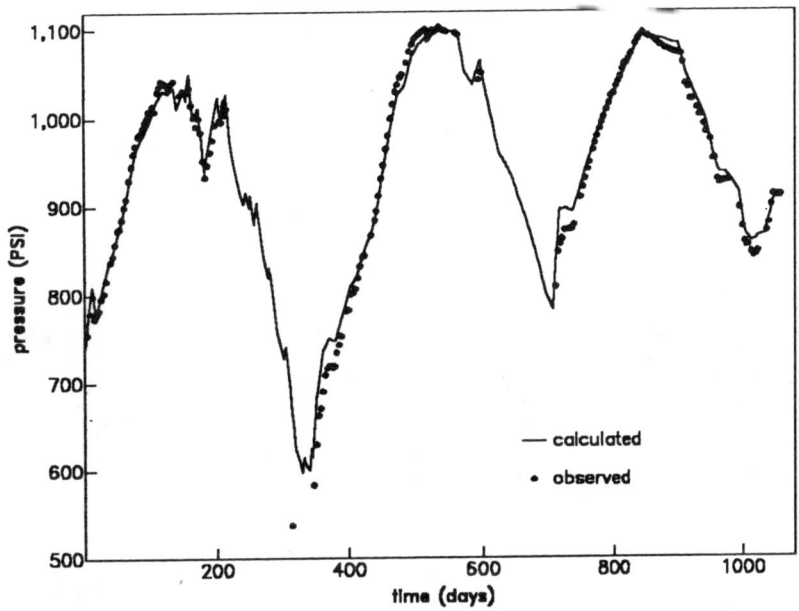

FIG. 1.   history matching

2.2   Transient draw-down test in a gas well

During 22 hours, a succession of different gas flow-rates was produced on a well located on the gas storage of CHEMERY (FRANCE); bottom well pressure was recorded at the same time. The different levels of flow-rate were at too near intervals to allow an analytical interpretation based on steady-state flow conditions. Our method shows greater efficiency and accuracy.

We represent the storage around the well by a layer of constant thickness and porosity. The grid size is smaller near the well in order to increase the precision of the results. The initial field of transmissivity has a constant value of 40 Darcy-meters. Figure 2 shows the optimized field of transmissivity: we can see a local reduction around the well (just a part of the used grid is represented on this figure) which squares with the fact that this well was recently drilled.

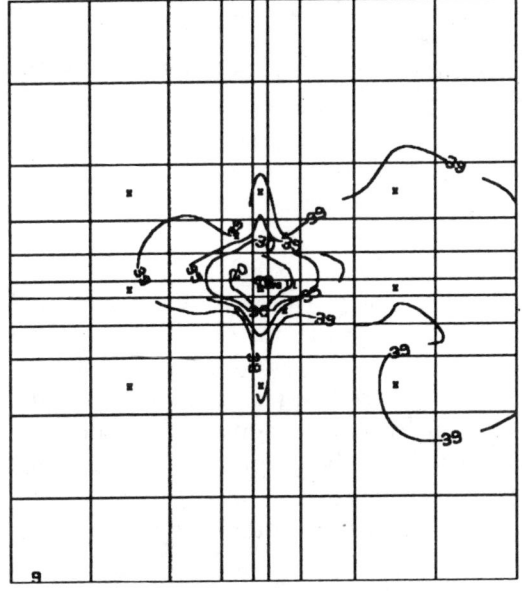

FIG. 2.   well test

Figure 3 shows the comparison between observed and calculated pressures with the initial field of transmissivity and with the adjusted one.

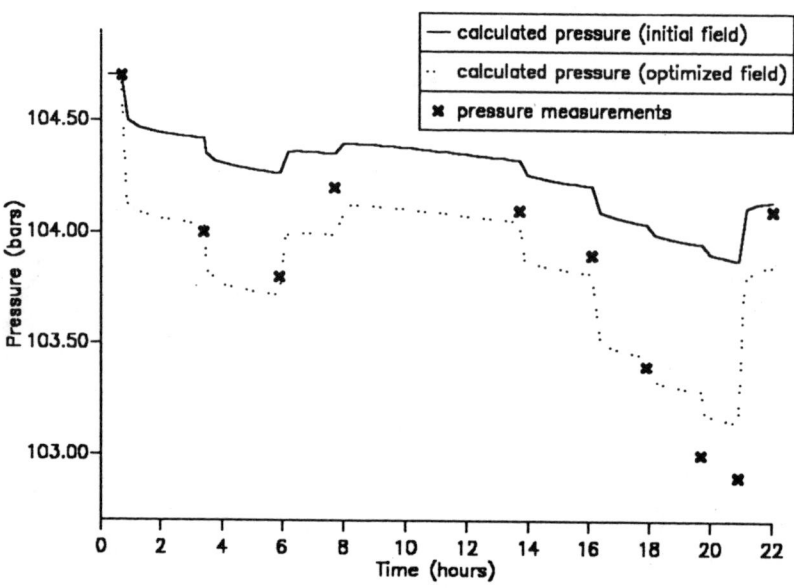

FIG. 3.   well test

CONCLUSION

- An inverse modeling procedure using a Lagrangian method to determine automatically the permeability field from historical record of pressures has proven its efficiency in the case of non-linear flow equations, especially in the case of gas-flow in porous medium.

- The use of geostatistical information combined with the pilot points method allows us to obtain an adequate and effective method by reducing the number of unknowns while keeping the geological consistency.

- It is now possible to interpret automatically gas well tests under unsteady-state conditions, or to develop a gas reservoir description from historical record of pressures and flow-rates when the water-drive of the aquifer is not too important, this description being needed for planning future development.

- An extension to three dimensions is already under development.

REFERENCES

de Marsilly, G. Lavedan, G., Boucher, M. and Fasanino, G., 1983, Interpretation of interference tests in a well using geostatisical techniques to fit the permeability distribution in a reservoir model, Geostatitical for Natural Resources Characterization, Part 2, NATO ASI Series, lecture n°49.

Chavent, G. and Lemonier, P., 1973, Identification de la non-linéarité d'une équation parabolique quasilinéaire, rapport de recherche n°45, INRIA, Rocquencourt, France, 63 p.

Fasanino, G., Molinard, J.E., de Marsilly, G. and Pelcé, V., 1986, Inverse modeling in gas reservoir, SPE 15592, 15 p.

Matheron, G., 1971,The theory of regionalized variables and its application, cahier n°5, Centre de Morphologie Mathématique, Ecole des Mines, Fontainebleau, 250 p.

2nd European Conference on the Mathematics of Oil Recovery
© D. Guérillot, O. Guillon (Editors) and Éditions Technip, Paris 1990, pp. 335-337
27 rue Ginoux, 75015 Paris

Computer Geological Simulation in Oil Recovery

V. A. Badyanov[1]

At the Siberian Scientific-Research Institute for the Oil Industry we have developed some methods of solving geological problems. They are realized in the form of a computer system GEOPAK-2. The system can solve three classes of problems providing construction of a producing formation geological model according to the type of a filtration model used: a detailed geological correlation of well profiles; an estimation of the object geological structure complexity parameters; geometrization of pools and differentiated calculation of reserves. The first of them is basic for solving the others. The results of solving the second one are used in oil recovery estimation procedures based on one-dimensional statistical models of filtration. The solution of the third-class problems is necessary when using two- and three-dimensional determined models of filtration.

The initial data in wells are bottom-hole coordinates, altitudes, elongations, deths of markers and of all permeable layers boundaries, physical parameters of rock, oil and gas and position of water-oil and gas-oil contacts. One can use indirect information, such as seismic survey data.

An objective basis to solve the problem of geological correlation is the presence to this or that degree a latent rhythm of geological objects structure which is a genetic consequence of rhythmical sedimentation on a rather large area on the background of disturbance factors of geological conditions.

A disturbance background suppression and an exposure of a systematic component (i.e., rhythm) is done by synthesizing a great number of particular well profiles into one generalized geological-statistical profile (GSP).

If the GSP has a rhythmical character, it is dismembered into rhythms. Each well is compared with the GSP. The layers are assigned to one of the rhythms or two and more rhythms simultaneously (mergence)

---

(1) The Siberian Scientific-Research Institute for the Oil Industry, 50 Lyet Coktyabrya ST, 118, 625016 Tyumen, USSR.

according to the criteria of spatial proximity.As a result,dismemberment of a producing horizon into strata is executed and their one-to-one correspondence in wells is set.

The results of correlation are used in the following classes of problems.

In technological calculations on the basis of one-dimensional filtration model,two aspects of objects geological structure complexity are usually taken into account: a natural reservoir inner geometry complexity (morphological complexity) and reservoir permeability heterogeneity.

Morphological complexity influences horizontal (lateral) and vertical fluid filtration.The first phenomenon is a result of strata discontinuity which decreases oil recovery.When predicting within one-dimensional filtration models, such a decrease is allowed for by sweep efficiency that is a relation of oil recovery from a discontinuous stratum to oil recovery from a continuous one,all the other conditions being equal.

As a result of correlation problem solving,a matrix desription of the object is obtained.Every element of the three-dimensional matrix has areal coordinates (wells) X and Y and a time coordinate Z (paleodepth) with the prescibed logical function "reservoir-non-reservoir".An integral(variability frequency)and differential (average value) characteristics of "reservoir-non-reservoir" pattern are determined at every paleodepth level.A sweep function,i.e.sweep efficiency as a function of the distance between production and injection zones is determined by matching the above-mentioned characteristics with the help of analytical relationships obtained by

the Monte-Carlo method on two-dimensional mathematical models.Averaging in Z provides a sweep function of the object.Sweep efficiency for any regular well pattern is calculated using an averaged sweep function with regard for distances between production and injection zones.

In the problem of accounting for fluids vertical crossflows in the process of development of water-oil and gas-oil zones it is important to define a contacting oil-saturated thickness fraction. An oil-saturated reservoir is considered to be contacting when it immediately adjoins a water- or gas-bearing stratum or when it is separated from them with clay layers with a less total thickness than the given one.For this purpose,a geological profile is generated on the basis of each well data;on this profile a fraction of contacting oil-saturated thickness is determined by analytical geometry methods for any value of oil-saturated thickness.

By averaging for the wells, a contacting thickness fraction for the whole object and for oil-saturated thickness classes is obtained.

Layer-by-layer permeability heterogeneity of the object is calculated as the square of the layers permeability variation factor in wells,zonal heterogeneity is calculated as the square of variation factor between mean permeabilities of adjacent wells.

A number of other heterogeneity values are also calculated.

The problem of oil pools geometrization is connected with reconstruction of a sufficient number of various geological characteristics fields on the wells data basis.A field is expressed as geological characteristic values determined in a dense square network of points.

The problem layout is the following.For the calculated object, mean values of porosity and oil saturation factors and oil characteristics are calculated in wells.

The object top and bottom marks fields are reconstructed including usage of indirect information (e.g.,seismic survey data),oil-water contact,gas-oil contact.With the help of arithmetic and logical operations on the obtained fields,the field of the pool total thickness with outer and inner oil and gas pool outline fixation is calculated,thus reconstructing the pool outer and inner oil and gas pool outline fixation is calculated,thus reconstructing the pool outer geometry.

To further obtain the oil-saturated thickness field,the total thickness field is multiplied by the oil-saturated oil thickness fraction field (the oil-saturated thickness fraction of the total pool thickness) and the lithological field "reservoir-non-reservoir".

The geological reserves field is obtained when multiplying the oil-saturated thickness field by the fields of oil saturation,porosity,and oil characteristics.At that,the oil saturation field can be reconstructed allowing for vertical zonation,i.e.,oil saturation lowering towards the oil-water contact.

Besides the fields to calculate reserves,the system can reconstruct some other fields of interest for hydrocarbon development,where initial data are available,for example,permeability fields,gas,oil

and water parting clay bands thickness fields and so on.

Differential calculation of reserves both for algorythmically obtained zones (i.e.,zones of various fluids contact,characteristics classes,rock types and so on) and for zones with the a priori assigned boundaries (i.e., reserves categories,development elements and so on) is made by the reserves field integration within the zones.

The system GEOPAK-2 generates the results in the form of tables, geological profiles,maps in isolines produced by a graph plotter, and numerical pool models on magnetic medium for further simulation of a filtration process.

2nd European Conference on the Mathematics of Oil Recovery
© D. Guérillot, O. Guillon (Editors) and Éditions Technip, Paris 1990, pp. 339-342
27 rue Ginoux, 75015 Paris

A New Formulation for Generalised Compositional Simulation

R. E. Mott and C. L. Farmer[1]

1   BASIC EQUATIONS

An important recent trend in reservoir simulation has been the development of multiple application simulators which can model both black oil and compositional processes within a single program. A number of different mathematical formulations have been proposed for solving the fluid flow equations in these simulators. With an IMPES solution scheme there are two basic approaches; Newton-Raphson methods, such as that of Young and Stephenson (1983), and the volume balance method of Watts (1986). Various fully implicit schemes have also been described. Although these methods all solve the same underlying equations, the relationship between them is not easily understood.

This paper presents a new formulation, which makes use of the best features of several previous methods. This formulation incorporates both IMPES and fully implicit solution methods. Although the method is based on a Newton-Raphson approach, it includes the volume balance method as a special case.

The formulation has been used in the development of a general purpose simulator which includes black oil and equation of state (EOS) compositional models. This simulator forms part of an integrated software system for geoscience and reservoir engineering applications.

We first describe the equations which are solved in a general purpose, multicomponent, multiphase, isothermal finite difference simulator. The starting point is a continuity equation for each component m.

$$\frac{\partial}{\partial t} \left(\phi \, a_m \right) = \theta_m \qquad (m = 1, \, n_c) \qquad (1)$$

where ϕ is the porosity, a_m is the density of component m (in a spatially uniform region, a_m is the number of units of component m in unit pore volume), and θ_m is the rate of accumulation of component m.

There are n_c continuity equations for the $n_c + 1$ unknowns a_m ($m = 1, \, n_c$) and the pressure P. An additional equation is needed to complete the system, and we derive this from an equation of state for the fluid.

An equation of state relates pressure, volume, temperature and component amounts, and can be written as

$$V = v (N, P, T) \qquad (2)$$

where the amount of each component in the fluid is given by the vector N. (The term "equation of state" is used here in its most general sense to denote any equation of the form given in (2), even if the fluid properties are calculated from a black oil model).

The volume in equation (2) is a homogeneous function of N, satisfying

$$\lambda v (N, P, T) = v (\lambda N, P, T) \qquad (3)$$

for all values of λ. Further details may be found in Farmer (1986).

---

(1) AEA Petroleum Services Winfrith, Dorchester, Dorset, DT2 8DH, UK.

Hence

$$v(a, P, T) = 1 \qquad (4)$$

which is assumed to hold when a is a function of position and time. Equations (1) and (4) are a set of $n_c + 1$ equations to be solved for a and P.

The finite difference forms of (1) and (4) are obtained by integration over a grid block i. We express these equations in terms of residuals which must be reduced to zero for convergence of the solution at a timestep. This gives a mass balance equation for each component over a timestep from time t to $t + \Delta t$:

$$r_{mi} = N_{mi}^{t+\Delta t} - N_{mi}^{t} - \Delta t \, \theta_{mi} = 0 \qquad (5)$$

where N_{mi} is the number of units of component m in block i, and θ_{mi} is the net flow rate of component m into block i from neighbouring blocks, wells and aquifers.

Equation (4) can be integrated to give

$$r_{Vi} = V_{Ti} - V_{Pi} = 0 \qquad (6)$$

where V_{Ti} is the total fluid volume in block i, given by

$$V_{Ti} = v(N_i, P_i) \qquad (7)$$

and V_{Pi} is the pore volume of block i. This is the volume balance equation of Watts (1986). It is now clear that the volume balance equation is the finite difference form of an equation of state, a point which has not been emphasised in earlier discussions of this equation.

2 SOLUTION SCHEMES

Equations (5) and (6) are solved by a Newton Raphson solution scheme with P and N_m ($m=1, n_c$) as primary variables. The accumulation term in (5) can be written as

$$\theta_{mi} = \sum_j T_{ji} \sum_\alpha \frac{k_r^\alpha}{\mu^\alpha} a_m^\alpha (P_j^\alpha - P_i^\alpha + \rho^\alpha g \Delta z) \qquad (8)$$

where T_{ji} is the transmissibility between block i and a neighbouring block j, a_m^α is the density of component m in phase α, k_r^α is the relative permeability of phase α, μ^α is the viscosity of phase α, ρ^α is the mass density of phase α, and where well and aquifer terms are omitted for simplicity.

First, consider an IMPES scheme where the block pressures in θ_{mi} are evaluated at time $t + \Delta t$, and all other terms are evaluated at time t. The changes in the primary variables over an iteration are given by

$$\delta N_{mi} - \Delta t \sum_j T_{mji} (\delta P_j - \delta P_i) = -r_{mi} \qquad (9)$$

$$\left(\frac{\partial V_T}{\partial P} - \frac{\partial V_P}{\partial P} \right)_i \delta P_i + \sum_m \left(\frac{\partial V_T}{\partial N_m} \right)_i \delta N_{mi} = -r_{Vi} \qquad (10)$$

where

$$T_{mji} = T_{ji} \sum_\alpha \frac{k_r^\alpha}{\mu^\alpha} a_m^\alpha \qquad (11)$$

Elimination of δN_{mi} from (9) and (10) yields the IMPES pressure equation for grid block i :

$$\left(\frac{\partial V_T}{\partial P} - \frac{\partial V_P}{\partial P} \right) \delta P_i$$
$$+ \Delta t \sum_m \left(\frac{\partial V_T}{\partial N_m} \right)_i \sum_j T_{mji} (\delta P_j - \delta P_i)$$
$$= -r_{Vi} + \sum_m \left(\frac{\partial V_T}{\partial N_m} \right)_i r_{mi} \qquad (12)$$

On the first iteration of a timestep, the initial estimates of P and N_m at time $t + \Delta t$ will be the corresponding values at time t. In this case

$$r_{mi} = -\Delta t \, \theta_{mi}^t \qquad (13)$$

Substituting into (12) gives

$$\left(\frac{\partial V_T}{\partial P} - \frac{\partial V_P}{\partial P}\right)_i \; \delta P_i$$

$$+ \Delta t \sum_m \left(\frac{\partial V_T}{\partial N_m}\right)_i \sum_j T_{mji} \; (\delta P_j - \delta P_i)$$

$$= -(V_T - V_P)_i^t - \Delta t \sum_m \left(\frac{\partial V_T}{\partial N_m}\right)_i \; \theta_{mi}^t \quad (14)$$

which can be re-arranged to give

$$\left(\frac{\partial V_P}{\partial P} - \frac{\partial V_T}{\partial P}\right)_i \; \left(P_i^{t+\Delta t} - P_i^t\right)$$

$$= (V_T - V_P)_i^t + \Delta t \sum_m \left(\frac{\partial V_T}{\partial N_m}\right)_i \; \theta_{mi}^{t+\Delta t} \quad (15)$$

Equation (15) is the finite difference form of the pressure equation in the volume balance method, equation (31) of Watts (1986).

Thus the volume balance method corresponds to the first iteration in a Newton-Raphson solution scheme. This was first pointed out by Wong et al. (1989), but the preceeding analysis provides a much simpler demonstration of the relationship between the two methods.

In most practical simulations, a single Newton iteration will reduce the volume balance errors to an acceptable level ($r_v/V_T < 0.001$). In this case the Newton-Raphson method described here is identical to the volume balance method. However, for a simulator implementation, the Newton-Raphson method gives the extra flexibility to reduce the volume error by iteration if needed; and it also simplifies the inclusion of both IMPES and fully implicit solutions within a single program.

In a fully implicit solution scheme, all terms in θ_{mi} (equation 8) are evaluated at time level $t + \Delta t$. Equation (9) is replaced by

$$\delta N_{mi} - \Delta t \sum_j \sum_{k=1}^{n_c+1} \left(\frac{\partial \theta_{mi}}{\partial X_{ki}} \delta X_{ki} + \frac{\partial \theta_{mi}}{\partial X_{kj}} \delta X_{kj}\right)$$
$$= -r_{mi} \quad (16)$$

where X_{ki} ($k=1, ...n_c + 1$) denote the primary variables (P and N) in block i.

Equations (10) and (16) give a set of $(n_c + 1)$ linear equations in each block, to solve for the $(n_c + 1)$

unkowns δX_{ki}. However, equation (10) is local (it only involves δX for a single block) so it can be used to eliminate one of the unkowns from equation (16) before entering the linear solver. This reduces the set of linear equations to n_c per block. The fully implicit solution scheme is similar to that of Chien et al. (1985).

3 FLUID PROPERTIES

The formulation presented in this paper is completely general, in that there is no limit to the number of components or the number of phases, and any component can exist in any phase. It can be used for black oil, compositional, miscible flood and chemical flood simulation.

The inclusion of a number of fluid models within a single simulator is possible because it is easy to isolate the calculation of fluid PVT properties. For each fluid model (e.g. black oil, compositional) a set of routines is provided to calculate the PVT properties listed in Table 1. The requirements for IMPES and fully implicit solutions are identical except in the derivatives of properties such as viscosity or density.

The fluid property models can be divided into two categories, those where a flash calculation requires an iterative solution (such as an equation of state compositional model), and those where no iteration is needed (such as a black oil model). The solution of the flash calculation in an iterative model is discussed in the next section, in the context of an EOS compositional model.

TABLE 1
Requirements for Fluid PVT
Properties in Each Grid Block

Input   Primary Variables (P, N_m)

Output   Total Fluid Volume
           Derivatives of Total Fluid Volume
           Phase Saturations
           Phase Viscosities
           Phase Densities
           Phase/Component Concentrations
           Derivatives of Properties (fully implicit only)

4 FLUID PROPERTY CALCULATIONS IN AN EOS COMPOSITIONAL MODEL

In an EOS compositional model, a flash calculation involves the solution of a set of non-linear equations, representing the equality of liquid and vapour phase fugacities for each hydrocarbon component.

In the simulator described in this paper, the fugacity equations are solved exactly at each call to the fluid property routines, using a Newton-Raphson iteration scheme based on that of Michelsen (1982). In the following discussion, we refer to iterations on the mass and volume balance equations (5) and (6) as "outer" iterations, and to iterations on the fugacity equation as "inner" iterations.

In a typical IMPES compositional simulation, almost all of the timesteps converge in a single outer iteration. The inner iterations converge in a single iteration in some blocks, and in two iterations in the other blocks. It is unusual for the flash calculation to need more than two inner iterations.

As an example, the 3rd SPE Comparative Solution problem models a gas condensate reservoir produced by lean gas re-cycling followed by depletion. A simulation of this problem used 190 timesteps and 192 outer iterations for 15 years simulation. During the gas re-cycling period, about 50% of the grid blocks required one inner iteration, and the remaining blocks converged in two inner iterations. During the depletion phase, the proportion of blocks converging in one inner iteration increased to about 75%.

These results indicate that "coupled" formulations (eg Young and Stephenson 1983), where each iteration involves a Newton update on both the mass balance and phase equilibrium equations, are less efficient than "decoupled" formulations which use inner iterations to converge the flash calculation at each outer iteration. A coupled formulation will usually take at least two iterations per timestep, involving additional work in the flash calculation, because the same number of iterations must be taken in all grid blocks.

It is also worth noting that the first inner iteration does not require the Jacobian matrix to be recalculated. The derivatives of the total fluid volume include the effects of changes in phase compositions, so their calculation requires the derivatives of phase compositions with respect to P and N_m. On the next call to the fluid property routines, these derivatives can be used on the first inner iteration to estimate the change in phase compositions due to the changes in P and N_m.

5 SUMMARY AND CONCLUSIONS

A new formulation for generalised compositional simulation has been described. The important features of the method are

(a) Both IMPES and fully implicit techniques are incorporated

(b) The basic equations are solved by Newton-Raphson iterations, but the IMPES solution includes Watts' volume balance method as a special case

(c) In EOS compositional simulation, the flash calculation is converged at each Newton iteration.

ACKNOWLEDGEMENTS

This work was funded by the Robertson Group plc, ARCO British Ltd, Enterprise Oil plc, Shell UK Ltd and UK Department of Energy, as part of the TIGRESS project. We thank W.R. Rodwell for helpful discussions.

REFERENCES

Chien, M.C.H. et al., 1985,. A New Fully Implicit Compositional Simulator, SPE 13385, 8th SPE Symposium on Reservoir Simulation.

Farmer, C.L., 1986, Formulations of the Equations for Multi-Component Multi-Phase Flow through Porous Media. UKAEA Report No. AEEW - M 2376.

Michelsen, M.L., 1982, The Isothermal Flash Problem, Fluid Phase Equilibria, **9**, 21-40.

Watts, J.W., 1986, A Compositional Formulation of the Pressure and Saturation Equations, SPEJ, May 1986, 243-252.

Wong, T.W. et al., 1989, The Relationship of the Volume Balance Method of Compositional Simulation to the Newton - Raphson Method, SPE 18424, 10th SPE Symposium on Reservoir Simulation.

Young, L.C. and Stephenson, R.E., 1983, A Generalised Compositional Approach for Reservoir Simulation. SPEJ, October 1983, 747-742.

2nd European Conference on the Mathematics of Oil Recovery
© D. Guérillot, O. Guillon (Editors) and Éditions Technip, Paris 1990, pp. 343-346
27 rue Ginoux, 75015 Paris

P3D Modeling of Vertical Hydraulic Fracture Growth

P. Valko and B. Pertik[1]

ABSTRACT

Recent short cut design procedures for hydraulic fracturing treatment are based on two dimensional geometric description: either a penny shape or constant height is supposed. Unless the minimal horizontal stress is constant along the vertical direction or a high confining stress acts in the overburden and underburden these assumptions cannot be justified.

A more detailed approach predicts the variation of fracture height with lateral coordinate and time taking into account the minimal horizontal stress variation. According to the pseudo three dimensional (P3D) concept the height is determined by decoupling the vertical cross sections. In this work a framework model (the extension of the Nordgen – Kemp system) is proposed as a unified approach for P3D models. Within this framework several height determination concepts published in the literature are investigated. A critical comparison is made between vertical pressure drop and constant stress intensity approaches. A new quasi steady state height equation is proposed using a simple vertical – lateral pressure profile similarity assumption. The new model can treat with a continuously varying stress field and gives penny shape or constant height geometries in the limiting cases. Any known typical wellbore pressure behavior can be reproduced.

1 THE P3D FRAMEWORK MODEL

By framework model we mean a system of equations, which will be closed only after the way of height determination is specified. Our framework model is the generalization of Nordgen's two dimensional model (Nordgen, 1972, Kemp, 1987). The basis is the continuity equation

$$\frac{\partial A_c u}{\partial x} + \frac{2 C_1 (H^U + H^L)}{(t - \tau(x))^{1/2}} +$$

---

(1) Hungarian Hydrocarbon Institute, Szazhalombatta POB 32, 2443 Hungary.

$$+ \frac{2 S_p \, \partial(H^U + H^L)}{\partial t} + \frac{\partial A_c}{\partial t} = 0$$

Two boundary conditions are required. One at the wellbore, expressing that the known inflow $Q_o(t)$ (for 1 wing) is flowing into the fracture

$$A_c(0,t) \, u(0,t) = Q_o(t)$$

and one for the leading edge (Kemp, 1987):

$$\frac{d \, L(t)}{dt} = u(L,t)$$

where $L = L(t)$ is the actual half length (Stefan condition).

By the help of the pressure drop equation the linear velocity u, the width w and the pressure drop is connected:

$$- \frac{dp}{dx} = 2^{n+1} \left(\frac{3n+1}{n}\right)^n K \, u^n \, w^{-(n+1)}$$

The width equation is taken from the Perkins-Kern-Nordgen model :

$$w(0;x) = (H^U + H^L)[p(0;x) - \sigma(0)]/(2\bar{E})$$

where $\bar{E}$ is the width opening modulus, which can be expressed by the help of the shear modulus G, Young modulus E and Poisson ratio ν :

$$\bar{E} = G / [2(1-\nu)] = E / [4(1-\nu)^2].$$

Since at a given location x the cross section consists of half ellipses, $w(0;x)$, H^U and H^L determines A_c.

2 PREVIOUS HEIGHT EQUATIONS

In the MIT model (Settari and Cleary, 1984, 1986) height is increasing starting from a minimal value. The driving force of the growth is the deviation of the pressure from the horizontal model and the average minimal horizontal stress. In the decoupled vertical cross sections a KZGD model is solved, resulting in the height growth rate. Unfortunately this model is too complicated from the conceptual and computational point of view and no attempt has been made to reproduce it.

On the contrary in the ORU model (Palmer and Luiskutty, 1985) the height is determined by a simple concept: in every vertical cross section the computed stress intensity factor at the top is taken equal to the critical stress intensity of the rock. The previous one is computed from the pressure corresponding to the x location (and taken as constant in the vertical direction), from the given height and from the minimum horizontal stress. Hence an equation is constructed for the height, which is derivated and the resulting differential equation is solved backward starting from a leading edge height.

The computational advantages of the ORU model are obvious. Unfortunately there is a deep contradiction in the concept itself:
The pressure is decreasing toward the edge of the fracture and hence the greater the distance from the wellbore is the bigger the resulting height is if we want to keep the stress intensity constant. Such a fracture shape, however, contradicts to experience and common sense.

3 A NEW CONCEPT FOR HEIGHT DETERMINATION

It is supposed that the minimum

horizontal stress is known as a function of the z coordinate: $\sigma(z)$. The basic assumptions of the model are as follows:

(1) The pressure at the top of the fracture is equal to the minimum horizontal stress at the same height. Applying this principle at the wellbore we obtain

$$p(\ H^U(0)\ ;\ 0\ )\ =\ \sigma(\ H^U(0)\ )\ ,$$

an equation for the upper height. A similar equation is obtained for the $H^L(0)$ lower height at the wellbore.

(2) The vertical pressure profile we need to know is similar to the horizontal pressure profile we know from the continuity equation, since the same fluid is flowing. The mathematical form of this similarity is expressed by stating that the the vertical profile is obtained by a linear stretch operation: $p(z;0) = p(0;a^U z)$.

(3) The third assumption corresponds to the stretch factor a^U. Since it should be approximately equal to the ratio of the vertical and horizontal component of the average flow in the upper quarter wing we postulate $a^U = H^U(0)\ /\ L$.
(A similar equation gives the stretch factor for the lower part.)

(4) The height is decreasing toward the tip. The decrease is according to an elliptic rule reaching the a priori given leading edge height at the tip and resulting in a convex fracture shape.

The above assumptions make the framework model closed. A suitable numerical method is proposed to solve the resulting system of partial differential and algebraic equations which is essentially a Stefan's problem. A fully implicit scheme is applied. The resulting finite difference equations are solved by the Newton-Raphson technique with a problem specific convergence promoter. Since the algorithm produces a new mesh point at every time step a periodic mesh reduction is needed to keep the computation time within minutes on a PC.

4 APPLICATION

The input data for a fracture projected in South-West Hungary is given in Table 1. In this case the upper stress contrast is 100 % larger than the lower one.

Table 1. Fracture design input data

$\bar{E}$	$7.4\ 10^9$ Pa
n	0.51
K	2.2 Pa s$^{0.51}$
C_l	0.00008 m/s$^{1/2}$
S_p	0 m
stress field: σ_0 any value	
if z > 45 m: $\sigma = \sigma_0 + 1E6$ Pa	
if z < -45 m: $\sigma = \sigma_0 + 5E5$ Pa	
injection:	
rate: 0.045 m^3/s / 1 wing	
time: 126 min	
leading edge (U/L): 11 m/11 m	

The resulting fracture shape at the end of the treatment is shown on Fig. 1. The fracture penetrates both into the upper and lower layers but to a different extent. At the end of the treatment the maximum width at the wellbore is

10.54 mm . The upper height is 54.6 m while the lower one is 90 m. The half length is 129.8 m and the pressure drop from the wellbore to the tip is 1.079E6 Pa . The fluid efficiency is 26.3 % .

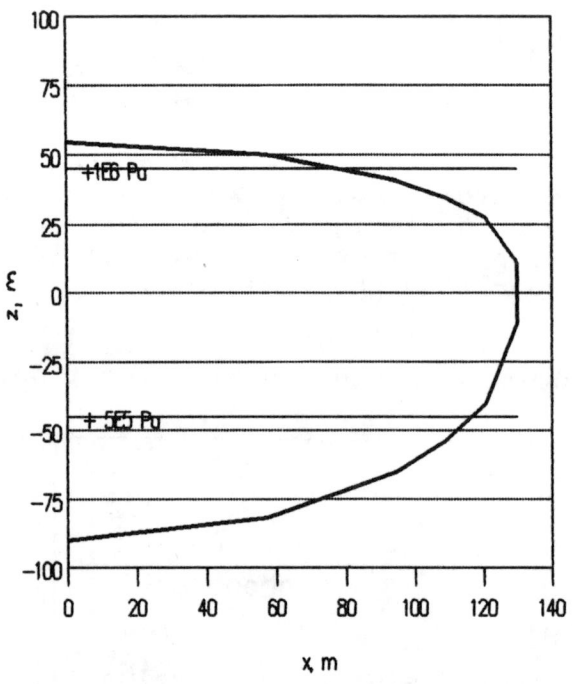

Fig. 1 Fracture shape at 126 min

LITERATURE

Settari,A. - Cleary,M.P.: Three-Dimensional Simulation of Hydraulic Fracturing. JPT pp 1177-90 (1984 Sept).

Settari,A. - Cleary,M.P.: Development and Testing of a Pseudo Three- Dimensional Model of Hydraulic Fracture Geometry. SPEPE pp 449-66 (1986 Nov) .

Palmer, I.D. Luiskutty, C.T.: A model of the Hydraulic Fracturing Process for Elongated Vertical Fractures and Comparison of Results With Other Models. SPE paper 13864 presented at the SPE/DOE Low Permeability Gas Reservoirs Meeting, Denver, Colorado, May 19-22 (1985).

Nordgen,R.P.: Propagation of Vertical Hydraulic Fractures. SPEJ, Trans AIME 253, pp 306-14 (1972 Aug).

Kemp,L.F.: Study of Nordgen's Equation of Hydraulic Fracturing. SPE paper 18959 presented at SPE Joint Rocky Mountain Regional / Low Permeability Reservoir Symposium and Exhibition. Denver, Colorado, March 6-8. (1989).

NOTATION

$A_c(x)$, m^2	cross section
a^U, a^L	stretch factors
C_l, m/s$^{1/2}$	leakage coeff.
S_p, m	spurt loss coeff.
E, Pa	Young modulus
$\bar{E}$, Pa	width opening modul.
G, Pa	shear modulus
H^U, $H^L(x)$, m	up./low. height
$\sigma(z)$, Pa	min. horiz. stress
n,	fluid. behav. index
K, Pa s^n	consist. ind.
L, m	fra. length (1 wing)
$p(z;x)$, Pa	pressure
t, s	time
u, m/s	linear velocity
u_l, m/s	linear leak. veloc.
Q_o, m^3/s	inj. rate (1 wing)
$w(x)$, m	width at z=0
x,y,z	coordinates
ν	Poisson ratio
$\tau(x)$, s	opening time

2nd European Conference on the Mathematics of Oil Recovery
© D. Guérillot, O. Guillon (Editors) and Éditions Technip, Paris 1990, pp. 347-350
27 rue Ginoux, 75015 Paris

Irregular Averaging of Filtration Transfer Processes in Heterogeneous Media

M. B. Panfilov[2]

ABSTRACT

The modification of the asymptotic method of averaging operators with rapidly oscillating coefficients (Sanchez-Palencia, 1980, Bakhvalov et al., 1984) is suggested for the media with heavy heterogeneity (great differences in permeability of separate areas). Several new models of the filtration processes generalizing a known phenomenological simulator (Barenblatt et al., 1960) are constructed with the aid of this method.

The composite has been selected as a physical model, the components of which are a connected high-conductive system and periodically scattered isolated compact blocks. The system period l is much less than macroscale L, so $\varepsilon = l/L \ll 1$.

1 STATEMENT OF PROBLEM

Let us consider a problem for pressure $p(x,t)$ in the process of filtration of a weakly compressible fluid which adheres to Darcy's law in each point of the medium:

$$b\left(\frac{x}{\varepsilon},\nu\right)\frac{\partial p}{\partial t} - \frac{\partial}{\partial x_i}\left(a\left(\frac{x}{\varepsilon},\omega\right)\frac{\partial p}{\partial x_i}\right) = f(x,t;\omega) \qquad (1)$$

at $x \in \Omega_1 \cup \Omega_2$, $t>0$, $\Omega_1, \Omega_2 \subset \Omega \subseteq \mathbb{R}^s$, $s \geq 2$

under the initial condition of $p(x,0)=p^{\circ}$, the condition of the solution quick damping towards p° in the infinity, and:

$$[a]_\Gamma = 0, \qquad [a\partial p/\partial n]_\Gamma = 0 \qquad (2)$$

Here, Γ – is a surface separating Ω_1 from Ω_2, n – is a normal to Γ, $[\cdot]$ – is a function jump symbol; Ω_1, Ω_2 are subregions occupied by blocks and fractures. Parameter ν is a ratio of porosities of blocks and fractures, while ω is a ratio of their permeabilities: $a_1 = a_2\omega$, $b_1 = b_2\nu$, $f_1 = f_2\omega$; where $a_i \equiv a(x)$, $x \in \Omega_i$, etc.

The condition of heavy heterogeneity is taken: $\omega \ll 1$.

In practice, within one deposit, the collector porosity varies within much narrower limits than the permeability does, so $\omega \ll \nu \leq 1$. It is accepted that $\nu \sim 1$.

Average equations for (1)-(2) are derived by means of construction of asymptotic expansion of problem at $\varepsilon, \omega \to 0$.

2 CLASSIFICATION OF MEDIA

Let us write down equation (1) in expanded coordinates $y_i = x_i/\varepsilon$, $i=1,\ldots,s$ on cell $Y=\{-1/2 < y_i < 1/2, \forall i\}$ consisting of

---

(1) Institute of Oil and Gas Problems, USSR Academy of Sciences, GSP-1, 117917, Moscow, USSR.

subregions Y_1, Y_2, which meet the block and fracture criteria:

$$\varepsilon_p b_2 \frac{\partial p}{\partial t} - \frac{\partial}{\partial y_i}\left(a_2 \frac{\partial p}{\partial y_i}\right) = \varepsilon^2 f_2, \quad y \in Y_1 \qquad (3\text{-}a)$$

$$\varepsilon^2 b_2 \frac{\partial p}{\partial t} - \frac{\partial}{\partial y_i}\left(a_2 \frac{\partial p}{\partial y_i}\right) = \varepsilon^2 f_2, \quad y \in Y_2 \qquad (3\text{-}b)$$

Parameter $\varepsilon_p \equiv \varepsilon^2 \nu/\omega$ has a meaning of a ratio of a separate block perturbation coverage time to the macro process specific time. Their variation changes qualitatively the type of solutions of equation (3-a) and causes four classes of heavily heterogeneous media which averaged models occurs importantly different.

(1) $\varkappa$-homogeneous media: $\varepsilon_p \sim \varepsilon^2$. In spite of heavy heterogeneity in permeability, the medium proves to be homogeneous in piezoconductance $\varkappa = a/b$ ($\varkappa_1 \sim \varkappa_2$), so the perturbation propagates through blocks and fractures with the same speed. The system conducts itself like a homogeneous one, the pressures in blocks and fractures are practically equal.

(2) Weakly $\varkappa$-heterogeneous media: $\varepsilon^2 < \varepsilon_t < 1$, ($1 > \varkappa_1/\varkappa_2 > \varepsilon$). Time boundary layer $\sim \varepsilon_p$ arises, within the limits of which the block-process is nonstationary. A difference of pressures between blocks and fractures is distinguishable, but has an order of ε_p.

(3) $\varkappa$-heterogeneous media: $\varepsilon_p \sim 1$. The process is always nonstationary on the block, as the time of covering the block with perturbation gets expanded through the entire period. Difference between the pressure values is about one.

(4) Heavily $\varkappa$-heterogeneous media: $\varepsilon \gg 1$, ($\varkappa_1/\varkappa_2 \ll \varepsilon^2$). For the entire period only narrow spatial boundary layer sizing $\sim \sqrt{\varepsilon_p}$ operates on the block near the boundary of the contact with the fracture.

3 METHOD OF IRREGULAR AVERAGING

Ordinary averaging methods (Sanchez-Palencia, 1980; Bakhvalov et al., 1984)

require some modification for the heavily heterogeneous media because a conception of an average through a cell of heterogeneity becomes non-constructive or loses its meaning at all (Panfilov, 1989).

Assume that operations of ordinary and conditional averaging: $<f> = \int_Y f dy$, $<f>_i = \int_{Y_i} f dy$ are determined for function $f(x, y, t)$ continuous through $y \in Y$. Then a irregular average is $<\chi(\omega)f> = \omega <f>_1 + <f>_2$, where the indicator function is $\chi(\omega) = \omega$, $y \in Y_1$; $\chi(\omega) = 1$, $y \in Y_2$.

The principle of the irregular averaging method is that the asymptotic expansion at $\varepsilon, \omega \to 0$ of solution (1)-(2) will be found in form (Panfilov, 1989):

$$p = p_{00}(x, t; \varepsilon_p) + p_{01}(x, y, t; \varepsilon_p) +$$

$$+ \sum_{k=1}^{\infty} \varepsilon^k [p_{k0}(x, t; \varepsilon_p) + p_{k1}(x, y, t; \varepsilon_p)]$$

where $p_{00} + \varepsilon p_{10} + \varepsilon^2 p_{20} + \ldots$ is the irregular average from p. In particular, for $\varkappa$-heterogeneous media the asymptotic main term falls to the functions of quick variables on blocks and the non-oscillating functions on fractures. The next stage of the method is averaging the block component of asymptotic in subregion Y_1.

4 MODEL FOR WEAKLY $\varkappa$-HETEROGENEOUS MEDIA

Assume that $P_1 = <p>_1/\alpha$, $P_2 = <p>_2/(1-\alpha)$ are average pressures on a block and fracture, where α is a volume fraction of blocks in the system.

Two stages are distinguished: 1 - at $t < \varepsilon_p$, the effective simulator for which coincides with (7), and 2 - at $t > \varepsilon_p$. By using the method described above, it is possible to obtain for the second stage the follows averaged simulator:

$$\begin{cases} A_{ij}^{(2)} \frac{\partial^2 P_2}{\partial x_i \partial x_j} - B \frac{\partial P_2}{\partial t} + \lambda_* \frac{\partial^2 P_2}{\partial t^2} = -<f_2>_2, \\[2mm] P_1(x, t) = P_2(x, t) - \tau_* \frac{\partial P_2}{\partial t}, \end{cases} \qquad (4)$$

where

$$A_{ij}^{(2)} = \left\langle a_2(y)\left(\frac{\partial \psi_i^{(2)}}{\partial y_j}+\delta_{ij}\right)\right\rangle_2,$$

$$\tau_* \equiv \varepsilon_t b_2 \xi^2/15a_2, \quad \lambda_* \equiv \nu \tau_* b_2/\alpha,$$

$$B^{(1)}=\nu\langle b_2\rangle_1, \quad B^{(2)}=\langle b_2\rangle_2, \quad B\equiv B^{(1)}+B^{(2)},$$

ξ-is a dimensionless radius of the block. Equation (4) for P_2 is an elliptic one (but not hyperbolic). Along with the initial condition $P_2(x,0)=P^\circ$, it requires an additional condition: $|P_2|<\infty, \forall t$.

Local function $\psi_i^{(k)}(y)$ are determined as 1-periodical solutions of the following problems in subregions Y_k:

$$L_0\psi_k^{(2)}=-\partial a_2/\partial y_k, \quad y\in Y_2; \quad \langle\psi_k^{(2)}\rangle_2=0,$$

$$(\partial\psi_k^{(2)}/\partial y_i+\delta_{ij})n_i\big|_s=0; \qquad (5)$$

where $L_0=\partial(a_2\partial/\partial y_i)/\partial y_i$, S - is a surface separating Y_1 and Y_2.

Relation (4) for P_1 is an equation of simple kinetics that characterizes the weakly non-equilibrium system.

5 MODEL FOR $\varkappa$-HETEROGENEOUS MEDIA

The following takes place with an acuracy of order ε:

$$P_1(x,t)=P_2(x,t)-\int_0^t\frac{\partial P_2(x,\theta)}{\partial\theta}K(t-\theta)d\theta, \qquad (6)$$

$$A_{ij}^{(2)}\frac{\partial^2 P_2}{\partial x_i\partial x_j}-B^{(2)}\frac{\partial P_2}{\partial t}=-B^{(1)}\int_0^t\frac{\partial P_2}{\partial\theta}K'(t-\theta)d\theta$$

where $\quad K(t)\equiv(6/\pi^2)\sum_{k=1}^\infty exp(-\pi^2 k^2 t/\lambda)/k^2$,

$$K'(t)\equiv dK/dt, \quad \lambda=\varepsilon_p b_2\xi^2/a_2.$$

Thus, the averaged system possesses memory. Exchange kernel $K(t)$ determines the intensity of return flows between the blocks and cracks.

The media with heavy $\varkappa$-heterogeneity operates merely as purely fractured one.

6 THE MISCIBLE MULTICOMPONENT FILTRATION

The initial system includes relation (1) for the pressure field and equation:

$$\beta b(\frac{x}{\varepsilon},\nu)\frac{\partial c}{\partial t} - a(\frac{x}{\varepsilon},\omega)gradPgradc=$$

$$=div\left(D(\frac{x}{\varepsilon},\omega_c)gradc\right) \qquad (7)$$

for concentration c of one of the mixture components with the convective diffusion taken into account. Here $\beta\sim 1$. Conditions for c have the form of: $c(x,0)=c^\circ$, $[c]_\Gamma=0$, $[D\partial c/\partial n]_\Gamma=0$. Coefficient D has such structure: $D_1=D_2\chi(\omega_c)$, ω_c - is a new parameter of problem. Let $\varepsilon_c=\varepsilon^2\nu/\omega_c$ is a measure of rate of diffusive waves expansion. Then it discovers a large quantity of systems that differs by character of asymptotic expansions. However they can be assembled in three groups according to correlation between the intensity of diffusive and convective transfers. We will present their examination for weakly $\varkappa$-heterogeneous media.

(1) C-systems (convective factor predominate on the blocks): $\varepsilon_c>1$ ($\omega_c<\varepsilon^2$). The concentration propagates through the single block in the form of a travelling wave with diffusive boundary layers near the limits of block and around the front.

(2) CD-systems (equilibrium between convective and diffusive factors on the block): $\varepsilon_c\sim 1$.

(3) D-systems (the diffusion predominates on the block): $\varepsilon_c<1$. The temporal boundary layer $\sim\varepsilon_c$, corresponding to the period of diffusive wave propagation, arise on the blocks.

7 AVERAGED MODEL FOR C-SYSTEMS

The resulting model for C-systems has not a specific form, but depends on the whole history of pressure change, especially on the factor $\gamma=sign(\partial P_2/\partial t)$.

It can be obtained for average concentrations C_i at $\gamma=1$:

$$\begin{cases} \dfrac{\partial C_1}{\partial t} = \dfrac{\partial P_2}{\partial t}\Big[C_2(x,t)-C_1(x,t)\Big], \\[2mm] \mathbb{D}_{ij}^{(2)}\dfrac{\partial C_2}{\partial x_i \partial x_j} - \beta B\dfrac{\partial C_2}{\partial t}+R_{km}^{(2)}\dfrac{\partial P_2}{\partial x_k}\cdot\dfrac{\partial C_2}{\partial x_m} = Q\equiv 0 \end{cases} \quad (8)$$

$$R_{km}^{(2)}= \left\langle a_2\left[\dfrac{\partial \psi_k^{(2)}}{\partial y_i}\dfrac{\partial \varphi_m^{(2)}}{\partial y_i} + \dfrac{\partial \varphi_m^{(2)}}{\partial y_k} + \dfrac{\partial \psi_k^{(2)}}{\partial y_m} + \right.\right.$$

$$\left.\left. +\delta_{km}\right]\right\rangle_2, \quad \mathbb{D}_{ij}^{(2)}= \left\langle D_2\left[\dfrac{\partial \varphi_i^{(2)}}{\partial y_j}+\delta_{ij}\right]\right\rangle_2$$

where $B=B^{(1)}+B^{(2)}$, functions $\varphi_i^{(k)}(y)$ are determined in the same way as $\psi_i^{(k)}(y)$ (5) with replacement of a_2 to D_2.

Since the coefficients of convective transfer $R^{(2)}$ in fractures depends upon local properties of diffusion process.

The coordinate of wave front (point of weakly jump) advances inside the block according to the following law:

$$\zeta_* = \exp\left(-[P_2(x,t)-P^\circ]/3\right), \quad \zeta\equiv r/\xi$$

where $\xi-$ is a radius of the block. The wave front becomes more abrupt during the time, and their rate decreases rapidly.

It has been received an interior solution near the block limit (at $1-\zeta\sim\varepsilon_c$), where the diffusive and convective terms are the same order, by aid of asymptotic expansions joint method:

$$c_1(x,\zeta,t)=C_2(x,t)+3\dfrac{\partial C_2}{\partial t}\left(\dfrac{\partial P_2}{\partial t}\right)^{-1}\left\{\zeta-1+\right.$$

$$\left.+\lambda_1\dfrac{\partial C_2}{\partial t}\left[1-\exp\left(-\dfrac{(1-\zeta)}{\lambda_1}\dfrac{\partial P_2}{\partial t}\right)\right]\right\},$$

where $\lambda_1\equiv 3D_2/\varepsilon_c b_*\xi^2$.

If $\gamma=-1$ (the pressure decreases), the fluid flow out the blocks into fractures, so the exchange term appears in (8):

$$\begin{cases} C_1(x,t) = c^\circ, \\[2mm] Q = -B^{(2)}\alpha(1-\alpha)^{-1}\dfrac{\partial P_2}{\partial t}\Big[C_2(x,t)-c^\circ\Big] \end{cases} \quad (9)$$

The real process of agent injection in beds is characterized by variation of γ in the course of time: $\gamma=1$ up to $t=t_1$, and $\gamma=-1$ in certain points of stratum later on. We obtain at $t>t_1$:

$$\dfrac{\partial C_1}{\partial t} = \dfrac{\partial P_2}{\partial t}\Big[\widetilde{C}_2(x,2\tau_1-\tau)-C_1(x,t)\Big] \quad (10)$$

and the right term Q of exchange in equation (8) for C_2 have a following form:

$$-B^{(2)}\dfrac{\alpha}{(1-\alpha)}\,(\partial P_2/\partial t)\Big[C_2(x,t)-\widetilde{C}_2(x,2\tau_1-\tau)\Big]$$

$$\tau_1(x)\equiv P_2(x,t_1)-P^\circ, \quad \widetilde{C}_2(x,\tau)\equiv C_2(x,t)\big|_{t=t(\tau)}$$

$$\tau(t;x)\equiv 2P_2(x,t_1)-P_2(x,t)-P^\circ.$$

The law of the front moving is: $\zeta_*=\exp[-(2\tau_1-\tau)/3]$.

The relation (10) presents an equation with delaying argument.

It is important that models (8)-(10) do not contain the new characteristic times, therefore it may be possible to state self-similar problems.

REFERENCES

Sanchez-Palensia, E., 1980, Non-homogeneous media and vibration theory, Berlin, Springer-Verlag, 398 p.

Bakhvalov, N.S., Panasenko, G.P., 1984, Averaging of processes in periodical media, Moscow, Nauka, 352 p.

Barenblatt, G.I., Zheltov, Y.P., Kochina, I.N., 1960, On basic representations of theory of filtration of homogeneous fluids in fractured rocks, Pricladnaya mathematica i mekhanica, **24**, 5, p. 852 - 864.

Panfilov, M.B., 1990, Averaged model of filtration in heavily heterogeneous media, Docladi AN SSSR, **311**, 2, p.313-317.

2nd European Conference on the Mathematics of Oil Recovery
© D. Guérillot, O. Guillon (Editors) and Éditions Technip, Paris 1990, pp. 351-354
27 rue Ginoux, 75015 Paris

History Matching Problems of Filtration Theory. Complex Adaptative Geological Models of Fields

B. Palatnik, I. Zakirov and G. Agaev[1]

To choose improved oil or gas field development it's necessary to create algorithm providing the possibility to calculate a field development performance during all the period of production for the different controlled influences on the bed. The simulation reliability and therefore the validity of the accepted project decisions depends on the degree of the adequacy of the mathematical models used to real physical processes as well as on the degree of the truthfulness of the initial geological-field information used at the simulation. It is the important peculiarity of the large oil and gas fields that they have common hydrodynamical connection between processes which take place in an aquifer, in a bed, in wells and fluid gathering, compressing, cleaning and treating and field pipeline transferring systems. Therefore to carry out long-term forecast of field development performance it is necessary to create the complex adaptive mathematical model which unites both reservoir and ground technological equipment (GTE) models.

To describe fluid moving in a geterogeneous bed we apply three-dimensional multiphase unsteady filtration model. The corresponding boundary problem describing both formation pressure and saturation dynamics in an actual field is written down relative to phase pressures. There's a need to note that we attach great significance to the aquifer simulation method. To match mathematical reservoir model history matching coefficient problem solution is used. This solution for both gas and oil reservoir and aquifer is formulated as follows

$$\nabla[k\, \sigma_l^* (\nabla P_l - \rho_l \bar{g})] = \sum_{j=1}^{2} m\, B_{lj}^* \frac{\partial P_j}{\partial t}$$

$$+ \sum_{i=1}^{n} q^l(x,t)\, \delta(x-x_i) + Q^l(x,t) \quad (I)$$

$$P_l(x,t) = P_{l,in}(t),\ x \in \mathbb{G} \quad (2)$$

$$\frac{\partial P_l}{\partial n} = 0,\ x \in \partial\mathbb{G},\ l = I,2 \quad (3)$$

$$Au = f,\ u \in \mathbb{U},\ f \in \mathbb{F}, \quad (4)$$

$$A: \mathbb{U} \to \mathbb{F}$$

where

$$\sigma_l^* = \frac{k_l \rho_l}{\mu_l};$$

P_l - are the phase pressures.

(1) Institute of Oil and Gas problems, USSR Academy of Sciences, GSP-1, 117917, Moscow, USSR.

The right part f of the operator equation (4) presents the parameters being measured at some points of the field G with co-ordinates X_i (for example $S_f(X_i,t)$ or $P_{f_j}(X_i,t)$). The values of the operator A at some point $u = [k(X),m(X)]$, is the corresponding parameter (for example $S(X_i,t)$), which is provided at the point X_i from the solution of the problem (I)-(3) at $k(X)$ and $m(X)$. Let the values P_{1f}, P_{2f} and S_f at points be measured respectively. Then f and Au can be expressed, for instance, in the form:

$$f = [P_{1f}(X_{11},t),\dots,P_{1f}(X_{1,n_1},t),$$
$$P_{2f}(X_{21},t),\dots,P_{2f}(X_{2,n_2},t),$$
$$S_f(X_{31},t),\dots,S_f(X_{3,n_3},t)]^T \quad (5)$$
$$Au = [P_1(X_{11},t),\dots,P_1(X_{1,n_1},t),$$
$$P_2(X_{21},t),\dots,P_2(X_{2,n_2},t),$$
$$S(X_{31},t),\dots,S(X_{3,n_3},t)]^T \quad (6)$$

The problem of operator equation solving (4) is so-called ill-posed, and the solution itself doesn't have a property of steadiness towards initial data. In (Alifanov et al., 1988; Bakushinskiy et al., 1989) it's shown that iterative method of solving (4) for which is used a value of gradient creates the regularization succession. In this case an iteration number is used as a regularization parameter. Suppose the equation is given

$$Au = f \quad (7)$$

where $A: U \to F$ is an operator ha-

ving a Freshey's derivative; U, F - denotes the Hilbert's space. Let's examine the functional $J(u) = 0.5\|Au - f\|_F^2$. To solve the equation (7) we apply one of the gradient methods

$$u_{n+1} = u_n - \beta_n J'u_n, \quad (8)$$
$$n = 0, I, 2, \dots$$

where β_n - is a step length along the found direction.

The gradient $J'(u)$ of the functional $J(u)$ in accordance with the following expression is defined

$$J'u = (A'_u)^* (Au - f) \quad (9)$$

where A'_u - is a Freshey's derivative of the operator A at the point u; $(A'_u)^*$ - denotes an operator, adjoint to A'_u.

For the problem (I)-(4) we've got the following expressions for the gradient $J(u)$

$$J'u = [J'_k u, J'_m u]^T \quad (I0)$$

where

$$J'_k u = - \int_0^T \sum_{i=1}^2 \sigma'_i [\nabla\psi_i, \nabla P_i - \rho_i \bar{g}] \, dt$$

$$J'_m u = - \int_0^T \sum_{i=1}^2 \sum_{l=1}^2 B'_{il} \psi_i \frac{\partial P_l}{\partial t} \, dt$$

While the functions $\psi_i(x,t)$ are the solution of the boundary problem

$$\nabla[\sigma_l \nabla\psi_l] + \sum_{j=1}^{2} (\alpha_{lj}[\nabla P_j - \rho_j \bar{g}, \nabla\psi_j] +$$

$$\beta_{lj}\psi_j + \frac{\partial}{\partial t}[B_{jl}\psi_j]) =$$

$$-\sum_{i=1}^{n_1}(P_{1i} - P_{1f,i})\,\delta(x-x_{1i}) +$$

$$(-I)^l \gamma \sum_{i=1}^{n_2}(S_i - S_{fi})\,\delta(x-x_{2i}) \quad (II)$$

$$\psi_l(x,T) = 0, \; x \in G \quad (I2)$$

$$\frac{\partial\psi_l}{\partial n} = 0, \; x \in \partial G, \; l = I,2 \quad (I3)$$

To get the iteration parameter values in (8) it's necessary to solve one-dimension optimization task

$$\beta_{\alpha n} = \min_{\beta_\alpha} J(\alpha^n - \beta_\alpha J'\alpha^n), \; \alpha = k,m$$

Appraisals of $\beta_{\alpha n}$ can be obtained by applying a linear approximation of the problem (I)-(3) in order to express the functional variation $\delta J(u)$ by $\delta P_j(u)$ and $\delta S(u)$ ones. The optimal values of $\beta_{\alpha n}$ in this case must be defined

out by the system of the equations

$$a_1\beta_{kn} + b_1\beta_{mn} = c_1$$
$$a_2\beta_{kn} + b_2\beta_{mn} = c_2$$

where a_i, b_i, c_i are calculated by δP_j, δS and the deviations between calculated and observed values of the parameters.

There's a need to note also, that for speeding the iteration procedure convergence it's necessary that finite-difference approximation of the equations (I)-(3), (II)-(I3), expressions for calculation $J'u$ and a method of permeability weighting corresponded each other.

The GTE system of gas treatment has some peculiarity that's worth to mention. The GTE model represents unlinear algebraical equations system describing the gas moving in wells, gas gathering, compressing, treating and cleaning systems and some inequalities determining technological restrictions as well. The algorithm of joint functioning both the reservoir and GTE models are represented on fig.

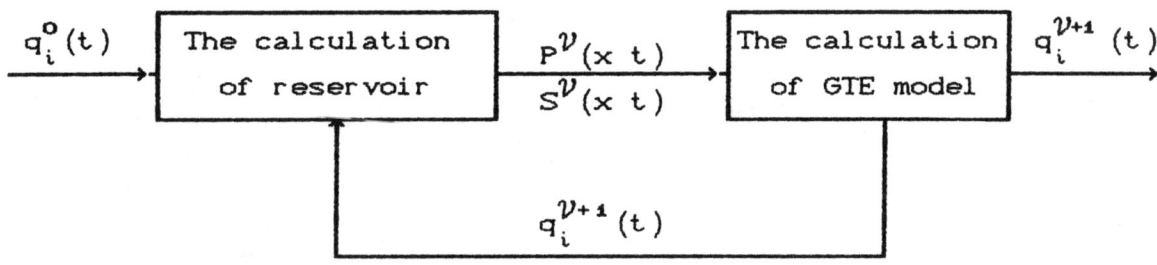

Fig. The algorithm of co-operation of reservoir and GTE models.

353

The reservoir and GTE models are connected by information and they function in an iterative cycle. When $(n+I)$-th time step is calculated the initial estimation of flow rate well values (taken for example from the previous time step) goes to the reservoir model entrance. The distribution of reservoir pressures and water saturations for $(n+I)$-th time step are the result of solving of the reservoir model.

The system of large gas-field construction, which is rather typical for largest northern gas fields consists of the separate gas fields (SGF), each of those consists of wells, gas gathering system (GGS), gas field compressor station (GFCS) and gas treating and cleaning system (GTCS), is connected each other by gas-field gathering main (GGM). At field exit a head gas compressor station (HGCS) with constant power N_{HGCS} is disposed, which have to provide determined exit pressure P_{HGCS}.

The movement of gas in the different elements of GTE system from a bed to the entrance point of HGCS station is described by the following system of the equations

$$\sum_{k=1}^{m_{ij}} \left(a_{ij}^k + \sqrt{b_{ij}^k + c_{ij}^k P_{ij}^2} \right) -$$

$$d_{ij}\sqrt{P_{ij}^2 - P_{en_i}^2} = 0 \ , \ j = I, l_i$$
$$i = I, M$$

$$\sum_{i=1}^{l_i} d_{ij}\sqrt{P_{ij}^2 - P_{en_i}^2} = Q_{iv}$$

$$\varepsilon_i = \varepsilon_i(Q_{im}(\rho), n_i)$$

$$N_i = N_i(Q_{im}(\rho), n_i, \rho_i)$$

$$\varepsilon_i = P_{i_{GTCS}} / P_{i_{GFCS}}$$

$$P_{i_{GTCS}}^2 - P_{i_{GGM}}^2 = \lambda_{i_{GTCS}} Q_{iv}^2$$

$$P_{l_{GGM}}^2 - P_{\nu_{GGM}}^2 = \lambda_{iv_{GGM}} q_{ev}^{i2}$$

$$\sum_{l=1}^{t_l} q_{lv} = 0$$

where ε_i – is a compression ratio of the turbocompressor unit respectively; respectively; n_i – is a number of revolutions; $Q_i(\rho)$ – is a volume flow rate; N_i – is a capacity i-th turbocompressor unit, i – is a number of GTCS, l_i – is a number of wells at j-th cluster. The following restrictions should be observed as well

$$n_i < n_{max} \ , \ Q_i(\rho) > Q_{min}$$

The solution of this system is fulfilled by Newton's method improved for faster convergence.

The algorithm of joint functioning both reservoir and GTE models provides complex simulation of oil and gas field development.

REFERENCES

I. Alifanov, O.M. et al., 1988, Ekstremalnie metody resheniya nekorektnyh zadach, Nauka, 285 s.

2. Bakushinskiy, A.B. et al., 1989, Iterativnie metody resheniya nekorektnyh zadach, Nauka, I26

3. Watson, A.T. et al., 1980, History matching in two-phase petroleum reservoirs , Soc. Pet. Eng. J., 52I-532, Trans., AIME.

4. Yang, P.H. et al., 1988, Automatic history matching with variable-metric methods , SPERE, Vol. 3, 3, 995-I00I.

2nd European Conference on the Mathematics of Oil Recovery
© D. Guérillot, O. Guillon (Editors) and Éditions Technip, Paris 1990, pp. 355-358
27 rue Ginoux, 75015 Paris

Multilevel Methods in Porous Media Flow

R. Teigland and G. E. Fladmark[1]

1 INTRODUCTION

The purpose of this paper is to describe the development and application of a cell-centered multilevel method applied to multiphase flow in porous media. The model that we consider is two-phase, 2-D compressible flow in a porous media.

In section 2 we set up the coupled set of nonlinear partial differential equations to be solved and discretize them using cell-centered differencing. A fully implicit scheme with water saturation and oil pressure as unknowns is used in the solution of the system.

The most time consuming part of a reservoir simulator is solving the set of linear equations that arises in each Newton iteration. Therefore it is clear that fast iterative methods that can handle jump discontinuities in the coefficients are needed. Cell-centered two-level methods have been widely used in nuclear reactor calculations. The method used was a two-step method based on a finite dimensional Galerkin technique. The Galerkin technique was first applied by E. L. Wachspress in 1965 for acceleration of the iterative solution of the equations in nuclear reactor calculations [7].

In section 3 we discuss the use of a linear multilevel scheme in order to solve the set of linear equations.

For smooth problems, multilevel methods will theoretically provide convergence in $\mathcal{O}(N)$ operations where N is the number of unknowns. The technique used in this paper is similar to the one used by Wesseling & Khalil [8]. In section 4 we present numerical results.

2 MODEL EQUATIONS

The model equations are derived by combining the mass conservation equations for the two components and Darcy's law. The resulting standard model equations are given below.

$$\nabla \cdot (\lambda_w(\nabla P_o - \nabla P_c - \gamma_w \nabla D)) = \qquad (1)$$
$$\frac{\partial}{\partial t}(\phi b_w S_w) + q_w$$

$$\nabla \cdot (\lambda_o(\nabla P_o - \gamma_o \nabla D)) = \frac{\partial}{\partial t}(\phi b_o S_o) + q_o \quad (2)$$

where $\lambda_l = \frac{K k r_l b_l}{\mu_l}$, $l = o, w$ and the b_l's are the formation volume factors defined as:

$$b_l = \frac{V_{l_{stc}}}{V_{l_{rc}}}$$

rc stands for reservoir conditions and stc stands for stock tank conditions. V is the volume of a fixed mass.

(1) Department of Applied Mathematics, University of Bergen, Allegaten 55, N5007 Bergen, Norway.

Cell-centered differencing is one of the simplest ways to approach the discretization of the flow equations, (See [2] for details about the discretization). Note that the cell-centered grid does not adversely affect the convergence properties of the discretization. It was thought for some time that uniform cell-centered grids were only $\mathcal{O}(\delta x)$, while uniform point-distributed grids where $\mathcal{O}(\delta x^2)$, (See Aziz and Settari [2]). The cell-centered approach is obviously more useful in modelling physical problems, where cell edges can be aligned with physical discontinuities in the problem. It is now known that the $\mathcal{O}(\delta x^2)$ convergence property applies to both uniform and non-uniform cell-centered grids (see Weiser and Wheeler [9]). The fully implicit method produces a set of $N \times n_{ph}$ nonlinear algebraic equations (where n_{ph} is the number of coupled equations per grid node and N is the number of grid nodes). The system of equations is solved using Newton iterations. The Jacobian matrix associated with the nonlinear system is highly unsymmetric.

3 MULTILEVEL METHODS

A class of iterative methods which currently plays an important role in reservoir simulation is formed by Conjugate Gradient and related methods. These solvers are often highly effective when used in combination with a suitable preconditioning technique.

Another class of iterative methods is that of multilevel methods. The advantage of multilevel methods is that they provide convergence in $\mathcal{O}(N)$ operations, whereas other iterative methods such as ICCG need $\mathcal{O}(N^{\alpha})$, $\alpha > 1$. Multilevel methods have been developed for problems with jump discontinuities in equation coefficients in the vertex-centered case and recently in the cell-centered case [1,3,8]. In the vertex-centered case the multigrid method can be expensive in terms of preliminary work and storage, especially in the three-dimensional case. Equation dependent interpolation operators are needed in this case [1,3]. Because of its greater simplicity, the recently developed cell-centered multigrid method [8] promises greater efficiency.

The multilevel method requires a representation of the differential operator on a sequence of increasingly coarser (or finer) grids. The basic multilevel idea is that it uses an efficient technique for eliminating high frequency error components on each grid. The other frequencies are eliminated on other grids. Thus, all modes are eliminated efficiently. The same idea of using a coarser grid in order to get a correction to the fine grid equations has been used by researchers in the field of reactor physics where they termed the method "coarse mesh rebalancing technique" [7]. Their construction of the matrices on coarser levels is similar to the Galerkin approximation suggested by Nicolaides [6] and Alcouffe et al [1].

Let a sequence of grids G^k (k=1(1)M) be defined with $h_1 > h_2 > \ldots > h_M$, where h_k is the mesh size on grid G^k. Denote the finite difference operator on G^M by A^M. The coarse-grid operator A^{k-1} is defined in terms of the fine-grid operator A^k and prolongation and restriction operators P^k_{k-1} and R^{k-1}_k:

$$A^{k-1} = R^{k-1}_k A^k P^k_{k-1}$$

This definition is called the Galerkin approximation. There are some advantages in using the Galerkin approach:
Irregular difference formulae at points near the boundary are automatically generated. If A^k is symmetric (positive definite) and if $R^{k-1}_k = (P^k_{k-1})^*$, then $A^{k-1} = R^{k-1}_k A^k P^k_{k-1}$ is also symmetric (positive definite). it seems natural to use a variational formulation in the construction of the coarse-grid operator in the case of discontinuous coefficients. This automatic prescription would only require the representation of the differential operator (or stencil) on the finest grid, the rest is automatically set up by the multilevel solver. Disadvantages of the Galerkin approach are:
The definition of A^{k-1} depends on A^k. A preprocessing phase is needed to compute A^{k-1}, (and $A^{k-2}, \ldots, A^1$) for the multilevel iterations.

The full-approximation scheme (FAS) that does not require linearity could be applied directly to the nonlinear equations, requiring only local linearization. One would like to use a Galerkin approach on this nonlinear MG scheme. The major limitation with the Galerkin approach for nonlinear schemes (as noted by Mc.Cormick [5]) stems from the fact that explicit use of the definition of $A^{k-1}(u^{k-1})$ (now a nonlinear operator) involves grid h^k computations. For the defect computations on grid G^{k-1}, $A^{k-1}(u^{k-1})$ may be evaluated by interpolating u^{k-1} to grid G^k, adding u^k, applying A^k, and transferring the result to grid h^k. The fact that in reservoir simulation we usually have a good initial guess when using values from the previous time step, Newton iterations will usually converge in a few iterations. This led us to use the linear multilevel scheme and not the nonlinear FAS approach. In the construction of the coarse-grid matrix A^{k-1} we are using the Galerkin-approximation since that is the only reliable approach when the equation coefficients are discontinuous. In multilevel methods P^k_{k-1} and R^{k-1}_k must satisfy the following requirements. Let m_P-1, m_R-1 be the maximum degree of polynomials that are interpolated exactly by sP or tR^* respectively for some real value of s or t. Then we must have $m_P + m_R > 2m$ with $2m$ the order of the partial differential equation to be solved. It is sufficient here to take $m_P = 1$ and $m_R = 2$ as suggested by Wesseling & Khalil [8]. Prolongation is chosen as constant interpolation, and restriction as scaled adjoint of linear interpolation in triangles, this preserves 7-point stencils. We refer to [8] for further details.

In the homogeneous case where K_x and K_y are of the same order we use the collective point Gauss-Seidel scheme (We use collective Gauss-Seidel schemes since we are treating the system of equations as a coupled system). In the anisotropic case we have three options in our code. Collective x-line Gauss-Seidel for $K_x \gg K_y$ denoted XLGS. Collective y-line Gauss-Seidel for $K_y \gg K_x$ denoted YLGS. Alternating line Gauss-Seidel if both kinds of anisotropies appear, denoted ALGS. The reason for choosing XLGS in the case of $K_x \gg K_y$ is that the points in the x-direction are strongly coupled, whereas the points in the y-direction are weakly coupled. We have also tried to accelerate the scheme using the multilevel scheme as a preconditioner to ORTHOMIN.

4   NUMERICAL EXPERIMENTS

We have tested the multilevel linear solver with our black-oil simulator. The following cases were considered. Ω is a square of constant thickness, $\Omega = (x,y) \mid 0 \leq x,y \leq 1000\,ft$. Water is injected in the lower left corner of Ω, and oil is produced in the upper right corner.

Pb 1. $K_x = K_y = 100\,mD \; \forall (x,y) \in \Omega$

Pb 2. $K_x = 1000\,mD, K_y = 10\,mD \; \forall (x,y) \in \Omega$

Pb 3. $K_x = 1000\,mD \qquad 0 \leq x \leq 200.\,ft$,
$K_x = 1000\,mD \;\; 200.0 < x \leq 1000.0\,ft$
$0 \leq y \leq 1000\,ft$
$K_y = 10\,mD \qquad \forall (x,y) \in \Omega$

Pb 4. $K_x = K_y = 10.0\,mD, 0 \leq x \leq 500\,ft$
$K_x = K_y = 500.0\,mD, 500.0 < x \leq 1000\,ft$
$0 \leq y \leq 1000\,ft$

Let ν_1 and ν_2 denote the number of pre- and post-smoothing steps respectively. $W(\nu_1, \nu_2)$- $V(\nu_1, \nu_2)$- and $F(\nu_1, \nu_2)$-cycles refers to the cycling strategy, see [4]. We define the average reduction factor by

$$\bar{\kappa} = \left(\frac{\|r^\nu\|_2}{\|r^0\|_2} \right)^{\frac{1}{\nu}}$$

where r^ν is the residual at the iteration ν, $r^\nu = f - Au^\nu$. The tabulated average reduction factors are averages over 20 timesteps.

h	$\nu = 10$		
	$pb.1$	$pb.2$	$pb.3$
1/16	.35	.33	.25
1/32	.36	.33	.32
1/64	.36	.30	.40

Table 1: results with $m_P + m_R = 2$.

Notice that the multilevel scheme is not efficient in this case. The reason is that prolongation and restriction are not accurate enough for a second order problem. Results with $m_P + m_R = 3$:

h	$\nu = 10$		
	$pb.1$	$pb.2$	$pb.3$
1/16	.07	.05	.065
1/32	.055	.04	.075
1/64	.06	.06	.08

Table 2: results for $W(2,1)$-cycles.

	$\nu = 10$		
h	$pb.1$	$pb.2$	$pb.3$
1/16	.09	.07	.08
1/32	.13	.09	.09
1/64	.13	.09	.11

Table 3: results for $V(2,1)$-cycles.

	$\nu = 10$		
h	$pb.1$	$pb.2$	$pb.3$
1/16	.07	.06	.06
1/32	.09	.08	.07
1/64	.10	.09	.09

Table 4: results for $F(2,1)$-cycles.

Smoother for test problems 1,2 and 3 is x-line Gauss-Seidel. In the next table we have tabulated results for test problem 4. Average reduction factors with a W(2,1)-cycle and the same cycle as preconditioner to ORTHOMIN (labeled orto-mg). Smoother is alternating line Gauss-Seidel.

	$\nu = 7$	
h	mg	$ortho - mg$
1/16	.06	.009
1/32	.06	.01
1/64	.08	.03
1/128	.08	.03

Table 5: results for test problem 4.

5   CONCLUDING REMARKS

We have implemented a cell-centered multilevel linear solver in a reservoir simulator. The method uses geometric restriction and prolongation operators. The multilevel solver can easily be implemented on existing simulators, requiring only stencils on the finest grid. Numerical experiments show that W- and F-cycles seem marginally better than V-cycles. Accelerating the multilevel scheme with ORTHOMIN reduces the total number of multilevel cycles. Implementation in three dimensions is easier in the cell-centered case than in the vertex-centered case. The convergence rates that we achieved are typical for those found when the multilevel method is used.

References

[1] R. E. Alcouffe, R. E. Brandt, J. E. Dendy, J. W. Painter. The multi-grid method for the diffusion equation with strongly discontinuous coefficients.
SIAM J. Sci. Statist. Comp. 2, p. 430-454.

[2] K. Aziz, A. Settari. Petroleum reservoir simulation.
Elsevier Applied Science Publishers 1979.

[3] A. Behie, P. Forsyth Jr. Comparisons of fast iterative methods for symmetric systems.
IMA Journal of Numerical Analysis 1983, **3**, p. 41-63

[4] A. Brandt. Multi-level adaptive solutions to boundary-value problems.
Math. of Comp. **31**, p. 333-390.

[5] S. F. McCormick. Multilevel Adaptive Methods for Partial Differential Equations.
Frontiers in applied mathematics. vol. 6. SIAM. Philadelphia 1989.

[6] R. A. Nicolaides. On some theoretical and practical aspects of multi-grid methods. Math. of Comp., vol. 33, p. 933-952.

[7] E. L. Wachspress. Iterative Solution of Elliptic Systems and Applications to the Neutron Diffusion Equations of Reactor Physics.
Prentice-Hall, Englewood Cliffs, New Jersey 1966.

[8] M. Khalil, P. Wesseling. Vertex-centered and cell-centered multigrid methods for interface problems.
Report 88-42, Delft University of Technology.

[9] A. Weiser, M. F. Wheeler. On convergence of block-centered finite differences for elliptic problems.
SIAM j. Numer. Anal. vol. 25, No. 2, p. 351-375, April 1988.

ACHEVÉ D'IMPRIMER
EN AOÛT 1990
PAR L'IMPRIMERIE NOUVELLE
45800 SAINT-JEAN-DE-BRAYE
No d'éditeur : 822
Dépôt légal 1990, no 11015
IMPRIMÉ EN FRANCE